THE ANTARCTIC CONTINENTAL MARGIN: GEOLOGY AND GEOPHYSICS OF OFFSHORE WILKES LAND

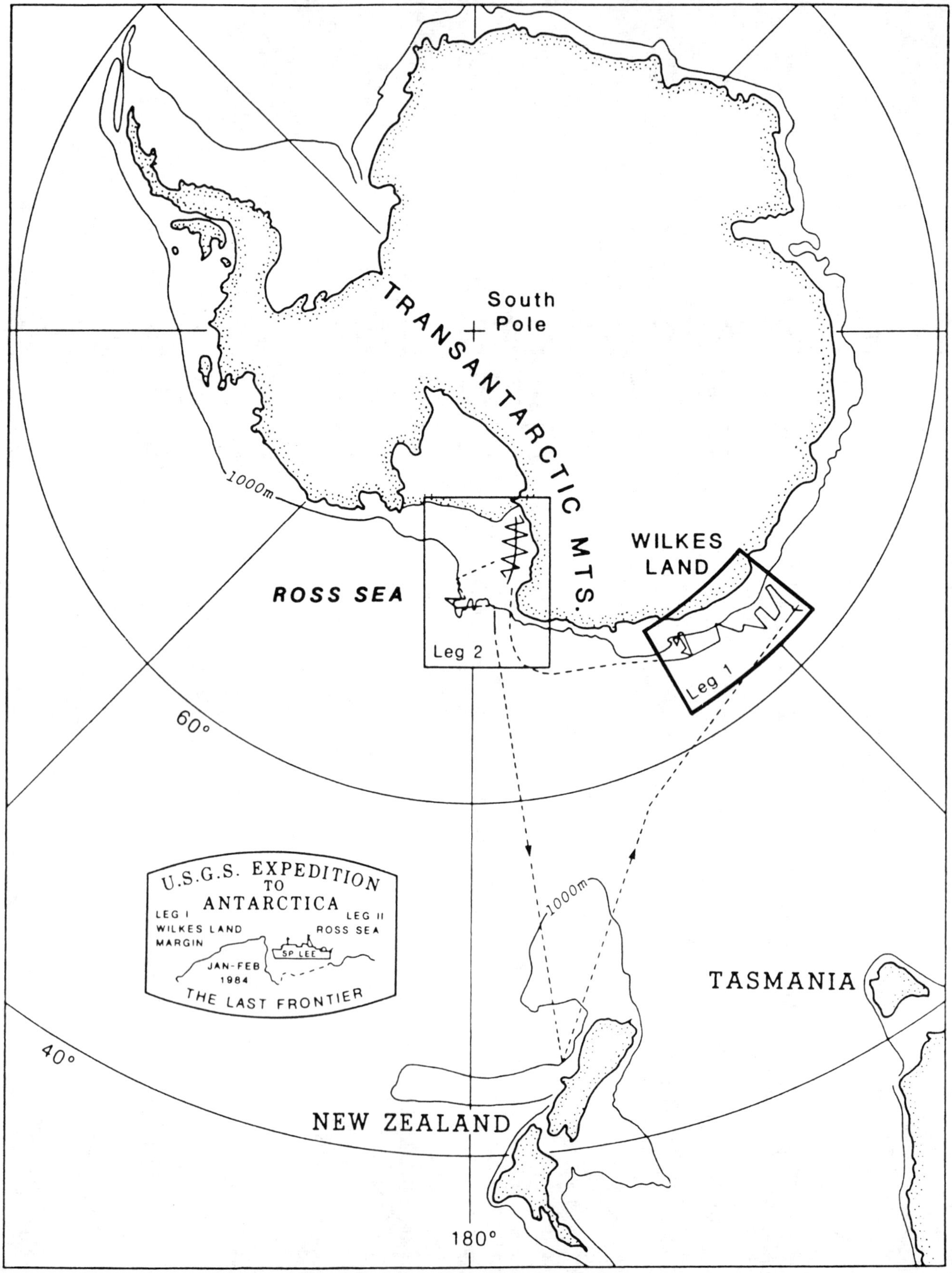
South Pole
TRANSANTARCTIC MTS.
WILKES LAND
ROSS SEA
1000m
Leg 2
Leg 1
60°
U.S.G.S. EXPEDITION TO ANTARCTICA
LEG I WILKES LAND MARGIN
LEG II ROSS SEA
SP LEE
JAN-FEB 1984
THE LAST FRONTIER
1000m
TASMANIA
40°
NEW ZEALAND
180°

Circum-Pacific Council for Energy and Mineral Resources
Earth Science Series, Volume 5A

THE ANTARCTIC CONTINENTAL MARGIN: GEOLOGY AND GEOPHYSICS OF OFFSHORE WILKES LAND

Edited by
Stephen L. Eittreim
and
Monty A. Hampton

Published by the Circum-Pacific Council
for Energy and Mineral Resources
Houston, Texas, U.S.A.

Published 1987
Manuscript received May 1986
Printed in the United States of America

Library of Congress Cataloging-in-Publication Data

The Antarctic continental margin.

(Earth sciences series/Circum-Pacific Council for Energy and Mineral Resources; v. 5A)
Bibliography: p.
1. Geology–Antarctic regions–Wilkes Land.
2. Geophysics–Antarctic regions–Wilkes Land.
I. Eittreim, Stephen. II. Hampton, Monty A.
III. Circum-Pacific Council for Energy and Mineral Resources. IV. Series: Earth science series (Houston, Tex.)
QE350.A567 1987 559.8′9 87-5184
ISBN 0-933687-04-4

Cover Design: Ben Servino, U.S. Geological Survey

Cover Illustration: Stephen Eittreim
Photograph of the U.S. Geological Survey Research Vessel *S.P. Lee* in McMurdo Sound. The seismic section below crosses the boundary between continental crust (left) and oceanic crust (right), showing the progressive seaward thinning of continental crust above Moho (seen at bottom of seismic record). High-amplitude reflectors near the continent–ocean boundary are believed to be volcanic, associated with the Cretaceous Gondwana breakup.

CPCEMR Earth Science Series Project Editors: F. L. Wong and H. G. Greene

Circum-Pacific Council for Energy and Mineral Resources
Earth Science Series

1. Tectonostratigraphic Terranes of the Circum-Pacific Region
   edited by D. G. Howell
2. Geology and Offshore Resources of Pacific Island Arcs—Tonga Region
   compiled and edited by D. W. Scholl and T. L. Vallier
3. Investigations of the Northern Melanesian Borderland
   edited by T. M. Brocher
4. Geology and Offshore Resources of Pacific Island Arcs—Central and Western Solomon Islands
   edited by J. G. Vedder, K. S. Pound, and S. Q. Boundy

5A. The Antarctic Continental Margin: Geology and Geophysics of Offshore Wilkes Land
   edited by S. L. Eittreim and M. A. Hampton

5B. The Antarctic Continental Margin: Geology and Geophysics of the Western Ross Sea
   edited by A. K. Cooper and F. J. Davey

Circum-Pacific Council publications are available from:
The AAPG Bookstore • P.O. Box 979 • Tulsa, Oklahoma, 74101 • USA
and its international distributors.

FOREWORD

The Earth Science Series of the Circum-Pacific Council for Energy and Mineral Resources (CPCEMR) is designed to convey the results of geologic research in and around the Pacific Basin. Topics of interest include framework geology, environmental geology, volcanology, oceanography, tectonics, geophysics, geochemistry, geothermal energy, hard minerals, petroleum geology, and applications of renewable energy. The CPCEMR supports and publishes results of scientific research that will advance the knowledge of energy and mineral resources potential in the Circum-Pacific region. The Earth Sciences Series is specifically designed to publish papers that include new data and new maps, report on CPCEMR-sponsored symposia and workshops, and describe the results of onshore and marine geological and geophysical investigations.

This volume reports the results for Leg 1 of the 1984 U.S. Geological Survey (USGS) marine investigation across the Wilkes Land and Ross Sea segments of the Antarctic continental margin (see frontice map). The results for Leg 2 are given in a companion volume of the Earth Science Series. Geophysical and geological data were collected aboard the USGS research vessel *S.P. Lee* during the USGS Operation Deep Sweep, which included other nations of Tonga, Fiji, Vanuatu, the Solomon Islands, and Papua, New Guinea.

Michel T. Halbouty

Chairman and President
Circum-Pacific Council for
Energy and Mineral Resources

PREFACE

The U.S. Geological Survey has had a long-standing commitment to Antarctic research in cooperation with the National Science Foundation. The 1984 Antarctic cruise of the R/V *S.P. Lee* was a milestone in this research program and represents the first marine multichannel seismic-reflection investigation in the Antarctic by the United States. This project also involved cooperation of scientists from government and academia in New Zealand, Australia, West Germany, France, and Japan, as well as the United States.

The multichannel seismic-reflection data collected by the USGS in the Wilkes Land and Ross Sea areas are a major contribution toward resolving the geologic structure and the potential resources of these parts of the Antarctic continental margin. In accord with the Antarctic Treaty provisions, the USGS multichannel seismic-reflection data have been made available to the international scientific community through the World Data Center in Boulder, Colorado.

The 1984 Antarctic cruises were the southernmost legs of a Pacific traverse by the *S.P. Lee* during the USGS Operation Deep Sweep. The scientific findings of these and other segments of Operation Deep Sweep are being released through volumes such as this one for the Earth Science Series of the Circum-Pacific Council for Energy and Mineral Resources.

The USGS marine operations in the Antarctic would not have been possible without the logistic support of the National Science Foundation, the U.S. Coast Guard, and the U.S. Navy. We appreciate their ongoing support for major scientific ventures in the difficult and hostile frontier of Antarctica. Geology is truly a global science, and results of the investigations reported here not only will provide new ideas and data to explain the earth's crust in Antarctica, but also will aid in unraveling the nature and history of similar continental margins around the world.

Dallas Peck
Director, U.S. Geological Survey

EDITORS' FOREWORD

Antarctica has been the focus of wide-ranging geologic and geophysical investigations over the past several decades, as a result of the Antarctic Treaty and the cooperative international scientific programs that the treaty has fostered. Yet the geologic framework of this vast continent and adjacent offshore margins is still poorly understood because of the extensive ice sheet and sea ice that masks underlying rocks. In January and February 1984, the U.S. Geological Survey (USGS) conducted multidisciplinary studies aboard its research vessel R/V *S.P. Lee*, including multichannel seismic-reflection surveys, of the Wilkes Land and Ross Sea continental margins to examine the geologic framework of the Antarctic offshore regions. This and a companion volume of papers presents the results of the 1984 USGS marine expedition. In addition, we include several invited papers about ongoing investigations in these areas, studies not directly related to the 1984 *Lee* survey.

The choice of the Wilkes Land (Leg 1) and Ross Sea (Leg 2) regions was influenced by both scientific and logistic criteria. The Wilkes Land continental margin offered a look at an Antarctic passive-rift margin, a conjugate to the well-studied Australian continental margin. The Ross Sea presented the opportunity to examine geologic structures within an area of suspected rifting and volcanism adjacent to the Transantarctic Mountains, the transcontinental mountain range dividing East and West Antarctica. Since 1980, West German, French, and Japanese scientists have conducted multichannel seismic-reflection surveys in these areas, and the 1984 USGS surveys complement the earlier work. In both of these areas, windows through the sea ice usually open during the period from December to March, making surveys of the margins possible. McMurdo Station in the Ross Sea offered a resupply port and a degree of protection associated with the U.S. Coast Guard icebreaker operations based there.

The Wilkes Land survey commenced at Christchurch (Lyttleton Harbor), New Zealand, on January 5, 1984, and ended at McMurdo Station, Antarctica, on February 2, 1984. The principal study areas were between longitudes 130° E and 150° E. The new results described in the following papers reflect not only our good fortune with excellent weather but also the dedicated work by the expedition crew and scientists.

Some highlights of the new results from the Wilkes Land survey include a clear seismic-stratigraphic look at this mid-Cretaceous margin formed by the breakup with Australia. Our recording of deep structures, including Moho reflections, exceeded in quality those normally achieved in passive margin settings. The origin of distinctive unconformities in the section can be tied to events associated with the rifting and subsidence of this margin. The more recent history is dominated by a period of active fan building, canyon cutting, and deposition of sediment from geostrophic contour currents.

The scientific findings in this volume present a preliminary and tantalizing look at the geologic features of the Wilkes Land continental margin, and we hope they will constructively add to our understanding of continental rifting processes that have acted since the breakup of Gondwana.

S. L. Eittreim
M. A. Hampton

PARTICIPANTS

Listed below are names and affiliations of the scientists and technicians who participated in the U.S. Geological Survey 1984 marine expedition aboard the R/V *S.P. Lee* off the Wilkes Land coast.

Stephen L. Eittreim	Chief Scientist	USGS
Monty A. Hampton	Co-Chief Scientist	USGS
Peter W. Barnes	Geologist	USGS
Barbara J. Blubaugh	Navigator	USGS
Jonathan R. Childs	Geophysicist	USGS
Margaret Golan-Bac	Geochemist-Navigator	USGS
David J. Hogg	Electronic Technician	USGS
Kris H. Johnson	Geologist-Navigator	USGS
Lawrence D. Kooker	Electronic Technician	USGS
Joseph Kravitz	Geologist	NOAA
Keith A. Kvenvolden	Geochemist	USGS
Alan McClenaghan	Captain	USGS
George W. Moore	Geologist	USGS
Gerald E. O'Brien	Mechanical Technician	USGS
Bruce M. Richmond	Geologist	USGS
Gregory L. Smith	Geophysicist	USGS
Howard M. J. Stagg	Geophysicist	BMR
John J. Veevers	Geologist	UM
Hal Williams	Mechanical Technician	USGS

BMR—Bureau of Mineral Resources, Canberra, Australia
NOAA—National Oceanic and Atmospheric Administration, Rockville, Maryland
UM—University of Macquarie, Sydney, Australia
USGS—U.S. Geological Survey, Menlo Park, California

CONTENTS

Eittreim, S. L., and Hampton, M. A., 1987, *The Antarctic Continental Margin: Geology and Geophysics of Offshore Wilkes Land*, CPCEMR Earth Science Series, v. 5A: Houston, Texas, Circum-Pacific Council for Energy and Mineral Resources

An Interpretation of the Multichannel Seismic Reflection Profiles across the Continental Margin of the Dumont d'Urville Sea, off Wilkes Land, East Antarctica

M. Tanahashi[1]
T. Saki
Japan National Oil Corporation
Tokyo, Japan

N. Oikawa
Japan Petroleum Exploration Co., Ltd.
Tokyo, Japan

S. Sato
Japan National Oil Corporation
Tokyo, Japan

The seismic stratigraphy and geological structure of the continental margin of the Dumont d'Urville Sea off Wilkes Land, near the 140°E meridian, are interpreted based on our marine geological and geophysical survey (TH82) in the 1982–1983 Antarctic summer season. Two sedimentary units are observed in the continental rise of this area. The upper unit (sequences DS-2, DS-3, and DS-4) is undeformed and well stratified, and probably represents the post-breakup marine Tertiary sequence. The maximum thickness is 2.0 sec (two-way travel time) in the center of the continental rise. The unit onlaps the lower one in the outer marginal high and inner continental slope side of the rise. The lower unit (sequences DS-5 and DS-6) is slightly deformed and poorly stratified. This unit is believed to be Cretaceous pre-breakup nonmarine to shallow-marine intracontinental sediments. Maximum thickness of the unit is 3.0 sec in the outer marginal high of the rise.

The acoustic basement below the continental rise is characterized by a severely block-faulted discontinuous reflection with extremely high reflectivity. It is probably composed of stretched and intruded continental igneous-metamorphic rocks of early Paleozoic and Precambrian ages.

Our interpreted tectonic history of the area is summarized as follows: Crustal extension and igneous intrusion occurred and made a broad inland prerift basin in East Gondwanaland during the mid-Mesozoic. Thick lacustrine, deltaic, and shallow-marine sedimentary sequences of the lower unit were deposited in the rift basin. Rapid regional subsidence of the rift graben occurred after the uplift of the central part of the rift basin (at present, the outer margin of the rise). This subsidence was caused by the initiation of breakup of East Gondwanaland and the seafloor spreading between Antarctica and Australia in the early Paleogene. After the breakup, thick marine sedimentary sequences of the upper unit filled the tectonically stable sedimentary basin of the continental rise.

INTRODUCTION

Since 1980, we have conducted general marine geological and geophysical surveys, including multichannel seismics, around the Antarctic Continent using R/V *Hakurei-maru*. Brief descriptions of previous surveys in the following areas have been published: Bellingshausen Sea (TH80, Kimura, 1982); Weddell Sea (TH81, Okuda et al., 1983); and Dumont d'Urville Sea and Ross Sea (TH82, Sato et al., 1984). The TH83 cruise off Wilkes Land was accomplished in 1983–1984. In this report, we describe and discuss the seismic stratigraphy and the geological structure of the Dumont d'Urville Sea area, off French Dumont d'Urville station on the coast of Wilkes Land, East Antarctica, based on seismic-reflection and refraction data, and other data obtained during the TH82 cruise.

[1]Present address: Marine Geology Department, Geological Survey of Japan, 1-1-3, Higashi, Yatabe, Tsukuba, Ibaraki 305, Japan.

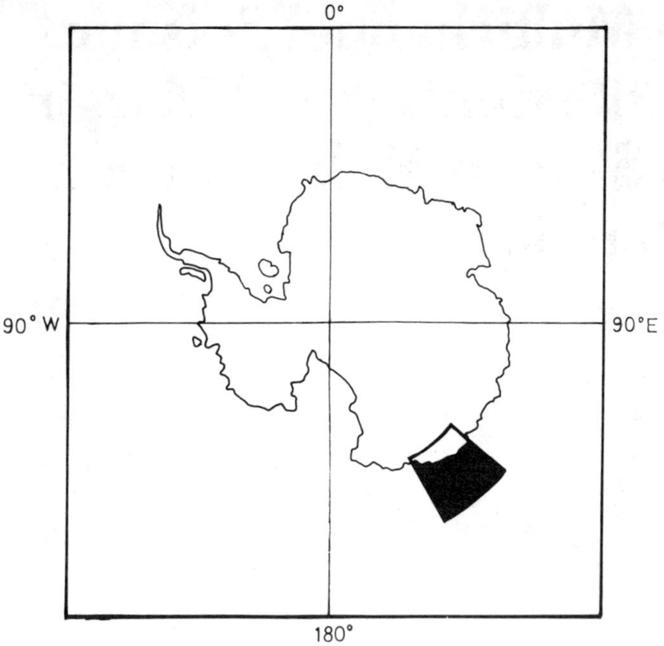

Figure 1—Index map of the study area, Dumont d'Urville Sea, off Wilkes Land, East Antarctica.

In the following description and discussion, the depth of reflectors and the thickness of seismic units are represented by two-way travel time in seconds.

PREVIOUS INVESTIGATIONS OF THE DUMONT D'URVILLE SEA AREA

The study area (Fig. 1) has had prior surveys. Research vessel U.S.N.S. *Eltanin* gathered abundant seismic, magnetic, and gravity data, and many bottom samples in the area (Houtz and Markl, 1972; Weissel and Hayes, 1972). Houtz (1974) inferred crustal structure by sonobuoy-refraction data. In 1973, *Glomar Challenger* drilled a hole down to late Oligocene at 958 m below the sea floor on site 269, 140°04.21′E, 61°40.57′S in the abyssal plain area (4285 m in water depth) during Deep-Sea Drilling Project (DSDP) Leg 28 (Hayes and Frakes, 1975).

Many sedimentological and paleontological studies have been carried out on samples recovered by the Australian Antarctic Expedition and the United States Deep Freeze 1979 in the continental shelf and slope areas (Anderson et al., 1980; Domack et al., 1980; Domack and Anderson, 1983; Truswell, 1982, 1983). Paleogeographic reconstruction by identification of geomagnetic anomaly lineations were discussed by Weissel et al. (1977) and Cande and Mutter (1982). Paleoenvironments were estimated by sedimentological analysis of the continental margin area of southern Australia (Deighton, 1977; Deighton et al., 1976).

In 1982, a 48-channel (24-fold) multichannel seismic-reflection survey was carried out by the Institut Francais du Petrol (IFP) in this area using Prakla-Seismos M/S *Explora* (Behrendt, 1983).

The TH82 cruise was carried out by Japan National Oil Corporation (JNOC) also during the 1982–1983 season. Dumont d'Urville Sea and Ross Sea were surveyed in the early

Table 1—Summary of obtained data in the early half of TH82, 1982–1983, in the Dumont d'Urville Sea area (Sato et al., 1984). © 1984 Memoir of National Institute of Polar Research.

Items	Total amount of data
Survey period	11 days
Seismic-reflection survey	820 km
multichannel survey	680 km
single-channel survey	140 km
Seismic-refraction survey	3 lines
Magnetic and gravity	4000 km
Piston coring with HF	3
Smith-McIntyre grab sampling	1
Dredge haul	2

and latter parts of the cruise, respectively. This paper is based upon the results of the early part of the cruise.

During the 1983–1984 season, the TH83 cruise was carried out off Wilkes Land area, 105°–140° E, also by JNOC (Tsumuraya et al., 1985). The eastern end of the survey lines of TH83 crosses the survey lines 3SMG and 8SMG of the TH82. In addition, the U.S. Geological Survey (USGS) surveyed this area by geophysical and geological methods with a 24-channel seismic reflection survey in leg 1 of their R/V *S.P. Lee* cruise in 1984 (Eittreim et al., 1984).

INSTRUMENTATION AND DATA COLLECTION IN TH82

The survey vessel used was R/V *Hakurei-maru* (overall length 87 m; 1821 GRT; one diesel engine; without ice strengthening) chartered from the Metal Mining Agency, Japan.

Total amount of collected data is summarized in Table 1. Survey tracks and sampled points are shown in Figure 2 with rough bathymetric contours based on the GEBCO bathymetric map.

Equipment and operating conditions in the survey were as follows: The seismic source was a Bolt PAR 1500C air gun with a 550 in.^{3} (9.2 l) firing chamber with wave shaping kit, operated at 1900 psi (133 kg/cm^2) air pressure. A small water gun, Seismic Systems Inc. S80 with 80 in.^{3} (1.3 l) firing chamber, was used in a single-channel survey line (5SMG) on the shallow continental shelf. High-pressure air was supplied by a Norwalk APS 120 air compressor with maximum capacity of 120 standard ft^3/min.

The hydrophone streamer used was a 24-channel Seismic Engineering Co. ministreamer with a polyuletan jacket. Channel intervals were 25 m with 32 hydrophone elements per channel. The multichannel streamer was used on lines 2SMG, 3SMG, and 8SMG; the single-channel streamer was used on line 5SMG for high-resolution survey in the shallow shelf area.

The data recorder was a Texas Instruments DFS-V digital recorder with record length set to 5 sec with delay-time control in the deep-sea area. Digital sampling interval was set at 4 msec.

Three sonobuoy-refraction measurements, with a Japan Radio Co. NRE-8A receiver and an Oki Electric OC-01 radio sonobuoy, were carried out intermittently during the seismic-reflection survey.

The vessel speed during the seismic-reflection survey was maintained at about 4 knots, and the firing interval was fixed at 22 sec in the constant-time firing mode giving distances between the nearest shot points of roughly 50 m. Because of the 25-m channel interval, the recordings are sixfold and the spacing of the common midpoints (CMP) is about 12.5 m.

All the seismic-reflection data were processed after the cruise by a geophysical contractor, Japan Petroleum Exploration Co. (JAPEX). A conventional stacking process sequence was applied to all the data, and in addition a wave equation (F-K) migration process was applied to selected data.

Magnetic and gravity measurements were made along all survey tracks using a Geometrics G801 proton magnetometer and a Lacoste–Romberg S-63 sea gravimeter, respectively.

Navigation was done with a Magnavox Navy Navigation Satellite System (NNSS) using the Transit Satellites. A Nippon Electronic Co. 12-kHz depth sounder NE16 system and a Raytheon 3.5-kHz subbottom profiling system with CESP-II signal correlater were used for bathymetric and subbottom profiling. Digital bathymetric data, from a Raytheon PDD, were transmitted into a data logging system of NNSS and recorded on magnetic tape with the navigation, magnetic, and gravity data.

Geologic samples were collected at several locations in the area by piston corer with thermister sensors for terrestrial heat flow measurements, a Smith–McIntyre grab sampler with Benthos deep-sea camera, and a chain bag dredge.

TOPOGRAPHY OF THE SURVEY AREA

The bathymetric map by GEBCO is the most useful as a basic topographic map of the survey area (Fig. 2). The outer part of the continental shelf shows a gentle inclination toward the continent along profiles 3SMG and 5SMG. This feature is likely caused partly by gravitational subsidence of the ice-covered continent and partly by development of the northwest–southeast-trending erosional trough possibly carved by glaciers (Johnson et al., 1982). The water depth of the trough is up to 700 m, and the relative depth difference from the surrounding normal shelf floor is about 400 m. The shelf has a width of about 120 km, with the shelf edge at depths of about 300 m on our survey lines. Apparent terrace-like features can be observed in the middle part of 5SMG line.

The continental slope shows a steep inclination (13°) down to a depth of about 2200 m from the shelf edge.

The continental rise is about 100–120 km wide. It has a very gentle slope less than 1° on the line 3SMG and deepens from about 2200 m at the foot of the slope to 3800 m at the outer edge. The sea floor of the shallower portion of the continental rise is cut by many submarine canyons to a depth of as much as 400 m, whereas the lower half is very smooth.

A characteristic topographic high is observed on the outer marginal part of the continental rise. Its relative height is about 750 m, with a width of 35 km on the line 3SMG. The high is west-northwest–east-southeast trending according to the GEBCO bathymetric map.

A very flat abyssal plain is present north of the high. The depth increases slightly from 3900 m at the northern end of the high to 4400 m at the junction of lines 2SMG and 3SMG. It decreases again to the north end of line 2SMG to 4200 m. Neither seamounts nor knolls can be observed on our profiles in the abyssal plain area.

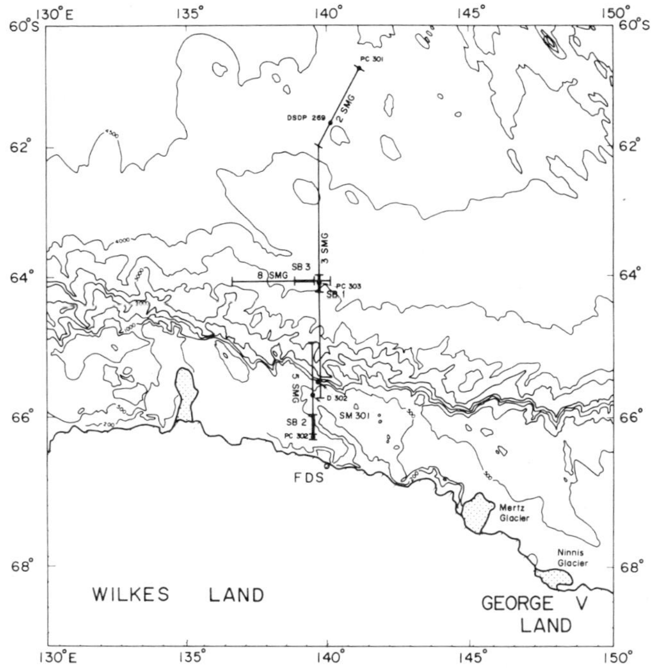

Figure 2—Survey track chart of the multichannel seismic-reflection survey lines from cruise TH82 (2SMG, 3SMG, 8SMG) and the single-channel seismic survey line (5SMG). SB shows the refraction survey lines using sonobuoys. PC shows sampling points by a piston corer with thermister sensors for a heat-flow measurement. D and SM mean the positions of dredge haul and Smith-McIntyre grab sample. Bathymetric contours are based on the General Bathymetric Chart of the Ocean (GEBCO). Contour interval is 500 m with an extra 200 m contour. FDS shows the location of French Dumont d'Urville station.

SEISMIC STRATIGRAPHY AND STRUCTURE ACROSS THE CONTINENTAL MARGIN ALONG THE 140° E MERIDIAN

Line drawing interpretations of the seismic sequences on the conventionally processed seismic profiles of the survey area are shown in Figure 3. Six seismic sequences (DS-1 to DS-6) can be identified, and they are grouped into upper and lower geological units of the continental rise by a prominent unconformity.

The survey results of the refraction seismics and gravity and magnetics are taken into consideration in the interpretation. The solution of the sonobuoy refraction measurements are shown on Table 2 after Sato et al. (1984). Gravity, magnetic, and bathymetric profiles are shown on Figure 4.

Three examples of processed seismic profiles are presented in Figures 5, 6, and 7.

Continental Shelf Area

Three seismic sequences are identified in the continental shelf area (Fig. 3C, 3E): (1) An upper sequence in the terrace-like topographic high in the middle part of the shelf; (2) a middle sequence in all profiles of the shelf; (3) a lower sequence in the shelf margin.

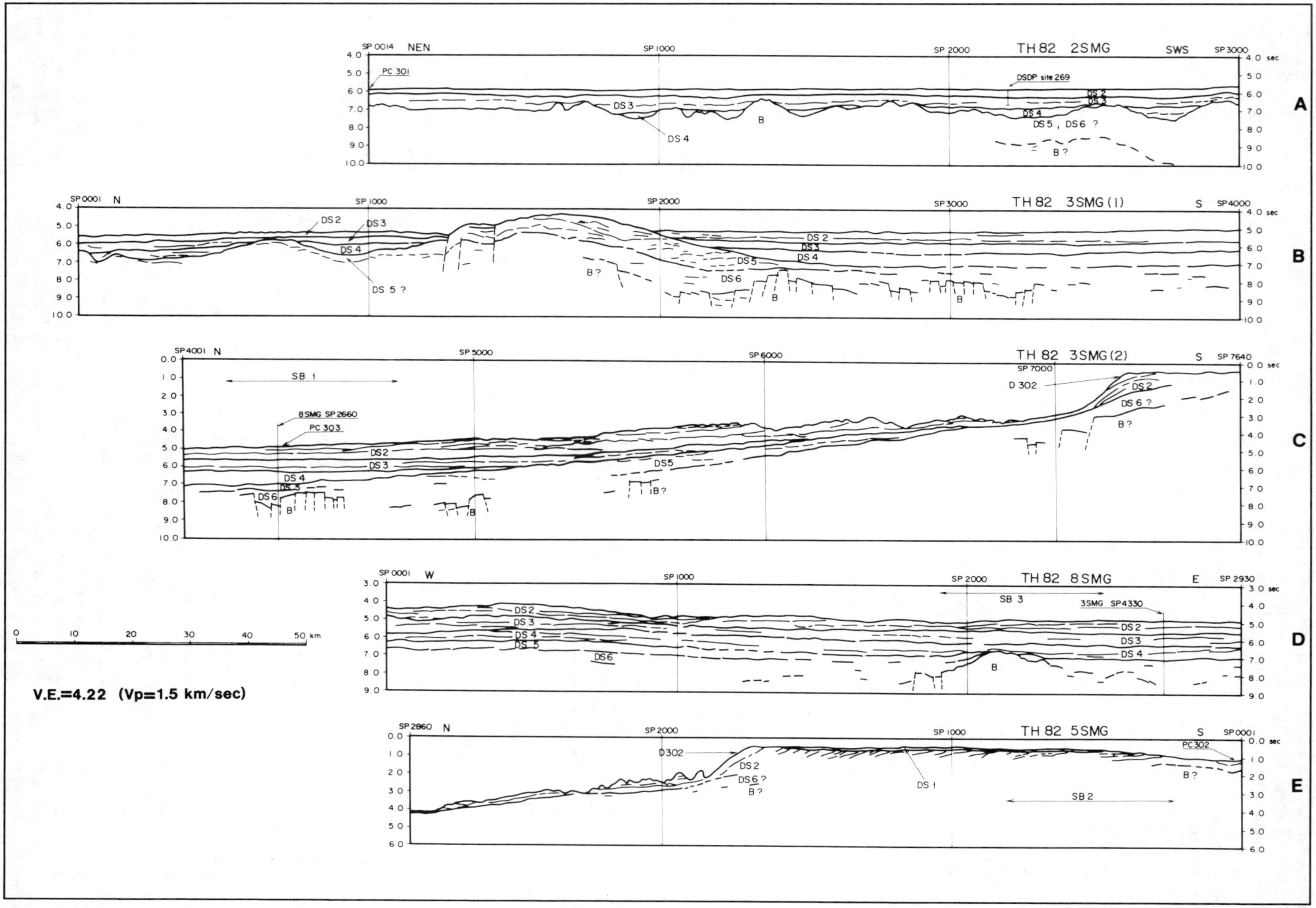

Figure 3—Interpretative line drawings of the multichannel (A, B, C, D) and single-channel (E) seismic reflection profiles that were obtained in TH82 in the Dumont d'Urville Sea.
A. 2SMG in the abyssal plain. B. 3SMG(1) mainly in the lower continental rise. C. 3SMG(2) in the mid-lower to upper continental rise and the outer part of the continental shelf. D. 8SMG east to west along strike of the continental rise. E. 5SMG in the continental shelf and upper rise.

Table 2—Sonobuoy refraction solutions in the Dumont d'Urville Sea in TH82 (Sato et al., 1984). © 1984 Memoir of National Institute of Polar Research.

	V_P (km/sec)	Thickness (km)	Depth (km)
SB-1	(1.5)		
			3.6
	1.7	1.3	
			4.9
	3.0	1.1	
			6.0
	4.0	3.0	
			9.0
	5.0		
SB-2	(1.5)		
			0.5
	1.6	0.1	
			0.6
	2.2	0.8	
			1.4
	3.0	0.4	
			1.8
	3.9	1.9	
			3.7
	4.7		
SB-3	(1.5)		
			3.5
	1.7	0.7	
			4.2
	2.3	0.9	
			5.1
	2.9	0.9	
			6.0
	3.5	0.1	
			6.1
	5.8		

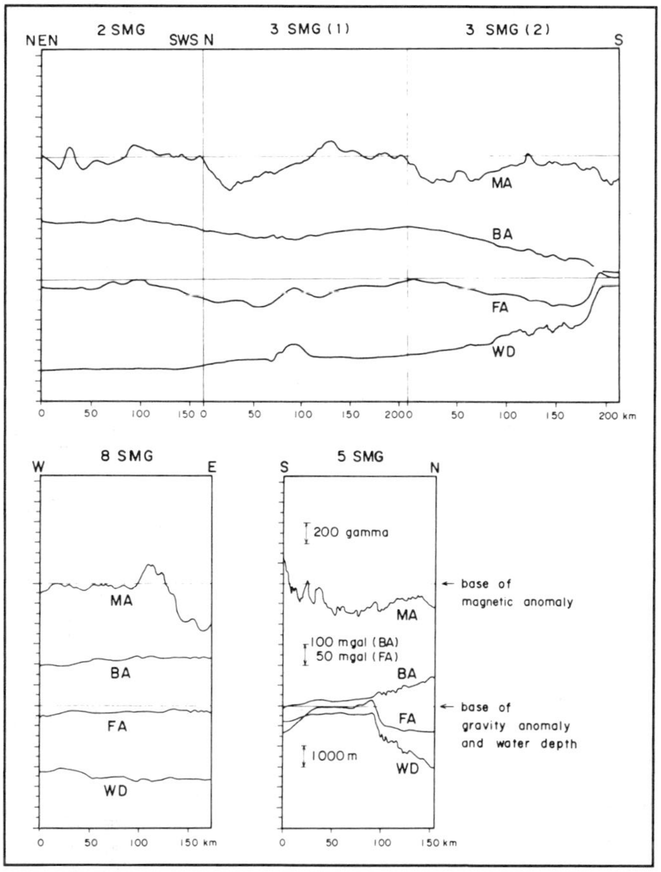

Figure 4—Geophysical profiles of the Dumont d'Urville Sea. MA, magnetic anomaly; FA, free air gravity anomaly; BA, Bouguer anomaly; WD, water depth. Free air anomaly and magnetic anomaly are calculated based on the IGSN67 and the IGRF1975 reference fields, respectively. Daily variations of the magnetic field have not been corrected for. Simple Bouguer anomaly is calculated using the density estimation of the water and the crust as 1.03 and 2.67 g/cm^3, respectively.

The upper sequence, designated as DS-1, is finely stratified and overlies the middle sequence by a discordant unconformity. Analyses of samples taken from the survey area (Anderson et al., 1980; Domack and Anderson, 1983; Sato et al., 1984) indicate that this sequence is composed mainly of glacial marine sediments, probably deposited after maximum glaciation. It cannot be identified in the continental rise and abyssal plain area. Its maximum thickness is about 0.2 sec.

The middle sequence, designated as DS-2, has reflectors characterized as tangential oblique progradational. In the southern part of the continental shelf, it concordantly overlies the lower sequence or acoustic basement. Its lower limit cannot be positively identified. In the middle part of the shelf, the lower part of it is masked by strong multiple reflections, and in the shelf margin it downlaps onto a lower sequence. This sequence apparently is continuous with the uppermost sequence of the continental rise. The maximum thickness is about 1.0 sec.

In the continental shelf margin, DS-2 probably consists mainly of Neogene and older strata, as determined from laboratory analysis of a carbonaceous siltstone sample taken from the upper part of the sequence at a depth of about 550 m. The sample contains diatom flora, possibly of early Miocene (D302; Sato et al., 1984).

A Lower Cretaceous sequence containing abundant palynomorphs occurs beneath Quaternary sediments in the possibly eroded trough off the George V coast (Domack et al., 1980; Domack and Anderson, 1983). Also, abundant reworked Paleogene and Upper Cretaceous palynomorphs have been recovered from bottom sediments off the George V coast (Truswell, 1982, 1983). This paleontological evidence suggests that DS-2 is composed of Cretaceous to lower Neogene, nonmarine to marine sediments from south to north.

The lower sequence, designated as DS-6, occurs in the shelf margin between DS-2 and the acoustic basement. It is characterized by homogeneous dispersed internal reflections and has a maximum thickness of about 1.0 sec. This sequence may continue in the southern part of 5SMG where the lower boundary of sequence DS-2 cannot be identified. The geology of DS-6 cannot be positively described, but by comparison with the stratigraphy of southern Australian basins, it is probably composed mainly of Lower Cretaceous nonmarine strata.

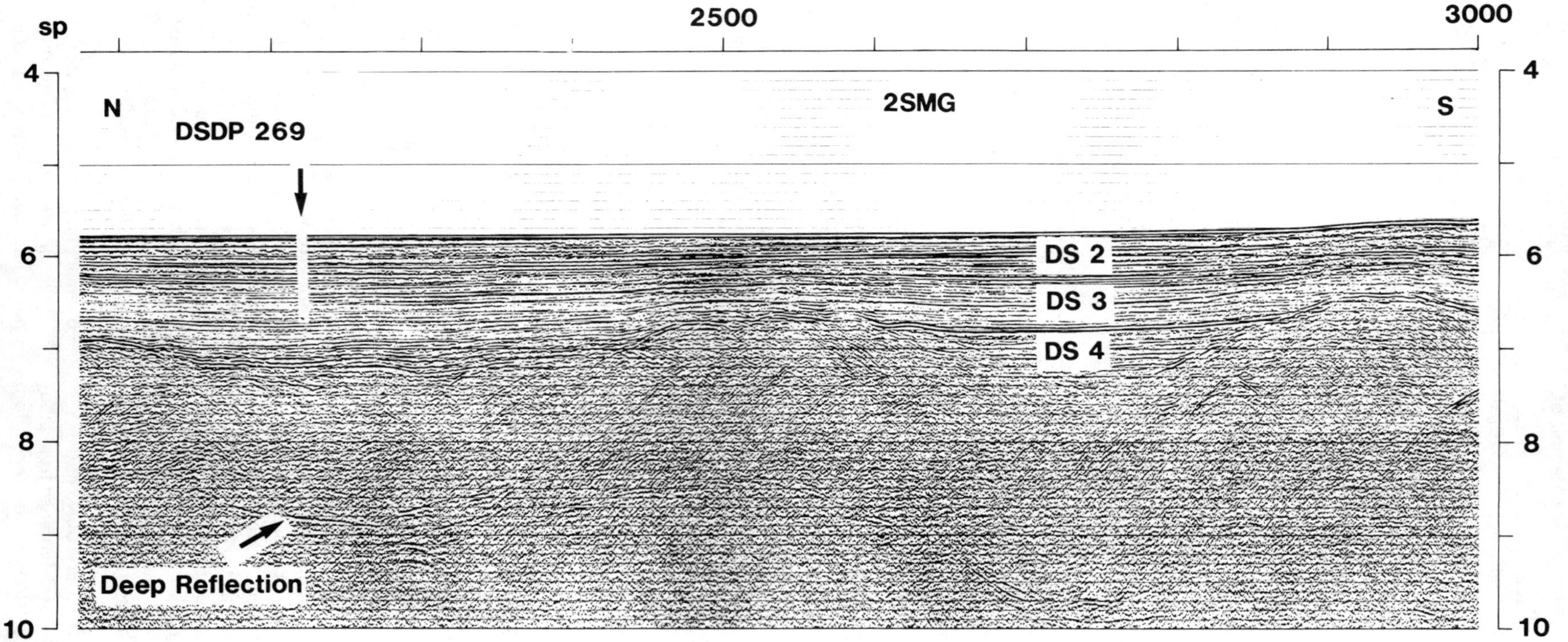

Figure 5—Part of the conventionally processed seismic profile, 2SMG, around site 269 of DSDP Leg 28. The acoustic basement under DS-4 resembles normal oceanic basaltic basement. However, the deep reflections at about 9 sec may show the existence of the fragmental continental complex.

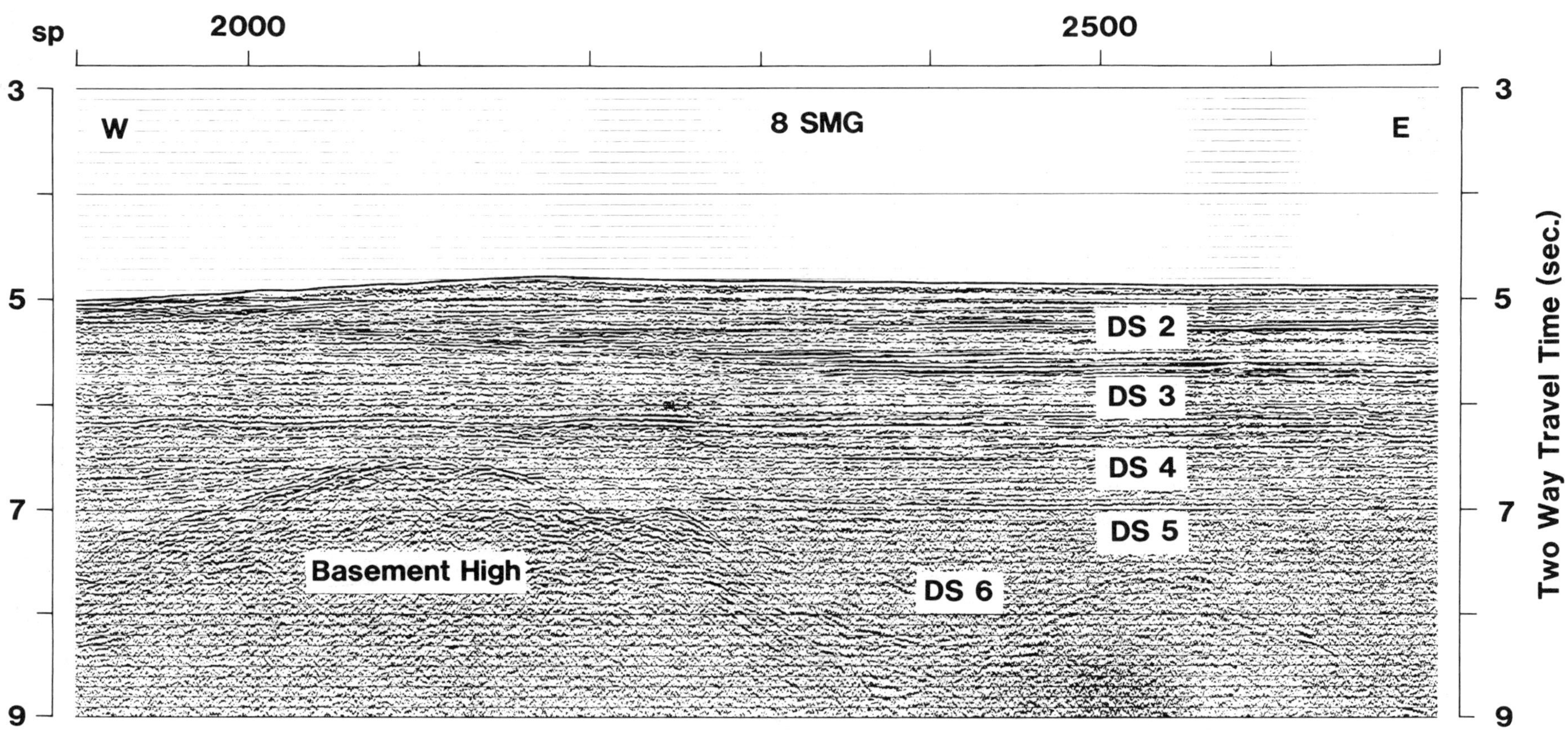

Figure 6—Part of the conventionally processed seismic profile, 8SMG, across a basement high.

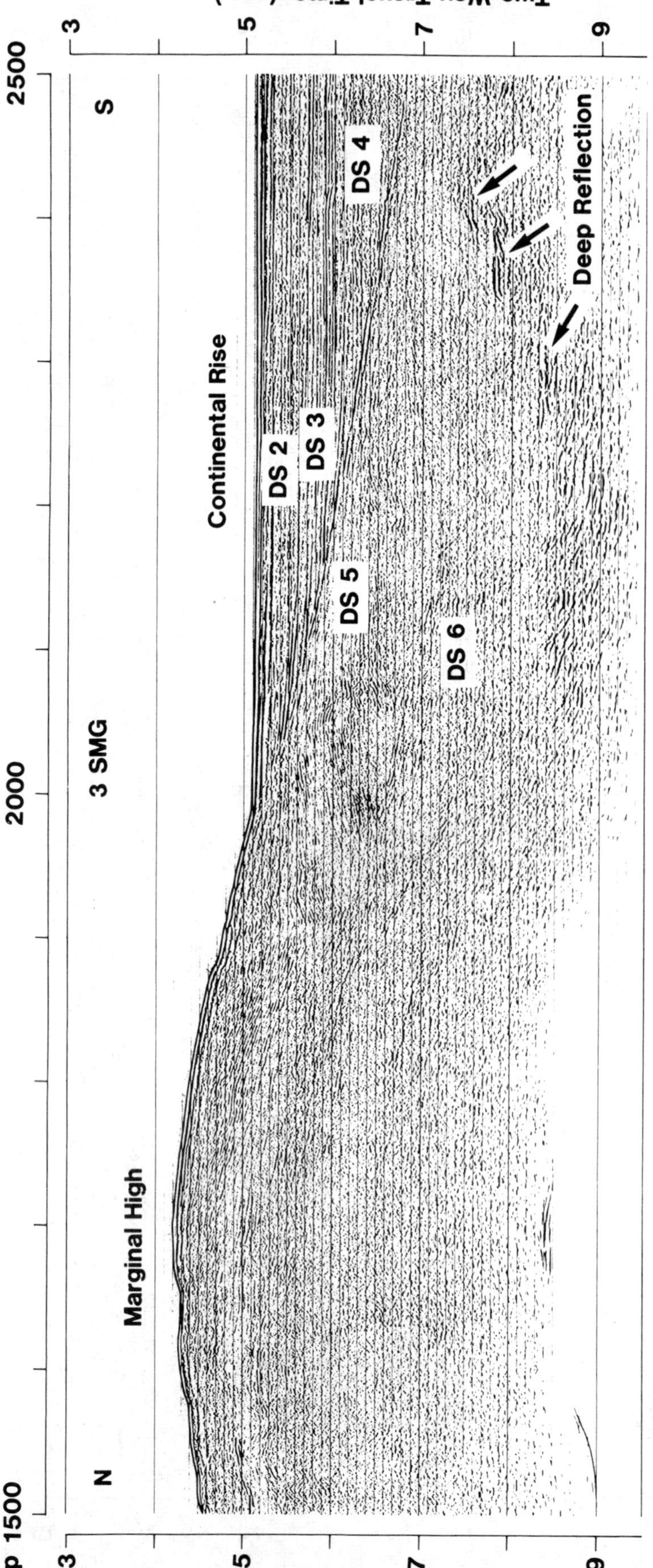

Figure 7—Part of the time-migrated seismic profile, 3SMG, processed by wave equation (F–K) migration, across the marginal topographic high of the continental rise.

Acoustic basement is interpreted as the early Paleozoic to Precambrian igneous–metamorphic complex that is exposed on the Antarctic coastal region.

Continental Slope Area

The steep continental slope is thought to be of mainly depositional origin. Sequence DS-2 on the shelf downlaps to sequence DS-6 on the continental slope (Fig. 3C), showing the northward outbuilding of the shelf edge during the deposition of sequence DS-2. The upper portion of sequence DS-6 has some strong continuous reflectors, whereas the lower portion is characterized by homogeneous acoustic character.

Continental Rise Area

In descending order, there are five sedimentary sequences, DS-2, DS-3, DS-4, DS-5, DS-6, and the acoustic basement, in the continental rise (Fig. 3B, 3C, 3D). They are grouped into two geological units by the prominent unconformity below sequence DS-4 (See Fig. 7).

The Upper Unit

The upper unit (sequences DS-2, DS-3, and DS-4) fills the sedimentary basin whose depocenter lies in the central part of the rise. Sequences DS-2, DS-3, and DS-4 are discriminated by strong reflectors, and their interrelations (DS-2/DS-3 and DS-3/DS-4) are considered as nearly parallel unconformities. The unit shows little deformation, and its maximum thickness is about 2.0 sec. The unit onlaps sequences DS-5 and DS-6 at the outer marginal high and the inner continental slope side of the rise (Fig. 3B, 3C). Sequences DS-3 and DS-4 show limited distribution and are not present in the southern part of the rise and the continental shelf and slope areas, whereas sequence DS-2 appears to continue to the continental shelf. The geologic ages of the unit are roughly estimated from the DSDP site 269 data.

The upper part of sequence DS-2 generally shows very clear stratifications in the lower continental rise, but it shows some chaotic reflectors under the rough sea floor of the upper rise. This chaotic nature probably shows the strong influence of submarine slumping and submarine channel activities in the area. These activities probably began with the Miocene glacial advance, and they are probably still active.

The maximum thickness of sequence DS-2 is about 1.2 sec in the middle part of the rise. Analysis of PC303 (Sato et al., 1984) shows that the uppermost part of sequence DS-2, in the center of the rise, is composed of siliceous silt and sandy silt. The depositional age of sequence DS-2 is estimated as Neogene to Quaternary, based on the correlation with DSDP 269.

Sequence DS-3 has a stratified nature similar to the lower part of sequence DS-2. Its maximum thickness is about 0.75 sec. Sequence DS-3 is estimated to be Oligocene marine strata, based on the correlation with DSDP 269.

Sequence DS-4 shows fewer continuous stratifications compared with the upper sequences DS-2 and DS-3. The lower surface of sequence DS-4 cannot be identified in the central part of the rise. Sequence DS-4 is interpreted as shallow-marine deposits after its reflector's less continuous character. The DS-4 maximum thickness is about 0.8 sec. The depositional age of sequence DS-4 is speculated as mainly Paleocene and Eocene by extrapolation from the seismic record at DSDP 269 in the abyssal plain.

The Lower Unit

The lower unit consists of seismic sequences DS-5 and DS-6. The boundary between DS-5 and DS-6 and the lower boundary of sequence DS-6 are not obvious. The maximum thickness of the unit is about 3.0 sec in the outer marginal high of the rise. The unit fills and covers the rough surface of the basement and is exposed or only thinly covered at the outer marginal high of the rise. The unit has been eroded and deformed, especially in the marginal high. The deformation is concordant with the basement deformation. The depositional ages of sequences DS-5 and DS-6 are roughly estimated as Late Cretaceous and Early Cretaceous, respectively, after comparison with stratigraphic sequences in the southern continental margin of Australia (Falvey and Mutter, 1981).

Sequence DS-5 occurs in the continental rise except in the southernmost part, where it onlaps to sequence DS-6. It shows stratifications with poor continuity. Sequence DS-5 shows characteristic acoustic features in the outer marginal high of the rise (Fig. 7). The upper part shows clear stratifications, and the lower part, which overlies the lower sequence DS-6 with a discordant unconformity, shows chaotic reflectors. The high variability of the seismic character may have been produced by nonmarine to shallow-marine, high-energy sedimentary environments during the deposition of sequence DS-5 in the outer marginal high. Sequence DS-5 has been deformed into anticlinal folds. Its thickness is about 2.0–2.8 sec. The upper surface of sequence DS-5 in the marginal high area shows erosional truncation.

The configuration of sequence DS-6 is not well observed on our profiles. Sequence DS-6 is thinner, is less stratified, and is a more completely homogeneous seismic sequence in the middle and landward part of the rise than in the marginal high. The maximum thickness is estimated as 1.0–1.5 sec in the marginal high area. The thickness of sequence DS-6 in the middle part is about 0.5 sec, and it thickens landward up to about 1.0 sec or more where its lower boundary cannot be observed. The homogeneous seismic character of sequence DS-6 suggests a nonmarine intracontinental sedimentary origin. It is probably correlated with the sequence between DS-2 and the basement in the continental shelf area.

Acoustic Basement

Acoustic basement is characterized by extremely strong, discontinuous reflectors on profile 3SMG (Fig. 7). The high contrast in acoustic impedances with sequence DS-6 probably suggests that the acoustic basement is composed of Paleozoic–Precambrian igneous–metamorphic continental basement. Some deep reflections are possibly block-faulted flat segments (Fig. 7). It probably supports the interpretation that at least some portions of the acoustic basement are composed of continental basement. But, in contrast, a basement high observed on the profile 8SMG suggests a possible igneous intrusion (Fig. 6). In some places, there are stratified segments in the acoustic basement. The interpreted continental basement probably continues to the basement under sequence DS-6 in the continental shelf area.

Abyssal Plain Area

The survey line crosses DSDP site 269, which penetrated possible late Oligocene clay and silty clay at a depth of 958 m (Hayes and Frakes, 1975). Three sedimentary sequences,

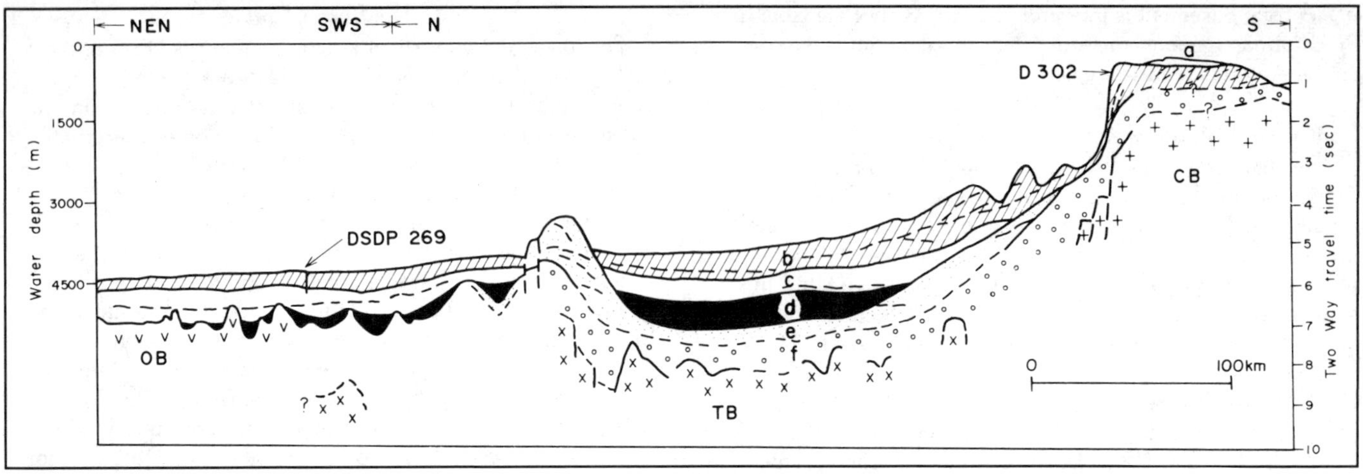

Figure 8—Simplified and estimated geologic section across the continental margin in the Dumont d'Urville Sea area. a, upper Neogene glacial and glacial marine sequence (seismic sequence DS-1); b, Neogene marine sequence in the continental rise, mainly Paleogene nonmarine to marine sequence in the shelf (DS2); c, upper Paleogene marine sequence (DS3); d, lower Paleogene marine and shallow-marine sequence (DS4); e, Upper Cretaceous mainly shallow-marine sequence (DS5); f, Lower Cretaceous continental sequence (DS6); CB, continental basement; TB, transitional basement; OB, oceanic basement.

which show little deformation, and the rough acoustic basement can be observed on our profile (Fig. 3A). The age and the depositional character can be determined from the DSDP results. The upper sequence (DS-2) is well stratified and probably correlates with the Miocene to Quaternary turbidite and silt sequence. The middle sequence (DS-3) is less stratified and is correlated with the clay and silty clay sequence estimated as mainly Oligocene. The lower sequence (DS-4) is stratified, fills the relief of the acoustic basement, and pinches out northward. It is speculated as mainly Eocene marine strata from the formation age of the oceanic volcanic basement estimated by remnant magnetic anomaly identification (Weissel and Hayes, 1972; Cande and Mutter, 1982). It shows weak deformation apparently concordant with the acoustic basement roughness.

The acoustic basement on the outer part of profile 2SMG resembles typical oceanic basaltic basement. But on the southern part of the profile, a distinct deep reflector, which resembles the basement in the continental rise area, can be observed (Fig. 5). It may show the upper surface of the continental fragment. It has a depth below the sea floor of about 3.0 sec and could also be interpreted as the Mohorovicić discontinuity. The typical oceanic basaltic basement occurs continuously only in the area north of site 269.

DISCUSSION OF THE STRUCTURE AND THE GEOLOGICAL DEVELOPMENT OF THE DUMONT D'URVILLE SEA AREA

Six seismic sequences, DS-1, DS-2, DS-3, DS-4, DS-5, and DS-6, can be identified in the continental rise and shelf. They can be grouped into upper and lower units by a prominent unconformity between sequences DS-4 and DS-5 in the continental rise area. Sequence DS-1 can be identified only in the shelf area. Sequences DS-2, DS-3, and DS-4 can be observed in the continental rise and abyssal plain areas. Sequences DS-3 and DS-4 onlap lower sequences on both margins of the continental rise basin to the north and south. They cannot be observed in the shelf area. Sequence DS-4 pinches out northward in the abyssal plain area. Sequences DS-5 and DS-6 can be observed in the marginal high of the continental rise. DS-5 is truncated under the upper continental rise, and DS-6 probably continues to the continental shelf. Both DS-5 and DS-6 may exist in the southern part of the abyssal plain area.

The upper and lower units in the continental rise area are probably correlated to the upper stratified sequence and the lower unstratified sequence, respectively, which are identified on the USGS profiles (Eittreim et al., 1984).

Figure 8 shows the sedimentary basin configuration on the TH82 profiles across the continental margin off Dumont d'Urville station, on the basis of interpretation of the seismic-reflection profiles described above.

Basement Structure

The basement in the continental rise and shelf is possibly fragmented continental igneous–metamorphic complex of the early Paleozoic–Precambrian age, probably intruded in places by igneous bodies. The regional, gentle, high, free-air anomaly over the deep continental rise sedimentary basin (Fig. 4) may be caused by the crustal thinning under the rise. The continental rise in our survey area corresponds to the so-called magnetic quiet zone (Weissel and Hayes, 1972; Cande and Mutter, 1982), which probably shows that the crust beneath the continental rise is a transitional crust between continental and oceanic as in the southern continental margin of Australia (Talwani et al., 1979). Even in the abyssal plain area, the southern part may contain some continental fragments (Fig. 5).

The characteristic marginal high of the continental rise has been interpreted as an oceanic–volcanic and volcanoclastic ridge by Wannesson (personal communication). However, our profile suggests that the marginal high is probably composed mainly of stratified and contorted sedimentary sequences (Fig. 7) and apparently shows anticlinal folding. Hence, we suggest

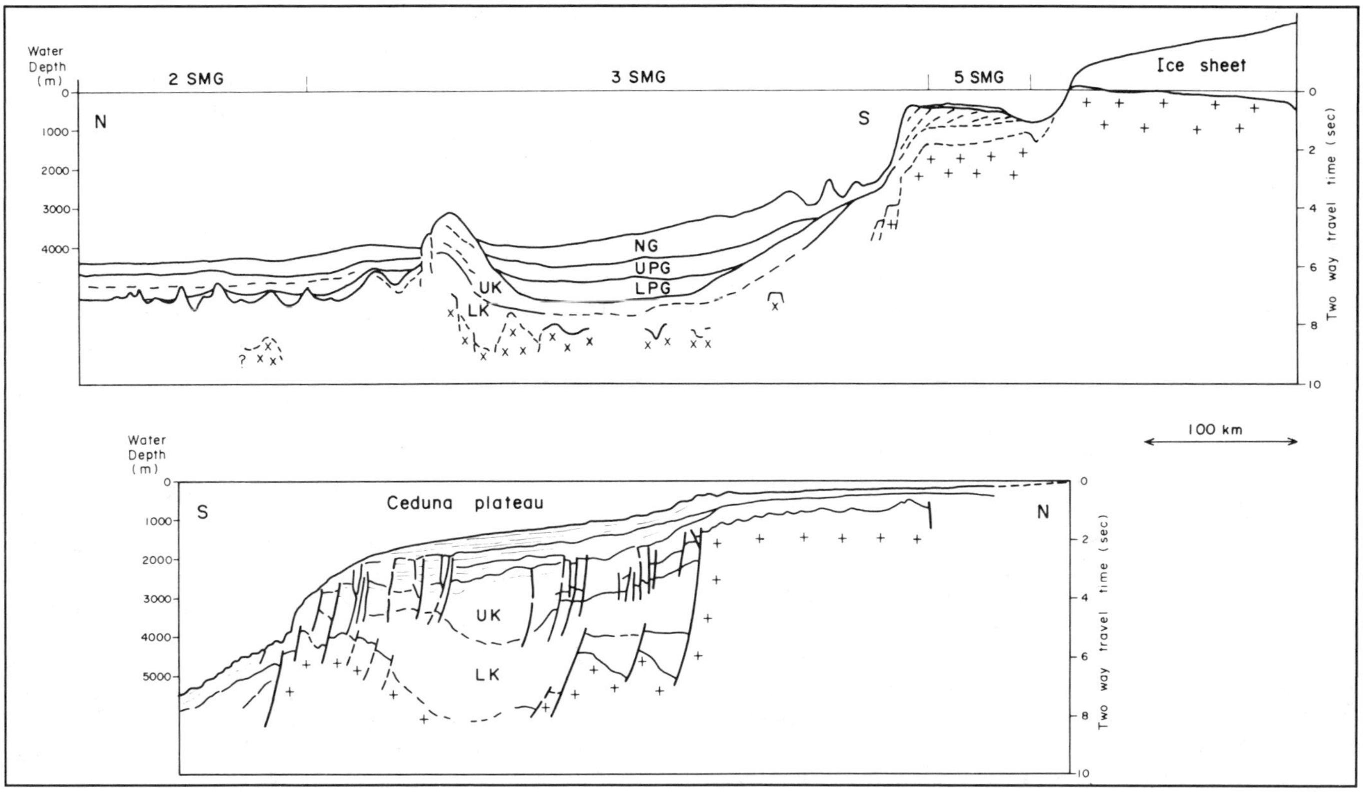

Figure 9—The comparison of the two continental margins, the Dumont d'Urville Sea area, off Wilkes Land, East Antarctica, and the Ceduna Plateau region, Great Australian Bight Basin, Southern Australia (Falvey and Mutter, 1981). NG, Neogene; UPG, upper Paleogene; LPG, lower Paleogene; UK, Upper Cretaceous; LK, Lower Cretaceous.

that the marginal high consists of anticlinally folded continental and shallow-marine sedimentary sequences, although this does not exclude the possibility of volcanic activity and igneous intrusion as inferred by Wannesson.

The initial breakup time is uncertain, because of the confusion of the magnetic anomaly identifications (Cande and Mutter, 1982). The magnetic quiet zone of the survey area probably is not due to slow spreading in the Cretaceous long magnetic quiet epoch before anomaly 34 as suggested by Cande and Mutter (1982). It is probably due to the transitional nature of the crust between oceanic and continental. This interpretation is supported by the block-faulted basement relief, gravity anomaly, and seismic facies. Initial breakup time may have been during early Paleogene, if we accept the identification by Weissel and Hayes (1972). If the interpretation of the deep strong reflectors on the southern part of 2SMG as continental fragments (Fig. 3A) is true, the transitional crust is probably developed over the continental rise and the southern part of the abyssal plain, and the initiation of normal seafloor spreading is probably middle Eocene at 100 km north of the present outer marginal topographic high.

Our heat flow data on PC301 in the abyssal plain and PC303 in the continental rise are 2.78 and 1.60 HFU, respectively (Sato et al., 1984). There are too few data for discussion of the tectonic interpretation, but we comment that the value at PC303 is not inconsistent with both hypotheses by Cande and Mutter (1982) and by this paper.

It is important to consider the geomagnetic anomaly data with the constraints of the basement structure to make a more plausible tectonic reconstruction of the area in the future.

Geologic History

The counterpart of the continental margin in the Dumont d'Urville Sea is the Ceduna Plateau region in the Great Australian Bight Basin, on the southern margin of Australia (Mutter, 1978; Weissel et al., 1977). The comparison of both profiles is shown in Figure 9 at the same scale using a profile shown by Falvey and Mutter (1981).

An extensive unconformity has been observed between infrarift (e.g., Early Cretaceous Otway Group in the Otway Basin) and riftogenic (e.g., Late Cretaceous Sherbrook Group in the Otway Basin) sequences in the southern continental margin of Australia (Falvey and Mutter, 1981). This unconformity is probably correlated to the unconformity between sequences DS-5 and DS-6. Some stratified segments, which can be observed in the acoustic basement (e.g., Fig. 7), may be Jurassic sediments deposited before the development of the rift activity.

Sequence DS-6 in the continental rise was possibly deposited in the Early Cretaceous and is inferred to be intracontinental deltaic or lacustrine deposits from its unstratified seismic character. The basin filled by sequence DS-6 may continue to the ice-covered Wilkes Basin in Wilkes Land (Drewry, 1976), which is partly exposed on the coast east of Ninis Glacier (Domack and Anderson, 1983), and has possibly been connected to the Murray Basin in Southern Australia (Veevers, 1982).

The faults, which severely deformed the basement, scarcely influence sequence DS-6 except in the area of the marginal high. It may show that sequence DS-6 was deposited in a formerly stretched, less active, broad rift basin. Sequence DS-6 shows weak deformation over the profile. The upper part of sequence DS-6 is probably erosional.

The deposition of sequence DS-5 followed the basin subsidence in about Late Cretaceous. Its center was probably located at the present northern margin of the continental rise. The sedimentary environment is interpreted as deltaic and shallow marine based on the chaotic nature of reflectors in the marginal high. Sequence DS-5 cannot be observed in the continental shelf and slope areas, possibly because of erosion during the time of continental breakup in Paleogene or by erosion preceding deposition of sequence DS-2 at the expansion of the continental glacier in the Tertiary.

There is less deformation in sequences DS-5 and DS-6 in the Antarctic margin, compared with the Upper and Lower Cretaceous sequences on the profile of the Australian margin.

The discordant unconformity between sequences DS-4 and DS-5 is probably correlated to the breakup unconformity of early Paleogene age that is observed on the southern continental margin of Australia (Falvey and Mutter, 1981).

The upper continental rise shows the highly active gravity flows and submarine channel processes of Neogene time. In the lower continental rise, hemipelagic silt and turbidites have been deposited in that same time. The post-breakup marine Tertiary is thicker here than that of the southern continental margin of Australia.

The prograding nature of sequence DS-2 at the shelf margin shows the northward progress of the continental shelf edge and continental slope. The prograding of the post-breakup marine sequences is also widely observed on the southern continental margin of Australia (Boeuf and Doust, 1975; Denham and Brown, 1976; Fraser and Tilbury, 1979; Falvey and Mutter, 1981). The very weak deformation of sequence DS-2 probably shows the stable nature of the post-breakup tectonism in the rise area. In contrast, some deformation of DS-2 on the southernmost part of the shelf area is observed (Fig. 3E), which may show the glacial subsidence and isostatic rebounding of the Antarctic continent in the post-breakup period.

SUMMARY

Despite the limitations of coverage by the survey lines and the small amount of geologic age data and lithologic evidence, we can give some concluding remarks on the geologic development of the continental margin of the Dumont d'Urville Sea area as follows:

1. The basement of the continental rise and shelf is possibly composed of early Paleozoic to Precambrian igneous–metamorphic continental strata. This basement has been extremely stretched and intruded and modified to a so-called transitional crust between continental and oceanic crusts, possibly during the early to middle Mesozoic. Distribution of the transitional crust is probably shown by the so-called magnetic quiet zone in this area.
2. The basement subsided slightly to form a rift-stage sedimentary basin possibly in the Early Cretaceous where the intracontinental sedimentary sequence DS-6, fluviatile, lacustrine, and deltaic sediments, filled especially the marginal part of the basin. Succeeding minor erosion of sequence DS-6, deltaic and shallow-marine sediments (sequence DS-5), were deposited transgressively from the central part of the basin. The DS-5 principal deposition site can be observed at the present marginal topographic high of the continental rise.
3. The present continental rise subsided quickly, after uplift and erosion, and made a continental rise sedimentary basin of Tertiary shallow-marine to marine sequences DS-4, DS-3, and DS-2. These sequences are distinguished from the lower sequences DS-5 and DS-6 by the unconformity associated with the breakup of East Gondwanaland possibly during early Paleogene age. They onlap both southern and northern sides of the continental rise basin, and they have been deposited in the gradually subsiding sedimentary basin of the continental rise.
4. Seafloor spreading is estimated to have started at the north of the outer marginal topographic high in Middle Eocene.
5. The marginal high that occurs along our survey line at the foot of the lower continental rise is believed to be composed mainly of anticlinally folded continental and shallow-marine sequences DS-5 and DS-6.
6. Sequence DS-2 shows prograding on the outer part of the continental shelf. The lower landward portion of sequence DS-2 in the shelf area is probably composed mainly of a nonmarine sequence. The sedimentation of sequence DS-2 possibly occurred during the Neogene.
7. Deformation of sequence DS-2 on the continental shelf area was possibly influenced by Quaternary glacial activities. DS-1, a flat stratified sequence of possible postglacial, glacial-marine compositions, was deposited on the eroded sequence DS-2 on the shelf.
8. The chaotic sedimentary nature in the upper part of sequence DS-2 on the upper continental rise probably indicates the increase in influence of glacial sedimentation during the Neogene.
9. The irregular low subsidence rate inferred by Domack and Anderson (1983) for the Lower Cretaceous sequence may not directly reflect the difference of the subsidence history between two continental margins of southern Australia and east Antarctica. Sequence DS-6, which probably contains nonmarine Lower Cretaceous sequences correlative with those in southern Australia, shows large subsidence, which probably amounts to 7 km or more in the continental rise area. In addition, it shows 3 km of subsidence even in the continental shelf margin.
10. The characteristics of the structure of the continental margin of the Dumont d'Urville Sea compared with the southern continental margin of Australia are as follows:
    a. Thinner and less deformed infrarift and rift-stage sedimentary sequences (DS-5 and DS-6).
    b. Thicker post-breakup marine Tertiary in the continental rise.
    c. Large gradual subsidence after the continental breakup.
    d. Steeper continental slope.

Although some differences exist, the framework of the geological development of the continental margins is quite similar between these two continents (cf. Boeuf and Doust, 1975).

It is important that a closer analysis should be made of the tectonics in this area and to compare results of other survey cruises in this area, i.e., TH83, IFP, and USGS. Scientific deep-sea drilling in the Ocean Drilling Program in the near future is expected to study the tectonic and sedimentological development of this continental margin, which has the characteristic young breakup passive margin features. Future detailed studies of the area will enable more realistic constraints on the tectonic development of Atlantic-type passive continental margins in the plate tectonic framework.

ACKNOWLEDGMENTS

We wish to thank the shipboard scientists, officers, and crew of R/V *Hakurei-maru* for their assistance during the TH82 survey cruise. Thanks are also extended to Mr. Y. Kaneda and Mr. I. Mizukoshi, Technology Research Center, Japan National Oil Corporation (TRC, JNOC), for their kind discussions of and suggestions for this study. We are greatly indebted to Mr. N. Asakura, former Director of the Geophysical Laboratory, TRC, JNOC, for promoting and supporting the *Hakurei-maru* Antarctic research cruises. The manuscript was improved through critical review by Drs. A. Mizuno and Y. Okuda of the Geological Survey of Japan.

REFERENCES CITED

Anderson, J. B., D. D. Kurtz, E. W. Domack, and K. M. Balshaw, 1980, Glacial and glacial marine sediments of the Antarctic continental shelf: Journal of Geology, v. 88, p. 399–414.

Behrendt, J. C., 1983, Are there petroleum resources in Antarctica?, *in* J. C. Behrendt, ed., Petroleum and mineral resources of Antarctica: U.S. Geological Survey Circular 909, p. 3–25.

Boeuf, M. G., and H. Doust, 1975, Structure and development of the southern margin of Australia: the Australian Petroleum Exploration Association Journal, v. 15, pt. 1, p. 33–43.

Cande, S. C., and J. C. Mutter, 1982, A revised identification of the oldest sea-floor spreading anomalies between Australia and Antarctica: Earth and Planetary Science Letters, v. 58, p. 151–160.

Deighton, I., 1977, The Paleogene stratigraphy of the southern Australian margin—an environmental approach to the general theory of rifted margin development: M. S. Thesis, University of Sydney, Sydney, Australia.

Deighton, I., D. A. Falvey, and D. J. Taylor, 1976, Depositional environments and geotectonic framework; southern Australia continental margin: the Australian Petroleum Exploration Association Journal, v. 16, pt. 1, p. 25–36.

Denham, J. I., and B. R. Brown, 1976, A new look at the Otway Basin: the Australian Petroleum Exploration Association Journal, v. 16, pt. 1, p. 91–98.

Domack, E. W., and J. B. Anderson, 1983, Marine geology of the George V continental margin; combined results of Deep Freeze 79 and the 1911–14 Australasian Expedition, *in* R. L. Oliver, P. R. James, and J. B. Jago, eds., Antarctic earth science: Canberra, Australian Academy of Science, p. 402–406.

Domack, E. W., W. W. Fairchild, and J. B. Anderson, 1980, Lower Cretaceous sediment from the East Antarctic continental shelf: Nature, v. 287, p. 625–626.

Drewry, D. J., 1976, Sedimentary basins of the East Antarctic craton from geophysical evidence: Tectonophysics, v. 36, p. 301–314.

Eittreim, S. L., A. K. Cooper, and Scientific Staff, 1984, Marine geological and geophysical investigations of the Antarctic continental margin, 1984: U.S. Geological Survey Circular 935, 12 p.

Falvey, D. A., and J. C. Mutter, 1981, Regional plate tectonics and the evolution of Australia's passive continental margins: BMR Journal of Australian Geology and Geophysics, v. 6, n. 1, p. 1–29.

Fraser, A. R., and L. A. Tilbury, 1979, Structure and stratigraphy of the Ceduna Terrace region, Great Australian Bight: the Australian Petroleum Exploration Association Journal, v. 19, pt. 1, p. 53–65.

Hayes, D. E., and L. A. Frakes, 1975, General synthesis—Deep-Sea Drilling Project, Leg 28, *in* D. E. Hayes, ed., Initial reports of the Deep-Sea Drilling Project: Washington, DC, U.S. Government Printing Office, v. 28, p. 919–942.

Houtz, R. E., 1974, Continental margin of Antarctica; Pacific-Indian sectors, *in* C. A. Burk, ed., The geology of continental margins: New York, Springer-Verlag, p. 655–658.

Houtz, R. E., and R. G. Markl, 1972, Seismic profiler data between Antarctica and Australia, *in* D. E. Hayes, ed., Antarctic oceanology. II. The Australian-New Zealand sector: American Geophysical Union, Antarctic Research Series, v. 19, p. 147–164.

Johnson, G. L., J. R. Vanney, and D. Hayes, 1982, The Antarctic continental shelf, *in* C. Craddock, ed., Antarctic geoscience: International Union of Geological Science, Series B, v. 4, p. 995–1002.

Kimura, K., 1982, Geological and geophysical survey in the Bellingshausen Basin, off Antarctica: Nankyoku Shiryo (Antarctic Record), n. 75, p. 12–24.

Mutter, J. C., 1978, An analysis of the Mesozoic–early Cenozoic rifting history of Australia and Antarctica: M. S. Thesis, University of Sydney, Sydney, Australia.

Okuda, Y., T. Yamazaki, S. Sato, T. Saki, and N. Oikawa, 1983, Framework of the Weddell basin inferred from the new geophysical and geological data: Memoir of National Institute of Polar Research, Special Issue 28, p. 93–114.

Sato, S., N. Asakura, T. Saki, N. Oikawa, and Y. Kaneda, 1984, Preliminary results of geological and geophysical surveys in the Ross Sea and in the Dumont d'Urville Sea, off Antarctica: Memoir of National Institute of Polar Research, Special Issue 33, p. 66–92.

Talwani, M., J. Mutter, R. Houtz, and M. König, 1979, The crustal structure and evolution of the area underlying the magnetic quiet zone on the margin south of Australia, *in* J. S. Watkins, L. Montavert, and P. W. Dickerson, eds., Geological and geophysical investigations of continental margins: American Association of Petroleum Geologists Memoir 29, p. 151–175.

Truswell, E. M., 1982, Palynology of seafloor samples collected by the 1911–14 Australasian Antarctic Expedition; implications for the geology of coastal East Antarctica: Journal of the Geological Society of Australia, v. 29, p. 343–456.

———, 1983, Geological implications of recycled palynomorphs in continental shelf sediments around Antarctica, *in* R. L. Oliver, P. R. James, and J. B. Jago, eds., Antarctic earth science: Canberra, Australian Academy of Science, p. 394–399.

Tsumuraya, Y., M. Tanahashi, T. Saki, T. Machihara, and N. Asakura, 1985, Preliminary report of the marine geophysical and geological surveys off Wilkes Land, Antarctica in 1983–1984: Memoirs of National Institute of Polar Research Special Issue 37, p. 48–62.

Veevers, J. J., 1982, Australian-Antarctic depression from the mid-ocean ridge to adjacent continents: Nature, v. 295, p. 315–317.

Weissel, J. K., and D. E. Hayes, 1972, Magnetic anomalies in the Southeast Indian Ocean, *in* D. E. Hayes, ed., Antarctic oceanology. II. The Australian-New Zealand sector: American Geophysical Union, Antarctic Research Series, v. 19, p. 165–196.

Weissel, J. K., D. E. Hayes, and E. M. Herron, 1977, Plate tectonic synthesis: The displacements between Australia, New Zealand and Antarctica since the late Cretaceous: Marine Geology, v. 25, p. 231–277.

Eittreim, S. L., and Hampton, M. A., 1987, *The Antarctic Continental Margin: Geology and Geophysics of Offshore Wilkes Land*, CPCEMR Earth Science Series, v. 5A: Houston, Texas, Circum-Pacific Council for Energy and Mineral Resources

Seismic Sequences and Their Distribution on the Wilkes Land Margin

S. L. Eittreim
G. L. Smith
U.S. Geological Survey
Menlo Park, CA

Four sedimentary seismic sequences, bounded by unconformities, and three volcanic seismic sequences near the boundary with oceanic crust are recognized on the Wilkes Land margin of Antarctica. Using multichannel seismic data collected by the U.S. Geological Survey (USGS) in 1984, we define three unconformities, the two deepest of which are associated with rifting and continental breakup. A *breakup* unconformity, separating synrift sediments below from postrift above, terminates at the oldest oceanic crust, and a deeper unconformity, perhaps a *rift-onset* unconformity, erosionally truncates strata of the deepest of the four sedimentary sequences. The deepest sequence is well stratified and block faulted and is interpreted as prerift continental strata. A shallow-dipping reflector under the block faults may represent a décollement surface; it intersects the Moho near the oceanic crust boundary. The morphology of the Moho can be followed in detail across the transition from oceanic crust to continental crust. Under oceanic crust, the Moho is rough and undulant, while under the adjacent thinned continental crust, it is generally smoother, parallel-laminated reflectors are sometimes observed, and the Moho surface exhibits offsets that may be faults or shear zones.

A marginal rift basin, about 40 km wide, lies adjacent to the edge of oceanic crust and is interpreted as a remnant rift valley. The floor of the marginal rift basin is about 3 km deeper than the adjacent oceanic crust, and the basin is filled with steeply prograded clastic sediments. Using the stretching and thinning model of D. P. McKenzie, the present observed depth of the prerift surface, and an age of breakup of 95 Ma, we calculate the extent of crustal thinning across the transition zone. Under the midcontinental rise the model infers thinning by a factor of 1.7, and in the region of the marginal rift basin, thinning is inferred to be a factor of 5 and greater. The thickness of prerift crust, measured from the *rift-onset* unconformity to Moho, ranges from 20 km down to as little as 5 km, consistent with the above values of thinning calculated from the McKenzie model.

INTRODUCTION

In order to better resolve the geometry and tectonics of continental rifting and to better understand the sedimentary sequences deposited at passive continental margins, numerous recent studies have focused on the transition from continental crust to oceanic crust. The region of transition from a normal 35 km thick continental crust to oceanic crust 10 km thick is characterized by a seaward-thinning wedge thought to be attenuated continental crust (e.g., Montadert et al., 1979; Keen and Barrett, 1981; Le Pichon and Sibuet, 1981; Sawyer, 1985). Extensional thinning of continental lithosphere, and its resultant subsidence due to thinning and to thermal decay after an initial heat pulse, have been modeled by McKenzie (1978), and this model has been used extensively to analyze subsidence rates at passive margins. Some have argued, however, that the seaward-thinning wedge of transitional crust is not attenuated continental crust but is entirely new volcanic material formed at the time of breakup (LaBrecque and Zitellini, 1985; Bohannon, 1986).

Prerift, synrift, and postrift sedimentary sequences have various expressions on different margins. Key unconformities such as *rift-onset* and *breakup* unconformities (Falvey, 1974) marking stages in the evolution of the breakup of continents are not always obvious in seismic records, and even their existence is locally uncertain (Austin and Uchupi, 1982). The volume of volcanic deposits associated with continental breakup is variable. Hinz (1981) and Mutter et al. (1982) have identified sequences of seaward-dipping seismic reflectors of high-velocity material that have been identified by drilling (Geotimes, 1982) as basalt flows intercalated with volcaniclastics and other sediments and associated with a kind of volcanism best described as subaerial seafloor spreading. These seaward-dipping lava sequences appear in some places to form the "weld" joining continental to normal oceanic crust and are clearly important to understanding the dynamics of continental breakup and the transition to seafloor spreading.

On the Wilkes Land margin, cruises 37 and 41 of the R/V *Eltanin* obtained single-channel seismic lines as part of wide-ranging surveys in the Southeast Indian Ocean (Hayes et al.,

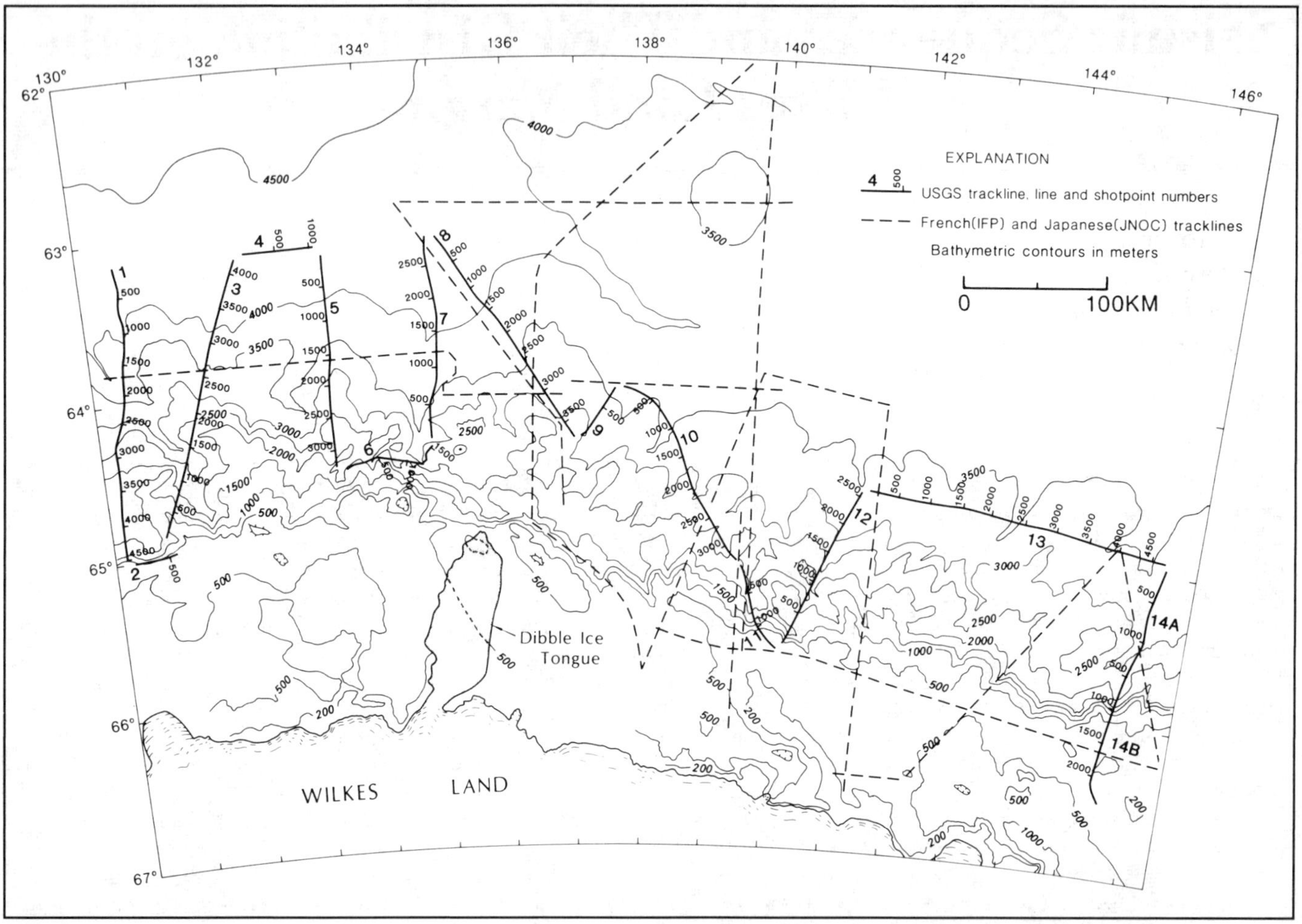

Figure 1—*S.P. Lee* cruise 1-84 multichannel seismic tracks with common depth points (CDP) indicated along tracks. Bathymetry from Chase (this volume). Tracks of Institut Francais du Petrol (IFP) and Japan National Oil Company (JNOC) shown dashed.

1975, 1976). König (this volume) discusses the *Eltanin* data as well as other geophysical data in the region between Australia and Antarctica. More recently French (Institut Francais du Petrol, IFP) and Japanese (Japan National Oil Company, JNOC) surveys have been carried out with multichannel seismic equipment on this margin (Wannesson et al., 1985; Tsumuraya et al., 1985; Tanahashi et al., this volume). On the Australian margin, extensive seismic and well data define the margin stratigraphy and structure in detail; Veevers (this volume) compares these data with the Wilkes Land margin.

The USGS survey was concentrated in the eastern part of offshore Wilkes Land (Fig. 1). The geology here is dominated by the Wilkes subglacial basin that lies parallel to the western side of the Transantarctic Mountains. This ice-filled valley, with depths to the ice-sediment interface reaching 1500 m below sea level, opens toward the coast on the north. The Wilkes subglacial basin is filled with an unknown thickness of flat-lying sediments according to radio-echo sounding data of Steed and Drewry (1982). Gradients in the magnetic field indicate depths to magnetic basement of several kilometers in places (Behrendt, 1983). Although East Antarctica is thought of as a tectonically stable Precambrian shield, the rugged sub-ice morphology (Steed and Drewry, 1982) suggests that this northeastern sector adjacent to the Transantarctic Mountains and including Victoria Land may be anomalous. The youngest widespread sedimentary deposits recognized in this part of Antarctica are those of the Beacon Supergroup (Barrett, 1982), a series of terrigenous clastics of mostly fluvial origin, ranging in age from Devonian to Jurassic, and intruded by flat-lying Jurassic sills (Ferrar Dolerites). Steed and Drewry (1982) speculated that beds of the Beacon floor the Wilkes subglacial basin. Barrett (1982) correlated the Beacon rocks to similar beds in Tasmania, which implies their existence in basinal areas of eastern Wilkes Land-northern Victoria Land.

The Gondwana breakup separating Australia from Antarctica occurred in the mid-Cretaceous according to interpretation of seafloor magnetic anomalies by Cande and Mutter (1982). From interpreted edge-effect anomalies at the edge of oceanic crust, Veevers (1986; and this volume) specified 95 Ma as the age of first generation of new oceanic crust. The earlier continental rifting phase, however, spans a time of uncertain length perhaps beginning in the Middle Jurassic with the Ferrar Dolerites. Cande and Mutter (1982) and Veevers (1986; and this volume) indicated times of slow, medium, and then fast spreading rates between the two continents. Wells on the southern Australian margin show a change from rapid to slow subsidence in the mid-Cretaceous that was postulated by Mutter et al. (1985) to reflect the change from continental crustal rifting to seafloor spreading.

The multichannel seismic data collected on R/V *S.P. Lee* cruise 1-84 in January 1984 cover the region from 130° to 146° E, with north–south lines spanning the region from the continental shelf-break northward onto oceanic crust (Fig. 1). The sedimentary sequences and reflectors dip northward in most records, and the upper sedimentary sequences of the continental rise lap onto oceanic crust. The line spacings, approximately 100 km, are not close enough to resolve the detailed distribution of structures and stratigraphic packages, but they do reveal significant trends in the major seismic sequences. This report defines the major seismic sequences and discusses their distribution over the survey area. Of particular interest here is the sedimentary and volcanic seismic stratigraphy revealed at the edge of oceanic crust. Seismic-reflection studies of passive margins are commonly hampered by acoustically opaque deposits such as evaporites or carbonate reefs that mask deep structures at the boundary with oceanic crust (Hutchinson et al., 1982). The seismic data presented here appear to be free of such acoustic problems.

MULTICHANNEL SEISMIC DATA

Recording and Equipment

The 1800 km of seismic-reflection data (lines 1 through 14B, Fig. 1) were recorded under ideal sea conditions with swells rarely above 1 m. Wave-induced noise was negligible, and the streamer cable was towed at a level depth of 12 m. Icebergs and sea ice limited the southern extent of the survey, permitting only one line extending far onto the continental shelf. Slight course changes were necessary on some lines to avoid icebergs, particularly along lines such as 6 and 11, which ran parallel to the pack ice. The recording parameters and equipment used are shown in Table 1. The data were recorded with a GUS model HDDR-4200 recording system, and the energy source was a tuned array of five air guns of volumes 148, 194, 194, 309, and 466 in.^{3}. Air guns were fitted with bronze sleeves to allow for better fit between pistons and sleeves in the cold water. No problems were encountered with icing of air lines to the air guns despite the fact that the sea water was at −1° C and no antifreeze was added to the compressed air.

Data Processing

The major aim in processing was the production of the best-stacked section possible to provide an accurate tool for structural and stratigraphic analysis. We have concentrated on accurate velocity estimations and have migrated data to properly locate dipping events. The basic processing steps are illustrated in Table 2. In addition to these basic steps, segments of data requiring special attention were analyzed with constant velocity stacks to attain the best stacking velocities and, on shallow-water portions, prestack deconvolution to attenuate multiple reflections.

Standard semblance-type velocity analyses were run at intervals of 2.5 km for each line using a summation of three adjacent common depth point gathers for each analysis. Because of the low noise content of the data, good quality velocity analyses were obtained in general (Fig. 2). Depth sections were computed for lines 1, 3, 5, and 7 using velocity models constructed from smoothed stacking velocity files.

Table 1—Recording parameters and equipment used during cruise L1-84-AN.

SOURCE:	BOLT AIR GUNS
AIR GUNS IN ARRAY:	5
NET VOLUME:	1311 CU. IN.
MANIFOLD PRESSURE:	1800 - 2000 PSI
GUN DEPTH:	10.5 M
SHOT INTERVAL:	50 M
STREAMER:	SEI MULTIDYNE.CHARGE COUPLED
GROUP INTERVAL:	100 M
AVERAGE DEPTH:	12.5 M
GROUP LENGTH:	100 M
PHONES/GROUP:	60
DEPTH CONTROLLERS:	SEI VARIABLE WING BIRDS
RECORDING:	GUS HDDR 4200. BINARY GAIN
SAMPLE INTERVAL:	2 MS
RECORD LENGTH:	10 S
RECORDING FILTER:	5-110 HZ
NUMBER OF CHANNELS:	24
NAVIGATION:	MAGNAVOX SYSTEM
PRIMARY:	"TRANSIT" SATELLITE
SECONDARY:	"GPS" SATELLITE

Table 2—Processing sequence.

1. DEMULTIPLEX:	
DESAMPLE:	4 MS
GAIN RECOVERY:	
REFORMAT:	PHOENIX I
2. TRACE SHOT EDIT:	
3. STATIC CORRECTIONS:	
RECORDING STATICS:	276 MS
DATUM:	SEA LEVEL
4. CDP SORT:	
5. VELOCITY ANALYSIS:	
WINDOW LENGTH:	60 MS
WINDOW INTERVAL:	4 MS
BAND PASS FILTER:	3-6-40-50 HZ
VELOCITY RANGE:	1400-4900 M/S
6. NMO CORRECTION:	
7. 24-FOLD STACK:	
8. BANDPASS FILTER:	HANNING
FILTER POINTS:	64
TIME WINDOW:	0.0 - 12.0 S
FREQUENCY:	4-8-50-60 HZ
9. AGC:	350 MS WINDOW

Bottom multiples are a severe problem on Antarctic shelves due apparently to the high velocity and reflectivity of the surficial, perhaps glacially overcompacted, sediments. Various types of deconvolution techniques were tested to attenuate these multiples. Among these were predictive and minimum entropy, applied before and after stacking. The minimum-entropy deconvolution, applied before stacking, appears to be the most effective.

Time migration after stack was applied to most of the long lines (as indicated in captions of record section figures) using the method of finite difference approximation to the wave equation, preceded by a band-pass filter. To smooth the velocity field laterally and vertically, every other RMS velocity function was eliminated before migration and then migration velocities were generated by reducing, by trial and error, the RMS velocities by 10–25%, increasing with time.

The record sections shown in Figures 3–11 have combined two traces into one by vertical 2:1 stack, thereby displaying one trace every 100 m. The resulting record has a vertical exaggeration of 7:1 (at the sea floor) and accentuates the

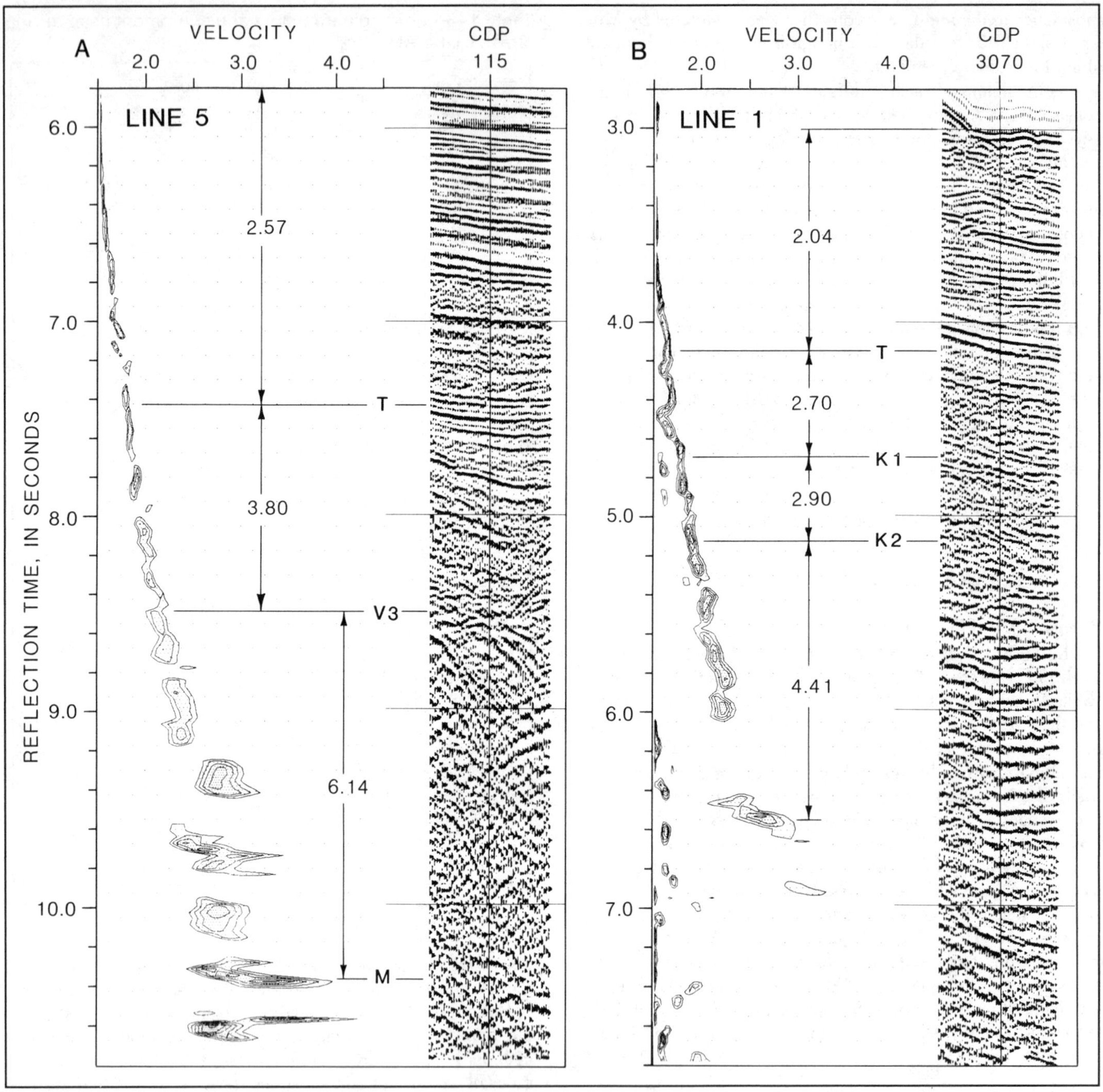

Figure 2—Semblance velocity plots, with contours representing relative coherency, and accompanying seismic-reflection record for the locations: A. On oceanic crust overlain by mostly deep marine sequence A. T is the unconformity between sequences A and B, M is Moho reflection, and V3 is oceanic basement or Layer 2. B. On continental crust of the mid-continental rise. K1 and K2 are the unconformities that divide sequences B, C, and D on continental crust.

regional stratigraphic and structural trends at the expense of detailed resolution and allows the 200+ km lines of this survey to be fit on reasonable-size foldout figures for this report. Figures 12–14 give blow ups of record sections with Figures 13 and 14 having low vertical exaggeration to better resolve details of sequence characteristics and boundaries. Digital tapes of the demultiplexed or stacked data can be obtained from the National Geophysical Data Center in Boulder, Colorado.

SEISMIC SEQUENCES AND SEQUENCE BOUNDARIES

Four major seismic sequences, bounded by unconformities, are interpreted as sedimentary sequences and termed sequences A through D. These sequences could be further subdivided locally, but the A through D breakdown has the greatest regional application over the whole survey area. From the bottom up, unconformities K2, K1, and T separate the four major sequences (Figs. 3–11, 15). Falvey (1974) proposed a

stratigraphic model of passive margin deposition incorporating two unconformities associated with continental breakup, termed *rift-onset* and *breakup*, that are caused by vertical tectonic effects occurring at the initiation of rifting and seafloor spreading, respectively. The two unconformities are separated by synrift sediments deposited during the unknown time span (presumably on the order of 10 Ma or more) between these two events. We will argue below that K2 and K1 represent Cretaceous rift-onset and breakup unconformities, respectively, and that T represents a postrift Tertiary event related either to relative sea-level change or a local tectonic event. On the northern ends of the lines, distinctive reflector sequences are found that are interpreted as volcanic units; we have termed these V1, V2, and V3. Figures 15 and 16 give descriptive definitions of each reflector sequence and sequence boundary. In the following sections, we describe the seismic data, defining the individual units starting with the deepest.

Basement

Oceanic crust can be recognized on the north ends of some lines (Figs. 3–7) by: (1) a reversal in the generally observed seaward slope of basement reflectors, and (2) a marked change in basement character. At the change in slope, normally accompanied by a step-up to the north, the basement surface increases in roughness, expressed as overlapping hyperbolic reflectors in unmigrated seismic data. We interpret these rough reflectors as the top of layer 2 of oceanic crust. Two different volcanic sequences are recognized as oceanic crust; the northernmost, termed V3, is associated with normal oceanic crust seaward of continental margins (Shipley et al., 1978). V3 has an extremely rough upper surface and no internal reflectors. The more southern of the two oceanic volcanic sequences, termed V2, has a rough upper surface but contains internal reflectors that dip generally but not exclusively seaward. These sequences are elongated east–west, parallel to the oceanic crustal magnetic anomalies.

Landward of oceanic crust beneath the continental rise, no reflectors are observed that are easily identifiable as continental crystalline basement. The deepest reflectors that we consistently observe beneath the continental rise are part of a highly stratified, block-faulted sequence with relatively high velocities averaging 4.2 km/sec (see Fig. 17). We term this deepest unit sequence D. Between the generally northward-dipping surface of sequence D forming the sloping continental margin on the south and the oceanic crust on the north, lies a basin with depths from 8 to 9 sec (6 km subbottom depth) that we term a marginal rift basin.

Sequence D

Sequence D is best displayed on lines 1 and 3 with its upper boundary defined as an erosionally truncated surface, K2 (see line 1, CDP 2000 to 2300, Fig. 3). The bottom boundary of D is not well defined; the reflections die out at depths from 7 to 8 sec. In a few locations (e.g., CDP 2300, line 1, Fig. 3), a base of sequence D can be defined by a weak cross-cutting reflector; this reflector may represent a Paleozoic–Precambrian basement, or may simply be a décollement surface associated with crustal extension. The strata of D are moderately to highly reflective, continuous, and parallel. Occasional high-amplitude beds with extreme reflectivity and uniform thickness occur that are suggestive of intrusive sills (line 1, CDP 2900 at 6.9 sec, Fig. 3). No divergence of bedding is discerned except near its erosional top, where small 200-m half-graben basins with divergent bedding lie adjacent to fault scarps (e.g., line 5, CDP 1950, Fig. 5). Therefore, it appears that the normal faulting postdates the deposition of D sediments. With the exception of drag folds near some faults, the beds are relatively undeformed. The widths of blocks between normal faults average about 5 km, and fault dips range from vertical to 25° and average about 50°. Some of the faults flatten out with depth to become parallel to what may be the deep décollement surface mentioned above (line 1, CDP 2500, Fig. 3). In places, as much as 1 km of strata in the upper part of sequence D is erosionally truncated (Fig. 3, CDP 2150; Fig. 12), and the thickness of eroded beds probably is more than 1 km locally, judging by the elevation of the lowermost D strata over the arch in the middle of lines 1 and 3. As we will argue below in the *Discussion* section, sequence D is believed to be sediment of a prerift intracratonic basin.

Sequence C

The top of sequence C is marked by an increase in stratal reflectivity and by a strong reflector and unconformity, K1. K1 is a generally nonconformable surface with significant topography and with fault-associated cuestas that truncate subcropping strata (Figs. 3, 4) and suggest an erosional surface. Normal faults are more difficult to trace in sequence C than in D because the sediments of sequence C are less reflective and the reflectors are less continuous and are chaotic in places. K1 is generally the upper limit of normal faults associated with crustal extension (Figs. 3, 4), and K1 terminates on the north at the southern edge of oceanic crust. The base of sequence C, K2, is also marked by a distinct change in stratal reflectivity, with higher amplitude and lower frequency reflectors characterizing the strata of D below. Northward downlapping of C strata onto K2 is observed locally, but directions of downlap are not always consistent. Thickening of C into the marginal rift basin appears to be due to addition of younger strata on top to the north.

A steeply dipping, prograded series of reflectors constitutes the upper portion of sequence C in the marginal rift basin on line 3 (Fig. 13). These prograding strata, with dips averaging 10°, terminate below at a high-amplitude, low-frequency reflector, termed V1, interpreted as a volcanic unit. In a few places, commonly in small basins on the downthrown sides of fault scarps, the strata of sequence C are divergent, indicating syndepositional faulting (see line 3, CDP 3100, Fig. 4). Some of the throughgoing faults of C have larger displacement at the base of the sequence than at the top, also indicating growth faulting. Sequence C is believed to be synrift sediment, deposited in the marginal rift basin and adjacent areas, prior to the initiation of seafloor spreading.

Sequences B and A

The upper two sequences, interpreted to be postrift, are discussed in detail by Hampton et al. (this volume). B is a relatively unstratified, low-reflectivity sequence. The strata that can be distinguished in B dip north, diverge, and downlap onto K1 near the edge of oceanic crust. T, the boundary between B and A, is a high-amplitude reflector onto which the overlying strata of A locally onlap southward. Sequence A is a well-stratified sequence with variable stratal configurations, reflect-

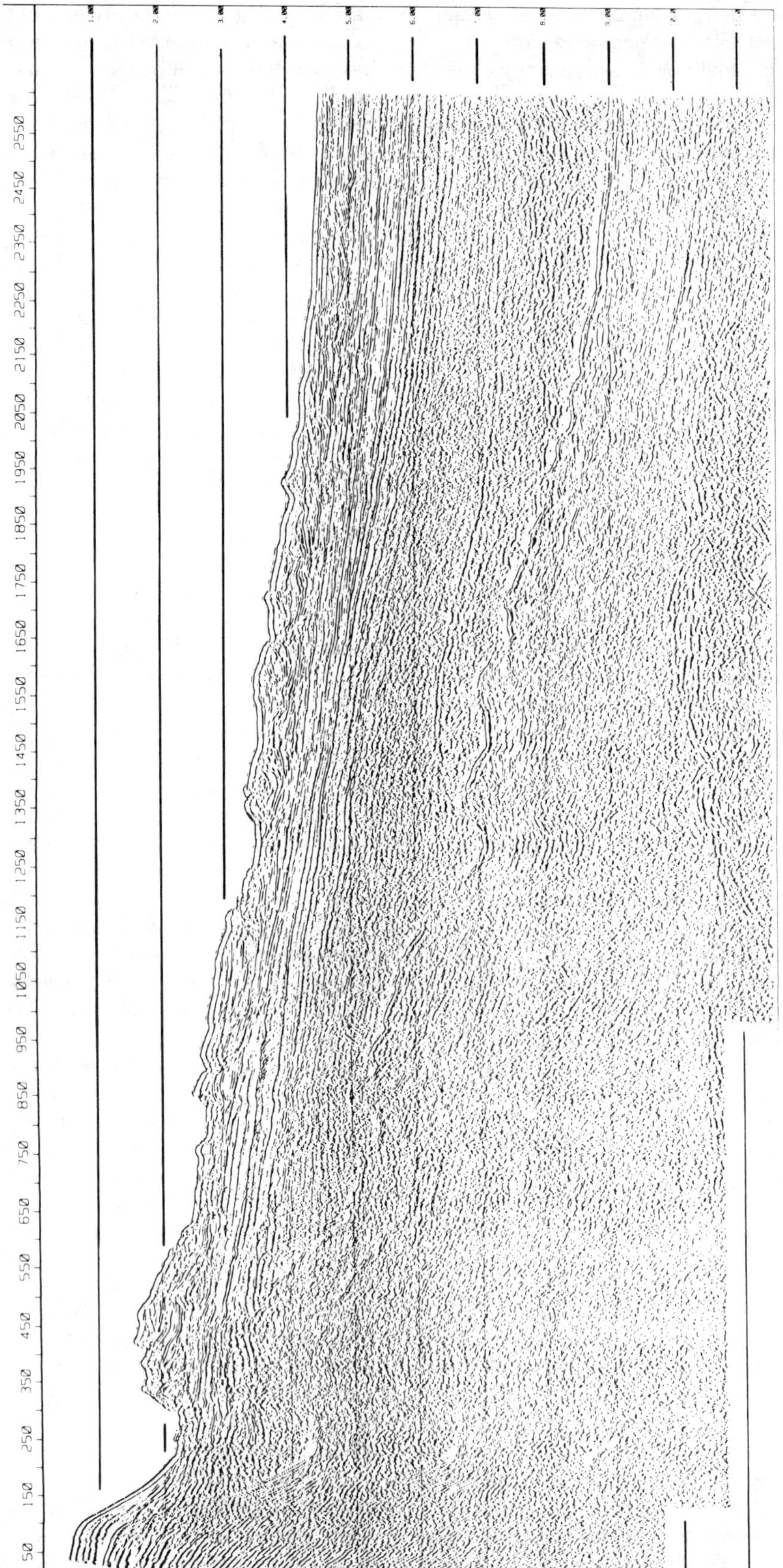

50
150
250
350
450
550
650
750
850
950
1050
1150
1250
1350
1450
1550
1650
1750
1850
1950
2050
2150
2250
2350
2450
2550

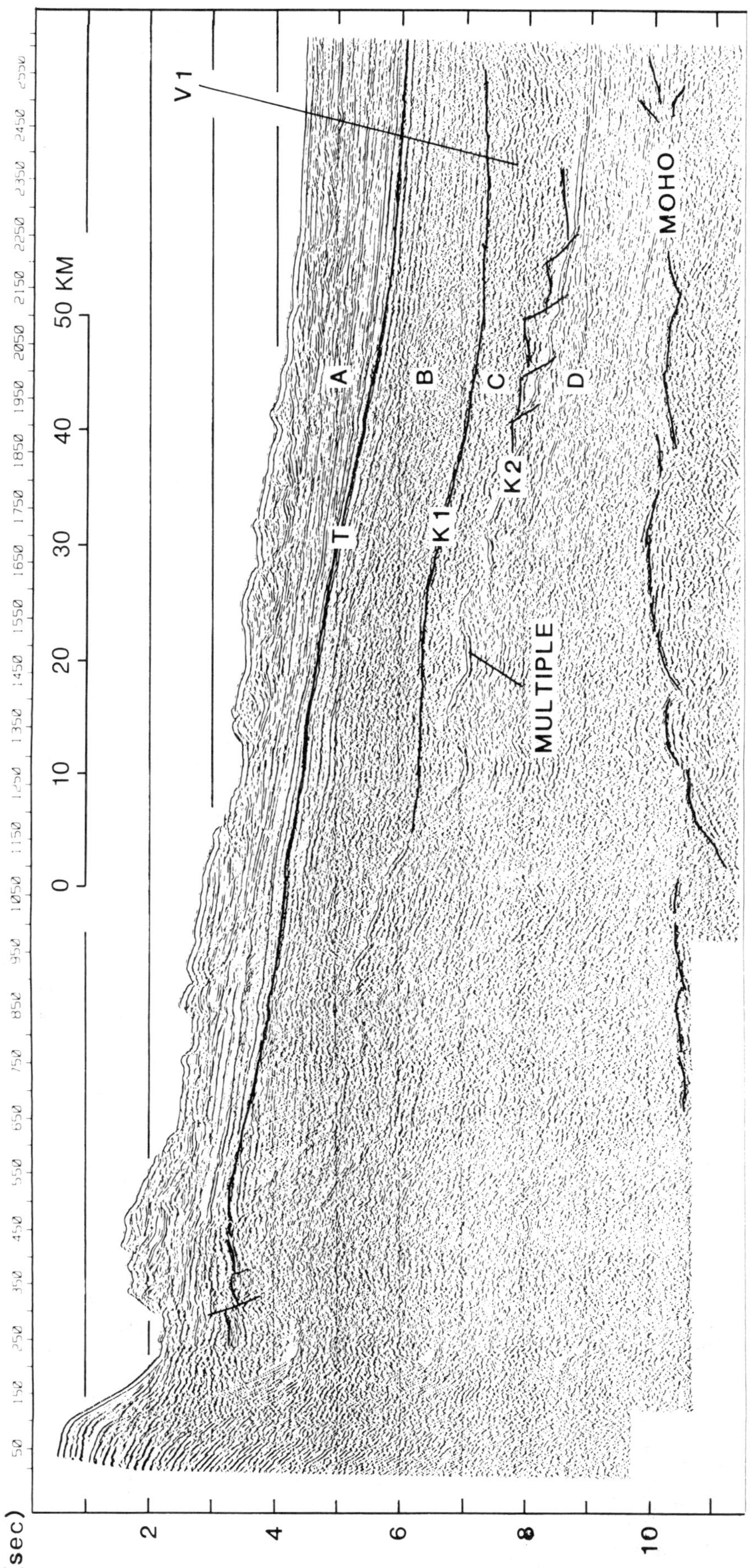

Figure 9—Time section, line 12 (not migrated). 7:1 vertical exaggeration at sea floor (uninterpreted seismic line on facing page).

ing the present continental rise depositional processes of contourite, turbidite, and mass-flow deposition. Wannesson et al. (1985) reports on a multichannel seismic line obtained through DSDP drill site 269 at 61°41′ S, 140°04′ E, which dated the middle portion of sequence A, at a depth of 958 m, as Oligocene or older.

CONTINENT-OCEAN BOUNDARY

The term continent–ocean boundary (COB) refers here to the landward edge of oceanic crust, which is also the northern boundary of the marginal rift basin, that 5 km deep basin lying adjacent to oceanic crust. Figure 18 shows line drawings of depth sections of lines 1, 3, 5, and 7 registered along the interpreted COB locations, Figure 13 includes a line drawing of the line 3 COB crossing at 1:1 vertical exaggeration, and in Figure 19 the COB is mapped over the survey area. Because the region landward of the oceanic crust edge is one of transition to normal thickness continental crust (30–35 km), it has been called the continent–ocean transition zone. Although the general view is that this transition zone crust is thinned continental crust (Montadert et al., 1979; Keen and Barrett, 1981; Le Pichon and Sibuet, 1981; Klitgord and Behrendt, 1979), some have preferred the use of the term *rift-stage* crust (Hutchinson et al., 1982) or *anomalous* crust (König and Talwani, 1977). Alternatively, some authors view this zone as a wedge of landward-thickening oceanic crust (LaBrecque and Zitellini, 1985; Bohannon, in press). For the reasons given in the following sections we interpret this transition zone to be occupied by continental crust, thinned in the rifting process.

Wannesson et al. (1985), reporting on the IFP survey of this margin that overlaps the eastern portion of our survey, defined a continent–ocean boundary as the seaward extent of layered, block-faulted continental basement. The boundary that Wannesson et al. (1985) described is 50 km to the south of our COB interpretation (Fig. 19). Our determination of the COB in the eastern half (east of line 7) is poor; this region may have great tectonic complexity because large-offset fracture zones impinge on the margin there (Veevers, this volume), and we have less seismic data coverage. Only lines 1, 3, 5, and perhaps 7, all in the western area, clearly show the edge of ocean crust and the associated volcanic and sedimentary sequences on the adjacent thinned continental crust. The line-to-line consistency gives us confidence in our interpretation of the COB location in the west. Our projection of this boundary to the east, using lines 8, 9, and 10, is less certain.

Oceanic Crust

The oldest identified seafloor-spreading magnetic anomaly, A34, occurs at the rough-surfaced basement ridge on line 3 (Fig. 18). This line is approximately 10 km west of the *Eltanin* 41 line from which, among others, the anomaly identifications were made (Cande and Mutter, 1982). A minor revision was made by Veevers (1986), who interpreted this anomaly as due to an edge effect of oceanic against continental crust in addition to the A34 polarity reversal. This latter interpretation provides an explanation of the anomalously broad positive magnetic feature that has been interpreted as due to A34. Magnetic data from our cruise confirmed the shape of this anomaly (Childs and Stagg, this volume). Therefore, there is little doubt, given the acoustic signature and magnetic field data, that the rough-surfaced basement north of CDP 3750 on line 3, and similar features on adjacent lines, represent oceanic crust.

Two volcanic seismic sequences have been identified with oceanic crust (Fig. 16). The southernmost sequence, V2, has a rough upper surface but contains a series of irregular internal reflectors that generally dip seaward. The superposition of strata (onlapping and overlapping) shows that the youngest V2 beds occur to the north. V3 is interpreted as normal oceanic crust with a rough upper surface and no internal reflectors, and it occurs on the northern end of lines 3 and 5. The upper reflectors of V3 onlap those of V2 (Fig. 18).

Marginal Rift Basin

As shown in Figures 3–6 and 18, a basin with a width averaging about 40 km is consistently observed landward of the edge of oceanic crust. The northern flank of this basin is an abrupt rise of about 2 km up to the surface of the oceanic crust whereas the southern flank is a more gentle incline of strata with some normal faulting. Beneath the basin at about 9 sec we observe a block-faulted, stratified sequence. This is observed on lines 1, 3, and 5. This stratified sequence is correlated to sequence D because of similar characteristics and a continuity with beds of sequence D on the south. The occurrence of sequence D beneath the marginal rift basin implies that the basin is floored by continental crust rather than new igneous crust of some anomalous type. The strata of this sequence dip gently (about 10°), and fault blocks are about the same size as in sequence D further landward where this sequence is well-defined. Overlying this, a high-amplitude sequence, termed V1, with rougher morphology consisting of from 2 to 5 reflectors, lies at about 8–8.5 sec. V1 has a distinctive, easily recognized signature on most of the lines in the vicinity of the oceanic crust edge, with generally low but variable frequency, very high amplitudes, and a moderately rough morphology, with interfingering, short, and segmented reflections. V1 occupies the central portion of the marginal rift basin; its roughness is clearly greater than the underlying planar-bedded, block-faulted strata of sequence D; and, except for the large throughgoing normal faults on lines 5 and 7, it has not been affected by, and is therefore younger than, the block-faulting in the beds below (Figs. 13, 14, 18). Overlying V1 on line 3 (Fig. 13), and less clearly so on line 1, we observe a sequence of sigmoid-prograding reflectors filling this basin. The prograding facies have unusually long and straight slope faces (1 km vertical relief) with rather abrupt slope changes at top and abrupt base-of-slope breaks or simple downlaps at bottom.

Superposition of Seismic Sequences at the COB

The sequence of deposition of volcanic and sedimentary units can be interpreted based on the superposition and onlap relations of strata, referring to Figure 13. The volcanic sequence V1 was deposited on a thin layer of lower sequence C, apparently on the floor of the marginal rift basin. The prograded sequence of the upper part of C was then deposited with northward onlaps onto V1. The 1-km relief on the prograded foresets indicate a basin of 1 km or greater depth. In turn the prograded beds are topped by the oldest beds of the V2 sequence that are roughly coeval with unconformity K1.

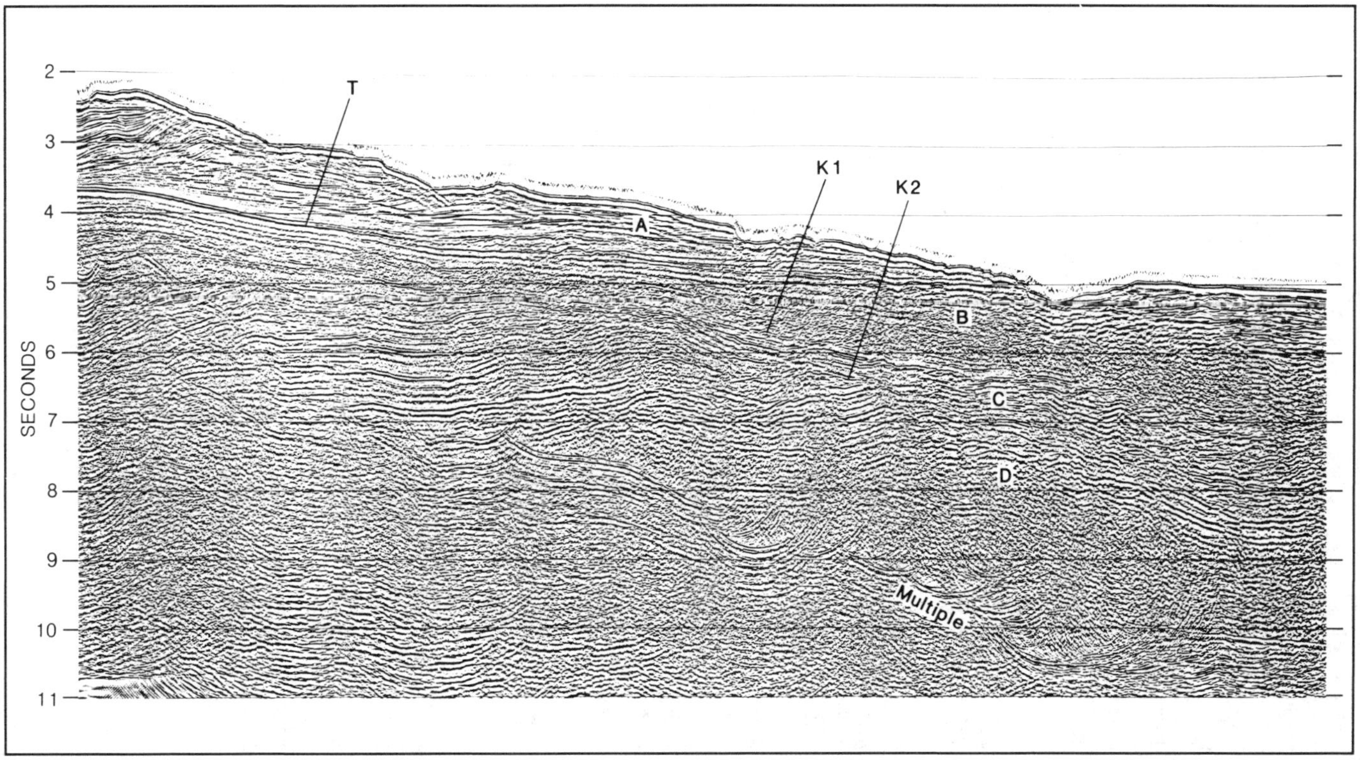

Figure 12—Portion of line 1 migrated seismic section, 7:1 vertical exaggeration at sea floor, showing principal sequences and sequence boundaries. See Figure 3 for location.

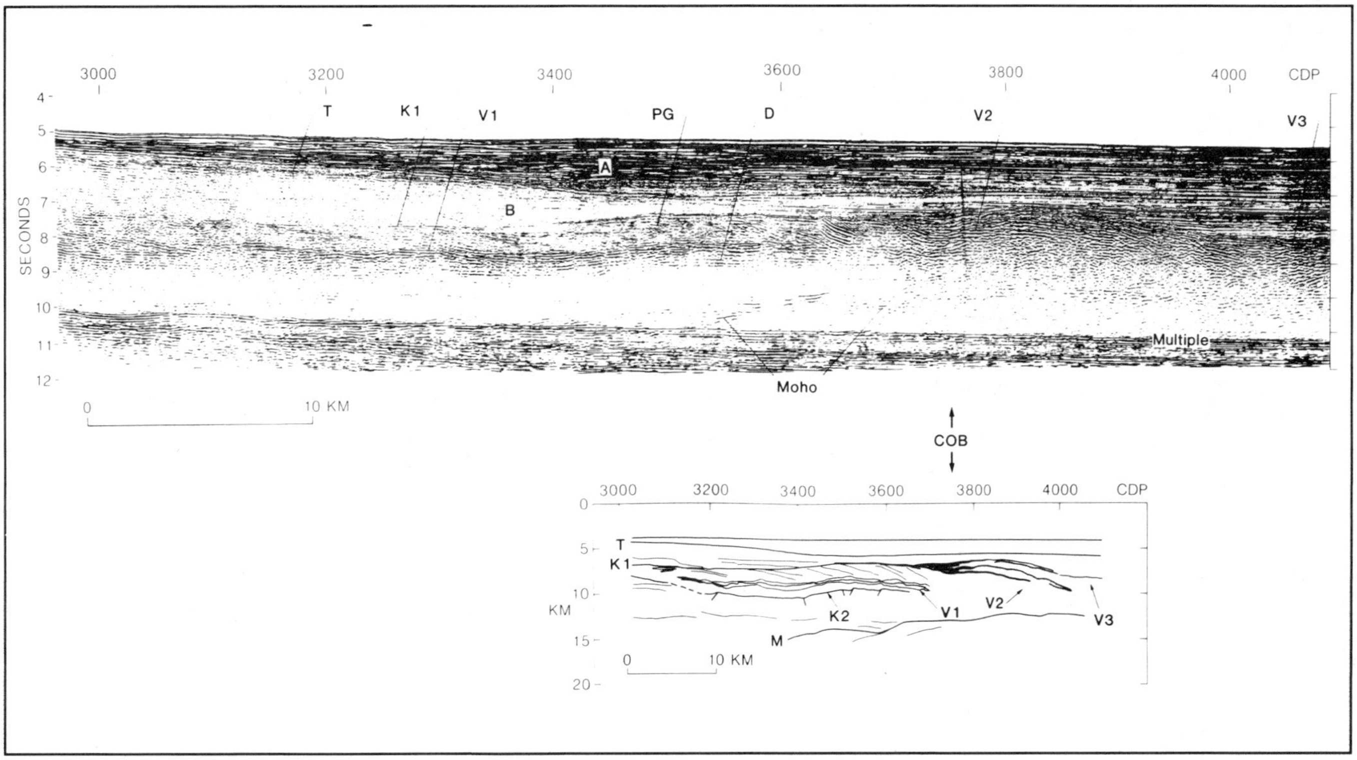

Figure 13—Portion of line 3 migrated seismic section, 2:1 vertical exaggeration at sea floor, showing unconformities T, K1, sequences A, B, D, prograded beds (PG) of sequence C in marginal rift basin, and volcanic sequences V1, V2, and V3. See Figure 4 for location. Line drawing is of same line, at 1:1 vertical exaggeration, for the portion across the continent-ocean boundary (COB).

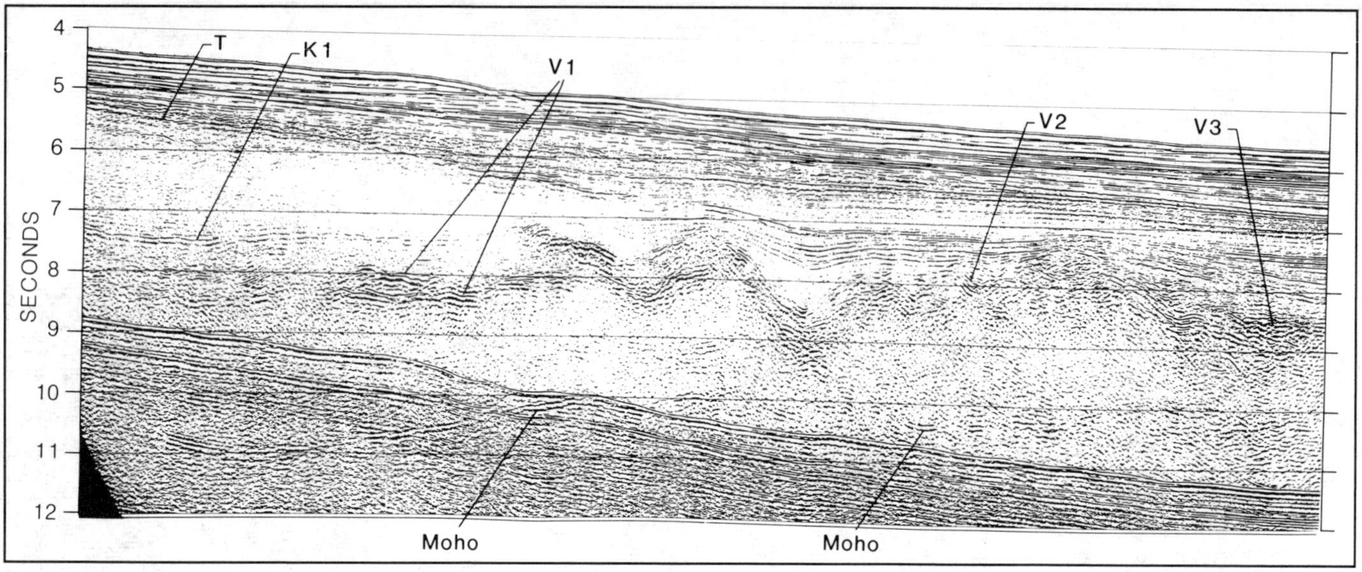

Figure 14—Portion of line 5 migrated seismic section, 2.5:1 vertical exaggeration at sea floor, showing volcanic seismic sequences V1-V3 associated with the edge of oceanic crust. See Figure 5 for location.

SEDIMENTARY SEISMIC-SEQUENCES

Velocity	Sequence	Unconformity	Description	Environment
2.0	A		well-stratified; reflectors continuous, variable configuration; flat at deep-water end; cut-and-fill, drifted deposits, slumps at shallow-water end	CONTINENTAL RISE & SLOPE
		T	onlap above, offlap or erosional truncation below; high amplitude	
2.7	B		very weak stratifications; poorly continuous & short discontinuous reflectors gently dipping and diverging north	CONTINENTAL SHELF
		K1	locally erosional; variable morphology surface	
3.7	C		faulted discontinuous strata, chaotic in places; high energy prograded strata at deep end, line 3	SYNRIFT BASIN
		K2	erosional	
4.2	D		well-stratified, planar beds; block faulted; occasional high-amplitude bed	TERRESTRIAL OR EARLY-SYNRIFT BASIN

Figure 15—Definitions of seismic sedimentary sequences derived from the records of Figures 3-11. Sequence boundaries are unconformities. The velocities shown are average interval velocities derived from line 3.

VOLCANIC SEISMIC-SEQUENCES		
	Description	Interpretation
V1 	High amplitude, low frequency, sinuous series of 2-5 reflectors sometimes faulted ; onlaps beds of lower C (synrift) sequence and constitutes part of upper C sequence near boundary with ocean crust	Flood basalt deposited on thinned continental crust
V2 	moderate roughness overlapping and generally seaward-dipping internal reflectors; higher elevation upper surface than V1	Stratoid, or Icelandic basalts of earliest oceanic crust, perhaps subaerial
V3 	myriad overlapping hyperbolae on unmigrated record; no internal crustal layering above Moho; seaward-most of 3 types; poor reflectivity	normal oceanic crust of pillow basalts and sheet flows at surface

Figure 16—Definitions of seismic volcanic sequences derived from the records of Figures 3–11.

The time correlation between oldest V2 and K1 is shown by the feather-edge of V2 that extends southward onto sequence C and becomes indistinguishable from K1 (Fig. 13). Younger V2 strata are added to the sequence northward, although in some places, such as on line 5, this northward progression to younger strata is irregular, with landward, as well as seaward, dips implying volcanic flows having sources moving in variable directions, rather than exclusively seaward. Finally, V2 strata are onlapped by beds of the oldest "normal" oceanic crust (with no internal reflectors), presumed to be typical pillow basalts and sheet flows overlying a sheeted dike complex of layer 2.

It is important to point out that the sequence of dipping reflectors that we interpret as prograded clastics (PG) are quite different from the seaward-dipping reflector (SDR) packages that have been mapped off Norway and other passive margins and interpreted as basalt flows (Hinz, 1981; Mutter et al., 1982; Mutter, 1985). Our PG unit is bounded at the base by a high-reflectivity unit, V1, angular changes are seen in the reflectors near the top that are characteristic of topset-foreset slope changes, and the PG reflectors are more closely spaced (about 80 m) than that observed in the SDR reflector packages (about 200 m).

Moho Reflections

Deep reflections, presumed to mark the M-discontinuity, were well recorded on all the major lines except 8, 10–11, and 14A and B. The clearest examples of Moho reflections are on lines 5 and 12 (Figs. 5, 9). Semblance velocity analysis, although not resolving velocities precisely above this reflector, give high average crustal interval velocities ranging from 4.5 to 7.0 km/sec, ruling out an origin for these events as side-reflected energy from shallower depths (e.g., Fig. 2). Sonobuoy refraction velocities confirm mantle velocities at the approximate depths of these reflectors on lines 1 and 8 (Childs and Stagg, this volume). Under oceanic crust, Moho reflections occur at about 10 sec (4.5 sec subbottom), and they deepen southward under transitional crust to 11 sec (7 sec subbottom) before loss of signal. As shown in Figure 18, corresponding depths, based on smoothed velocity models from the CDP stacking velocities, range from 12 to 20 km below sea level. Both laminated and single-event Moho reflections are seen. On line 5 (Fig. 5), the reflection changes character from a rough, overlapping series of reflectors beneath oceanic crust to a smooth, single-event reflector along a descent southward beneath transitional crust to two inflection points, one at CDP 1400 and one at CDP 1800. At the inflection points where the reflector changes dip, the character of the reflections changes to laminated. The highest amplitude Moho reflections occur on line 12, where the Moho character is a series of rough, overlapping multiple events with a duration of about a half-second.

DISTRIBUTION OF SEISMIC SEQUENCES

Figure 19 shows the major tectonic boundaries and disposition of the three volcanic facies marking the transition to oceanic crust. Large fluctuations in basement depth on the shore-parallel line, line 13, may be caused by numerous oblique crossings of the COB, or alternatively, may be caused by 50-km spaced fracture-zone offsets. Neither of the connecting lines 12 or 14 show clear crossings of the COB but may show indications of its approach at the north ends of the lines in the form of shallowing basement depths. The progression of volcanic sequences from V1, which floors the marginal rift basin, to V2, which forms the oldest oceanic crust, to V3, "normal" oceanic crust, is shown for the westernmost lines. This progression is best shown in line 3 (Fig. 13). The COB converges on the shelf break toward the east.

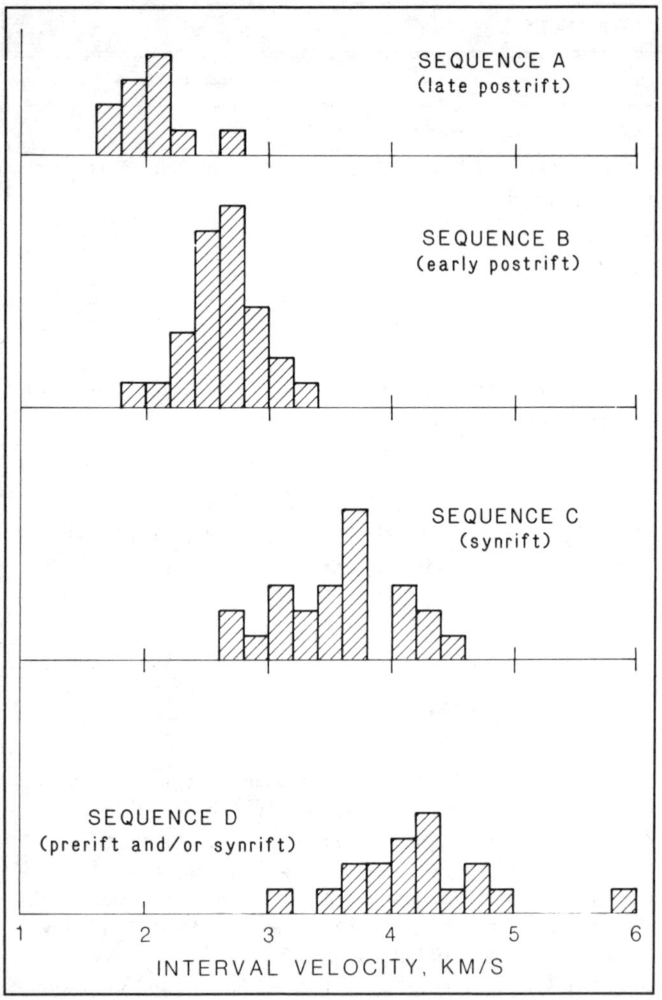

Figure 17—Interval velocities derived from best (i.e., least ambiguous) semblance velocity analyses of line 3, arranged by sequence.

A break in slope, or hingeline, of the K2 surface marks the major northward increase in thickness of synrift sediments which, in the case of lines 1 and 3 (Figs. 3, 4, CDP 2400 and 2000, respectively), occurs on the seaward side of an arch. Near the continental slope–shelf break, there is evidence of another major northward drop in K2 marking a much greater relief in block-faulted continental basement. This landward hingeline is observed in lines 11 and 14 (Figs. 8, 11, CDP 700 and 1000, respectively) and probably corresponds to a landward increase in crustal thickness associated with the magnetic trough (Veevers; König, both this volume). The magnetic trough is a conspicuous magnetic low observed on both the Australian and Antarctic margins at about the continental slope. König and Talwani (1977) postulate that it is caused by a magnetic edge effect associated with the juxtaposition of thick continental crust against a thinner, rifted crust of uncertain composition. Both the outer and inner hingelines appear to mark the most significant northward thinning of continental crust. The landward hingeline does not always occur at the continental slope-shelf break; on some lines, the shelf break appears as a purely sedimentary feature with no structural control.

Total Crustal Thickness

Total crustal thickness, from sea floor to Moho, ranges from about 10 km (4 sec) beneath oceanic crust to about 20 km (7 sec) beneath thinned continental crust (Fig. 20). The preliminary identification of Moho reflections in the data of IFP (Wannesson et al., 1985) are included in the map of Figure 20 and agree with our interpretation of Moho depths. The major increase in depth of the Moho beneath thinned continental crust occurs a few kilometers south of the COB, and this inflection point occurs at a crustal thickness of 12 km (5 sec). The inclination of the Moho along its steepest incline is about 7° (e.g., line 5, Fig. 18).

Total Sediment Thickness

The marginal rift basin contains the thickest sediment, greater than about 7 km (4 sec), and this zone of maximum thickness dominates the isopach map (Fig. 20). Another basin with thicknesses greater than about 5 km (3 sec) occurs between the outer and inner hingelines. This landward basin is poorly defined because of the generally poor data on the shelf, where a hard sea floor inhibits the penetration of acoustic energy and generates strong bottom multiples. In the far eastern area on lines 13 and 14, the total sediment thickness is greatest, reflecting a greater thickness of postrift (i.e., post K1) sediment. A greater sediment supply to the east may also be responsible for the fact that the shelf break is prograded farther northward, closer to the COB. The contours through the east–west line 13 are intended to be schematic only; the sediment thickness trends here may be controlled by basement ridges that have a general north–south trend parallel to fracture zones of the oceanic crust.

Sequence C

More than about 3 km (1.5 sec) of sequence C fills the marginal rift basin (Fig. 21). The landward basin, south of the outer hingeline, contains lesser thicknesses. Sediments of this landward basin show less stratification (Fig. 3), suggesting a shallower water, less-marine environment. The arch of continental basement about 80 km distant from the COB thus produces two parallel synrift basins, with the marginal rift basin, adjacent to oceanic crust, collecting far thicker, and probably more marine-rich sediment.

Sequence B

Sequence B is the weakly stratified unit deposited immediately on top of K1 (i.e., postrift) and forms wedges (Fig. 21) with maximum thicknesses greater than about 2 km (1.5 sec) at approximately 50 km landward from the COB on the westernmost three lines. The abrupt seaward thinning, versus the more gradual landward thinning, suggests that this packet is a shelf/slope wedge (Hampton et al., this volume).

Sequence A

Sequence A, the most recent, highly stratified sequence (Fig. 21), is thickest in the eastern part of the survey, with more than about 3 km (2.5 sec) reflecting presumably a greater sediment source in that area in the Tertiary. Locally, thicknesses are controlled by the canyon-channel systems and associated levees in detailed patterns that are not resolvable with the line spacing here. A minimum in this sequence occurs at the boundary between upper and lower rise, where there is a

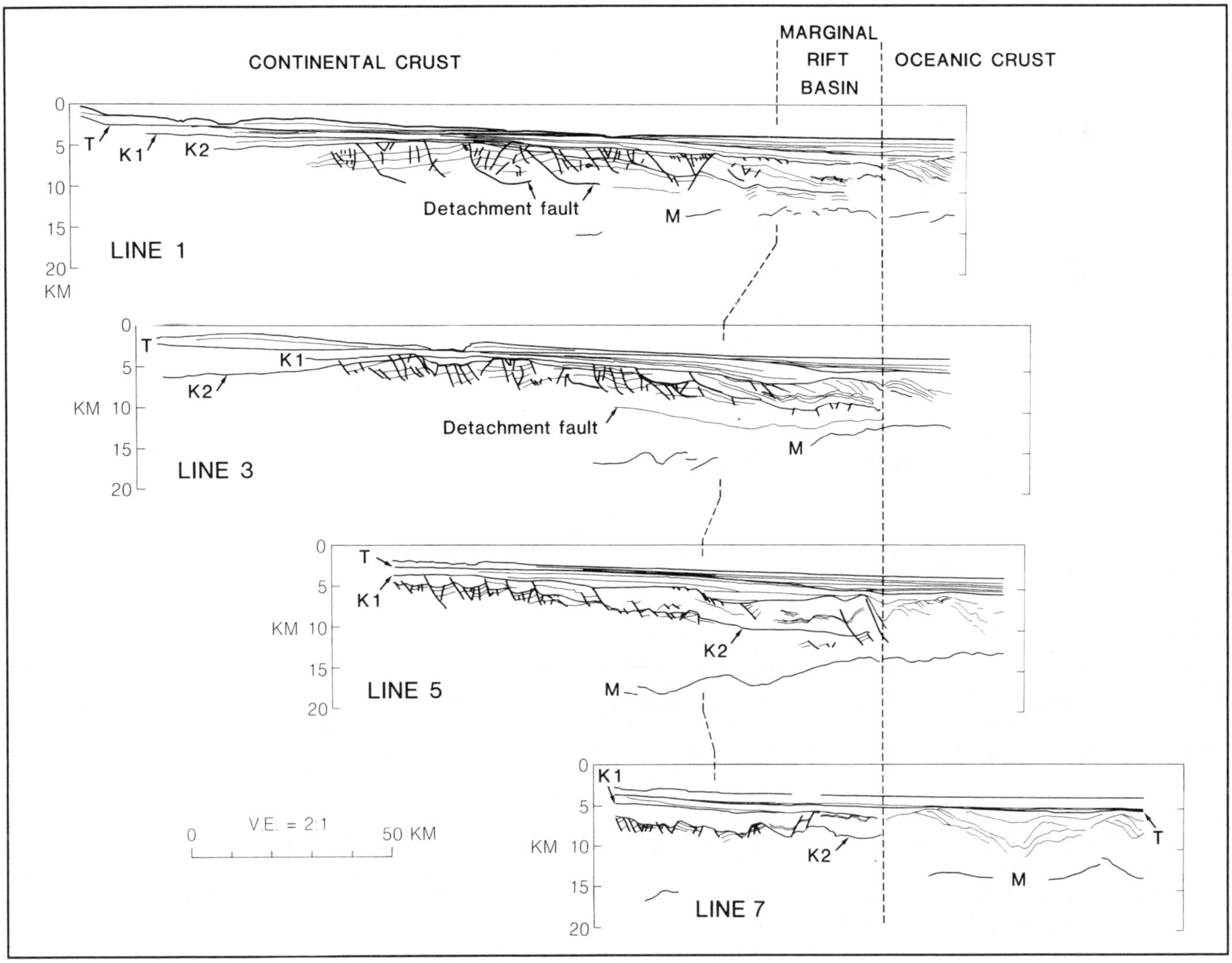

Figure 18—Line drawings of depth sections of four westernmost lines registered along the continent-ocean boundary (COB). Vertical exaggeration 2:1. Depth computation based on velocity model generated from smoothed stacking velocities.

break in slope separating the dissected-fan topography of the upper rise from the smooth slope of the lower rise.

SUBSIDENCE HISTORY

We believe, from the seismic evidence here and the evidence from other passive margins, that the transition zone is floored by thinned continental crust. Unconformity K2, the assumed top of prerift rocks, shows evidence of subaerial erosion, and today is at a depth of more than 8 km below sea level across large portions of the seismic lines. This subsidence of the margin is in response to thinning of the crust and postrift thermal contraction of the lithosphere, supplemented by sediment loading. Figure 22 shows subsidence curves for continental crust based on the McKenzie (1978) model of lithospheric stretching, heating, and cool-down. The model calls for an initial subsidence caused by thinning of the crust, then a prolonged period of thermal subsidence associated with contraction of the lithosphere. The principal feature of the subsidence curves is the exponential decay of subsidence with time, with the greatest subsidence occurring in the first few tens of millions of years after breakup. The largest uncertainties in the application of the McKenzie model appear to be near the time of rifting and breakup. Pre-breakup uplift along the rift shoulders can be a significant factor in subsidence analyses as discussed by Steckler (1985) for the Red Sea and Rowley and Sahagian (1986). There seems little doubt, however, that after a few tens of millions of years beyond breakup, the subsidence of passive margins fits the McKenzie model quite well based on available well data (Watts, 1981) and the model provides a basis for analyzing the Wilkes Land margin data. The question to be addressed is whether the present depth of surfaces that were originally at or near sea level can be explained by an existing model and if so, what the model implies in terms of the degree of crustal thinning.

Two methods are used to derive thinning, or β, factors. One is to measure directly the present crustal thickness (K2 to Moho), and divide that into an assumed initial thickness. Another is to assume that the present depth of K2 was

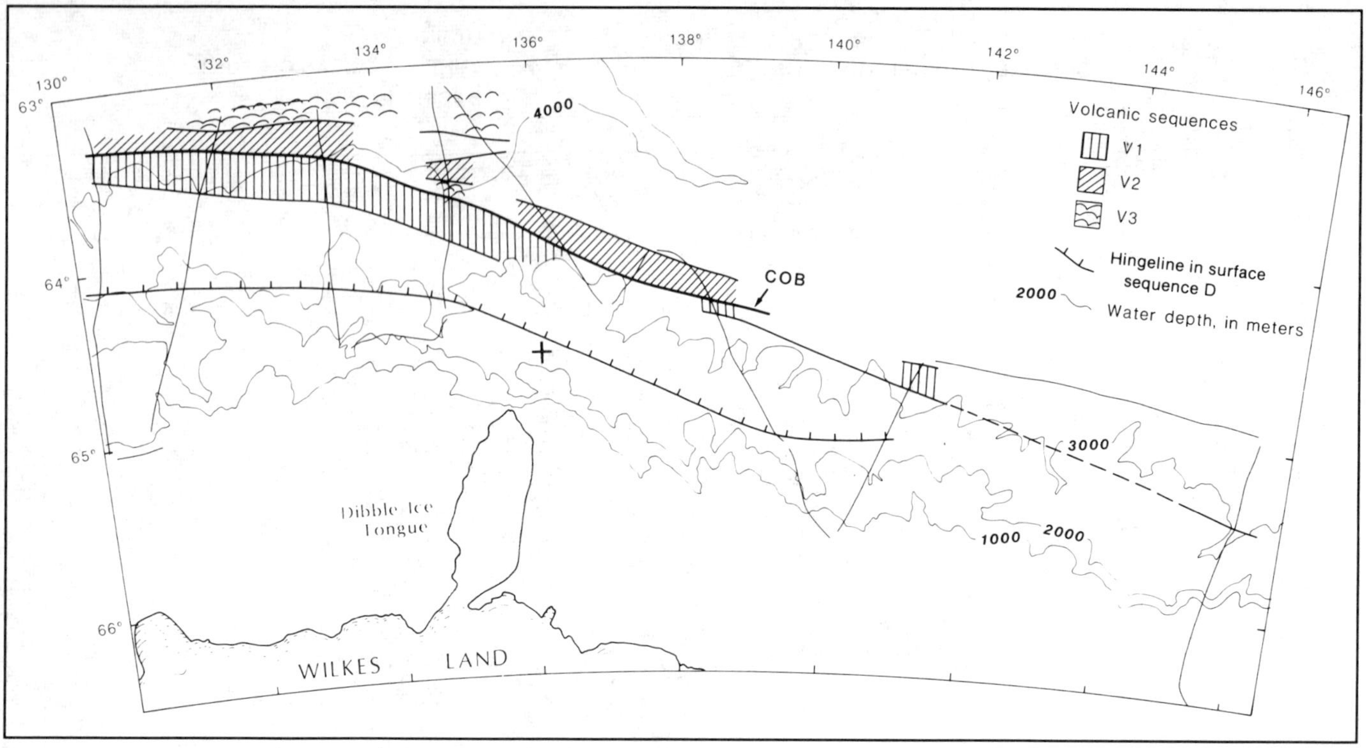

Figure 19—Tectonic map of area showing the three basement volcanic types associated with oceanic crust, the edge of oceanic crust (COB) and a hingeline, noted as a northward drop in surface of sequence D and significant increase in thickness of synrift sequence C. Also shown as (+) is the COB location as interpreted by Wannesson et al. (1985).

reached by postrift subsidence according to the McKenzie model and calculate what β factor would be required to produce such subsidence. The two methods are not completely independent, since both assume the same initial crustal thickness. To compensate for the depression of the crust caused by its sediment load, we have used the empirically derived Crough (1983) assumption of 600 m of depression per second of travel time of overlying sediment. This figure has been found to apply with less than 10% error for thicknesses of sediment of 3 sec or less, and smaller error for thinner sediments (Crough, 1983). At CDP 2100 on line 1, unconformity K2 displays clear indications of erosion by the truncation of reflectors, and thus provides a sea-level marker and will be used as a reference location to apply the McKenzie subsidence model. The same calculations should, in a more detailed analysis, be made all along the profiles where seismic packet boundaries are well defined. The intent in this report is a more limited analysis to test the model's applicability and to find out whether the degree of subsidence observed is physically reasonable. The unloaded depth of unconformity K2 at CDP 2100, line 1, is shown to correspond to a stretching factor, β, of 2.4 (Fig. 22), assuming the McKenzie model with initial crustal thickness of 35 km and an age of breakup of 95 Ma. If 110 Ma, instead of 95 Ma, is used for the breakup time and 30 km, instead of 35 km, is used for initial crustal thickness, the inferred β value would change by about 10%. Although some of the assumptions in the method may be in error, it is clear that the present depth of the prerift surface K2 can be explained by the crustal thinning and thermal contraction model, using reasonable β values.

Calculations of unloaded K2 depth were made along each of lines 1, 3, and 5, the intersections with the field of β curves at 95 m.y. past breakup was noted, and the inferred β values plotted in Figure 23. This calculated β value is compared with a "measured" β value, which is the observed K2-to-Moho thickness divided into 35 km, the same assumed prerift crustal thickness used in the model. Note that calculated β values under the landward portions of the profiles in the region of well-defined block faults of sequence D are about 1.8 and in the marginal rift basin adjacent to oceanic crust become large, generally above 4. Le Pichon and Sibuet (1981) pointed out that $\beta = 3.2$ is a critical value of stretching that corresponds to crustal subsidence of 2.5 km, the depth of midocean ridge crests. In general this value would not be expected to be exceeded, although these same authors pointed out that it may be locally greater in the Bay of Biscay. Hence the meaning of the large β values is not clear, but the general agreement between calculated and "measured" β values lends support to the idea that the McKenzie model is a good basis for explanation of basement depths observed for this region and suggests that an "over-thinning" of continental crust occurred beneath the marginal rift basin; or that the marginal rift basin is underlain by some kind of anomalous, so far inadequately modeled, crust.

In Figure 24, the seismic sequence boundaries K2, K1, and T are backtracked to their original position at sea level, for the location CDP 2100, on line 1, along curves consistent with the McKenzie (1978) model. The thicknesses of sequence B and C were corrected for compaction using the seismic velocities of the sequences, the Nafe and Drake (1963) curves of varia-

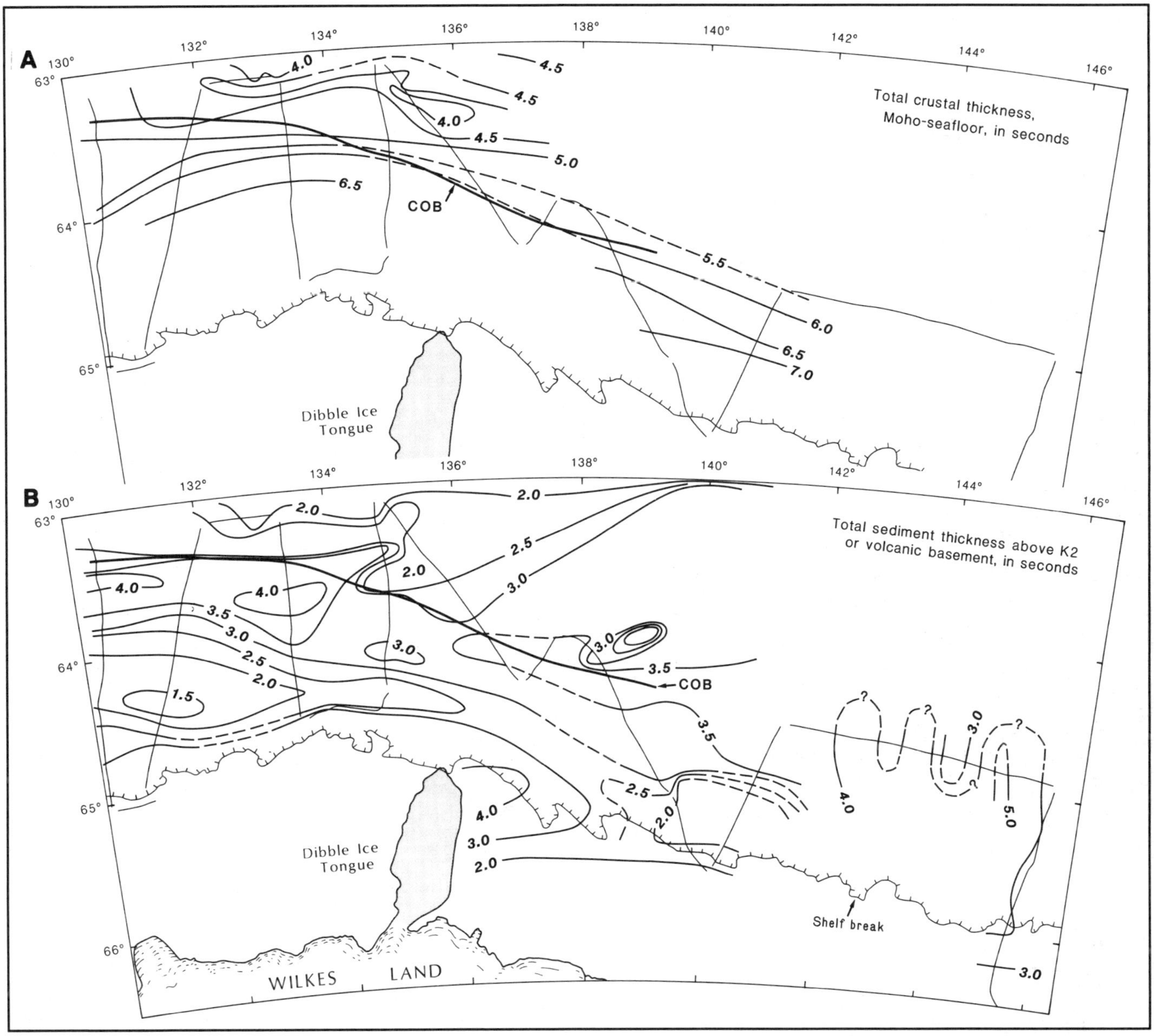

Figure 20—A. Total crustal thickness measured from sea floor to Moho in seconds. B. Total sediment thickness, in seconds, above basement. Basement is defined as one of the volcanic sequences on oceanic crust, or unconformity K2 on continental crust.

tions of density with velocity in sediments, and an initial assumed density of 1.9 g/cm^3. The corrected thicknesses were then used to fix the distance between sequence boundaries at the end of each depositional sequence, and the exponential curves were drawn fixed to the depth of burial at the present time. The greatest uncertainty in the subsidence history probably involves times near breakup, or the transition from rift to drift. The McKenzie model does not consider a finite rifting time length, and our knowledge of the tectonic events at this critical point is sketchy at best.

An age is implied for T of Late Cretaceous, or at the youngest, considering the uncertainties in initial subsidence time of K2, lowermost Tertiary. There could of course be a large time hiatus represented at T, with overlying sediments much younger. However, to be consistent with typical subsidence curves observed on other passive margins, which are well explained in general by the McKenzie model, Figure 24 shows that surface T, a highly reflective horizon interpreted as a shelf deposit (Hampton et al., this volume), must have sunk below wave-base not more than about 40 million years after breakup.

DISCUSSION AND CONCLUSIONS

The two unconformities, K2 and K1, both displaying erosional characteristics, may fit the model of Falvey (1974) as *rift-onset* and *breakup* unconformities, respectively. According

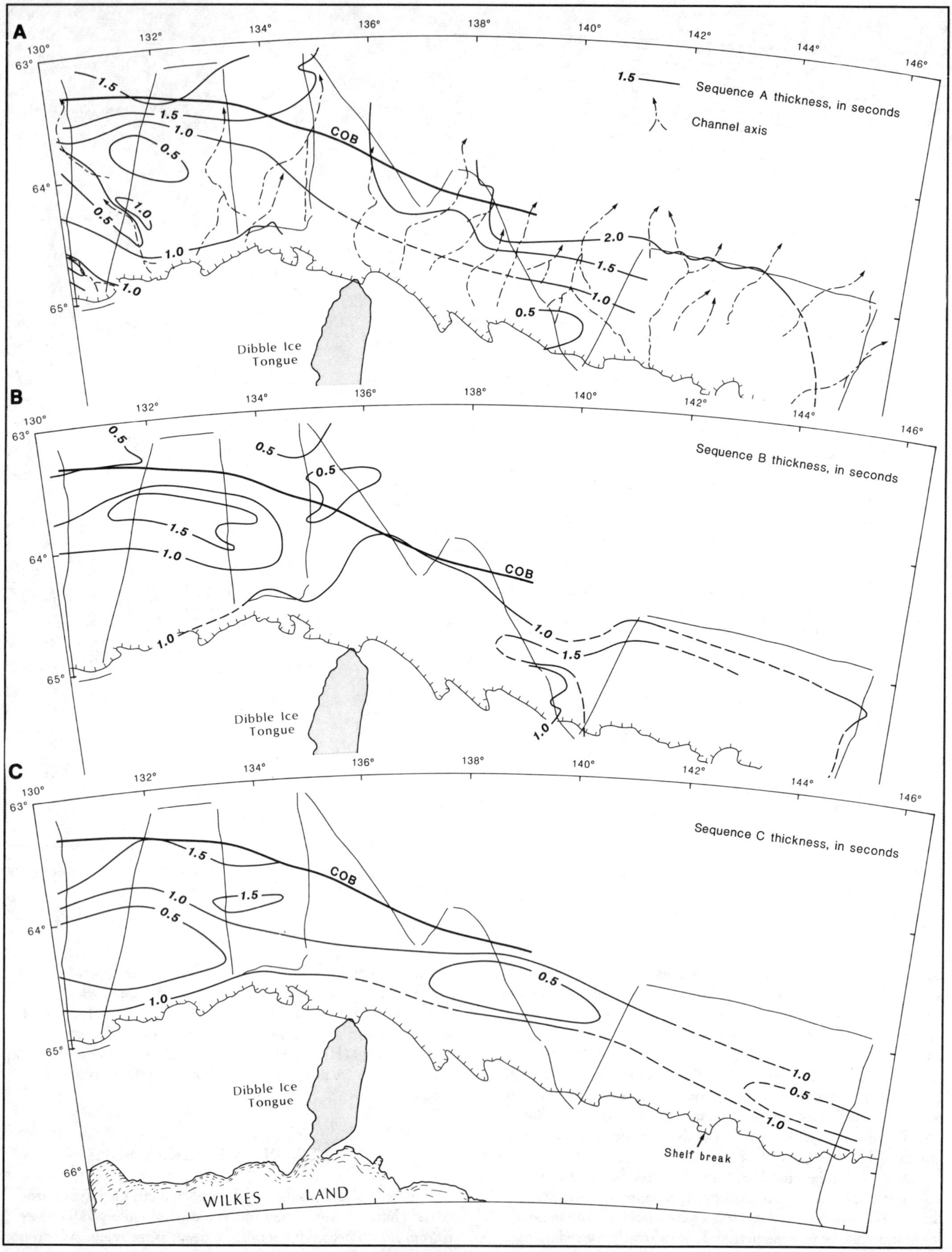
A
130°
132°
134°
136°
138°
140°
142°
144°
146°
63°
64°
65°
1.5
Sequence A thickness, in seconds
Channel axis
COB
1.5
1.0
0.5
1.0
0.5
1.0
1.0
2.0
1.5
1.0
0.5
Dibble Ice
Tongue
B
130°
132°
134°
136°
138°
140°
142°
144°
146°
63°
64°
65°
Sequence B thickness, in seconds
0.5
0.5
0.5
1.5
1.0
COB
1.0
1.0
1.5
1.0
Dibble Ice
Tongue
C
130°
132°
134°
136°
138°
140°
142°
144°
146°
63°
64°
65°
66°
Sequence C thickness, in seconds
1.5
COB
1.0
1.5
0.5
0.5
1.0
1.0
0.5
1.0
Dibble Ice
Tongue
Shelf break
WILKES
LAND

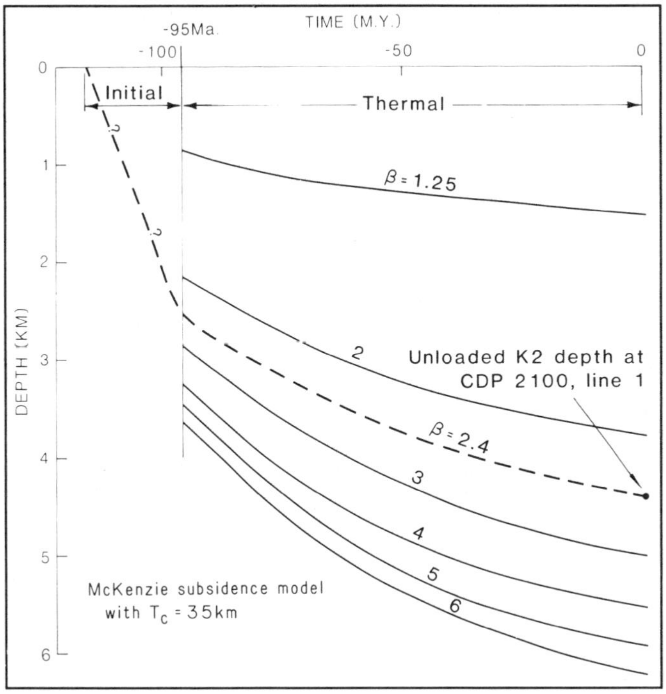

Figure 22—Theoretical crustal subsidence versus time according to the McKenzie (1978) model with an assumed initial crustal thickness of 35 km and an age since breakup of 95 Ma, for a range of different factors of stretching, β. Other initial crustal parameters are assumed same as in McKenzie (1978). The tectonic subsidence [corrected for sediment loading by the Crough (1983) assumption] of sequence boundary K2 is shown for the location CDP 2100 on line 1, and implies a $\beta = 2.4$. The subsidence history near and before the time of breakup during the continental rifting phase is uncertain as suggested by question marks.

to Falvey (1974) a *rift-onset* unconformity is caused by vertical movements associated with thermal doming and the block faulting of rift tectonics. A *breakup* unconformity is due to a pulse of uplift associated with the first generation of oceanic crust and deep convective upwelling of asthenospheric material. The period of rift-onset erosion would begin at the time of rifting, would extend into the synrift period, and would depend on relative local relief. The upper unconformity, K1, matches this model on three grounds: (1) it is the uppermost limit of extensional faults of the underlying sediments; (2) it is time-correlative with the oldest identified volcanics of oceanic crust; that is, sediments older than K1 have not been deposited on oceanic crust; and (3) it marks a significant change in sedimentation regime, as evidenced by the change in acoustic character across it and by its erosional nature. On the basis of (1), it marks the time of cessation of major extension of continental crust expected to coincide with the onset of separation of the continents by seafloor spreading. On the basis of (2), it is directly linked in time with this oldest oceanic crust.

The lower unconformity, K2, on the other hand, is less certain in its identity as a rift-onset unconformity. It prominently displays erosional characteristics and marks a change in sedimentary regime, but whether it is the deepest of such unconformities in the section cannot be determined with certainty partly because of the complex faulting deeper in sequence D. The best evidence of syndepositional faulting appears to be in sequence C, whereas sequence D shows parallel-bedded strata with no evidence of growth faults or divergence of strata, and on these grounds the prerift and synrift definitions seem to fit. In any case, K2 formed at the time of or just before the development of the marginal rift basin, with post-K2 sediments filling the basin. Because K2 is offset by normal faults, rifting must have continued after its formation, but some fault offsets can be found that are degraded by erosion with little fault movement evident posterosion. Since the marginal rift basin was a likely early development in the rifting history of the margin, K2 is certain to mark an early stage in the rifting history, and our preferred interpretation is that K2 is indeed a rift-onset unconformity that formed during the time of extensional faulting of continental crust.

Assuming K2 and K1 are rift-onset and breakup unconformities, respectively, sequence D is then prerift and C, synrift. Subaerial erosion of the upper beds of D took place across the arch or hingeline beneath the upper continental rise on lines 1 and 3. During rifting, subsidence of the top of D was greatest in the marginal rift basin, allowing deposition of 1–3 km of synrift sediments. It was minimal, zero, or negative (uplift) over the arch under the continental rise. Sequence D might be correlated with the Beacon Supergroup of the Antarctic continent (Barrett, 1982; La Prade, 1982), a series of fluvial clastic beds about 2.5 km thick, distinctively flat-lying and intruded by the Ferrar Dolerites in the Jurassic. Seismic velocities measured in the field for the Beacon rocks range from 4.1 to 4.4 km/sec (Crary, 1963; Bentley and Clough, 1972) and in the laboratory, from 2.2 to 4.4 km/sec (Barrett and Froggatt, 1978). Figure 17 shows an average velocity of sequence D of 4.2 with a range of about 3.5 to 5 km/sec. Paleocurrent transport directions in Beacon sediments of the Transantarctic Mountains indicate north-directed flows toward Tasmania (Barrett, 1982); thus, there is reason to expect a thickening of Beacon sediments in the direction of the northern Wilkes Basin on the eastern Wilkes Land margin. Sub-ice morphology studies (Steed and Drewry, 1982) show a flat sediment infilling of the subglacial Wilkes Basin, suggesting significant thickness of relatively young sediments, perhaps Beacon, there.

The COB is defined in this paper as the southernmost extent of oceanic crust as determined by seismic-reflection characteristics. Wannesson et al. (1985) interpreted a COB about 50 km south of our interpreted COB, based on location of the northernmost extent of clearly indicated block-faulted continental crust. Our two different definitions bracket the area of the deep marginal basin; the basin is flanked on the south by blocks of sequence D faulted down to the north and on the north by elevated rough-surfaced basement presumed to be oceanic crust. Therefore, the basic difference in interpretation involves the nature of the crust forming the marginal rift basin. Similar marginal rift basins have been found on other passive margins: e.g., northwest Atlantic (Klitgord and Behrendt, 1979), northeast Atlantic (Avendik et al., 1982), southwest Norwegian margin (Eldholm and Sundvor, 1980), and southern Australian margin (König and Talwani, 1977; Veevers, this volume). If our interpretation is correct that the marginal rift basin is floored by thinned continental crust, based on the

Figure 21—A. Sequence A (late postrift) thickness in seconds. B. Sequence B (early postrift) thickness in seconds. C. Sequence C (synrift) thickness in seconds.

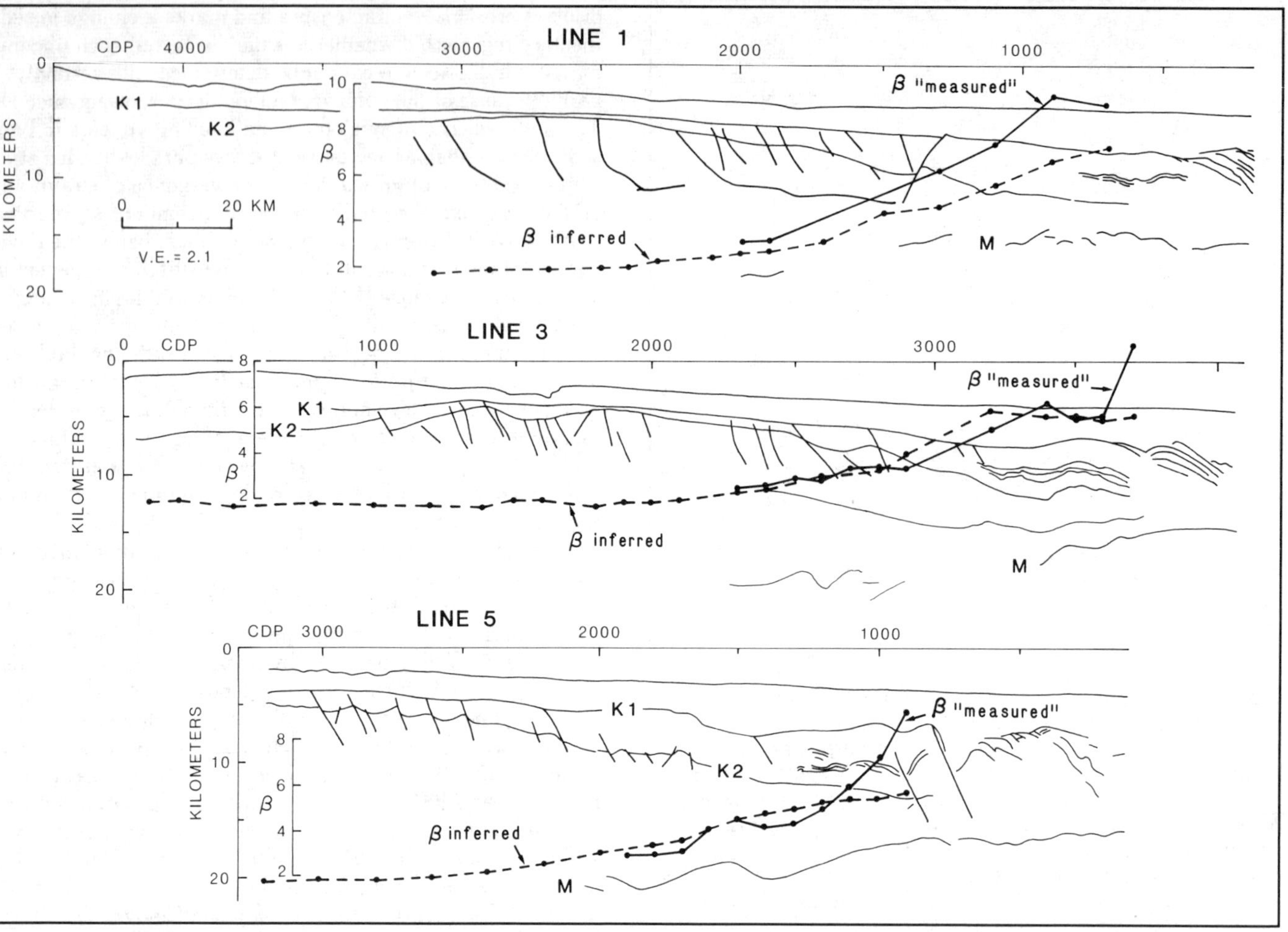

Figure 23—Depth sections of the three westernmost lines, with stretching factor, β, "inferred," by the McKenzie (1978) model (such as done in Fig. 22) and "measured" by taking the present observed K2-to-Moho thickness divided into the assumed initial thickness of 35 km. The latter calculation assumes that thinning equals stretching.

observation in lines 1, 3, and 5 of parallel-bedded block-faulted reflector sequences beneath the volcanic floor (V1) of this basin (see line 3, CDP 3200-3700), then an extreme thinning of continental crust is implied (to a minimum of about 5 km locally, measured from the top of K2). As shown in Figure 23 and discussed above, β factors of >5 were "inferred" and "measured" for this area. Such thinning cannot have been accomplished solely by simple extension of the whole crust, because the observed blocks of continental crust are rotated only mildly, approximately 10°. Rowden and Keen (1980), Keen and Barrett (1981), and Bott (1982) suggested varying β factors vertically, with more stretching in the lower crust to account for this discrepancy. Moore and Eittreim (this volume) discuss a similar mechanism of extension in which lower crustal slabs move out away from the rift axis from underneath largely intact upper layers, from which they are separated by detachment surfaces.

We interpret the marginal rift basin at the edge of oceanic crust as a former rift valley. The 40 km width and 2 km depth of the marginal rift basin is approximately the same as presently active rift valleys such as the East African Rift Valley (Baker et al., 1972). The prograded clastics with foreset beds of 1-km relief seen on line 3 require an environment of rapid deposition, presumed to be tectonically active. Similar deep, prograded sequences of synrift age are seen on the Atlantic margin near the edge of oceanic crust (Klitgord and Grow, 1980, fig. 7). Crustal elevation is commonly associated with the rift valleys of the earth, and the uplifts responsible for the erosion at K2, on the shoulders of this former rift valley, can be explained either by a particular cross-sectional geometry of the stretched zone in the lithosphere (Rowley and Sahagian, 1986), by secondary mantle convections associated with rifting (Steckler, 1985), or by curvature in detachment surfaces along which the upper blocks of continental crust move during extension (Davis, 1983).

Three different volcanic sequences are associated with oceanic crust and the COB region. The landward-most, V1, is relatively flat-lying, with moderate local roughness and very high acoustic impedance contrast with the surrounding sediments. V1 is deposited either on the oldest of the synrift deposits or directly on the prerift surface K2, the uncertainty being due to our inability to resolve whether there is a significant thickness of sediment between K2 and the lower boundary of V1. The size of this deposit ranges from 10 to 30 km in north–south width and several hundred meters to a kilometer in thickness. It is interpreted as a flood basalt that blanketed the

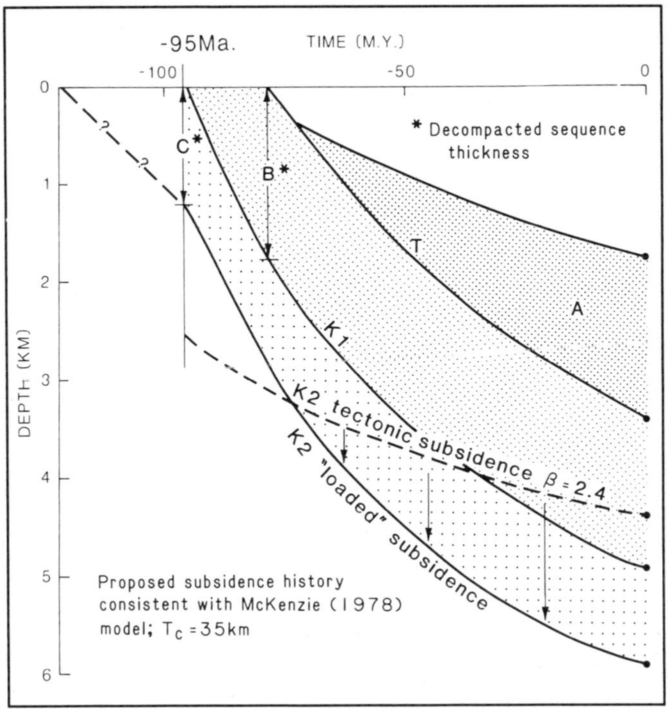

Figure 24—Proposed subsidence history at CDP 2100 line 1 assuming the McKenzie (1978) model, $\beta = 2.4$, modified by assuming that sequences B and C were deposited at or near sea level. The sequences were decompacted using the Nafe and Drake (1963) relation between velocity and density and the Crough (1983) assumption was used for the difference between unloaded and loaded depth of K2.

fault-block topography of sequence D below. The confinement of sequence V1 to the center of the marginal basin suggests that it may have been extruded on the floor of an already existing rift valley.

Volcanic sequence V2, the earliest of which was deposited on top of the prograded clastics of the rift valley, is interpreted as lava flows extruded from a generally northward-migrating extrusion center, thus producing the seaward dip by crustal loading and rotation according to the model of Mutter (1985). The occurrence of these seaward-dipping reflectors in the ridge at the edge of oceanic crust and at the location of magnetic anomaly 34 confirms the idea of Mutter (1985) that the bulk of these sequences are not constructed on continental crust, as initially proposed by Hinz (1981), but constitute new volcanic crust. However, in partial agreement with the Hinz model, the oldest of these deposits is a thin feather-edge, up to 100 m thick, that does overlie the marginal basin of thinned continental crust (line 3, CDP 3500-3700, Fig. 13).

The youngest and seaward-most volcanic sequence, V3, has characteristics typical of oceanic crust observed elsewhere (Shipley et al., 1978) with rough upper surface and no internal reflectors above the Moho. Our lines unfortunately traverse only short portions of this V3 crustal type where it is well defined (lines 3 and 5), but it may exist on other lines where the interpretation is more ambiguous (lines 7, 8, and 14).

Extrapolation of our stratigraphy onshore is difficult because of limited and poor quality continental shelf data (line 14, Fig. 11). Wannesson et al. (1985) and Tanahashi et al. (this volume) show continental basement (our sequence D) rising southward beneath the shelf from depths of from 2 to 4 sec at the shelf break to depths of from 0.5 to 2 sec beneath the inner shelf. Domack et al. (1980) reported Lower Cretaceous sediment cored at the bottom of an ice-scoured inner-shelf trough at 1400 m depth, showing that synrift sediments are at levels within about a kilometer of sea level at least in some locations along the Wilkes Land margin. The likely configuration then is a southward elevation of sequences. In places where crystalline basement rocks crop out at the coast such as Dumont d'Urville (140° E) there may be a fault along the coast at the inner margin of the inner-shelf trough (Domack et al., 1980).

Although very little can be stated with any certainty regarding hydrocarbon potential, because the age and nature of strata can only be guessed at, the following facts are relevant. On the conjugate Australian margin Cretaceous synrift and postrift deposits are shallow-water carbonaceous sandstone, siltstone, and marl (Fraser and Tilbury, 1979). Some of these sediments have been deposited in restricted environments, and coal beds occur in the Cenomanian. In our survey area, total sediment thicknesses range from 1.5 to 10 km above K2. The thickness of the Cretaceous synrift unit C ranges up to 6 km in the marginal basin near the COB and 3 km in basins and wedges on the flank of the seaward-dipping K2 under the upper continental rise. The thickness of sequence D below K2 adds at least another 4 km to the section thickness under the upper continental rise. A time–temperature model using the depth history of burial was generated using our depth sections, the assumed sequence ages based on a 95 Ma breakup time, and the assumed history of subsidence of Figure 24. The results show that there is enough depth of burial for much of the Cretaceous synrift section to be in the time–temperature oil-generating window (Waples, 1980) in both the areas of the marginal rift basin near the COB and in the basin landward of the hingeline. Whether the composition of these sediments would be conducive to hydrocarbon generation is of course unknown.

Our interpretation of the progression of events that produced the seismic stratigraphy discussed above is the following:

1. The Gondwana crust was extended in Jurassic and Cretaceous time, causing the normal faulting of sequence D and subsequent erosion of fault-block edges producing unconformity K2. Erosional products were deposited in local basins at the foot of fault blocks, and synrift sequence C was deposited on both sides of an arch along what is now the midcontinental rise. Extension continued during the synrift period with growth of the normal faults up through much of sequence C.
2. Approximately 2 km of subsidence occurred localized in a rift valley of observed width about 40 km, and a flood basalt, unit V1, was deposited onto the earliest beds of synrift sequence C on the rift valley floor.
3. Clastic sediments prograded northward across the floor of the rift valley, on top of V1, producing the latest portion of sequence C synrift deposits. The steep dip (10°) and long slip-face (1 km relief) argues for a very high energy sedimentation environment with high rates of deposition.

4. At 95 Ma (Veevers, 1986), volcanic flows, V2, were deposited on top of the prograded beds of upper C, with the first bed of V2 progressively thickening toward the north. Subsequent V2 deposits onlapped generally southward and were seaward rotated as a result of progressive subsidence northward from the volcanic crustal load as the center of extrusion migrated northward (Mutter, 1985). This earliest stage of new igneous crust generation (spreading) is assumed to have occurred subaerially, and it coincided with the development of K1, the breakup unconformity.
5. A few million years later, the surface of volcanic extrusion dropped below sea level for the first time, resulting in the production of pillow basalts and sheet flows typical of oceanic crust.
6. The margin thermally subsided and postrift sequences B and A were deposited. Southward migration of the shoreline of about 200 km occurred during this time (see Hampton et al., this volume).

ACKNOWLEDGMENTS

We thank D. Howell, R. Bohannon, J. Veevers, W. Dillon, and B. Tucholke for critically reviewing the manuscript and Jon Childs for help and guidance in MCS data processing. Captain A. McClenaghan and his crew skillfully and courageously guided the R/V *S.P. Lee* through sometimes difficult sea-ice conditions during the course of this survey. We thank the National Science Foundation, Division of Polar Programs, for logistic support and the United States Coast Guard for support with the icebreaker *Polar Sea*.

REFERENCES CITED

Austin, J. A., and E. Uchupi, 1982, Continental-oceanic crustal transition off Southwest Africa: Bulletin of the American Association of Petroleum Geologists, v. 66, p. 1328–1347.

Avendik, F., A. L. Camus, A. Ginsburg, L. Montadert, D. G. Roberts, and R. B. Whitmarsh, 1982, A seismic refraction and reflexion study of the continent–ocean transition beneath the North Biscay margin: Philosophical Transactions of the Royal Society of London, Series A, v. 305, p. 1–25.

Baker, B. H., P. A. Mohr, and L. A. Williams, 1972, Geology of the eastern rift system of Africa: Geological Society of America Special Paper 136, 67 p.

Barrett, P. J., 1982, History of the Ross Sea region during the deposition of the Beacon Supergroup 400–180 million years ago: Journal of the Royal Society of New Zealand, v. 11, p. 447–458.

Barrett, P. J., and P. C. Froggatt, 1978, Densities, porosities, and seismic velocities of some rocks from Victoria Land, Antarctica: New Zealand Journal of Geology and Geophysics, v. 21, p. 175–187.

Behrendt, J., 1983, Are there petroleum resources in Antarctica? *in* J. Behrendt, ed., Petroleum and mineral resources of Antarctica: U.S. Geological Survey Circular 909, p. 3–24.

Bentley, C. R., and J. W. Clough, 1972, Antarctic subglacial structure from seismic refraction measurements, *in* R. J. Adie, ed., Antarctic geology and geophysics: Oslo, Universitetsforlaget, p. 683–692.

Bohannon, R. G., 1986, Tectonic configuration of the western Arabian continental margin, southern Red Sea: Tectonics, p. 477–499.

Bott, M. H. P., 1982, Stress based tectonic mechanisms at passive continental margins, *in* R. A. Scrutton, ed., Dynamics of passive margins: American Geophysical Union, Geodynamics Series, v. 6, p. 147–153.

Cande, S. C., and J. C. Mutter, 1982, A revised identification of the oldest sea-floor spreading anomalies between Australia and Antarctica: Earth and Planetary Science Letters, v. 58, p. 151–160.

Crary, A. P., 1963, Results of United States traverses in East Antarctica 1958–61: IGY Glaciological Report Series 7, American Geographical Society, New York, 144 p.

Crough, S. T., 1983, The correction for sediment loading on the seafloor: Journal Geophysical Research, v. 88, p. 6449–6454.

Davis, G. H., 1983, Shear-zone model for the origin of metamorphic core complexes: Geology, v. 11, p. 342–347.

Domack, E. W., W. W. Fairchild, and J. B. Anderson, 1980, Lower Cretaceous sediment from the East Antarctic continental shelf: Nature, v. 287, p. 625–626.

Eldholm, O., and E. Sundvor, 1980, The continental margins of the Norwegian-Greenland Sea: Recent results and outstanding problems: Philosophical Transactions of the Royal Society of London, Series A, v. 294, p. 77–86.

Falvey, D. A., 1974, The development of continental margins in plate tectonic theory: Australian Petroleum Exploration Association Journal, v. 14, p. 95–106.

Fraser, A. R., and L. A. Tilbury, 1979, Structure and stratigraphy of the Ceduna Terrace region, Great Australian Bight Basin: Australian Petroleum Exploration Association Journal, v. 19, p. 53–65.

Geotimes, 1982, Leg 81 drills west margin, Rockall Plateau: September, 1982, p. 21–23.

Hayes, D. E., R. Houtz, M. Talwani, A. B. Watts, J. Weissel, and T. Aitken, 1975, Preliminary report of volume 23, U.S.N.S. Eltanin cruises 33–38, March 1968-May 1969: Lamont-Doherty Geological Observatory Technical Report No. CU-1-75, 214 p.

———, 1976, Preliminary report of volume 24, U.S.N.S. *Eltanin* cruises 39–45, June 1969-October 1970: Lamont-Doherty Geological Observatory Technical Report No. CU-2-76, 302 p.

Hinz, K., 1981, A hypothesis on terrestrial catastrophes, wedges of very thick oceanward-dipping layers beneath passive continental margins—their origin and paleoenvironmental significance: Geologisches Jahrbuch, Reihe E., Geophysik, Heft 22, p. 3–28.

Hutchinson, D. R., J. A. Grow, K. D. Klitgord, and B. A. Swift, 1982, Deep structure and evolution of the Carolina Trough, *in* J. S. Watkins and C. L. Drake, eds., Studies in continental margin geology: American Association of Petroleum Geologists Memoir 34, p. 129–154.

Keen, C. E., and D. L. Barrett, 1981, Thinned and subsided continental crust on the rifted margin of eastern Canada; crustal structure, thermal evolution and subsidence history: Royal Astronomical Society Geophysical Journal, v. 65, p. 443–465.

Klitgord, K. D., and J. C. Behrendt, 1979, Basin structure of the U.S. Atlantic margin, *in* J. S. Watkins, L. Montadert, and P. W. Dickerson, eds., Geological and geophysical investigations of continental margins: American Association of Petroleum Geologists Memoir 29, p. 85–112.

Klitgord, K. D., and J. A. Grow, 1980, Jurassic seismic stratigraphy and basement structure of western Atlantic magnetic quiet zone: Bulletin of the American Association of Petroleum Geologists, v. 64, p. 1658–1680.

König, M., and M. Talwani, 1977, A geophysical study of the southern continental margin of Australia: Great Australian Bight and western sections: Geological Society of America Bulletin, v. 88, p. 1000–1014.

LaBrecque, J. L., and N. Zitellini, 1985, Continuous seafloor spreading in Red Sea: An alternative interpretation of magnetic anomaly pattern: American Association of Petroleum Geologists Bulletin, v. 69, p. 513–524.

La Prade, K. E., 1982, Petrology and petrography of the Beacon Supergroup, Shackleton Glacier area, Queen Maud Range, Transantarctic Mountains, Antarctica, *in* C. Craddock, ed., Antarctic geoscience: Madison, University of Wisconsin Press, p. 581-590.

Le Pichon, X., and J. C. Sibuet, 1981, Passive margins: A model of formation: Journal of Geophysical Research, v. 86, p. 3708-3720.

McKenzie, D. P., 1978, Some remarks on the development of sedimentary basins: Earth and Planetary Science Letters, v. 40, p. 25-32.

Montadert, L., O. de Charpel, D. Roberts, P. Guennoc, and J. Sibuet, 1979, Northeast Atlantic passive continental margins: Rifting and subsidence processes, *in* M. Talwani, W. Hay, and W. B. F. Ryan, eds., Deep drilling results in the Atlantic Ocean, continental margins and paleoenvironment: American Geophysical Union, Maurice Ewing Series, v. 3, p. 154-186.

Mutter, J. C., 1985, Seaward dipping reflectors and the continent-ocean boundary at passive continental margins: Tectonophysics, v. 114, p. 117-131.

Mutter, J. C., M. Talwani, and P. L. Stoffa, 1982, Origin of seaward-dipping reflectors in oceanic crust off the Norwegian margin by "subaerial seafloor spreading": Geology, v. 10, p. 353-357.

Mutter, J. C., K. A. Hegarty, S. C. Cande, and J. Weissel, 1985, Breakup between Australia and Antarctica: A brief review in light of new data: Tectonophysics, v. 114, p. 255-279.

Nafe, J. E., and C. L. Drake, 1963, Physical properties of marine sediments, *in* M. N. Hill, ed., The sea; ideas and observations in the study of the seas, v. 3: New York, Interscience, p. 794-815.

Rowden, L., and C. E. Keen, 1980, Rifting processes and thermal evolution of the continental margin of eastern Canada determined from subsidence curves: Earth and Planetary Science Letters, v. 51, p. 343-361.

Rowley, D. B., and D. Sahagian, 1986, Depth-dependent stretching: A different approach: Geology, v. 14, p. 32-35.

Sawyer, D. S., 1985, Total tectonic subsidence: A parameter for distinguishing crust type at the U.S. Atlantic continental margin: Journal of Geophysics Research, v. 90, p. 7751-7770.

Shipley, T. H., R. T. Buffler, and J. S. Watkins, 1978, Seismic stratigraphy and geologic history of Blake Plateau and adjacent western Atlantic continental margin: Bulletin of the American Association of Petroleum Geologists, v. 62, p. 792-812.

Steckler, M. S., 1985, Uplift and extension at the Gulf of Suez: Indications of induced mantle convection: Nature, v. 317, p. 135-139.

Steed, R. H. N., and D. J. Drewry, 1982, Radio echo sounding investigations of Wilkes Land, Antarctica, *in* C. Craddock, ed., Antarctic geoscience: Madison, University of Wisconsin Press, p. 969-976.

Tsumuraya, Y., M. Tanahashi, T. Saki, T. Machihara, and N. Asakura, 1985, Preliminary report of the marine geophysical and geological surveys off Wilkes Land, Antarctica in 1983-1984: Memoirs of National Institute of Polar Research Special Issue 37, p. 48-62.

Veevers, J. J., 1986, Breakup of Australia and Antarctica estimated as mid-Cretaceous (95 $\pm$5 Ma) from magnetic and seismic data at the continental margin: Earth and Planetary Science Letters, v. 77, p. 91-99.

Wannesson, J., M. Pelras, B. Petitperrin, M. Perret, and J. Segoufin, 1985, A geophysical transect of the Adelie margin, East Antarctica: Marine and Petroleum Geology, v. 2, p. 192-201.

Waples, D. W., 1980, Time and temperature in petroleum formation: Application of Lopatin's method to petroleum exploration: Bulletin of the American Association of Petroleum Geologists, v. 64, p. 916-926.

Watts, A. B., 1981, The U.S. Atlantic continental margin: Subsidence history, crustal structure and thermal evolution: American Association of Petroleum Geologists Education Course Notes, Series 19, 75 p.

Weissel, J. K., and D. E. Hayes, 1972, Magnetic anomalies in the Southeast Indian Ocean, *in* D. E. Hayes, ed., Antarctic oceanology. II. The Australian-New Zealand sector: American Geophysical Union, Antarctic Research Series, v. 19, p. 165-196.

Eittreim, S. L., and Hampton, M. A., 1987, *The Antarctic Continental Margin: Geology and Geophysics of Offshore Wilkes Land*, CPCEMR Earth Science Series, v. 5A: Houston, Texas, Circum-Pacific Council for Energy and Mineral Resources

The Conjugate Continental Margins of Antarctica and Australia

J. J. Veevers
Macquarie University
New South Wales, Australia

The Southeast Indian Ridge opened at 96 Ma at slow rates of 4.5 mm/year to 48 Ma and 10 mm/year to 44 Ma, and then at a fast rate of 23 mm/year to the present. Three types of magnetic quiet zones were generated: (1) attenuated continental crust between a magnetic trough and the continent-ocean boundary (COB); (2) oceanic crust between the COB and A34 (Cretaceous quiet zone); and (3) east of the George V Fracture Zone, oceanic crust cut by a complex of fracture zones or affected by frequent ridge jumps.

The COB on both margins is marked by (1) the contrast in the seismic character of oceanic and continental crust and (2) by a prominent magnetic anomaly, modeled as generated at the truncated edge of oceanic blocks and less magnetic continental crust. The COB trend of the Wilkes Land margin matches its Australian counterpart in the reconstruction at breakup found by overlapping the two COB anomalies.

The breakup event defines the depositional sequences of both margins. A comparison of the Antarctic sequences with those of Australia indicates that the Antarctic sequence D is synrift (Early Cretaceous), and C to A postrift, with the unconformity between D and C (K2) the breakup unconformity.

The attenuated crust landward of the COB along transects across the Great Australian Bight and the conjugate Wilkes Land margin indicates a pre-breakup extension of 360 km ($\beta = 1.8$). This extension constrains a reconstruction of the individual subplates of mainland Australia, Tasmania, the South Tasman Rise, and Antarctica that underwent relative motion during an Early Cretaceous phase of continental extension.

REGIONAL COMPARISON

Previous Work

The continental margins of Antarctica and Australia are separated today by the 3000 km wide Southeast Indian Ocean (Fig. 1) that developed by seafloor spreading since continental breakup in the mid-Cretaceous. The Southeast Indian Ridge, paralleled by the conjugate margins, trends easterly along lat 50° S past the deep and intensely fractured Australian-Antarctic Discordance at 120°–128° E (Weissel and Hayes, 1974; Anderson et al., 1980; Vogt et al., 1983) to 139° E, whence it is offset southeastward by a zone of transform faults (George V) to 63° S, near its continuation past 162° E with the Pacific-Antarctic Ridge (Hayes and Conolly, 1972). Centered on the Australian-Antarctic Discordance and extending southward and northward into the continents is the Australian-Antarctic Depression, an area of 15×10^6 km^2 of anomalously deep oceanic lithosphere and adjacent continental lithosphere, whose correspondence with a negative satellite free-air gravity anomaly (Fig. 2) suggests that the depression is maintained dynamically by motions in the asthenosphere (Veevers, 1982).

The region was explored by Lamont-Doherty Geological Observatory on cruises of the R/V *Eltanin* and *Vema* from 1968 to 1976 (Weissel and Hayes, 1972, 1974); the Australian margin part of the region was described by König and Talwani (1977), Talwani et al. (1979), and König (1980), and the Wilkes Land margin by König (1980, 1981); the seafloor-spreading magnetic anomalies were revised by Cande and Mutter (1982), and the geophysical data were presented by Hayes et al. (1975a, 1976, 1977, 1978).

Weissel and Hayes (1972) subdivided the Southeast Indian Ocean into four longitudinal zones on the basis of contrasts in magnetic, morphologic, and seismic properties, and inferred that seafloor spreading started 55 Ma (magnetic anomaly 22) and proceeded through five stages from slightly different poles and rates, at times with marked asymmetry. Magnetic quiet zones along portions of the adjacent continental margins cut across the strikes of the oldest seafloor-spreading magnetic anomalies and were related to particular conditions that occurred during the initial rifting history of the continents. Studies of seismic profiles by Boeuf and Doust (1975) and extensive studies of crustal structure by Talwani et al. (1979) later confirmed that the Australian magnetic quiet zone was characterized by a rift-zone crust different from that of either the continent or ocean. Cande and Mutter (1982) revised the identification by Talwani et al. (1979) and König (1980) of the oldest seafloor-spreading magnetic anomalies, from A19–22 to A20–34, and showed that breakup of Antarctica and Australia probably occurred sometime between 110 and 90 Ma. Spreading took place at three half-rates: (1) between 90 and 48 Ma at the very slow half-rate of 4.5 mm/year, during which

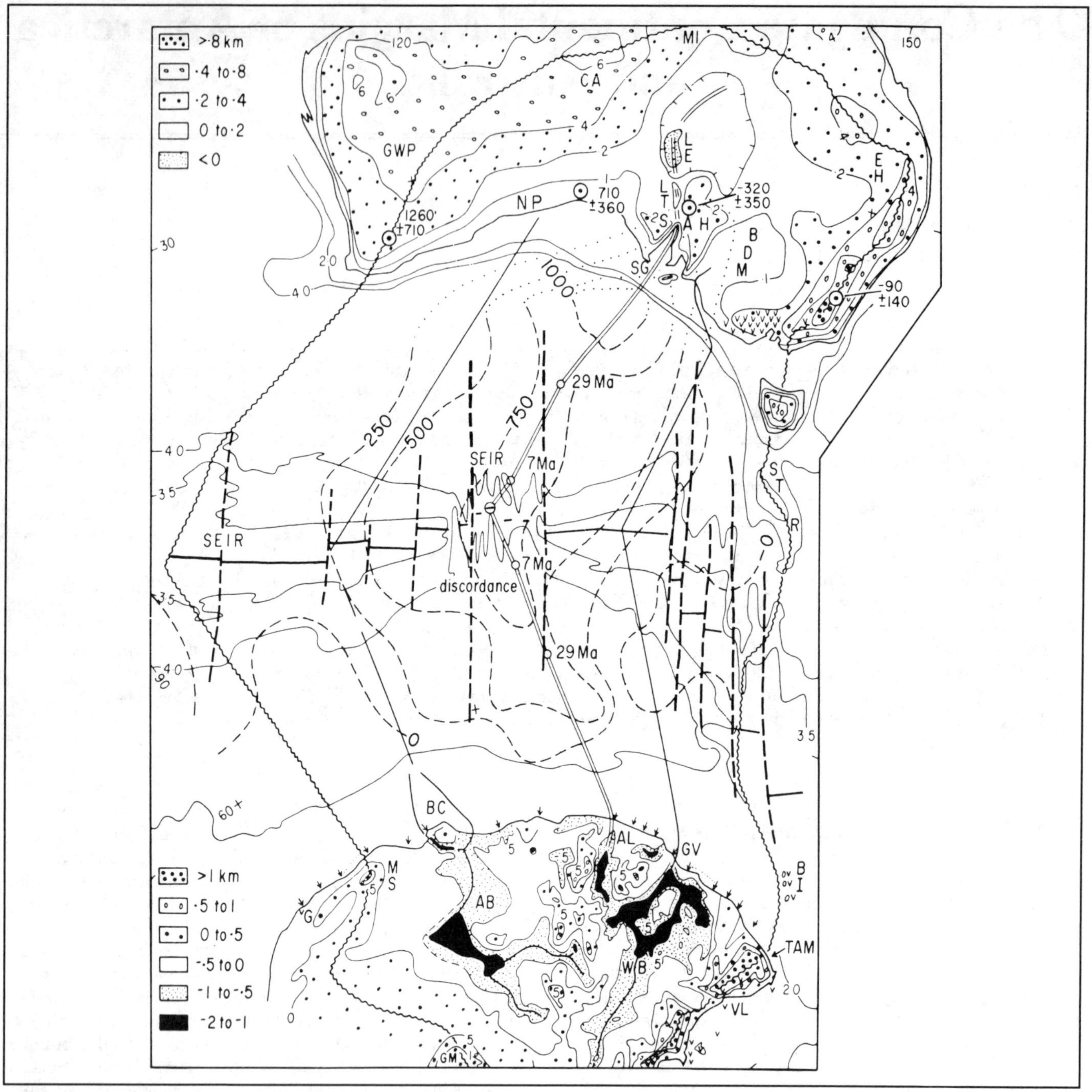

Figure 1—Morphology of the Australian-Antarctic Depression. Australia shown by contours (km) of 1° × 1° area mean rock altitude (Wellman, 1979) to afford direct comparison with the ice-smoothed topography of Antarctica; Southeast Indian Ocean by the 3.5 and 4.0 km isobaths and the crest of the Southeast Indian Ocean Ridge (SEIR) and fracture zones (Circum-Pacific Map Project, 1981); and Wilkes Land by sub-ice bedrock contours (km) (Drewry, 1983). Residual depth anomalies (m) of the Southeast Indian Ocean (broken lines) from Watts and Daly (1981), interpolated by dotted lines to the continental values across southern Australia (Wellman, 1982). Coastal outcrop in Wilkes Land west of the Transantarctic Mountains (James and Tingey, 1983) shown by arrows. Modern (<5 Ma) volcanics (Aziz-ur-Rahman and McDougall, 1972; Armstrong, 1978; Kyle et al., 1979) shown by Vs. Axis of depression shown by double lines, subsidiary axes by single lines, and flank crests by wavy lines. Circles along axis indicate datum points along isochrons. Lambert equal-area projection. Modified from Veevers (1984). AB, Aurora sub-ice basin; AL, Adelie Land; BI, Balleny Islands; BC, Budd Coast; CA, central Australia; EH, Eastern Highlands; GM, Gamburtsev Mountains; GV, George V Land; GWP, Great Western Plateau; LE, Lake Eyre; LT, Lake Torrens; MDB, Murray-Darling basin; MI, Mount Isa block; MS, Mount Sandow; NP, Nullarbor Plain; SAH, South Australian Highlands; SG, Spencer Gulf; STR, South Tasman Rise; TAM, Transantarctic Mountains; VL, Victoria Land; WB, Wilkes sub-ice basin.

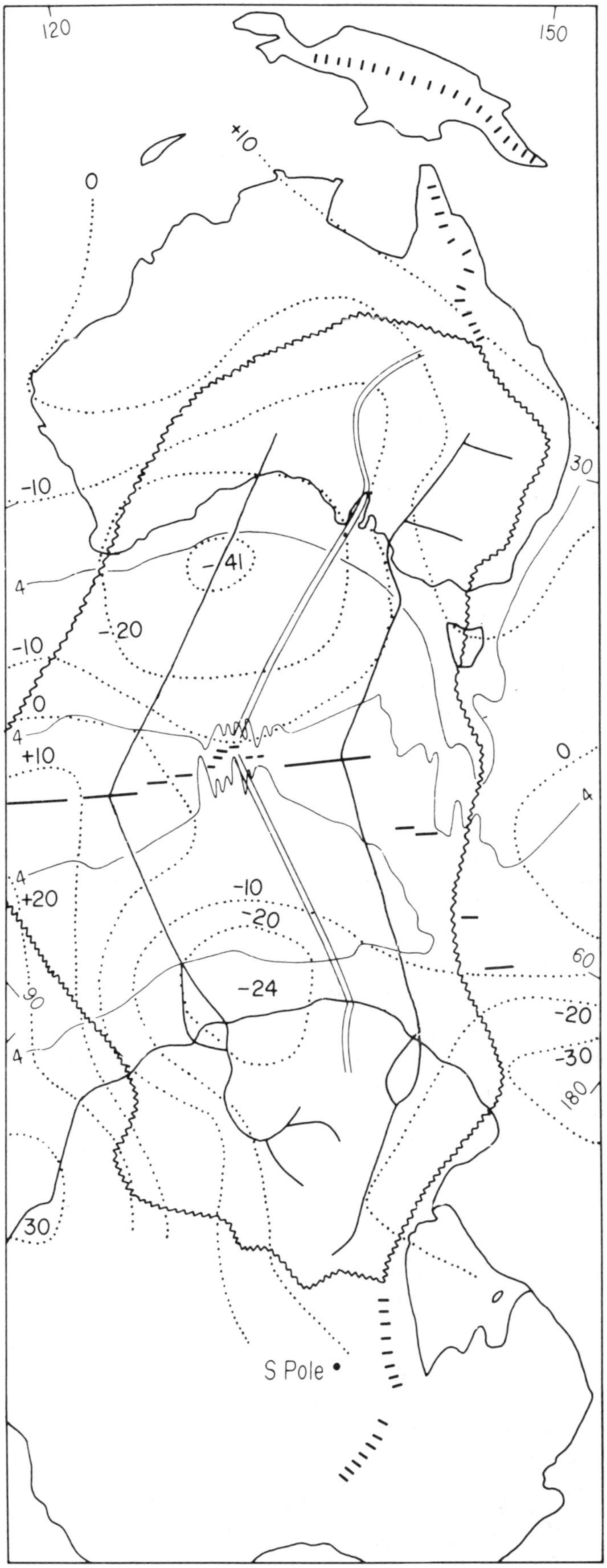

the rough topography of the Diamantina Zone was generated, (2) from 48 to 44 Ma at 10 mm/year, and (3) from 44 Ma to the present at 23 mm/year. A20 (44 Ma) marks the change between the early phase of slow opening and the late phase of rapid opening.

In his detailed study of the conjugate margins of Antarctica and Australia, König (1980) interpreted the cross-cutting of the oldest seafloor-spreading magnetic anomalies off southern Australia and off Wilkes Land eastward of 132° E (shown in Cande and Mutter's Figure 1, 1982) as due to a lag in the onset of seafloor spreading in the east, manifested by extension of continental crust in the east at the same rate as seafloor spreading in the west. König (1981) published a new finite pole and rotation angle for the closure of Australia and Antarctica based on the identification of the magnetic anomaly at the seaward edge of each marginal quiet zone as representing the landward limit of oceanic crust.

I reinterpreted the magnetic anomaly (Cande and Mutter's [1982] A34) at the edge of the marginal quiet zone as due to the edge effect at the continent–ocean boundary (COB) of less magnetic continental crust abutting strongly magnetic oceanic crust, and found an extrapolated age of 96 Ma, rounded to 95 ± 5 Ma (mid-Cretaceous), for continental breakup (Veevers, 1986). Younger oceanic crust from 80.5 to 49.0 Ma at the COB east of 131.2° E was interpreted as being emplaced by jumps of the ridge crest to Australia, leaving previously emplaced oceanic crust on the Antarctic plate. The breakup of Australia and Antarctica was thus postulated to have been isochronous at 95 ± 5 Ma (mid-Cretaceous) without any subsequent continental extension east of 131.2° E. New magnetic data, given below, support this model.

Veevers and Eittreim (in press) independently confirmed this model by using the identification of the COB in R/V *S.P. Lee* multichannel seismic profiles from the Wilkes Land margin (documented by Eittreim and Smith, this volume) augmented by magnetic anomalies to show that the Wilkes Land COB parallels that of its Australian counterpart, so that the oceanic crust of the Southeast Indian Ocean has the same width both east and west of 130° E; and we confirmed that the pole of closure found by König (1981), from joining the seaward edge of each marginal quiet zone, fitted the Australian and Antarctic COBs as identified in seismic profiles (Fig. 3). König's (1981) pole, slightly modified for the amount of closure, describes the line of breakup or the opening suture, and thus indicates the location of originally conjugate points (Table 1). Furthermore, we showed that along a transect across the Great Australian Bight and the conjugate Wilkes Land margin, the suture zone underwent 360 km of extension before breakup, and from this derived a pole of rotation that describes the Early Cretaceous separation of the four subplates of mainland Australia, Tasmania, the South Tasman Rise, and Antarctica (Table 2).

Figure 2—Map showing the correspondence of the negative satellite free-air gravity anomalies (dotted lines) with the Australian–Antarctic Depression. Gravity anomalies, in mgal, from Anderson et al. (1973) to 70° S, extended southward from Gaposchkin and Lambeck (1971). Modified from Veevers (1984). © 1984, Clarendon Press, Oxford.

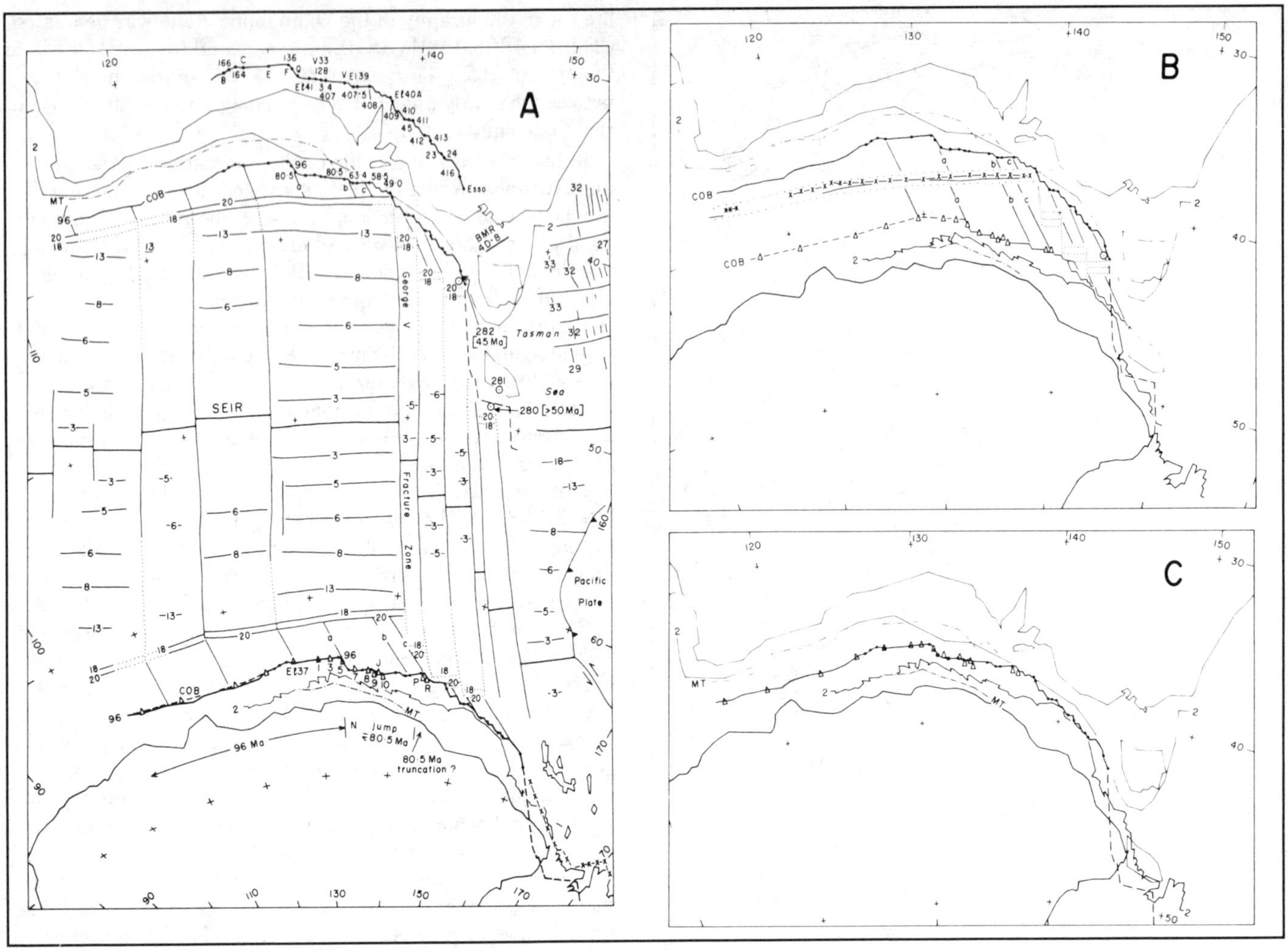

Figure 3—Reconstruction of Antarctica and Australia, by B. D. Johnson's computer program INTERMAP, computer runs by P. Fell. A. Present geography, showing coastlines and 2-km isobaths, and the Australian COB (data points labeled in box above) rotated (27.85°) about König's (1981) pole to Antarctica, to compare with the Antarctic COB determinations, as shown in Figure 8A. Also shown are magnetic trough (MT), selected seafloor-spreading magnetic anomalies from the ridge-crest anomaly to A20, and related fracture zones (e.g., a–c; inferred extensions shown by dotted lines). The oblique lines between A20 and the COB parallel the inferred transform-fault trend during the first phase of spreading (96–44 Ma). DSDP sites 280–282 and Tasman Sea spreading system also shown. B. Closure of the Southeast Indian Ocean to A18 (42 Ma), from Weissel et al.'s (1977) finite pole. C. Total closure of the Southeast Indian Ocean by a rotation of 27.85° (slightly modified from König's 28°) about König's pole (1.5° N, 37.0° E). Geographical coordinates refer to present-day Australia. Lambert equal-area projection. Note overlap of the COB of the western margin of Tasmania and the South Tasman Rise (broken line) on north Victoria Land and margin.

Comparative Regional Morphology, Structure, and Stratigraphy

Australia

According to its morphology and structure (Fig. 4A), the southern margin of Australia comprises three sectors:

Sector I extends to 125° E, is bounded on the north by the Archean Yilgarn Block enclosed by the Proterozoic Albany–Fraser Mobile Zone on the south and southeast, and by the Proterozoic Darling Mobile Zone and superimposed Phanerozoic Perth Basin on the west; on the south it is bounded by the Diamantina Zone of rugged east-trending oceanic ridges and troughs. The margin has a narrow shelf and slope, underlain by the Cretaceous–Cenozoic Bremer Basin and a rise that broadens to the east.

Sector II extends to 137° E, encompasses the reentrant of the Great Australian Bight with a broad shelf backed by the low-standing Nullarbor Plain of the Cretaceous–Cenozoic Eucla Basin in the west and center, here including the subsurface Permian Denman Basin, and by the eastern tip of the Permian and Jurassic–Cretaceous–Cenozoic Polda Basin in the east, and bordered by the Proterozoic and Archean Gawler Block; the slope expands into the narrow Eyre Terrace and the broad Ceduna Terrace on the east, and the rise narrows in front of the Ceduna Terrace; this sector is bounded oceanward by an abyssal plain that extends eastward to the projected line of the George V Fracture Zone.

Sector III trends southeast from 137° E, has a broad shelf indented on the west by the gulfs within the South Australian Highlands, and on the east has a narrow shelf underlain by the

Table 1—Conjugate points on the COB, in order from west to east.

Age (Ma) oldest sea floor Ant. Plate	Antarctica		Australia		Age (Ma) oldest sea floor Aust. Plate
	Long° E	Profile	Long° E	Profile	
				B	96
				166	
				164,405	
	127.2	El 37E	127.75	C	
	130.4	1	129.5	E, 407	
	131.95	3, El 41	130.3	136	
	133.35	5	131.1	F	
				Q	80.5
	134.9	7	132.1	132	
96				El 41	
				130	
	136.75	8	132.8	128, V33	
		El 37J			
	138.5	10	133.3	407-3,4	
				V	63.4
				407-5	
				El 39	
				408	58.5
				El 40A	49.0
		George V Fracture Zone			
80.5		El 50			

Table 2—A. Finite poles and angles, relative to Australia, for the reconstructions of Antarctica to Australia: (1) by closing the suture zone, (2) by closing the Southeast Indian Ocean by overlapping the Antarctic and Australian COBs (96 Ma), and (3) by closing part of the ocean by overlapping A20 (44 Ma) and (3′) A18 (42 Ma). Positive values indicate north latitudes, east longitudes, and counterclockwise rotations when viewed from above the pole.

Pole	Age (Ma)	Lat°	Long°	Angle°	Reference and remarks
1	160	5.68	38.36	30.64	Closes suture zone
2	96	1.5	37.0	27.85	Closes SE Indian Ocean to conjugate COBs. König (1981) modified from angle of 28°, for best fit of conjugate COBs
3	43.69(A20)	13.0	31.5	24.1	König (1980). Closes to A20, given by Cande and Mutter (1982)
3′	41.82(A18)	10.3	34.75	23.6	Weissel et al. (1977), table III, pole 6. Closes to A18

Table 2—B. Stage poles and angles, relative to Australia, for (a) the rift phase, (b) slow drift phase, and (c) fast drift phase.

Pole	Age (Ma)	Lat°	Long°	Angle°	Reference and remarks
a (1–2)	160–96 (COB)	42.03	40.58	−3.564	Extension in suture zone during rift phase
b (2–3)	96–43.69 (COB to A20)	−35.74	73.85	−6.82	Spreading during slow drift
c (3)	43.69–0 (A20 to A1)	13.0	31.5	−24.10	Spreading during fast drift

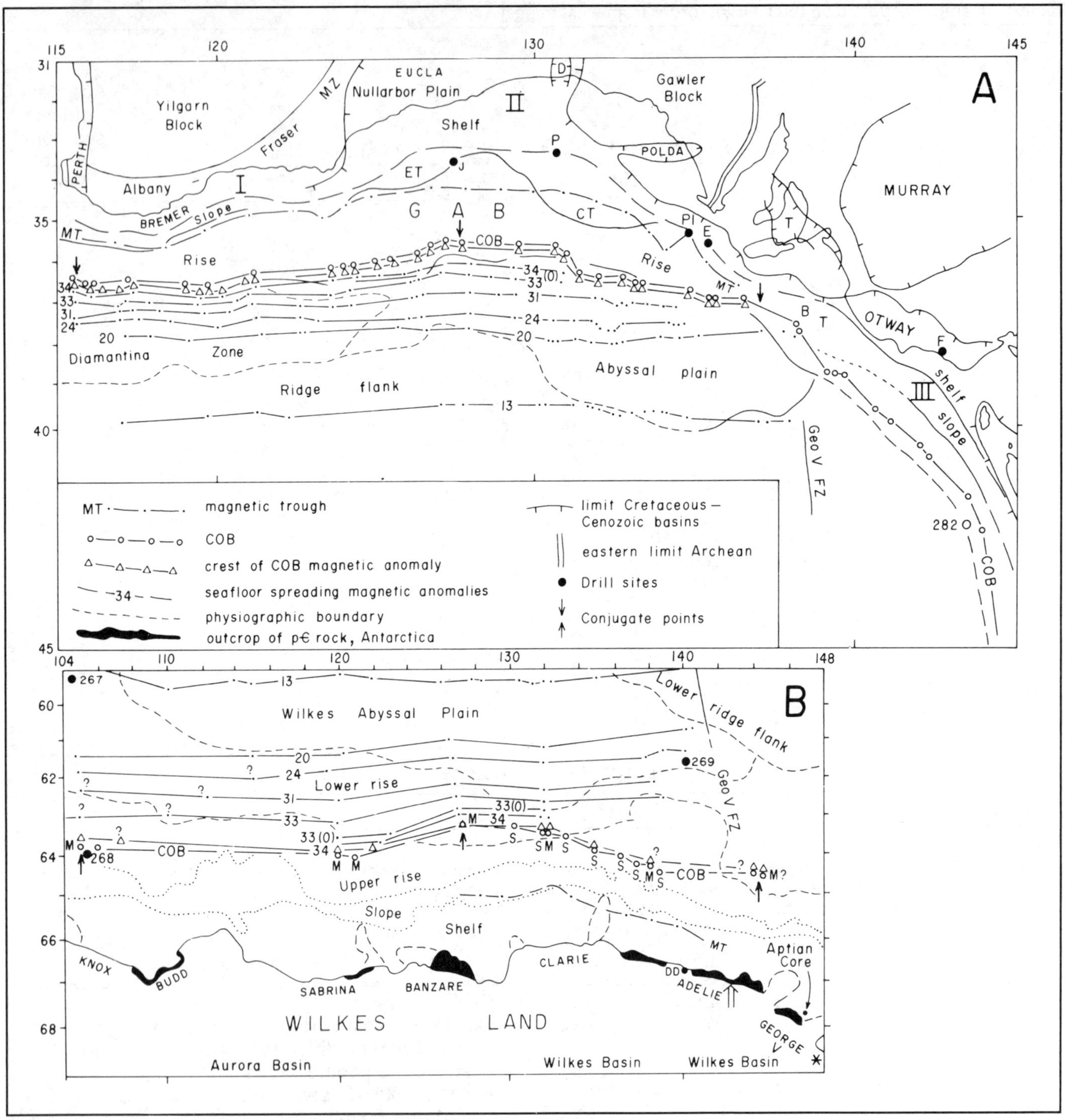

Figure 4—A. Southern Australian margin, 115°–145° E, showing sectors I–III; morphology and sedimentary basins (from Talwani et al., 1979), including the Great Australian Bight Basin (GAB); location of drillsites, from west to east: J, Jerboa; P, Potoroo; PI, Platypus; E, Echidna; F, Flaxmans; 282, DSDP site 282, the logs of which are given in Figure 5; MT, magnetic trough; COB and COB anomaly (Veevers, 1986, and this paper); and seafloor-spreading magnetic anomalies, added to and modified from Talwani et al. (1979), König (1980), and Cande and Mutter (1982). BT, Beachport Terrace; CT, Ceduna Terrace; ET, Eyre Terrace; D, Denman Basin; T, Troubridge Basin. Double line on Australia and double-shafted arrow on Antarctica mark the easternmost extent of known Archean rocks (Oliver et al., 1983). The thin arrows indicate the position on the COB of conjugate points. B. East Antarctic margin: morphology and magnetic trough (MT) (both from König, 1980), COB determined from seismic profiles (S) (Eittreim and Smith, this volume) and from the COB magnetic anomaly (M) (Veevers, 1986, and this paper); seafloor-spreading anomalies, added to and modified from König (1980) and Cande and Mutter (1982). DSDP sites 267–269 from Hayes et al. (1975b). Aptian core off George V Coast from Domack et al. (1980, 1983). The asterisk indicates exposed Permian-Triassic sedimentary rocks and Jurassic dolerite at Horn Bluff (Mawson, 1940).

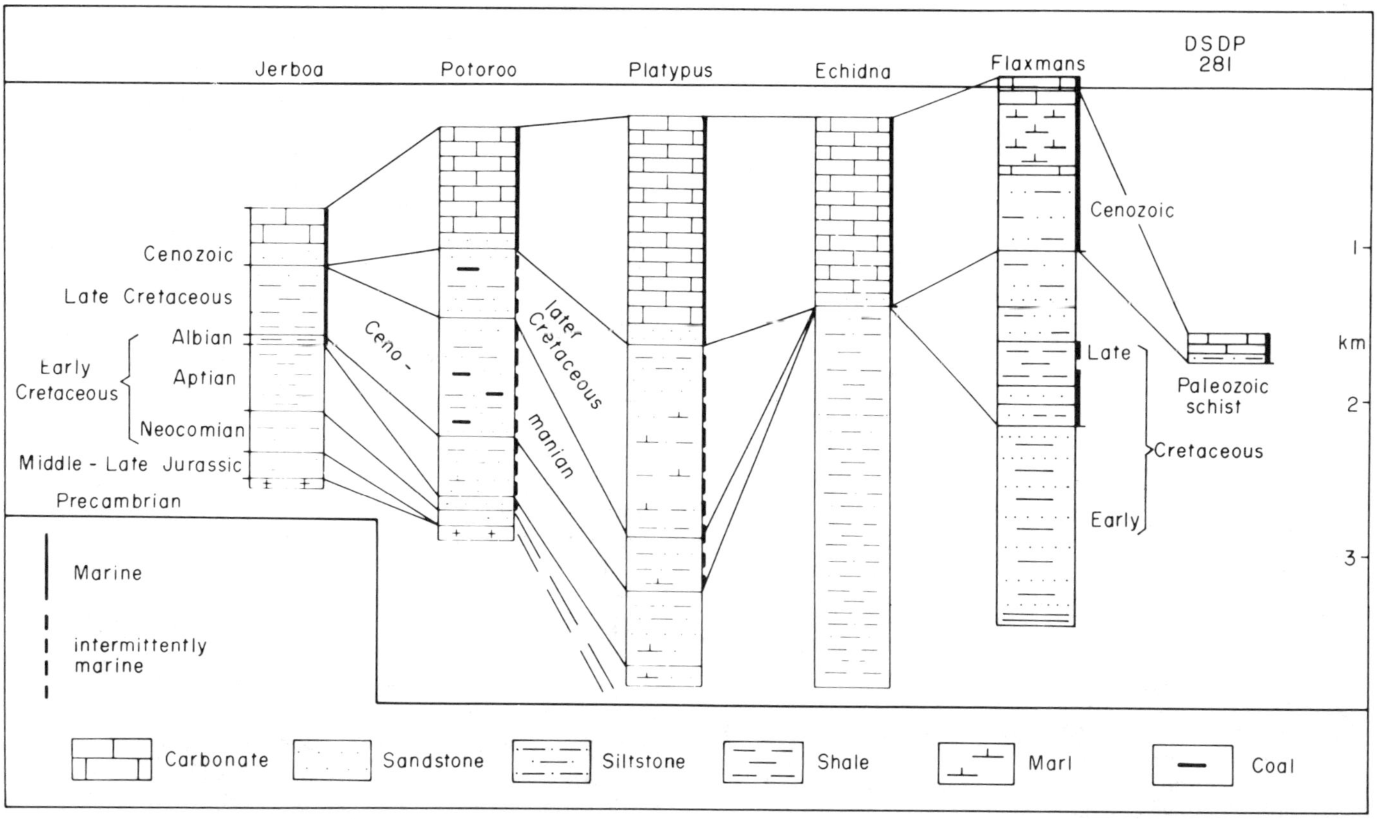

Figure 5—Well logs, Australian margin, located in Figure 4A, and DSDP 281 in Figure 3A. From Veevers (1984). Jerboa (Bein and Taylor, 1981), Potoroo and Platypus (Fraser and Tilbury, 1979), Echidna (Whyte, 1978), Flaxmans (Bain, 1965), DSDP 281 (Kennett et al., 1975). Depositional environment indicated by heavy line on side of column: solid line: marine; broken line: paralic; no line: nonmarine.

Otway Basin backed by the highlands of western Victoria and the Murray Basin on the north, and by King Island and Tasmania on the east. The slope and rise are together up to 200 km wide, and they narrow southward to a scarp on the western edge of the South Tasman Rise. Seaward of the COB, recognizable seafloor-spreading magnetic anomalies are lacking.

Except where it crosses the Ceduna Terrace, the magnetic trough (Talwani et al., 1979) marks the northern edge of the thick Late Jurassic and Cretaceous peripheral basin near the foot of the slope (Fig. 4A). The axis of the Australian–Antarctic Depression (Fig. 1) passes through the gulfs that split the South Australian Highlands, with subsidiary axes through the Great Australian Bight–Nullarbor Plain and Murray Basin.

The peripheral basins, as sampled by drilling on the shelf (Fig. 5) and at the edge of the South Tasman Rise (DSDP 281), rest on Precambrian granite and metamorphic rocks in the center and mid-Paleozoic metasedimentary rocks in the east, and comprise a sedimentary succession of mid-Jurassic to mid-Cretaceous faulted dominantly nonmarine detrital sediment unconformably overlain by intermittently marine Late Cretaceous detrital sediment, succeeded by Cenozoic shallow-marine carbonate. On the South Tasman Rise, only the Cenozoic sequence was found. The only older undeformed sediment known in the region is the very thin (75 m) Permian claystone of the Denman Basin and even thinner nonmarine glacial deposits of the Polda Basin and shallow-marine sediment and nonmarine glacial deposits of the Troubridge Basin. Traced seaward in seismic profiles into the axis of the peripheral basin, the Cenozoic (Fig. 6B) remains thin (<1 km) but the mid-Jurassic and Cretaceous sequence (Fig. 6A) thickens to an estimated maximum of 12 km in the Ceduna depocenter of the Great Australian Bight Basin and to 8 km in the Otway Basin, and then wedges out seaward. Seaward of the COB, the sediment is rarely 1 km thick (Fig. 6B); at Deep-Sea Drilling Project (DSDP) site 282, it comprises 300 m of Eocene and younger clay, silt, and calcareous ooze.

The pattern of magnetic anomalies is such that, with the notable exception of the Ceduna depocenter, most of the thick peripheral basin corresponds to the marginal magnetic quiet zone between the magnetic trough (traceable to 136° E) and the COB anomaly (Fig. 7C). On the other side of the COB anomaly (Fig. 8A) is a second kind of magnetic quiet zone comprising a strip of sea floor generated during the last 10 Ma of the Cretaceous long normal polarity interval (KLNPI) or Cretaceous magnetic quiet zone before the first reversed block at 86 Ma (A34). A third kind of magnetic quiet zone is underlain by the oceanic crust east of the projected George V Fracture Zone (at 138° E) and extends east across to Tasmania and south to the oldest recognized anomaly (A6) near the Southeast Indian Ridge. Unlike the other two zones, its origin is unknown, but it could be due to the crustal complexity associated with the narrow width of individual spreading segments

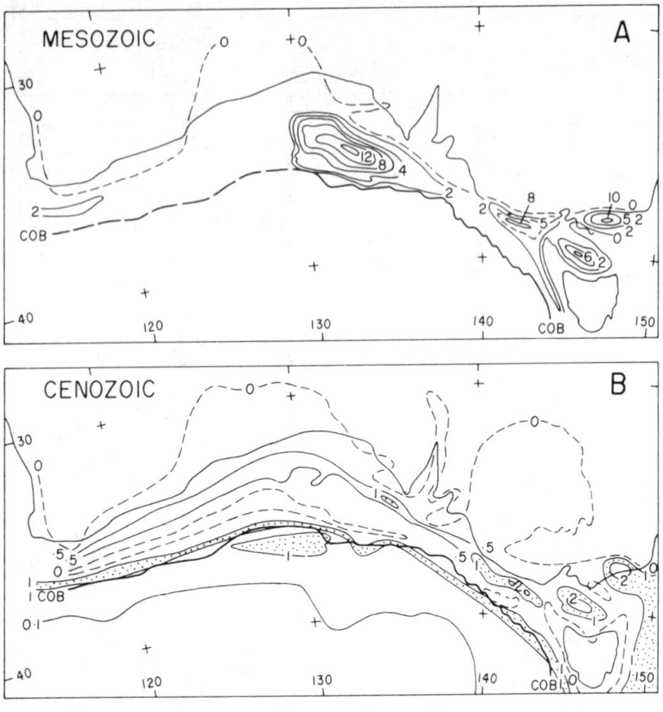

Figure 6—Isopachs (km) of Mesozoic (A) and Cenozoic (B) sediments in southern Australia. From Veevers (1984). © 1984, Clarendon Press, Oxford. Stipple in B indicates sediment thickness between 1 and 2 km.

(Roots, 1976) or repeated ridge jumps.

In summary, the peripheral trough along the southern margin of Australia is divided by structural saddles into three basins of thick mid-Jurassic and Early Cretaceous block-faulted dominantly nonmarine detrital sediment unconformably overlain by thick Late Cretaceous intermittently marine detrital sediment, with a thin layer of Cenozoic carbonate.

Antarctica

The coast of Wilkes Land (Fig. 4B) is rimmed by scattered outcrops of Precambrian metamorphic and granitoid rocks (James and Tingey, 1983; Oliver et al., 1983) between the ice sheet of the Aurora and Wilkes sub-ice basins to the south (Drewry, 1983) and ice shelves and the shelf to the north. The easternmost extent of Archean rocks is estimated by Oliver et al. (1983) to lie at 142°40′ E (broad arrow in Fig. 4B). The only known Phanerozoic rocks onshore are the Beacon Group and dolerite at Horn Bluff (asterisk in Fig. 4B), an exposed 150 m of sandstone, shale, and coal, presumably Permian or Triassic in age, intruded by a dolerite sill, presumably Jurassic in age (Mawson, 1940). Offshore the only known in situ sediment older than Cenozoic is a nonmarine siltstone cored in a 1406-m erosional depression at 147° E (Fig. 4B) that contains mid-Cretaceous (Aptian–Albian) spores (Domack et al., 1980, 1983). Reworked palynomorphs in the modern sediment offshore at 130°–135° E and 140°–150° E (Truswell, 1982, 1983) range in age from Early Cretaceous to early Cenozoic, with Cretaceous the dominant age. Reworked palynomorphs from farther west at 97° E, off the Shackleton Ice Shelf, contain in addition a Permian element, which is lacking to the east. The predominantly Cretaceous age of the reworked palynomorphs from that part of the Antarctic margin east of 105° E conjugate with Australia matches the predominantly Cretaceous age of the Australian margin deposits. The lack of reworked Permian palynomorphs on the Antarctic margin east of 97° E suggests the paucity if not absence of exposed Permian source rocks in this region.

DSDP drilling (Fig. 4B) penetrated sediment no older than late Oligocene (DSDP 268, 105° E) and middle Oligocene (DSDP 269, 140° E) in a sequence of silty clay with dropstones (Hayes et al., 1975b).

The Antarctic margin is notable for its wide complex shelf, with ice-cut depressions reaching depths of 1400 m and a shelf-break commonly at 600 m, a narrow slope, and an expanded width of the rise seaward of the COB, in particular at 105° E and 140° E (Fig. 4B); the rise, so much wider than its Australian conjugate, is attributable to the relative abundance of Antarctic Cenozoic sediment. East of 155° E, the margin trends southeastward; the shelf and slope are narrow, and they face a shallow rise that terminates at the Balleny Islands (Fig. 7B).

The three kinds of magnetic quiet zone found on and off the Australian margin appear here, too (Fig. 8B): a marginal quiet zone between the magnetic trough, traced to 141° E (Fig. 7A), and the COB anomaly; the Cretaceous long normal polarity interval between the COB anomaly and A34; and the area east of the George V Fracture Zone, although this area probably contains A31 and A20 (Fig. 8B).

Figure 7—A. Bathymetric profiles of the Antarctic margin (drawn from Vanney and Johnson, 1979). © 1979 Deutsches Hydrolishes Zeitscrift. South-trending profiles across the Wilkes Land margin, conjugate with those in facing Figure 7C, aligned along the magnetic trough (MT) from 127° to its easternmost extent at 141°, and then at the midshelf. Compared with the conjugate Australian margin, the magnetic trough is situated much shallower, from the shelf-break or upper slope to midshelf. The long even slope-rise lacks the terraces of conjugate Australia, but the western profiles (127°–130° E) resemble the Australian conjugates (128°–129°). B. East-trending profiles east of 155° E, from 67° S to 71° S, conjugate with those in facing Figure 7D, aligned on the midpoint of the slope-rise. A narrow slope passes downward into a rise only 2.5 km deep before it ascends toward the Balleny Islands. Apart from this difference, these profiles resemble their Australian conjugates off the west coast of Tasmania. C. Bathymetric profiles of the southern margin of Australia, all from B.M.R. (1979). © 1979 Australian Government Publishing. North–south profiles from 128° to 142° E, aligned on the magnetic trough (MT) from 128° to its easternmost extent at 136° E (in sector II), and then at the midpoint of the slope-rise. Short horizontal lines at top of each profile indicate sea level. The bulge of the Ceduna Terrace is bounded on either side by a short even slope that narrows at 137° E and then widens again at the Beachport Terrace. Farther east, the slope becomes smooth again. D. East-trending profiles across the southern margin east of 142° E, from 39° S to 46° S, aligned on the 3.5-km isobath. The long straight slope-rise in the north changes at 44° S to a scarp that marks the western edge of the South Tasman Rise.

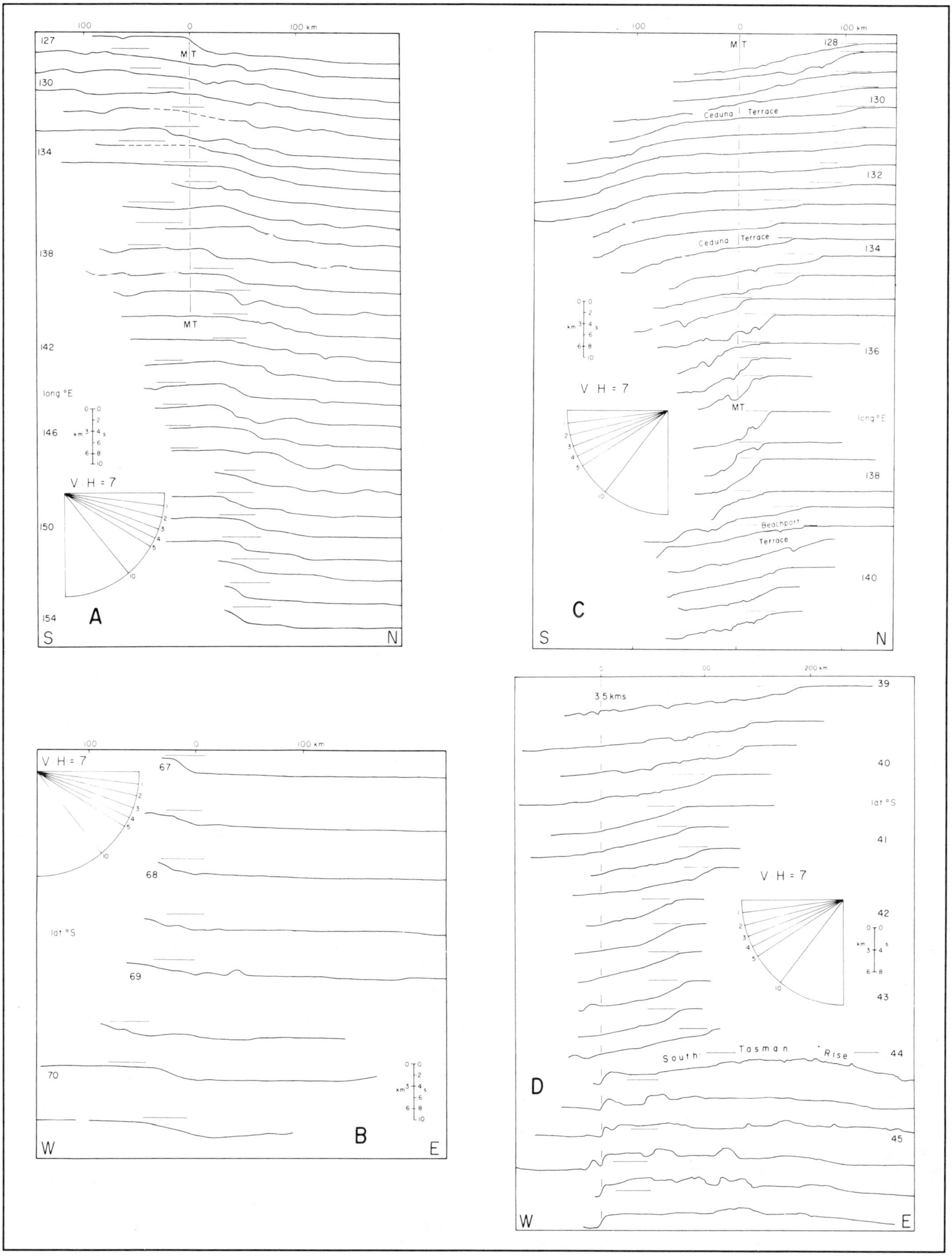
A
S
N
100
0
100 km
MT
127
130
134
138
142
146
150
154
long °E
V:H = 7
km
s
B
W
E
67
68
69
70
lat °S
C
128
130
132
134
136
138
140
Ceduna Terrace
Beachport Terrace
D
0
100
200 km
3.5 kms
39
40
41
42
43
44
45
South Tasman Rise

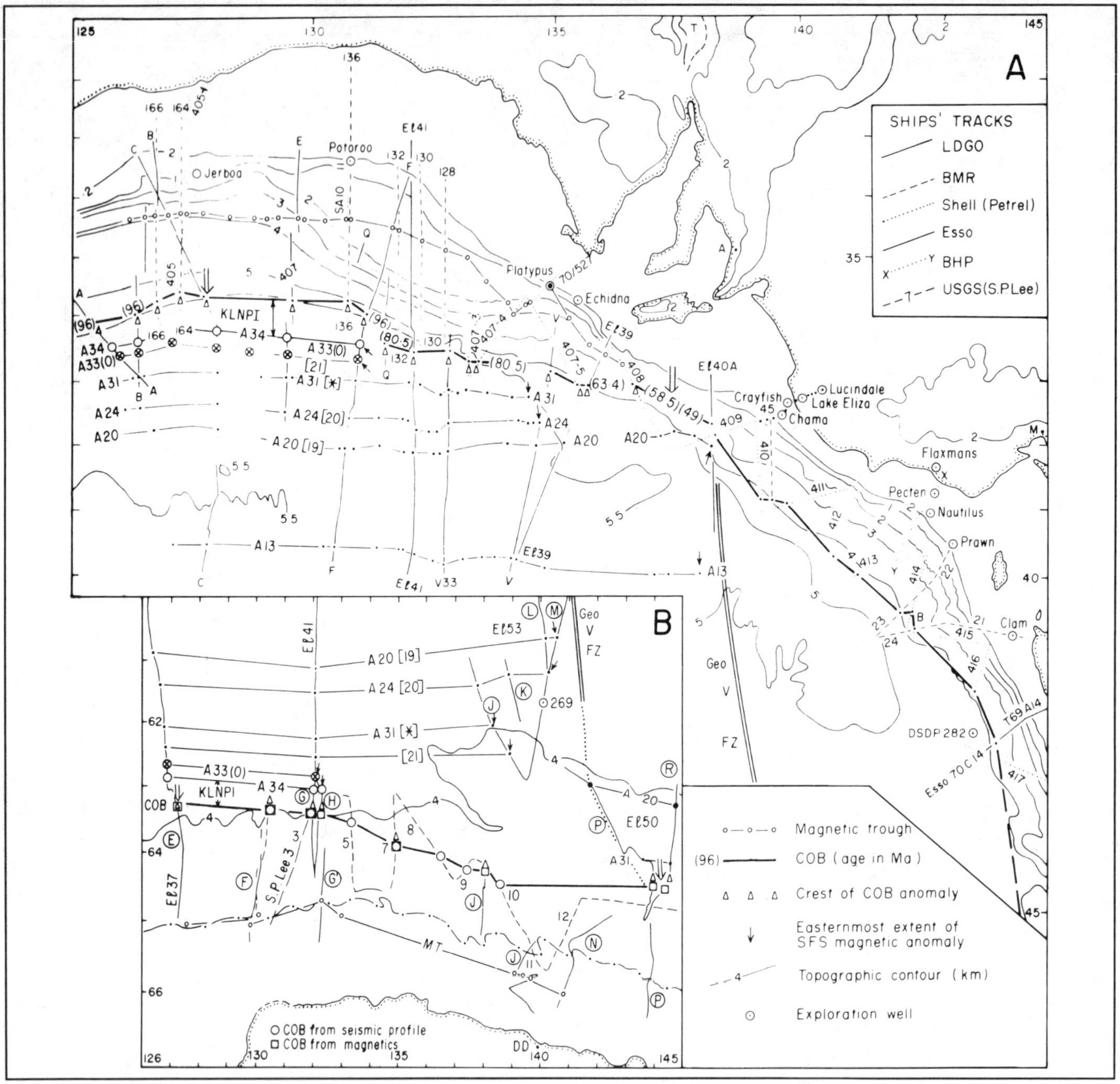

Figure 8—A. Southern margin of Australia between 125° and 145° E, showing ship's tracks (LDGO, from Talwani et al., 1979; Hayes et al., 1976; and König, 1980; BMR, from Willcox, 1978; B.M.R., 1979; and J. C. Branson, personal communication, 1984; Shell, from Boeuf and Doust, 1975; Esso, from Cooney et al., 1975 and personal communication, 1984; and B.H.P., from Denham and Brown, 1976); seafloor-spreading magnetic anomalies (from Cande and Mutter, 1982, with old identifications [in brackets] from König, 1980, with easternmost determinations shown by arrows); COB (heavy line) and crest of COB anomaly (triangle) from Figure 9B, and magnetic trough (from Cande and Mutter, 1982, after König, 1980, augmented by B.M.R., 1979); exploration wells (Deighton et al., 1976; Bein and Taylor, 1981; Fraser and Tilbury, 1979; Boeuf and Doust, 1975; and Kennett et al., 1975); bathymetry (from Hayes and Conolly, 1972). Broad arrows in A and B indicate conjugate points. Mercator projection. After Veevers (1986). B. Wilkes Land margin between 126° and 145° E, showing location of ships' tracks (R/V *Eltanin* 37E, F, G', N, and 41, from Hayes et al., 1976, and 50, 53, from König, 1980; R/V *S.P. Lee* from Eittreim and Smith, this volume), seafloor-spreading magnetic anomalies (from Cande and Mutter, 1982, with old determinations, by König, 1980, in brackets), and east of projected George V Fracture Zone, tentatively identified here; easternmost extent of anomalies west of George V Fracture Zone marked with arrow; COB (from *Lee* seismic profiles, from Eittreim and Smith, this volume, and from COB anomaly [crest located by triangles] on El 37 and 41, and *Lee* 1, 3, 7, from Fig. 9A), and magnetic trough (smaller circles, from Cande and Mutter, 1982, after König, 1980). Coastline, edge of continental terrace (0.8 ± 0.2-km isobath, dot-and-dashed line), and 4-km isobath, from Vanney and Johnson (1979). Mercator projection. After Veevers (1986). © 1986 Elsevier Science Publisher.

COMPARATIVE STRUCTURE AND STRATIGRAPHY OF THE SOUTHERN AUSTRALIAN (125°–145° E) AND WILKES LAND (130°–138° E) MARGINS

Southern Margin of Australia

A compilation of geophysical profiles of the Australian sector between 125° E and 145° E was made for comparison with the conjugate Wilkes Land margin. Ships' tracks are given in Figure 8A, magnetic profiles in Figure 9B, and magnetic profiles and line drawings of seismic profiles, projected on an azimuth of 180°, in Figures 10A through 10D.

Continent–Ocean Boundary (COB)

Boeuf and Doust (1975, Figure 3, profiles 405 and 413, shown here in the line drawings of Figs. 10A and 10D) recognized two types of acoustic basement in the R/V *Petrel* seismic profiles: (1) "Over much of the continental margin a flat and locally well defined bounding reflection, often displaced by normal faults, is taken to represent older rocks of the Precambrian or Palaeozoic metamorphic continental basement"; and (2) "Oceanwards on the section, a second type of basement appears usually as an envelope of diffraction curves. This diffraction envelope may be flat or may form pronounced hills or ridges which are generally expressed as magnetic anomalies... This second type of basement reflector is considered to represent the top of a basalt layer associated with the top oceanic crust." These types of basement, distinguished by stippling in the set of line drawings of seismic profiles of the Australian margin (Figs. 10A–D), provide the main constraint on the location of the COB. The boundary is not seen in those profiles with shallow seismic penetration, and the more widely applicable magnetic profiles were scrutinized for a magnetic signature of the COB. As noted by Boeuf and Doust (1975), west of 136° E the rugged topography of the oceanic crust near the COB is marked by magnetic anomalies. Compared to the oceanic crust, the continental crust is magnetically quiet to a distance of 50–200 km landward of the COB to the magnetic trough (König and Talwani, 1977; Talwani et al., 1979; König 1980, shown in Cande and Mutter, 1982, Figure 1), northward of which the magnetic profiles become jagged. König (1980) showed that the magnetic effect due to the rugged topography of the Diamantina Zone was not significant unless "a magnetic source body other than the commonly used magnetized layer of constant thickness" to model oceanic crust was used. Talwani et al. (1979) and König (1980, 1981) therefore recognized the oldest identified seafloor-spreading magnetic anomaly—their A22 equal to Cande and Mutter's (1982) A34 west of 131° E and A21 equal to A32–33 between 131° E and 136° E—as indicating the COB.

Later work (Fig. 9B) showed that the magnetic anomaly could be modeled better if generated by the contrast at the COB between the reversal sequence truncated at 96 Ma and a nonmagnetic continental crust (Veevers, 1986). The observed anomaly, in this interpretation, is the COB anomaly and not A34 [22], and the COB is located about halfway down the northern flank of the anomaly. A symmetrical set of anomalies at the conjugate Antarctic margin, found by König (1980), revised by Cande and Mutter (1982), and shown in Figure 9A, rules out the possibility of ridge jumping in the sector eastward to 134° E, so that continental breakup or the onset of spreading is dated by the truncation of the block model at 96 Ma. Rounded to 95 ± 5 Ma to allow for uncertainties, this age, at the Early to Late Cretaceous boundary, is that of the major (breakup) unconformity along the southern Australian margin between strongly faulted Early Cretaceous rocks, and lightly faulted to unfaulted Late Cretaceous and younger rocks (Veevers, 1984, p. 175). Between 131.2° and 139° E, the pattern of spreading is complicated by the convergence of the east-trending seafloor-spreading magnetic lineation with the east–southeast-trending COB (Fig. 8A); in Figure 9B, the COB anomaly is modeled as due to the same kind of effect as seen in the west, but with successively younger ages of truncation to 49 Ma, interpreted as indicating the youngest age of ridge crest jumping to the Australian COB.

These magnetic identifications of the COB are used in Figures 10A to 10D to align the combined set of seismic and magnetic profiles in the sector 125°–138.2° E. East of 138.2° E, seafloor-spreading magnetic anomalies near the margin are unknown, and the profiles are stacked along the seismically determined COB only. In the sector west of 138° E, six profiles contain a COB precisely determinable from seismic reflections (166, 164, 405, 407-3, -4, -5), and in each profile this coincides with or lies within 10 km of the magnetically determined COB; in the remaining profiles, the magnetically determined COB falls between widely spaced seismically determined oceanic and continental crust. As mapped in Figures 4A and 8A, the COB parallels the seafloor-spreading isochrons and, within each sector, is itself an isochron. The cross-cutting of the seafloor-spreading lineation by the COB, originally observed by Weissel and Hayes (1972), is most clearly marked by a 50-km southward offset of the COB between profiles F and Q (Fig. 9B), at 131.25° E (Fig. 8A), and by less clear offsets between sectors of each age to the east.

In many profiles (166, 164, 405 [Fig. 10A]; 407-3, 70-527, 410 [Fig. 10C]; 411, 413 [Fig. 10D]), the oceanic crust near the COB stands high in a hummock (Willcox, 1978).

Marginal Structure and Sequences

The structure of the southern Australian margin is dominated by normal faults (Fraser and Tilbury, 1979) of three kinds:

1. the major boundary faults that displace the Precambrian basement by about 6 km.
2. intrabasin faults, mostly synsedimentary, which displace the Lower Cretaceous horizons as much as 2 km.
3. near the outer margin, normal faults with displacement usually less than 0.2 km, and thrust faults, reported by Boeuf and Doust (1975, p. 41).

The intrabasin faults die out upward, within the Upper Cretaceous section; they grew fastest during the deposition of the Early Cretaceous section, which contains rollover anticlines, defining an epoch of crustal extension that terminated at the end of the Early Cretaceous, immediately after the deposition of a horizon called D by Fraser and Tilbury (1979), during the change from extension of continental crust to spreading of the sea floor. As in the Bass Basin to the east (Williamson et al., 1985), so in the Ceduna region, adjustment between fault blocks continued into the Late Cretaceous, but it was minor.

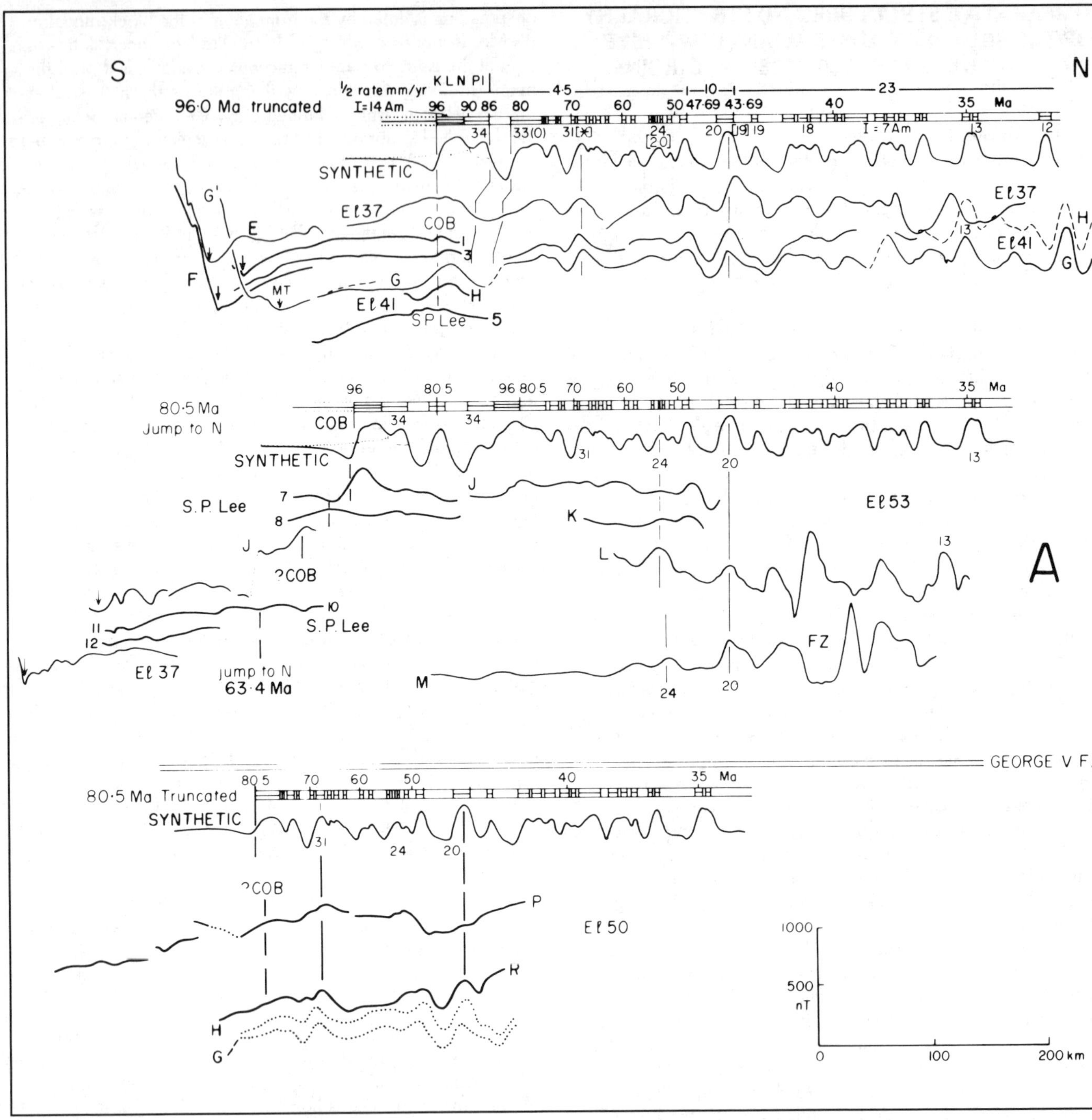

Figure 9—A. Magnetic profiles off Wilkes Land, located in Figure 8B, aligned along the COB, projected on an azimuth of 180° and compared with synthetic profiles. *Eltanin* 37 and 41 profiles from Hayes et al. (1976) and König (1980), *Eltanin* 50 and 53 profiles from König (1980) and Hayes et al. (1977, 1978); *Lee* profiles Childs and Stagg (this volume). Anomaly identifications modified from Cande and Mutter (1982), after old identifications of König (1980) [in brackets], and east of the projected George V Fracture Zone, by Veevers (1986). Magnetic trough shown by arrow. Magnetic-block models after model B of Cande and Mutter (1982), with the same parameters, except for a half-rate of 23 mm/year in place of 27 mm/year, which is probably a misprint. See Veevers (1986) for further details. B. Magnetic profiles of the Southern Australian margin and adjacent sea floor, located in Figure 8A, aligned along the COB, projected on an azimuth of 180°, and compared with synthetic profiles. After Veevers (1986). Magnetic-block models after Model B of Cande and Mutter (1982), with the same parameters, except the COB anomaly (interpreted by König, 1980, as A22, revised to A34 by Cande and Mutter, 1982) is modeled as generated by the contrast between a truncated oceanic block against nonmagnetic continental crust. The top set of profiles, from A to F (125° to 131° E) is matched by a synthetic profile generated from Cande and Mutter's (1982) model B truncated at 96 Ma, and with the older part of the last block, from 91 to 96 Ma, with double the magnetic intensity (or equivalently double the thickness). The dotted line corresponds to Cande and Mutter's last normal block of uniform intensity and width of 80 km. Further details in Veevers (1986).

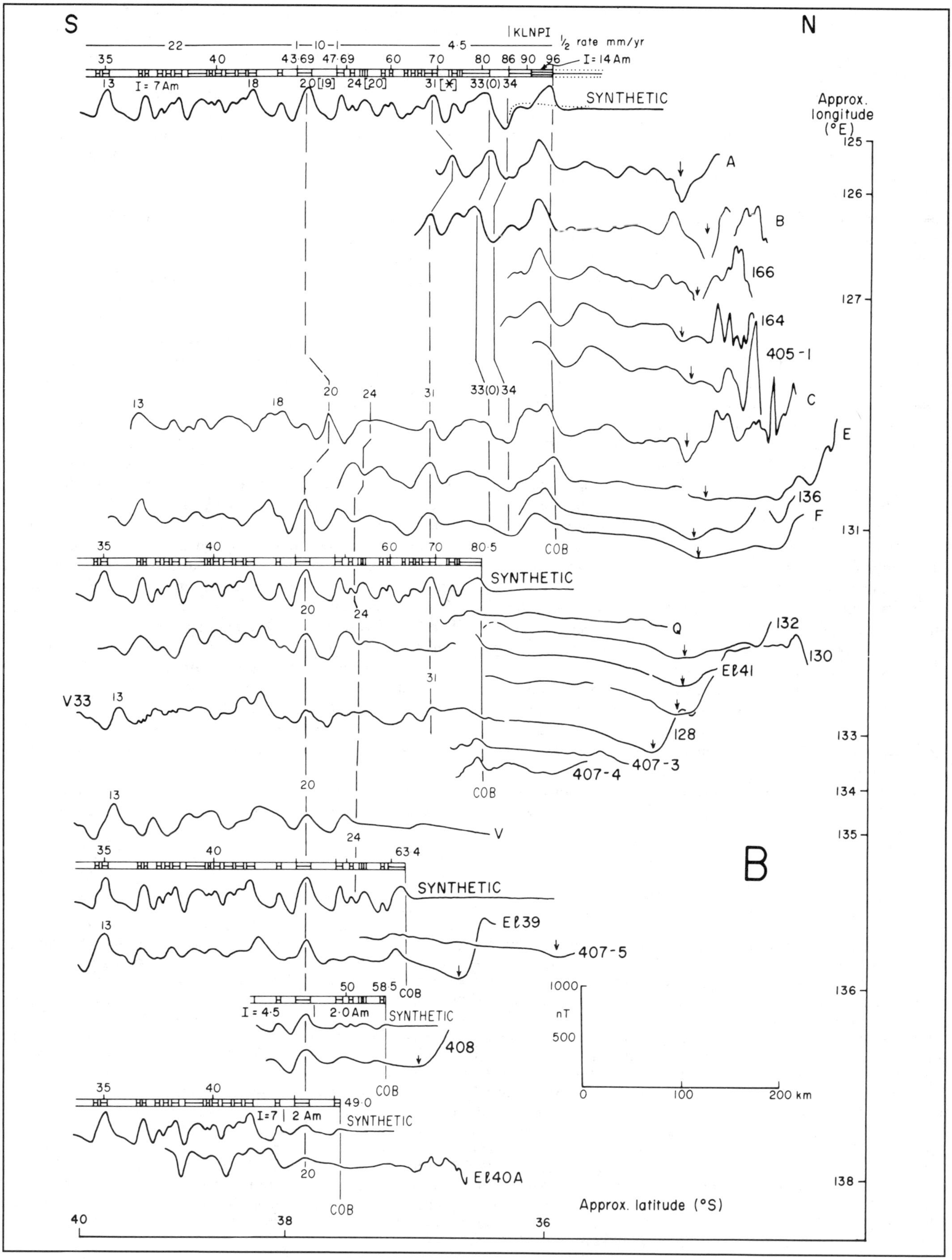
S
N
22
10
4·5
KLNPI
½ rate mm/yr
I = 14 Am
I = 7 Am
SYNTHETIC
Approx. longitude (°E)
A
B
166
164
405-1
C
E
136
F
COB
Q
132
130
El41
V33
128
407-3
407-4
V
El39
407-5
I = 4·5
2·0 Am
408
I = 7
2 Am
El40A
Approx. latitude (°S)
0
100
200 km
1000
nT
500
40
38
36

The boundary between the heavily block-faulted sequence below and the lightly faulted to unfaulted sequence above is interpreted (Veevers, 1984, p. 175) as the breakup unconformity (Falvey, 1974). In the Ceduna region, it is marked by horizon D of Cenomanian age (Fraser and Tilbury, 1979) and coincides with the magnetically determined age of breakup of 96 Ma. The unconformity is traceable throughout the rest of the southern margin, in the Otway, Bass, and Gippsland basins. Earlier attempts at correlating the breakup unconformity with the age of breakup, as indicated by the age of emplacement of the oldest adjacent sea floor, foundered on the interpretation by Weissel and Hayes (1972) of the oldest sea floor as A22 or only 53 Ma (Eocene), so that it was not until Cande and Mutter (1982) reinterpreted the oldest seafloor-spreading anomalies in the Southeast Indian Ocean as Early to mid-Cretaceous that valid correlation was possible.

Antarctic (Wilkes Land) Margin

Ships' tracks are given in Figure 8B, magnetic profiles in Figure 9A, and line drawings of seismic profiles (from Eittreim and Smith, this volume) together with magnetic profiles (from Childs and Stagg, this volume) in Figures 11A and 11B. Eittreim and Smith (this volume) located the COB from seismic evidence on lines 1, 3, 5, and 7, and tentatively on 10, and this position corresponds on lines 1, 3, and 5 with the landward flank of the magnetic anomaly at the seaward edge of the magnetic quiet zone, which is therefore recognized as the COB anomaly; on line 7, a well-marked magnetic anomaly 25 km south of Eittreim and Smith's pick is taken as more precisely marking the COB. In the case of line 10, the magnetic anomaly corresponds with a steep scarp in the oceanic crust (V3) and, as on the conjugate margin, is picked as the COB, 40 km north of Eittreim and Smith's pick.

Other magnetic profiles are available from El 37, 41, and 53 (König, 1980, this volume, revised by Cande and Mutter, 1982). Veevers (1986) found counterparts of the Australian model truncated at 96 Ma, west of 132° E. East of 132° E, the increasing distance of the COB from A20 confirms the implication of the model of ridge jumps to Australia leaving strips of "Australian Plate" sea floor on the Antarctic plate. In detail, and with the additional data from the R/V *S.P. Lee* (Childs and Stagg, this volume), I find (Fig. 9A) that the COB anomaly lies the following distances south of A20: 250 km in the segment from 127° E to 134° E at an age of 96 Ma; 330 km on *Lee* 7, implying a ridge jump to Australia at 80.5 Ma; 410 km on *Lee* 10, implying a jump to Australia at 63.4 Ma; and east of the southward projection of the George V Fracture Zone at 145° E, 140 km, implying a jump to Antarctica at 80.5 Ma.

Direct Comparison of the Conjugate Margins

Breakup is a synchronous event in the development of the conjugate Antarctic and Australian margins and is employed to define the depositional sequences of both margins—prebreakup, including before (prerift) and during (synrift) continental rifting or extension, and post-breakup (postrift). From the internal evidence of the Antarctic seismic records, Eittreim and Smith (this volume) define the lowermost sequence (D) as prerift and synrift, the next sequence (C) as synrift, principally from its stratigraphical relationship to the oldest oceanic crust and from indications of syndepositional faulting, and the upper two (B and A) as postrift. According to this interpretation, K2, a prominent erosional unconformity at the top of sequence D, is pre-96 Ma; and K1, another unconformity that marks the uppermost limit of faulting, is the breakup unconformity and hence ca. 96 Ma. No continental basement is recognized in the Antarctic seismic-reflection records. A comparison of the Antarctic sequences with those of Australia, known further by drilling, suggests rather that the Antarctic sequence D is synrift, and C to A postrift; that is, that the unconformity between D and C (K2) is the breakup unconformity.

K2 marks the upward limit of the *substantial* faults that express the phase of continental crustal extension that terminated at breakup. This feature is seen best at the outermost part of *Lee* 3 (Eittreim and Smith, this volume); the thinnest and hence most extended continental crust (only 2.5–5 km thick between the Moho and K2) is faulted along six faults that extend upwards to K2, which is overlain by the undeformed V1 and the set of progaders of sequence C. Here the considerable extension of the crust has been concluded *before*

Figure 10—A–D. Magnetic profiles and line drawings of seismic profiles of the Southern Australian margin, reduced to the same scale, and aligned on the COB. Stippling shows acoustic basement. The LDGO and BMR profiles are single-channel records. The *Petrel* profiles (405 to 413) are from processed records that show the nearest two traces gathered on a variable-area display. The Esso profile is from a 24-fold processed record. Profiles 407, 410, and YX are depth sections. Vertical exaggeration of the depth sections and of the sea floor of the seismic profiles is 7:1. Profiles west of the George V fracture zone projected on an azimuth of 180°; 409 and 411 are unrotated; and 413, YX, and 70C14-69A14 are rotated to 090°. Magnetic anomalies from Cande and Mutter (1982), old determinations by Talwani et al. (1979) and König (1980) shown in brackets. Crest of COB anomaly marked by old determinations of Talwani et al. (1979) and König (1980) placed in a box. Magnetic trough (MT) shown by broad arrow. Profiles A, B, E, and F from Talwani et al. (1979); E140A bathymetric profile only; 166, 164, 136, 132, 130, and 45 from Willcox (1978) and B.M.R. (1979); 405 and 405-1, 407 and SA 10-11, 407-3, 407-4, 407-5 and 70-527, 408, 410, 412, 409, 411, and 413 from Boeuf and Doust (1975) and from copies of profiles made available under the terms of the Submerged Lands Act by the designated authorities (Geological Survey of Western Australia, South Australian Department of Mines and Energy, Victorian Department of Minerals and Energy, and B.M.R., by courtesy of J. C. Branson and H. M. G. Stagg); YX from Denham and Brown (1976); and 70C14-69A14 from Cooney et al. (1975) and Esso (Australia) (personal communication, 1984). Line-drawings and interpretations of some of these profiles are given by Falvey and Mutter (1981) and Veevers (1981a). Sonobuoy determinations of sonic velocity from König and Talwani (1977) and Talwani et al. (1979); asterisk = poorly determined velocity. Potoroo and Platypus well-logs from Fraser and Tilbury (1979), and velocity profile of Potoroo from Talwani et al. (1979).

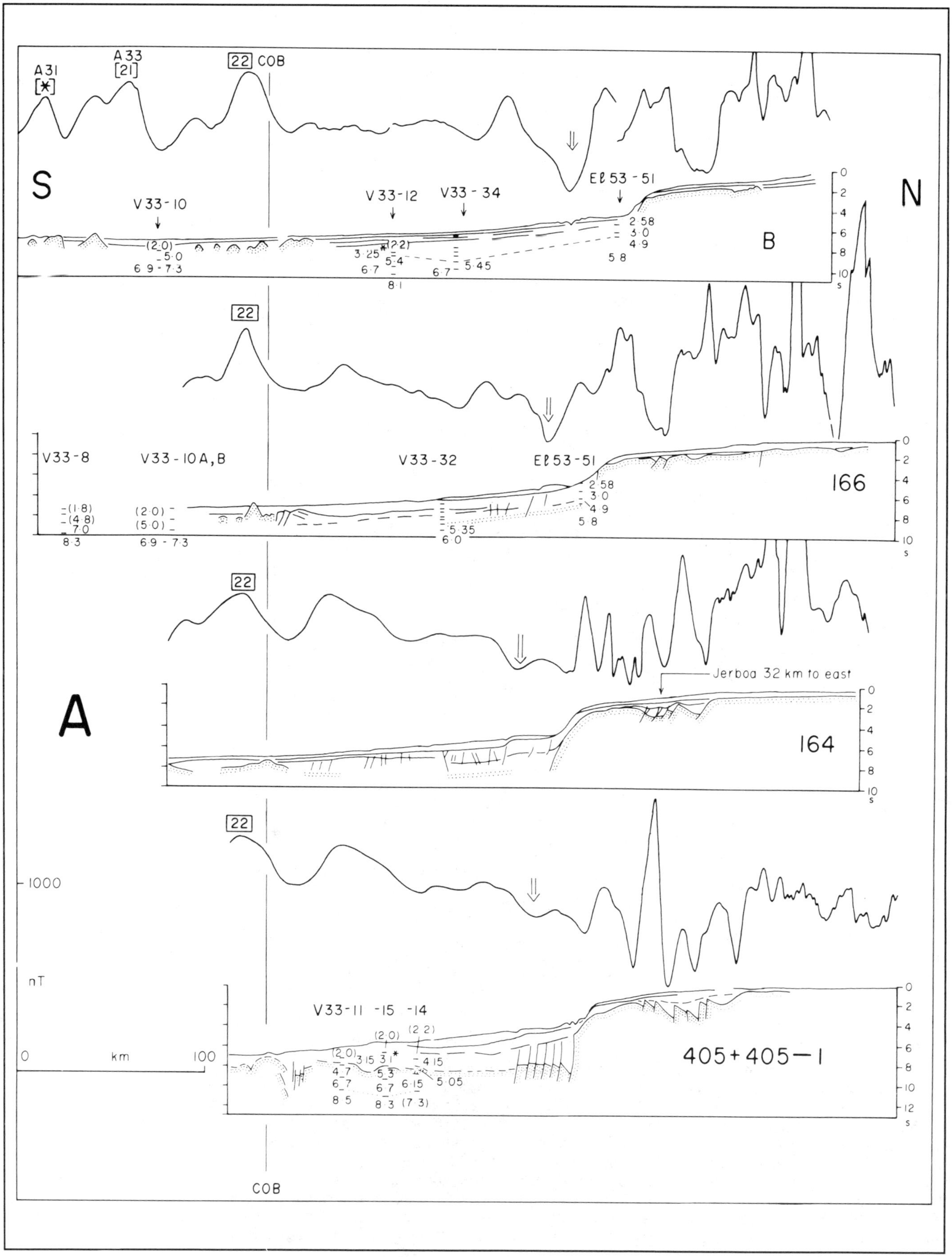

A31
[✱]
A33
[21]
22
COB
S
N
V33-10
V33-12
V33-34
El 53-51
B
(2.0)
5.0
6.9-7.3
3.25
(2.2)
5.4
6.7
8.1
6.7
5.45
2.58
3.0
4.9
5.8
V33-8
V33-10A,B
V33-32
El 53-51
166
(1.8)
(4.8)
7.0
8.3
(2.0)
(5.0)
6.9 - 7.3
5.35
6.0
A
Jerboa 32 km to east
164
1000
nT
0
km
100
V33-11 -15 -14
(2.0)
(2.2)
3.15
3.1
4.15
4.7
5.3
6.15
5.05
6.7
6.7
8.5
8.3
(7.3)
405+405—1
COB
s

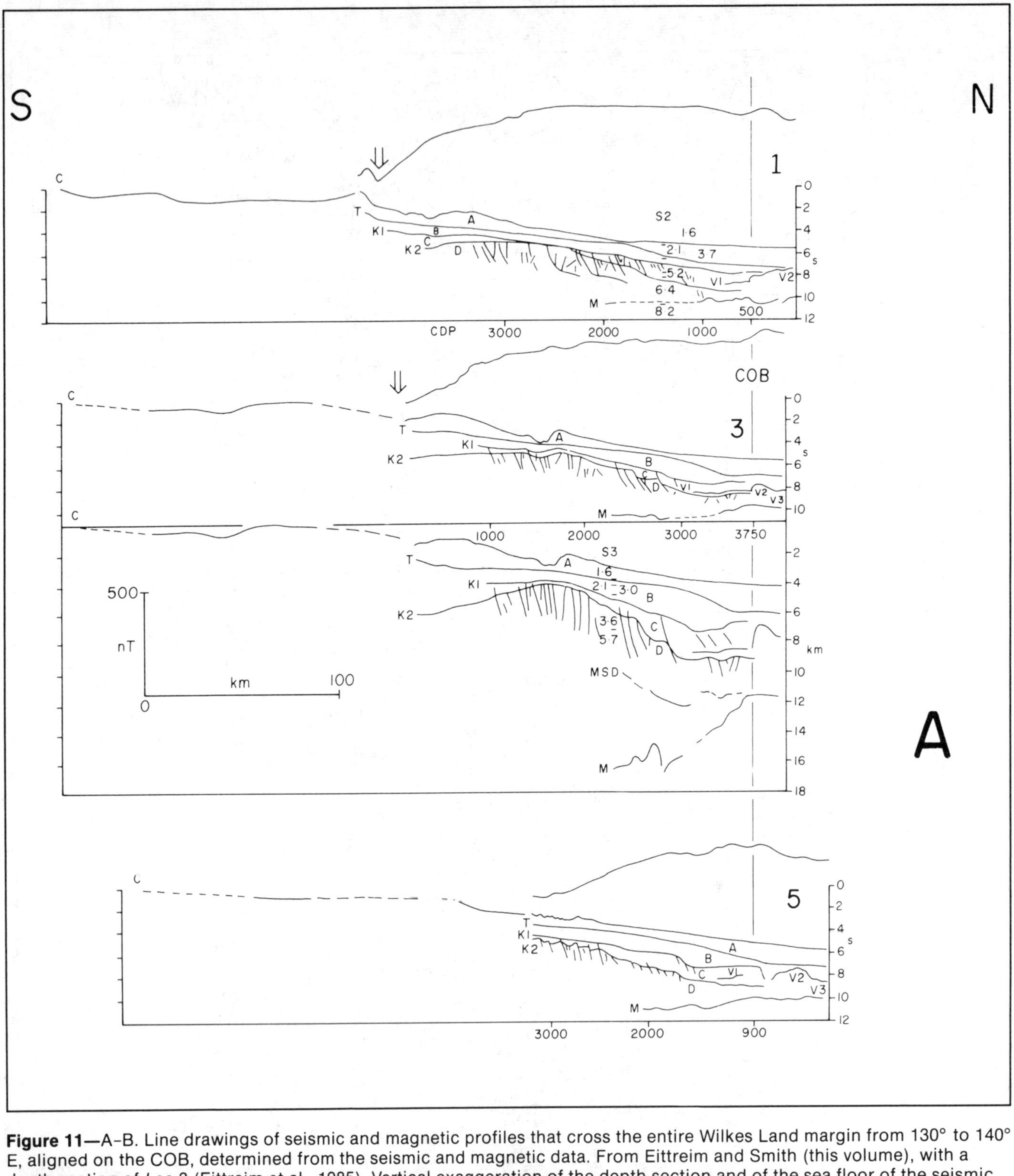

Figure 11—A-B. Line drawings of seismic and magnetic profiles that cross the entire Wilkes Land margin from 130° to 140° E, aligned on the COB, determined from the seismic and magnetic data. From Eittreim and Smith (this volume), with a depth section of *Lee* 3 (Eittreim et al., 1985). Vertical exaggeration of the depth section and of the sea floor of the seismic profiles is 7.35:1. Shown are sequences A to D, with bounding surfaces T, K1, and K2, the Moho (M), and volcanic sequences V3 (normal oceanic crust, myriad overlapping hyperbolae on unmigrated records), V2 (oceanic hummock, internal seaward-dipping reflectors), and V1 (high amplitude, rope-like in unmigrated records, series of 2 to 5 reflectors). In *Lee* 3, the outermost part of sequence C contains sediment progradera with apparent dip of 10°; in the depth section is shown MSD, a deep crustal reflector inferred by Eittreim et al. (1985) to be a décollement surface and a possible brittle-ductile boundary. Numerals at the bottom of each profile signify shot or common depth points. The sea floor beyond the southern end of each profile has been drawn to the coast (C) from bathymetric charts (Vanney and Johnson, 1979).

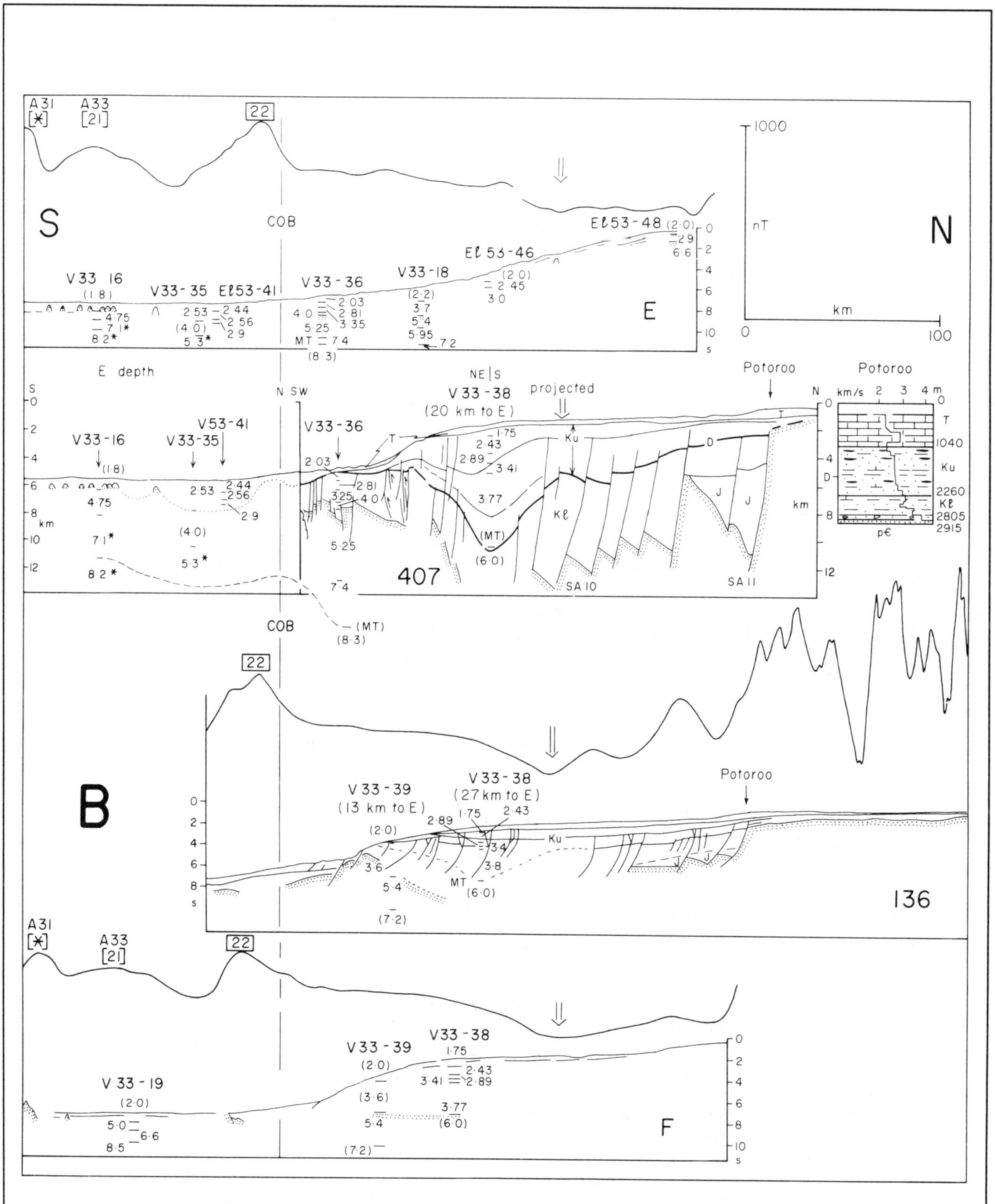

S
N
E
B
F
COB
nT
km
1000
0
100
A31
A33
22
V33-16
V33-35
V33-36
V33-18
V33-38
V33-39
V33-19
V53-41
E depth
projected
Potoroo
407
136
SA 10
SA 11
Ku
Kℓ
(MT)

Figure 10B

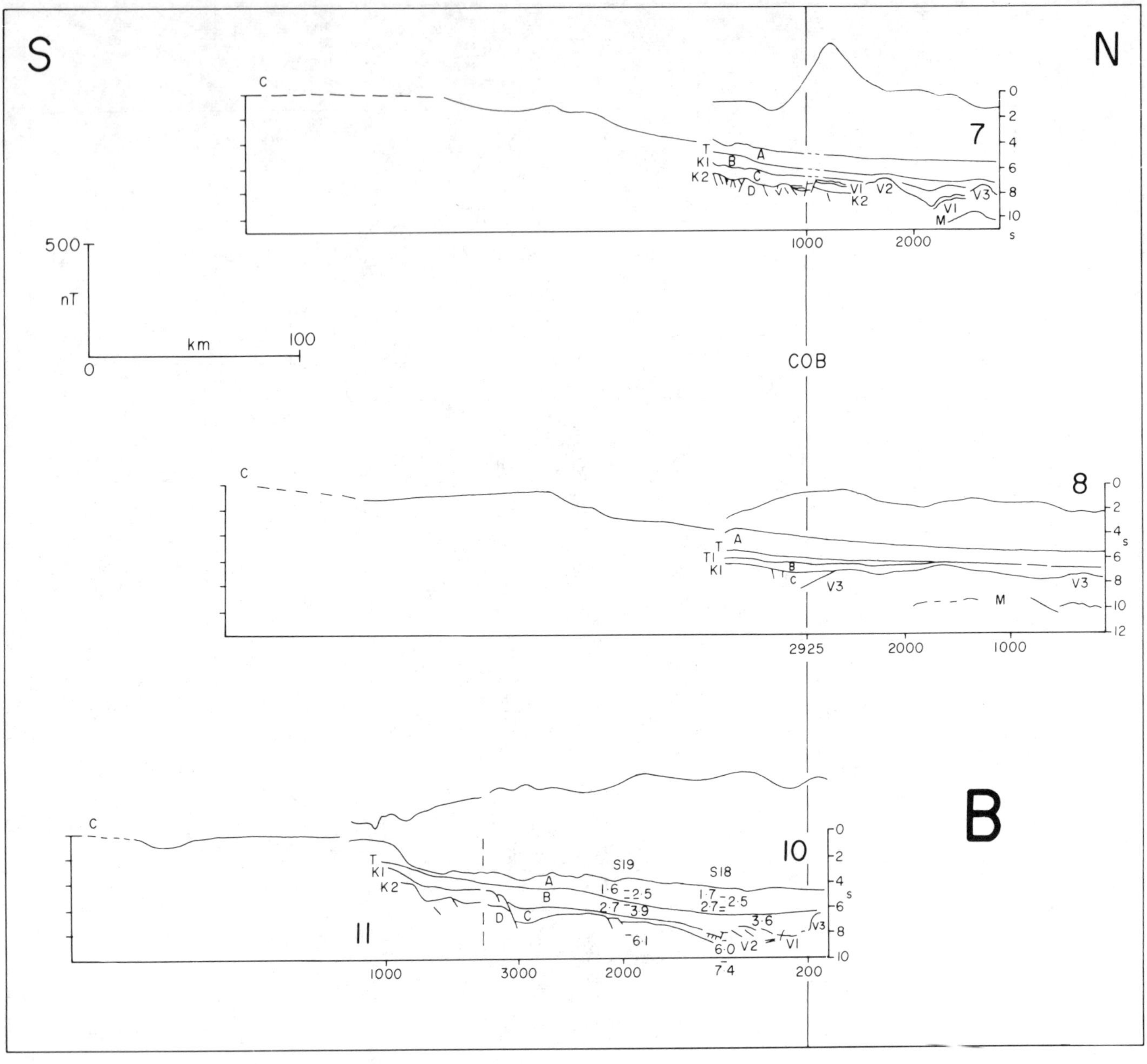

Figure 11B

the deposition of sequence C, indicating that K2 is the breakup unconformity. Table 3 is a tally of the various levels of the upward termination of the Antarctic faults interpreted by Eittreim and Smith (this volume) on *Lee* lines 1, 3, 5, 7, and 10/11, and of the Bass Basin (Australia) faults, including BMR line 4D-8 (Fig. 12), interpreted by Williamson et al. (1985). Of the 159 faults interpreted from the Antarctic records, 9.5% terminate within sequence D and 54% at or near its top; 10.5% terminate within sequence C and only 26% at or near its top. While it is true to say that K1 marks the uppermost limit of faulting, it is no less true to say that K2 marks the uppermost limit of substantial faulting—best indicated by the considerable relief on K2 compared with the subdued relief on K1.

A similar relationship is found on the conjugate Australian margin: a sequence (mid-Jurassic to mid-Cretaceous) terminating at Fraser and Tilbury's (1979) horizon D, overlain by a sequence (Late Cretaceous) cut in a few places and by a small throw by faults transmitted from below. As mentioned already, the main growth of these faults during the Early Cretaceous in Australia is shown by syndepositional structures, such as rollover anticlines and variations of thickness so that an Early Cretaceous epoch of major crustal extension by faulting can be contrasted with a Late Cretaceous and Cenozoic epoch during which the faults underwent only minor reactivation by movement between the old fault blocks, well-documented in the deep-penetration profiles of the Bass Basin to the east (Fig. 12; Williamson et al., 1985). Here the tally of the upward termination of faults (Table 3) is similar to that of the Antarctic faults: 47.5% terminate at the breakup unconformity, dated by drilling as mid-Cretaceous, 22% terminate within the overlying Late Cretaceous and Paleocene sequence C, and 30.5%

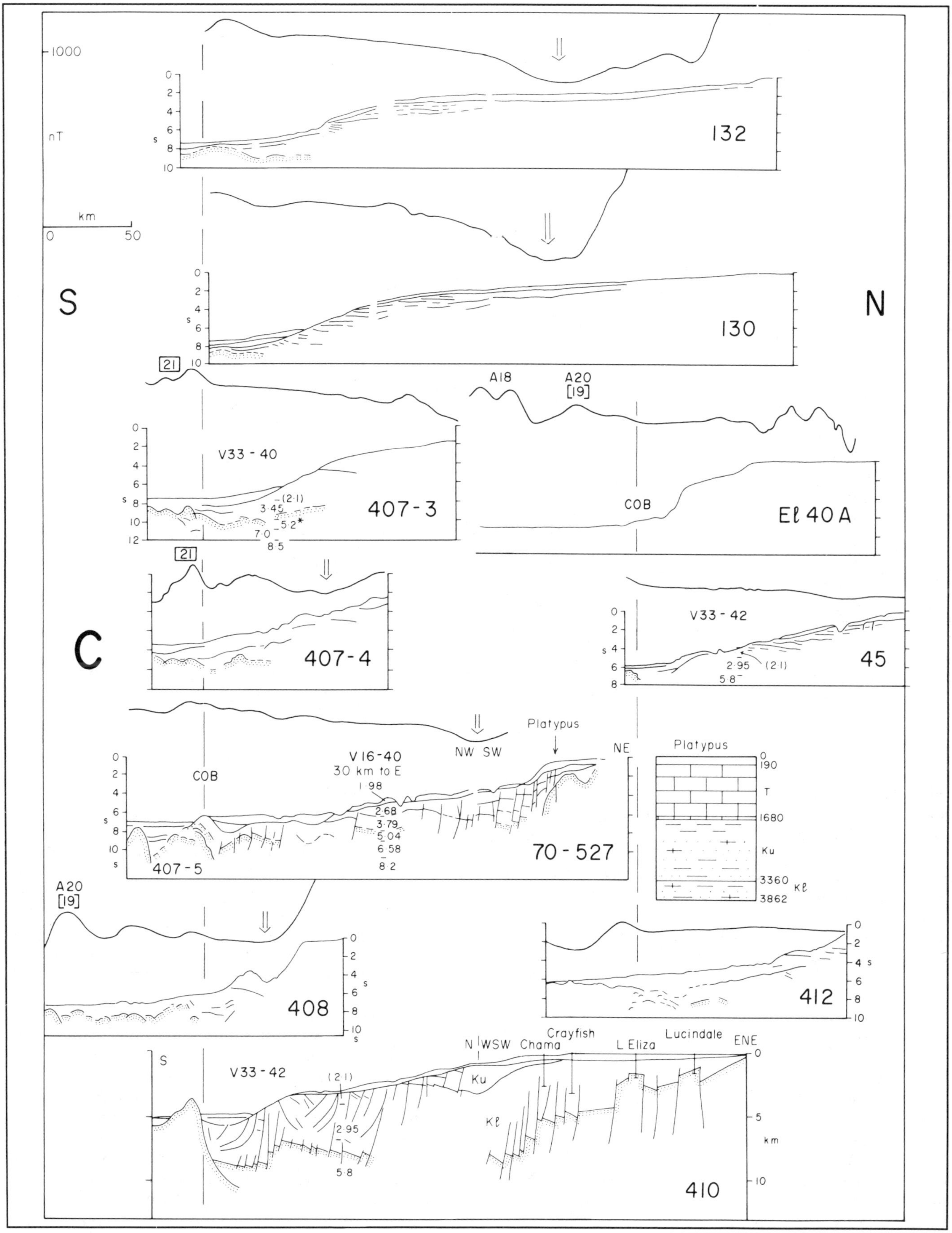

1000
nT
km
0 50
S
N
132
130
21
V33 - 40
407-3
A18
A20
[19]
COB
El 40A
407-4
C
V33 - 42
45
2·95
(2·1)
5·8
Platypus
V16-40
30 km to E
NW SW
NE
COB
70 - 527
407-5
190
T
1680
Ku
3360
3862
A20
[19]
408
412
Crayfish
N WSW Chama
L Eliza
Lucindale
ENE
V33-42
Ku
Kℓ
410
km

Figure 10C

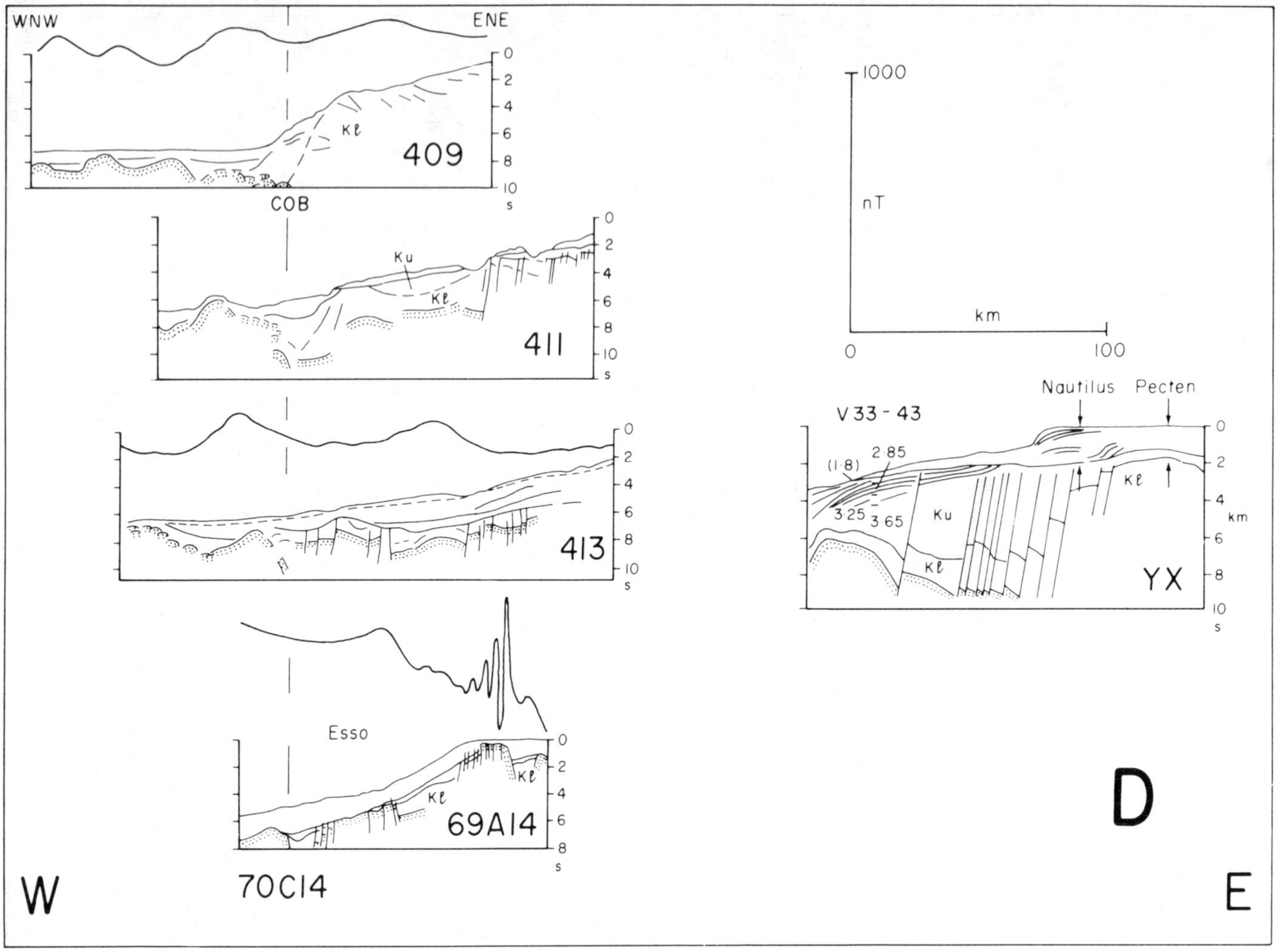

Figure 10D

terminate at or above the top of C. According to Williamson et al. (1985), "the Paleocene structure is influenced by the reactivation of normal faults developed at the time of the mid-Cretaceous rift unconformity and reflects drape over deeper features.... The mid-Cretaceous extensional phase constituted the last major phase of structure formation in the basin. Minor adjustment involving normal and reverse faulting continued after that time along with subsidence, but decreased with time so that little significant faulting is observed at the top of the Eastern View Coal Measures and only minor structural closures exist."

Syndepositional structure in Antarctica is difficult to find in the seismic records, except by inference of growth in the offset of faults from the bottom to the top of sequence C. From the observation that sequence D comprises highly stratified planar beds within the fault blocks, Eittreim and Smith (this volume) infer that no structural growth took place between the block faults and hence that substantial faulting took place no earlier than the age of the uppermost beds of sequence D. The observation of planar beds within each block is incontestable, but the scale of observation is so small—the average width of the blocks is 5 km—as to leave open the question of growth features manifested at a larger scale. Even so, the point is not critical: Most if not all of the faulting may have occurred immediately before the deposition of sequence C by rapid extension of continental crust in the style of deformation produced by the Late Triassic and Early Jurassic faulting of the thick, well-bedded interval G to F (Middle and Late Triassic Locker Shale and Mungaroo Sandstone) of the Exmouth Plateau (Exon and Willcox, 1980).

From the precise identification of the COB and of conjugate points (given in Table 1), each of the *Lee* seismic profiles is now compared with its Australian counterpart.

Lee *1 and Profile E (Figs. 11A and 10B)*

Profile E crosses the saddle between the Eyre Terrace, seen in the profiles on Figure 10A, and the Ceduna Terrace, seen in 407 (Fig. 10B). Significant differences aside (the thicker Cenozoic section, deep canyons on *Lee* 1), what little data are available in E match those of 1: a long, even slope and rise, similar width between the COB and magnetic trough (MT), similar oceanic crustal thickness. The oceanic crust hummock, seen in many southern Australian profiles, is seen in *Lee* 1, 3, 5, and 7, and comprises V2-type crust, with a rough upper surface and internal reflectors that generally dip seaward (Eittreim and Smith, this volume).

Table 3—Tally of upward termination of faults.

Wilkes Land Margin					
	Below K2 (within D)	At or near K2 (top of D)	Below K1 (within C)	At or near K1	Total
Lee 1	9	21	14	15	59
Lee 3	6	30	0	12	48
Lee 5	0	15	3	9	27
Lee 7	0	13	0	2?	15
Lee 10/11	0	7	0	3	10
Totals	15	86	17	41	159
Percent	9.5	54	10.5	26	100

Bass Basin					
	Within D	At or near BUU (top of D)	Within C	At or near top of C or above	Total
BMR 4D-8	0	11	13	12	36
BMR 18	0	6	0	3	9
BMR 17	0	11	0	3	14
Totals	0	28	13	18	59
Percent	0	47.5	22	30.5	100

Lee *3 and 407 (Fig. 13)*

By serendipity, the most revealing profiles, *Lee* 3 in Antarctica and profile 407 in Australia, are conjugate, so that this pair of profiles is the basis of a model cross section of the Antarctic–Australian suture zone. Profile 407 is a depth section made from a Shell multichannel seismic profile (Boeuf and Doust, 1975), projected in Figures 10B and 13 to an azimuth of 180°. The track of 407 has a southwest azimuth and crosses the contours of the Ceduna Terrace almost at a right angle; its seaward edge lies nearer the conjugate 1 than 3; on the shelf, it ties to Potoroo well (Fraser and Tilbury, 1979). Direct comparison is made in Figure 13A between the depth sections of 407 and *Lee* 3.

The following structural subdivisions can be made in Australia: (1) a landward platform, delimited by the major boundary fault from (2) the Ceduna depocenter, divided at the magnetic trough (MT) into a shallower inner part from the deeper axis, and (3) an outer ridge (the "outer high" of Boeuf and Doust, 1975); and in Antarctica, (1) a landward platform divided near the coast into an inner coastal part and an outer part beneath the shelf; and (2) seaward of the magnetic trough a depocenter divided by a median ridge into inner and outer parts.

The state of Antarctica and Australia at breakup (96 Ma), reconstructed by unloading the sediment and accounting for the thermal subsidence since 96 Ma, is shown in Figure 13B and applies to the other *Lee* lines. The section comprises a 600 km wide valley between boundary faults, with a 300 km wide median arch split by a 45 km wide median valley. Subsequent breakup along the COB took place not along the axis of the valley but asymmetrically along its northern edge so that the entire median valley remained on Antarctica. After breakup, Australia accumulated 8 km of Late Cretaceous detrital sediment in the rim basin behind the outer ridge from the main flow of drainage off the uplifted southeastern Australia (Veevers, 1984), including seaward progradersters, with a minor contribution from the eroding outer ridge. In Antarctica, a less copious source of sediment from continental Antarctica filled the inner depocenter, and local erosion of the median ridge (or from the side) supplied sediment that prograded into the lake or sea that filled the valley.

Lee *10 and 11 and Profiles 407-5 and 70-527 (Fig. 14)*

These profiles are not conjugate (the Australian profile 407-5 lies 150 km east of *Lee* 10 and 11 in the fit), but they contain a similar level of data. They have the same structural divisions as the section farther west, but 407-5 lies east of the Ceduna depocenter and so lacks the seaward axis of the depocenter. Otherwise, apart from the thickness of the K2 to seafloor fill, the profiles are similar, especially in their modern

Sequence	Lower boundary	SEISMIC CHARACTERISTICS					STRATIGRAPHIC INTERPRETATION		
		Configuration	Continuity	Amplitude	Frequency	Interval velocity range (km/s)	Maximum thickness (m)	Palaeoenvironment	Age
A	Concordant, onlapping at basin edges	Parallel	Continuous	Moderate to high	Moderate to high	1.6–2.9	2400	Restricted marine basin grading up into a carbonate-rich open marine shelf	Upper Eocene to Recent (Demons Bluff Formation and Torquay Group)
B	Concordant, onlapping at basin edges	Parallel	Continuous	Variable, but generally moderate	Moderate	2.5–3.1	1100	Non-marine basin: predominatly floodplain association including extensive coal swamps	Eocene (Eastern View Coal Measures)
C	Erosional truncation	Variable	Variable	Moderate to low	Moderate to low	3.1–5.5	6900	Non-marine basin: alluvial fan, floodplain and lacustrine associations	Late Cretaceous to Eocene (Eastern View Coal Measures)
D	–	Irregular	Discontinuous to moderately continuous	High	Low	–	–	Predominatly non-marine sediments unconformably on Palaeozoic and Precambrian basement	? Late Cretaceous and older (Otway Group and basement)

23/OV/64

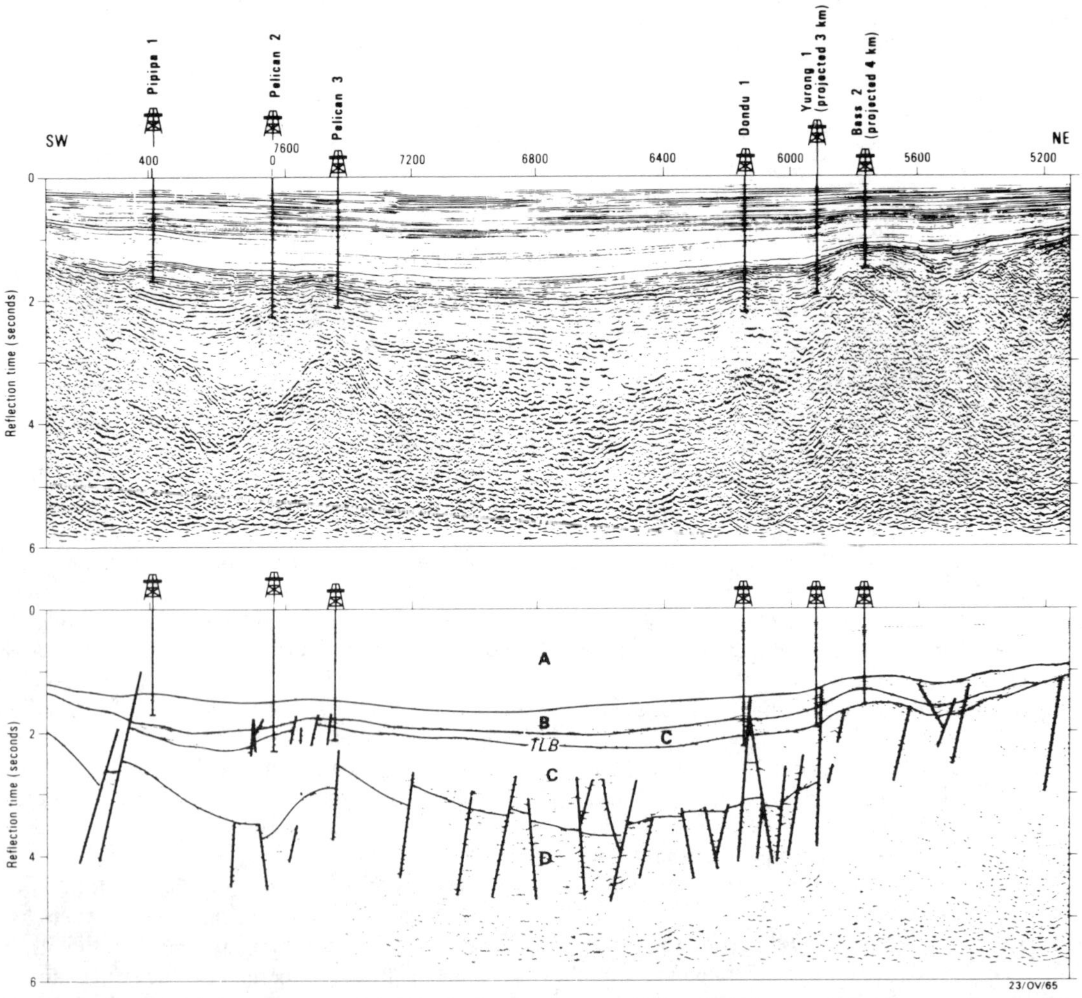

Figure 12—Bass Basin, seismic sequence table and interpretation of BMR Line 4D-8, located on Figure 3A. Line is 110 km long, and V:H = 8. From Williamson et al. (1985). © 1985 Australian Petroleum Exploration Association Journal.

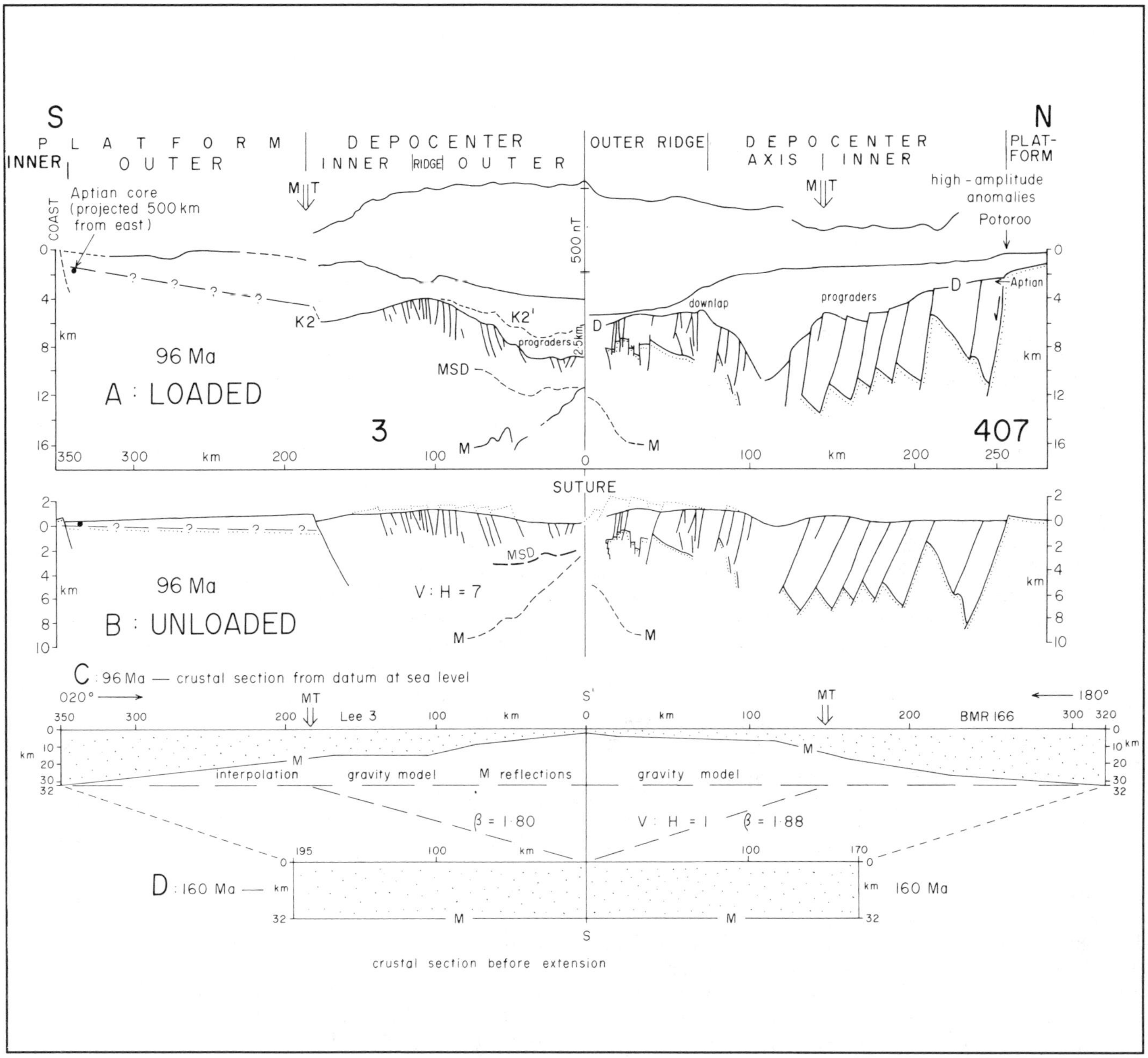

Figure 13—A. Depth sections and magnetic profiles of *Lee* 3 (Antarctica) and 407 (Australia) (projected along 180°) joined along the suture (COB), as at 96 Ma, but with present overburden included (loaded state). Sections show Moho (M), MSD (a deep crustal reflection, Eittreim et al., 1985), Precambrian basement (stippled), major faults extending upward to K2 (Antarctica) and to horizon D (Australia), position of known Aptian strata in Australia (in Potoroo well) and in Antarctica (projected 500 km from the east—Domack et al., 1980), the position of K2 isostatically adjusted to K2′ for the same thickness of overburden in Australia, the location of seaward-prograded strata and of downlapping strata, and the sea floor extended southward from the end of *Lee* 3 from bathymetry. In Antarctica, a structural scarp is inferred at the MT, a boundary fault at the coast, and a shallow K2 between. B. Conversion of this section to the unloaded state at breakup (96 Ma) by (1) rotating the COBs about a hinge near the coast through about 6 km from the initial elevation of the oldest sea floor to its present depth to allow for thermal subsidence since 96 Ma (Veevers, 1981b), and (2) complete adjustment for isostatic loading of the overlying sediment ($\rho = 2.2$ g/cm^3) and water. Dotted line above K2 and D indicates inferred surface since lost by erosion. C. Crustal section (no vertical exaggeration) at end of phase of extension (96 Ma) drawn from a datum at sea level to facilitate mensuration. From Veevers and Eittreim (in press). Antarctic section from *Lee* 3 (above) with downward termination of faults taken as the top of the crustal layer, and with Moho extended to end of *Lee* 3 by gravity modeling (Childs and Stagg, this volume), and thereafter by interpolation to assumed 32 km thick standard crust at the coast. Australian section is profile 166, with depth of Moho from gravity modeling constrained by seismic-refraction results (Willcox, 1977, Fig. 6) and surface of crustal layer from Figure 10A, again with a standard crust of 32 km at the coast. D. Restored crustal section (no vertical exaggeration) at 160 Ma immediately before the onset of rifting by extension, assuming a constant volume of crust.

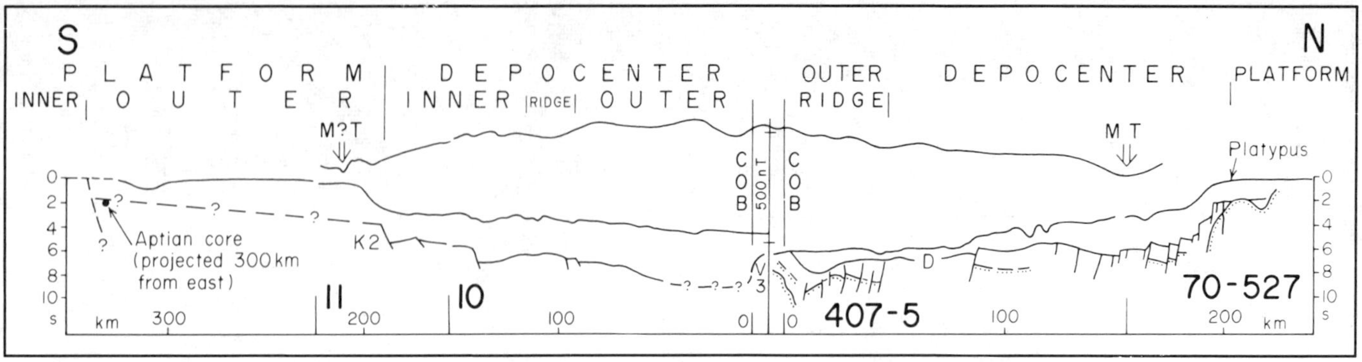

Figure 14—Seismic (time) sections and magnetic profiles of *Lee* 10 and 11 and 407-5 and 70-527 joined at points 5 km seaward of the COB (to show the equivalent oceanic hummocks) shortly after 96 Ma, but with present overburden (loaded state). Sections show Precambrian and oceanic basement in Australia (stippled) and oceanic basement in Antarctica (V3), major faults extending up to K2 (Antarctica) and to horizon D (Australia), position of known Aptian strata in Australia (in Platypus well—Whyte, 1978) and in Antarctica (projected 300 km from the east—Domack et al., 1980). Interpolation of K2 to the COB shown by queries.

morphology and magnetic symmetry.

As pointed out by Childs and Stagg (this volume), the Wilkes Land margin contains less sediment than its Australian counterpart; only east of a (north-trending) trending basement ridge at about line 7 is the sediment thickness (>7 km) in Antarctica greater than in conjugate Australia.

Summary

From the comparisons given above, I interpret K2 as the breakup unconformity and hence date sequence D as 95 ± 5 Ma and older; from comparison with Australia, it is Early Cretaceous and possibly also Late Jurassic (Fig. 15). Older Phanerozoic sediments are unknown along the southern Australian margin—very thin late Carboniferous/Permian sediment on the coast, in the Denman, Polda, and Troubridge basins, does not reach the margin where drilled—so there is no reason from comparison with Australia for inferring any older Phanerozoic sediment beneath the Wilkes Land margin. And from the Wilkes Land margin itself, off Cape Carr (130°–135° E) and in the Mertz Glacier area (140°–145° E), recycled palynomorphs recovered from surface samples comprise "an Early to mid-Cretaceous component and a Late Cretaceous through to early Tertiary suite. The Permian forms so prominent in the Shackleton Ice Shelf area (97° E) are absent" (Truswell, 1982). At the COB, K2 lies at a depth of 9 km beneath a sediment overburden of 4.7 km, 3.7 km thicker than the overburden on horizon D of the conjugate margin (Fig. 13). This extra overburden produces an isostatic loading of 2.5 km (= 2.2/3.3 × 3.7 km, using the ratio of sediment to mantle density); when this effect is removed, K2 is adjusted to K2′, which is within 1 km of the same depth at the COB as the conjugate horizon D (= breakup unconformity).

The only in situ Mesozoic sediment known from the Wilkes Land margin is nonmarine Aptian clay at a depth of 1407 m in the George V Basin (Domack et al., 1980), some 500 km west of *Lee* profile 3 (Fig. 13). This sediment, presumably beneath K2, suggests a boundary fault on the coast that offsets Precambrian rock at the coast to the greater depths offshore.

Sequence C, between K2 and K1, is correlated with the D to B interval of the Ceduna region, with K2 to V1 Cenomanian, and V1 to K1 Santonian–Campanian (Fig. 15). V1, interpreted by Eittreim and Smith (this volume) as flood basalt, represents a spill of lava from the young midoceanic ridge, comparable to the basalt at the breakup unconformity on the northwestern (Ashmore Volcanics) and western (Bunbury Basalt) margins of Australia (Veevers, 1984), and to the vesicular olivine basalt immediately at the mid-Cretaceous breakup unconformity in the Bass Basin, in Durroon-1 (Williamson et al., 1985). The prograded sediment above V1 was deposited in a 1.5 km deep lake or sea impounded behind the hummock of V2, as the prograded sediment of interval C to B of the Ceduna region was deposited in a restricted marine rim basin that deepened southward from paralic to bathyal (1.5 km) depths to onlap against the outer ridge or continental rim. The apparent downlap of the Upper Cretaceous strata on the outer ridge is due, in my opinion, to thermal subsidence after breakup of the ocean floor and coupled continental margin about a hinge 600 km north of the COB at the landward wedge-out of the Eucla Basin, so that the rim submerged below sea level only by 65 Ma, as shown by the initial overlap at the COB of the breakup unconformity by Tertiary sediment. Sequence B, between K1 and T, is inferred to extend from the Upper Cretaceous, possibly Maestrichtian, to the Eocene; and sequence A from the Oligocene to the present. The Tertiary sequence of the southern margin of Australia is dominantly carbonate and no thicker than 1 km; sequence A is up to 2 km thick, and probably represents detrital sediment, with a glacigene component, shed from the more copious source of Antarctica, and probably shaped by the Circum-Antarctic current system that has operated from the late Oligocene (Kennett et al., 1975). All three sequences, from the Late Cretaceous to the present, may overlap the oceanic hummock near the COB; indeed, off the southern Australian margin, some hummocks (profile 410, Fig. 10C) stand today above the seafloor sediment.

Comparison of Sonobuoy-Determined Velocities

The velocity spectrum of the conjugate Antarctic and Australian sequences can be compared by means of sonobuoy determinations of both margins but not by CDP-derived interval velocities, which are available for the Antarctic margins

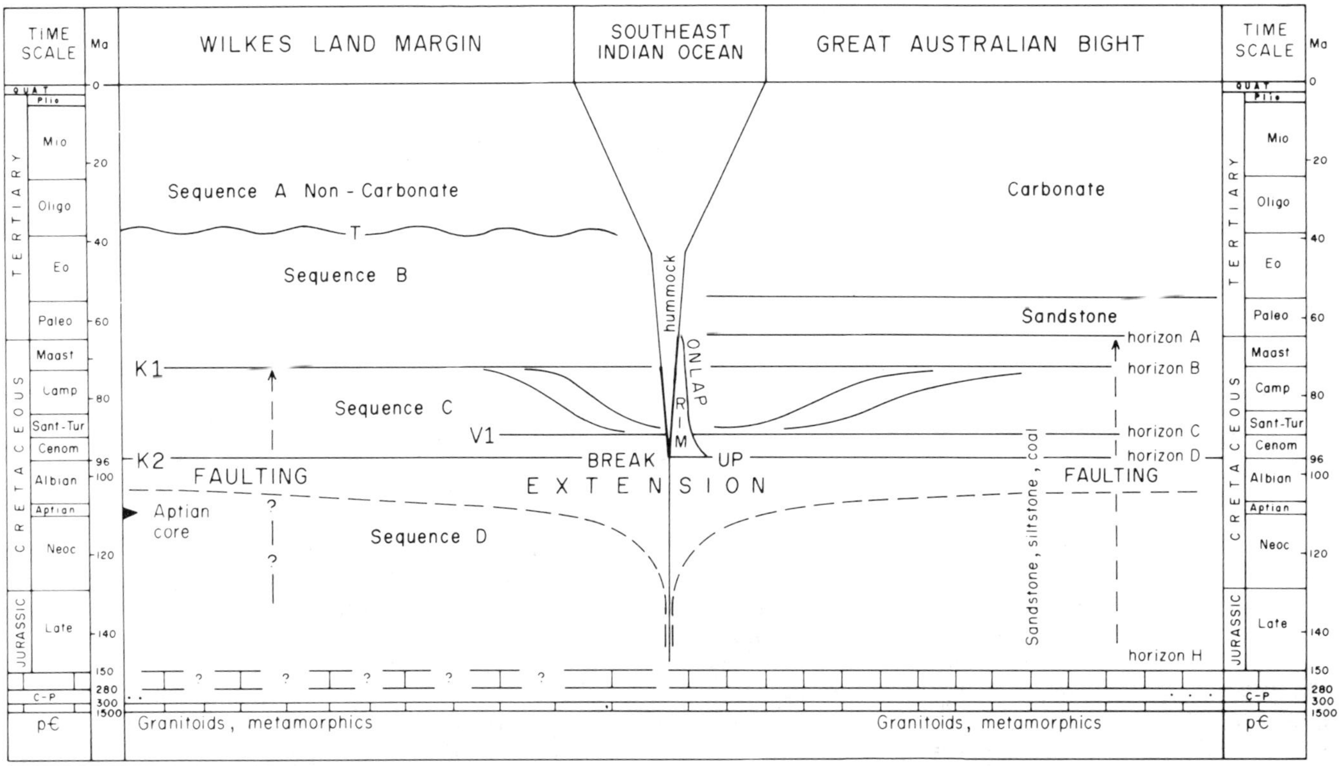

Figure 15—Schematic time-space diagram of the Wilkes Land-Great Australian Bight transect. Horizontal axis (distance) not to scale. The dots in the 300–280 Ma Carboniferous-Permian (C–P) interval represent the sediment at Horn Bluff on the coast of Wilkes Land, and the thin sediment of the Denman, Polda, and Troubridge basins on the Australian coast, which is unknown offshore.

only. Sonobuoys on *Lee* 1, 3, and 10 are situated sufficiently landward of the COB to provide unequivocal velocity determinations of the various continental sequences and crustal basement; the solutions of these sonobuoys (from Childs and Stagg, this volume) are superimposed on the seismic profiles in Figures 11A and 11B and listed, alongside comparable Australian data, in Table 4.

The velocities of the basement and of the Upper Jurassic to Early Cretaceous sequence of Australia match those of the Antarctic basement and sequence D; the wide range of the Australian Upper Cretaceous (2.43–3.77 km/sec) matches that (2.1–3.9 km/sec) of the Antarctic sequences B and C. The narrow range of the Australian Cenozoic sequence (1.75–1.98 km/sec) matches the lower range of the Antarctic sequence A (1.6–2.1 km/sec), and lacks the upper range of Antarctica (2.5–2.7 km/sec). This difference is attributable to the greater thickness of the Antarctic sequence A and its presumed clastic composition in contrast to the bathyal carbonate of Australia.

HISTORY OF THE AUSTRALIAN-ANTARCTIC SUTURE OR BOUNDARY ZONE BEFORE BREAKUP

A detailed reconstruction of Australia and Antarctica before the mid-Jurassic inception of the rift phase is given by Veevers and Eittreim (in press), from which the following summary is drawn.

The drift phase (<96 Ma) was preceded by a rift phase that in Australia lasted from the inception of rifting in the mid-Jurassic (160 Ma) or earliest Cretaceous (128 Ma) to breakup at 96 Ma. The rift phase in Australia is represented by the mid-Jurassic to mid-Cretaceous faulted sequence, and in Antarctica by the faulted terrane of sequence D.

In the Bass Basin (40° S, 145° E, Fig. 16A), Etheridge et al. (1984) found an extension ($\beta = 1.6$–1.8) in a south-southwest direction by Early Cretaceous movement on normal faults in the upper 10 km of the crust. Immediately east of the Bass Basin, a refraction section across the Bass Strait shelf (Johnson, 1973) showed that the main crustal layer ranges in thickness from about 35 km at the Victorian and Tasmanian coasts to a minimum of 15 km beneath the Gippsland Basin. Its cross-sectional geometry entails an extension of $\beta = 1.5$, similar to that of the adjacent Bass Basin. Bass Strait is only the northern part of the Australian–Antarctic suture zone. A complete transect across the suture zone is provided by the profiles of Figure 13A and 13B, from which the crustal sections of Figure 13C and 13D are derived. With crustal attenuation caused by extension during the rift phase, the cross-sectional geometry of the Great Australian Bight–Wilkes Land margin crust indicates 360 km of extension ($\beta = 1.84$), equivalent to 3.2° of latitude, along an azimuth of 035° at a right angle to the mean south-southeast trend of the Early Cretaceous faults of the Great Australian Bight and of the Otway, Bass, and Gippsland basins to the southeast (Denham and Brown, 1976; Etheridge et al., 1984; Threlfall et al., 1976; Fig. 16A). The rotation pole and angle of this extension is stage pole *a* (Table 2), and the closure of the suture zone about the pole is shown in Figure 16B. The distance to the pole is incalculable. I have chosen a pole (*a*) 45° distant from

Table 4—Sonobuoy-determined velocities (km/sec) of sequences.

	Antarctica				Australia		
Sequence	S2	S3	S18	S19	V33-38	V16-40	Age
A	1.6, 2.1	1.6	1.7, 2.5 2.7	1.6, 2.5	1.7	2.0	Cz
B	3.7	2.1, 3.0	3.6	2.7, 3.9	2.4, 2.9 3.4	2.7	UK
C	—	3.6	3.6	3.9	3.8		
D	3.7	3.6	—	3.9	3.8	3.8	LK UJ
Basement	5.2	5.7	6.0	6.1	(6.0)	5.0	pC

the Bight to the southeast on the great circle, to conform with the least coast-to-coast separation between the Otway Basin and the conjugate Horn Bluff region of Antarctica. A rotation of 4.5° about this pole effects a separation of 360 km at the Bight.

The 160-Ma reconstruction brings the estimated easternmost extent of Archean rocks in Antarctica opposite its Australian counterpart, but not precisely on trend. The Bowers Trough (BT), however, lies precisely on the trend of the equivalent Dundas Trough (DT). Neither of these features is intrinsically linear, but the trough could be expected to be straight between doglegs, such as the one in northern Tasmania, whereas the edge of the Archean terrane is not likely to be linear. Laird et al. (1977) reconstructed Tasmania and Antarctica by aligning the Bowers and Dundas troughs (since questioned by Jago, 1981). The reconstruction given here arrives at the same result, but as a by-product of a reconstruction made by eliminating the subsequent plate motions, not by an ad hoc alignment of features across the suture zone.

DISCUSSION AND CONCLUSIONS

1. The suture zone of extension between Australia and Antarctica was 3000 km long and 500–900 km wide. I earlier likened it to the modern rift divergence zone of the East African rift system where the 60 km wide fault-bounded Gregory rift valley of Kenya passes southward into a terrane of fault blocks 300 km wide in which crustal extension is distributed across a wide area (Veevers, 1984). But the width of the entire suture zone, taken with its structural style—Eittreim et al. (1985) liken the structure to that of the Basin and Range province of North America—suggest that the structure of the Basin and Range is the more appropriate modern analog.
2. On the west, Greater India was connected to Australia–Antarctica until breakup in the earliest Cretaceous, 128 Ma ago, when it diverged by seafloor spreading to the northwest (Veevers, 1984). Before breakup, continental extension between India and Australia–Antarctica and between Australia and Antarctica in the Jurassic must have been concurrent in a system whose arms met at a triple junction off the southwest tip of Australia unless the 360 km of extension between Australia and Antarctica took place after India's detachment at 128 Ma, as suggested by the concentration of syndepositional faulting on the southern margin of Australia in the Early Cretaceous.

   On the east, Lord Howe Rise–New Zealand was connected to Australia before breakup at 95 ± 5 Ma (Veevers, 1984). During the Early Cretaceous, the suture zone of extension either passed eastward into Lord Howe Rise–New Zealand or was offset by transform faults along what became the continental margin off New South Wales, which is parallel to the 027° azimuth of extension (C. McA. Powell, personal communication).
3. Extension between Australia and Antarctica took place in a Basin and Range zone approximated by the present coasts that mark the seaward limit of continental crust of standard thickness. Extension produced a zone of attenuated crust from 500 to 600 km wide, reaching 900 km wide in the middle part, equivalent to today's Australian–Antarctic Depression, and on the east, where the horst of Tasmania lies between the extended zones of Bass Strait and the area between Tasmania and north Victoria Land. During the Early Cretaceous, at least the eastern part of the zone (Otway and Gippsland basins) was dominated by deposition of alkaline-intermediate volcanogenic sediment (Otway and Strzelecki groups) and in the latest Jurassic by the Casterton Beds of the Otway Basin, lithic sandstone and other sediment with associated basaltic lava and pyroclastics (Douglas and Ferguson, 1976, p. 148). Some of the volcanogenic sediment may have originated from a magmatic arc east of the Gippsland Basin near the convergent boundary with the Pacific Plate, as argued by Jones and Veevers (in Veevers, 1984), or, as its alkaline affinity suggests (Gleadow and Duddy, 1981), from within the rift zone itself, as it does today in the analogous Basin and Range province of North America.
4. At the transect across the Great Australian Bight and Wilkes Land margin, the cross section was grossly symmetrical, with a 45 km wide median valley set within a 300 km wide arch flanked by depocenters, of which the Australian one (the Ceduna depocenter) is the thicker. Breakup at 96 Ma was not along the median line of the zone but on the northern side of the median valley, so that the outermost Antarctic margin is underlain by the median-valley depocenter, and the Australian margin by a rim or outer ridge, which subsequently impounded prograders in a thick

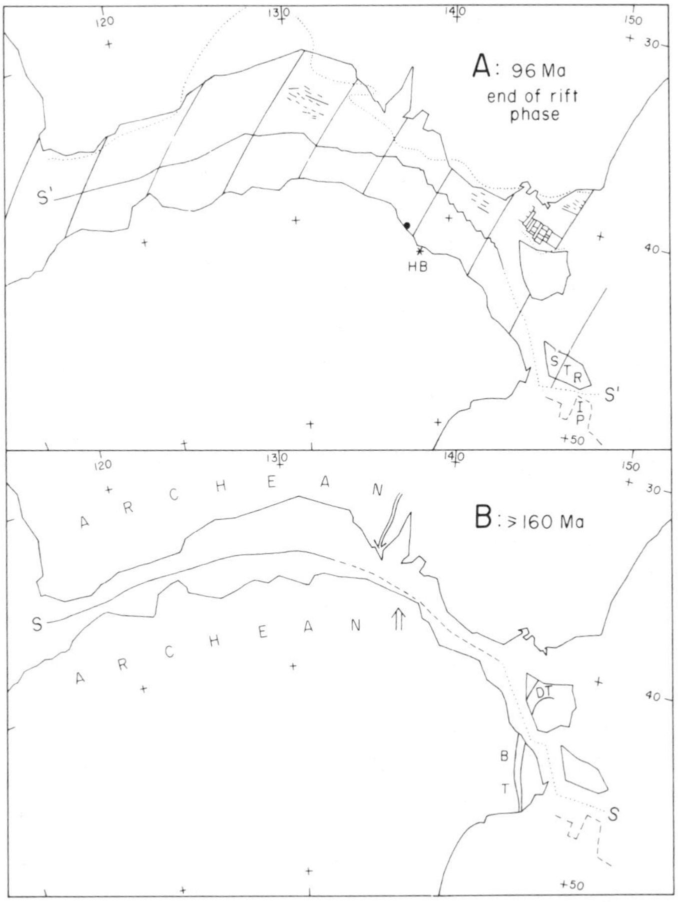

Figure 16—A. Reconstruction at the end of the rift phase at 96 Ma, with suture (S′) from Figure 3C (grid refers to present Australia), showing on Australia (1) azimuths of Early Cretaceous normal faults in the Great Australian Bight (Fraser and Tilbury, 1979), Otway Basin (Denham and Brown, 1976), Bass Basin (Etheridge et al., 1984) with, in addition, the orthogonal set of transfer faults, and Gippsland Basin (Threlfall et al., 1976), and (2) zero-line (dotted) of mid-Jurassic to mid-Cretaceous peripheral basin. Shown on Antarctica are the location of the Aptian core (solid circle), Horn Bluff (asterisk), and Iselin Plateau (IP) (broken line). The South Tasman Rise (STR) has been rotated by hand to remove the overlap shown in Figure 3C. Small circles about the rift-stage pole (*a* in Table 2) are drawn from coast to coast and crudely indicate the width of the zone of extension. See Veevers and Eittreim (in press) for details. B. Reconstruction before the inception of rifting 160 Ma ago, showing the positions on conjugate margins of the easternmost known Archean rocks (double lines and broad arrow) (Oliver et al., 1983) and of the Dundas Trough (DT) and Bowers Trough (BT) (Jago, 1981). S is the axis of the subsequent suture, as shown in Figure 13D, and, in the western half of the map, approximates the present position of the Australian magnetic trough. See Veevers and Eittreim (in press) for details.

rim basin in the Ceduna depocenter during the Late Cretaceous; the oldest sea floor was emplaced subaerially in a 1–2 km high volcanic hummock, comprising in Antarctica V2-type oceanic crust with internal seaward-dipping reflectors (Eittreim and Smith, this volume), which itself impounded progradеrs on the Antarctic margin, as seen in *Lee* 3, 7, and 8.

5. A few million years after breakup, the sea floor became emplaced at a lower elevation than that of the hummock and continued to spread, although at a faster rate after 44 Ma, within the regionally deeper Australian–Antarctic Depression to the present day. The tendency of the Southeast Indian Ridge to propagate eastwards across the southeast-trending Australian margin between 132° E and 138° E led to repeated ridge jumping to Australia, so that successively wider parts of the oceanic crust were transferred from the Australian to the Antarctic Plate. East of the George V Fracture Zone, ridge jumps to Antarctica and large offsets of the spreading ridge at closely spaced fracture zones gave rise to a magnetic quiet zone.

ACKNOWLEDGMENTS

I thank Steve Eittreim for inviting me to take part in the cruise of the R/V *S.P. Lee* to the Wilkes Land margin, and for arranging my subsequent visit to the U.S. Geological Survey, Menlo Park, and I express appreciation of the efforts of the officers and crew of the R/V *S.P. Lee* and of the scientists and technicians on board. Michael König kindly allowed me to cite information from his Ph. D. thesis, and various governmental authorities in Australia and Esso (Australia) made available seismic information. Sue Roots and Paul Fell were untiring in making computer runs of the reconstructions by David Johnson's program INTERMAP and in digitizing the bathymetric, seismic, and magnetic data for presentation at common scales. I thank Steve Eittreim and George Moore for valuable reviews of the manuscript, and Howard Stagg for discussion. Dr P. E. Williamson (Bureau of Mineral Resources) and the Australian Petroleum Exploration Association permitted the reproduction of a published figure.

REFERENCES CITED

Anderson, R. N., D. McKenzie, and J. G. Sclater, 1973, Gravity, bathymetry and convection in the earth: Earth and Planetary Science Letters, v. 18, p. 391–407.

Anderson, R. N., et al., 1980, The interrelation between variations in magnetic anomaly and amplitudes and basalt magnetisation and chemistry along the Southeast Indian Ridge: Journal of Geophysical Research, v. 85, p. 3883–3898.

Armstrong, R. L., 1978, K–Ar dating: late Cenozoic McMurdo Volcanic Group and dry valley glacial history, Victoria Land, Antarctica: New Zealand Journal of Geology and Geophysics, v. 21, p. 685–698.

Aziz-ur-Rahman, and I. McDougall, 1972, Potassium–argon ages on the Newer Volcanics of Victoria: Proceedings of the Royal Society of Victoria, v. 85, p. 61–69.

Bain, J. S., 1965, Summary of data and results, Otway Basin, Vic., Flaxmans No. 1 well of Frome-Broken Hill Co. Pty Ltd: Bureau of Mineral Resources, Petroleum Search Subsidy Acts Publication 62, 9 p.

Bein, J., and M. L. Taylor, 1981, The Eyre Sub-basin: Recent exploration results: Australian Petroleum Exploration Association Journal, v. 21, p. 91–98.

B. M. R., 1979, Australian continental margin survey: Canberra, Australian Government Publishing Service, 188 sheets. Scale 1:1,000,000.

Boeuf, M. G., and H. Doust, 1975, Structure and development of the southern margin of Australia: Australian Petroleum Exploration Association Journal, v. 15, p. 33–43.

Cande, S. C., and J. C. Mutter, 1982, A revised identification of the oldest sea floor spreading anomalies between Australia and Antarctica: Earth and Planetary Science Letters, v. 58, p. 151–160.

Circum-Pacific Map Project, 1981, Plate-tectonic map of the Circum-Pacific Region, Southwest Quadrant: Tulsa, American Association of Petroleum Geologists. Scale 1:10,000,000.

Cooney, P. M., P. R. Evans, and D. Eyles, 1975, Southern Ocean and its margins, *in* J. J. Veevers, ed., Deep Sea Drilling in Australasian Waters: Sydney, Challenger Symposium, p. 26–28.

Deighton, I., D. A. Falvey, and D. J. Taylor, 1976, Depositional environments and geotectonic framework: Southern Australian continental margin: Australian Petroleum Exploration Association Journal, v. 16, p. 25–36.

Denham, J. I., and B. R. Brown, 1976, A new look at the Otway Basin: Australian Petroleum Exploration Association Journal, v. 16, p. 91–98.

Domack, E. W., and J. B. Anderson, 1983, Marine geology of the George V continental margin: Combined results of Deep Freeze 79 and the 1911–1914 Australasian Expedition, *in* R. L. Oliver, P. R. James, and J. B. Jago, eds., Antarctic earth science: Canberra, Australian Academy of Science, p. 402–406.

Domack, E. W., W. W. Fairchild, and J. B. Anderson, 1980, Lower Cretaceous sediment from the East Antarctic continental shelf: Nature, v. 287, p. 625–626.

Douglas, J. G., and J. A. Ferguson, eds., 1976, Geology of Victoria: Geological Society of Australia Special Publication No. 5, 528 p.

Drewry, D. J., ed., 1983, Antarctica: Glaciological and geophysical folio: Cambridge, Scott Polar Research Institute, 9 sheets. Scale 1:6,000,000.

Eittreim, S. L., M. A. Hampton, and J. R. Childs, 1985, Seismic-reflection signature of Cretaceous continental breakup on the Wilkes Land margin, Antarctica: Science, v. 229, p. 1082–1084.

Etheridge, M. A., et al., 1984, Basin-forming structures and their relevance to hydrocarbon exploration in Bass Basin, southeastern Australia: BMR Journal of Australian Geology and Geophysics, v. 9, p. 197–206.

Exon, N. F., and J. B. Willcox, 1980, The Exmouth Plateau: Stratigraphy, structure, and petroleum potential: Australia, Bureau of Mineral Resources, Geology and Geophysics, Bulletin 199, 52 p.

Falvey, D. A., 1974, The development of continental margins in plate tectonic theory: Australian Petroleum Exploration Association Journal, v. 14, p. 95–106.

Falvey, D. A., and J. C. Mutter, 1981, Regional plate tectonics and the evolution of Australia's passive continental margins: BMR Journal of Australian Geology and Geophysics, v. 6, p. 1–29.

Fraser, A. R., and L. A. Tilbury, 1979, Structure and stratigraphy of the Ceduna Terrace region, Great Australian Bight Basin: Australian Petroleum Exploration Association Journal, v. 19, p. 53–65.

Gaposchkin, E. M., and K. Lambeck, 1971, Earth's gravity field to the sixteenth degree and station coordinates from satellite and terrestrial data: Journal of Geophysical Research, v. 76, p. 4855–4883.

Gleadow, A. J. W., and I. R. Duddy, 1981, Early Cretaceous volcanism and the early breakup history of southeastern Australia: Evidence from fission track dating of volcaniclastic sediments, *in* M. M. Cresswell and P. Vella, eds., Gondwana Five: Rotterdam, A. A. Balkema, p. 295–300.

Hayes, D. E., and J. R. Conolly, 1972, Morphology of the Southeast Indian Ocean, *in* D. E. Hayes, ed., Antarctic oceanology. II. The Australian–New Zealand sector: American Geophysical Union, Antarctic Research Series, v. 19, p. 125–145.

Hayes, D. E., et al., 1975a, Lamont-Doherty Survey of the World Ocean, Preliminary report of Volume 23, USNS *Eltanin*, cruises 33–38: LDGO Technical Report CU-1-75, 230 p.

———, 1975b, Initial reports of the Deep-Sea Drilling Project: Washington, DC, U.S. Government Printing Office, v. 28, 1017 p.

———, 1976, Lamont-Doherty Survey of the World Ocean, Preliminary report of Volume 24, USNS *Eltanin*, cruises 39–45: LDGO Technical Report CU-2-76, 220 p.

———, 1977, Lamont-Doherty Survey of the World Ocean, Preliminary report of Volume 25, USNS *Eltanin*, cruises 46–50: LDGO Technical Report CU-1-77, 170 p.

———, 1978, Lamont-Doherty Survey of the World Ocean, Preliminary report of Volume 26, USNS *Eltanin*, cruises 51–55A: LDGO Technical Report CU-1-78, 160 p.

Jago, J. B., 1981, Late Precambrian–Early Paleozoic geological relationships between Tasmania and northern Victoria Land, *in* M. M. Cresswell and P. Vella, eds., Gondwana Five: Rotterdam, A. A. Balkema, p. 199–204.

James, P. R., and R. J. Tingey, 1983, The Precambrian geological evolution of the East Antarctic metamorphic shield—a review, *in* Oliver, R. L., P. R. James, and J. B. Jago, eds., Antarctic earth science: Canberra, Australian Academy of Science, p. 64–68.

Johnson, B. D., 1973, A time term analysis of the data obtained during the Bass Strait Upper Mantle Project (Project BUMP): Australian Society of Exploration Geophysicists Bulletin, v. 4, n. 2/3, p. 15–20.

Kennett, J. P., et al., 1975, Initial Reports of the Deep-Sea Drilling Project: Washington, DC, U.S. Government Printing Office, v. 29, 1197 p.

König, M., 1980, Geophysical investigations of the southern continental margin of Australia and the conjugate sector of East Antarctica: Ph.D. Thesis, Columbia University, 311 p.

———, 1981, Geophysical data from the continental margin off Wilkes Land, Antarctica: Implications for pre-drift fit and separation of Australia–Antarctica: Eos, v. 62, p. 384.

König, M., and M. Talwani, 1977, A geophysical study of the southern continental margin of Australia: Great Australian Bight and western sections: Geological Society of America Bulletin, v. 88, p. 1000–1014.

Kyle, P. R., J. Adams, and P. C. Rankin, 1979, Geology and petrology of the McMurdo Volcanic Group at Rainbow Ridge, Brown Peninsula, Antarctica: Geological Society of America Bulletin, v. 90, p. 676–688.

Laird, M. G., R. A. Cooper, and J. B. Jago, 1977, New data on the lower Paleozoic sequence of northern Victoria Land, Antarctica, and its significance for Australian–Antarctic relations in the Paleozoic: Nature, v. 265, p. 107–110.

Mawson, D., 1940, Sedimentary rocks: Australasian Antarctic Expedition 1911–14: Sci. Reports, Series A, v. IV, pt. 11, p. 347–367.

Oliver, R. J., J. A. Cooper, and A. J. Truelove, 1983, Petrology and zircon geochronology of Herring Island and Commonwealth Bay and evidence for Gondwana reconstruction, *in* R. L. Oliver, P. R. James, and J. B. Jago, eds., Antarctic earth science: Canberra, Australian Academy of Science, p. 64–68.

Roots, W. D., 1976, Magnetic smooth zones and slope anomalies: A mechanism to explain both: Earth and Planetary Science Letters, v. 31, p. 113–118.

Talwani, M., et al., 1979, The crustal structure and evolution of the area underlying the magnetic quiet zone on the margin south of Australia: American Association of Petroleum Geologists Memoir 29, p. 151–175.

Threlfall, W. F., B. R. Brown, and B. R. Griffiths, 1976, Gippsland Basin, offshore: Australasian Institution of Mining and Metallurgy Monograph, v. 7, n. 3, p. 41–67.

Truswell, E. M., 1982, Palynology of seafloor samples collected by the 1911–14 Australasian Antarctic Expedition: Implications for the geology of coastal East Antarctica: Geological Society of Australia Journal, v. 29, p. 343–356.

———, 1983, Geological implications of recycled palynomorphs in continental shelf sediments around Antarctica, *in* R. L. Oliver, P. R. James, and J. B. Jago, eds., Antarctic earth science: Canberra, Australian Academy of Science, p. 394–399.

Vanney, J. R., and G. L. Johnson, 1979, The sea floor morphology seaward of Terre Adelie (Antarctica): Deutsches Hydrologisches Zeitschrift., v. 32, p. 77–87.

Veevers, J. J., 1981a, Australian rifted margins, *in* R. A. Scrutton, ed., Dynamics of passive margins: American Geophysical Union, Geodynamics Series, v. 6, p. 72–89.

______, 1981b, Morphotectonics of rifted continental margins in embryo (East Africa), youth (Africa–Arabia), and maturity (Australia): Journal of Geology, v. 89, p. 57–82.

______, 1982, Australian-Antarctic depression from the mid-ocean ridge to the adjacent continents: Nature, v. 295, p. 315–317.

______, ed., 1984, Phanerozoic earth history of Australia: Oxford, Clarendon Press, 418 p.

______, 1986, Breakup of Australia and Antarctica estimated as mid-Cretaceous (95 ± 5 Ma) from magnetic and seismic data at the continental margin: Earth and Planetary Science Letters, v. 77, p. 91–99.

Veevers, J. J., and S. L. Eittreim, in press, Reconstruction of Antarctica and Australia at breakup (95 ± 5 Ma) and before rifting (160 Ma).

Vogt, P. R., N. Z. Cherkis, and G. A. Morgan, 1983, Project Investigator. I. Evolution of the Australia–Antarctic Discordance deduced from a detailed aeromagnetic study, *in* R. L. Oliver, P. R. James, and J. B. Jago, eds., Antarctic earth science: Canberra, Australian Academy of Science, p. 608–613.

Watts, A. B., and S. F. Daly, 1981, Long wavelength gravity and topography anomalies: Annual Review of Earth and Planetary Sciences, v. 9, p. 415–448.

Weissel, J. K., and D. E. Hayes, 1972, Magnetic anomalies in the Southeast Indian Ocean, *in* D. E. Hayes, ed., Antarctic oceanology. II. The Australian–New Zealand sector: American Geophysical Union, Antarctic Research Series, v. 19, p. 165–196.

______, 1974, The Australian–Antarctic discordance: New results and implications: Journal of Geophysical Research, v. 79, p. 2579–2587.

Weissel, J. K., D. E. Hayes, and E. M. Herron, 1977, Plate tectonic synthesis and the displacements between Australia, New Zealand and Antarctica since the Late Cretaceous: Marine Geology, v. 25, p. 231–277.

Wellman, P., 1979, On the isostatic compensation of Australian topography: BMR Journal of Australian Geology and Geophysics, v. 4, p. 373–382.

______, 1982, Australian seismic refraction results, isostasy and altitude anomalies: Nature, v. 298, p. 838–841.

Whyte, R., 1978, Shell's offshore venture in South Australia: Australian Petroleum Exploration Association Journal, v. 18, p. 44–51.

Willcox, J. B., 1977, Some gravity models of the continental margin in the Australian region: Australian Society of Exploration Geophysicists, v. 8, p. 118–124.

______, 1978, The Great Australian Bight: A regional interpretation of gravity, magnetic, and seismic data from the Continental Margin Survey: Australia, Bureau of Mineral Resources, Geology and Geophysics, Report 201, p. 111.

Williamson, P. E., et al., 1985, Pre-Eocene stratigraphy, structure and petroleum potential of the Bass Basin: Australian Petroleum Exploration Association Journal, v. 25, p. 362–381.

Eittreim, S. L., and Hampton, M. A., 1987, *The Antarctic Continental Margin: Geology and Geophysics of Offshore Wilkes Land*, CPCEMR Earth Science Series, v. 5A: Houston, Texas, Circum-Pacific Council for Energy and Mineral Resources

Post-Breakup Sedimentation on the Wilkes Land Margin, Antarctica

M. A. Hampton
S. L. Eittreim
B. M. Richmond
U.S. Geological Survey
Menlo Park, California

The post-breakup seismic-stratigraphy displayed in multichannel and high-resolution profiles on the Wilkes Land margin records the accumulation of stratified sediment along the edge of the Antarctic continent after it separated from Australia. The seismic facies directly above the breakup unconformity consist of nonmarine/continental shelf, slope, and rise deposits that prograded across the subsiding margin. A major unconformity at the top of these deposits signals a possible subaerial exposure of the shelf, followed by a major rise in relative sea level. Deposits above the unconformity comprise deep-water fan, contourite, and hemipelagic facies. Contour currents sweep the modern slope and rise sea floor, degrading it in some places and depositing contourites in others.

INTRODUCTION

Divergent (passive) continental margins form as a direct consequence of continental fragmentation and the creation of new ocean basins. The transition between the rifting and drifting stages of continental fragmentation, termed *breakup*, is punctuated by the first generation of new oceanic crust and the initiation of seafloor spreading. An important change in tectonic style commences at the time of breakup; namely, the margins of the newly separated continental masses subside at a logarithmically decreasing rate as they drift away from the spreading axis. The subsidence fundamentally is driven by the cooling and contraction of lithosphere as the heat anomaly associated with rifting decays, but it is augmented by the deposition of sediment. Ideally, post-breakup prograding shelf and slope sediment accumulates above the faulted and eroded continental basement and is transitional seaward to deep-water rise and abyssal-plain deposits (e.g., Falvey, 1974). The sedimentary deposits along any particular margin are individualized as a consequence of several factors that include the lithology and relief of the source terrain, the climate, the position of sea level, and the proximity of major drainage systems.

The Wilkes Land continental margin (Fig. 1) originated when the Antarctic and Australian continents separated, probably in mid-Cretaceous time 90–110 Ma (Cande and Mutter, 1982), although the initial rifting began in the mid-Jurassic (Veevers, this volume). Deep seismic-reflection profiles across the margin clearly show that the basement consists of a south-to-north succession of block-faulted continental crust, thinned and intruded continental crust, and oceanic crust with a boundary ridge at its southern edge (Fig. 2) (Eittreim et al., 1985; Wannesson et al., 1985; Eittreim and Smith, this volume). Above basement is a series of depositional sequences separated by regional unconformities. The lowest sequence, present only above continental basement, was deposited during the rifting stage and is termed the *synrift sequence* (Eittreim et al., 1985). A major unconformity at the top of the synrift sequence is termed the *breakup unconformity* (Falvey, 1974; Eittreim et al., 1985). Younger sequences, present above both continental and oceanic basement, were deposited after breakup. These younger sequences are the subject of this paper.

Two geologic phenomena, in addition to thermally driven subsidence, can be identified as important influences on the Wilkes Land margin post-breakup sedimentation, especially with regard to the position of the shoreline. The first is climate, particularly the conditions that culminated with the formation of the Antarctic ice sheet. Significant climatic cooling began at about 38 Ma (Wolfe, 1978), the first ice-rafted sediment delivered to the ocean by local glaciers appeared at about 25 Ma (Hayes and Frakes, 1975), and the main ice sheet formed at about 14 Ma (Drewry, 1975; Shackleton and Kennett, 1975a, 1975b). The climate determines the amount of sediment that accumulates on a continental margin through its influence on the rate of erosion of the adjacent landmass, and sea level is eustatically lowered when ice sheets form. Rapid accumulation of sediment and rapid eustatic lowering of sea level can cause shoreline regression, even on a tectonically subsiding margin (Pitman, 1978).

The second phenomenon is the reorganization of plate motions and the associated change in the rate of seafloor spreading. World-wide plate reorganizations occurred at about 85 and 43 Ma (anomalies 34 and 19, respectively), and relatively slow rates of seafloor spreading (4.5–10 mm/year) occurred on the Southeast Indian Ocean Ridge (the spreading

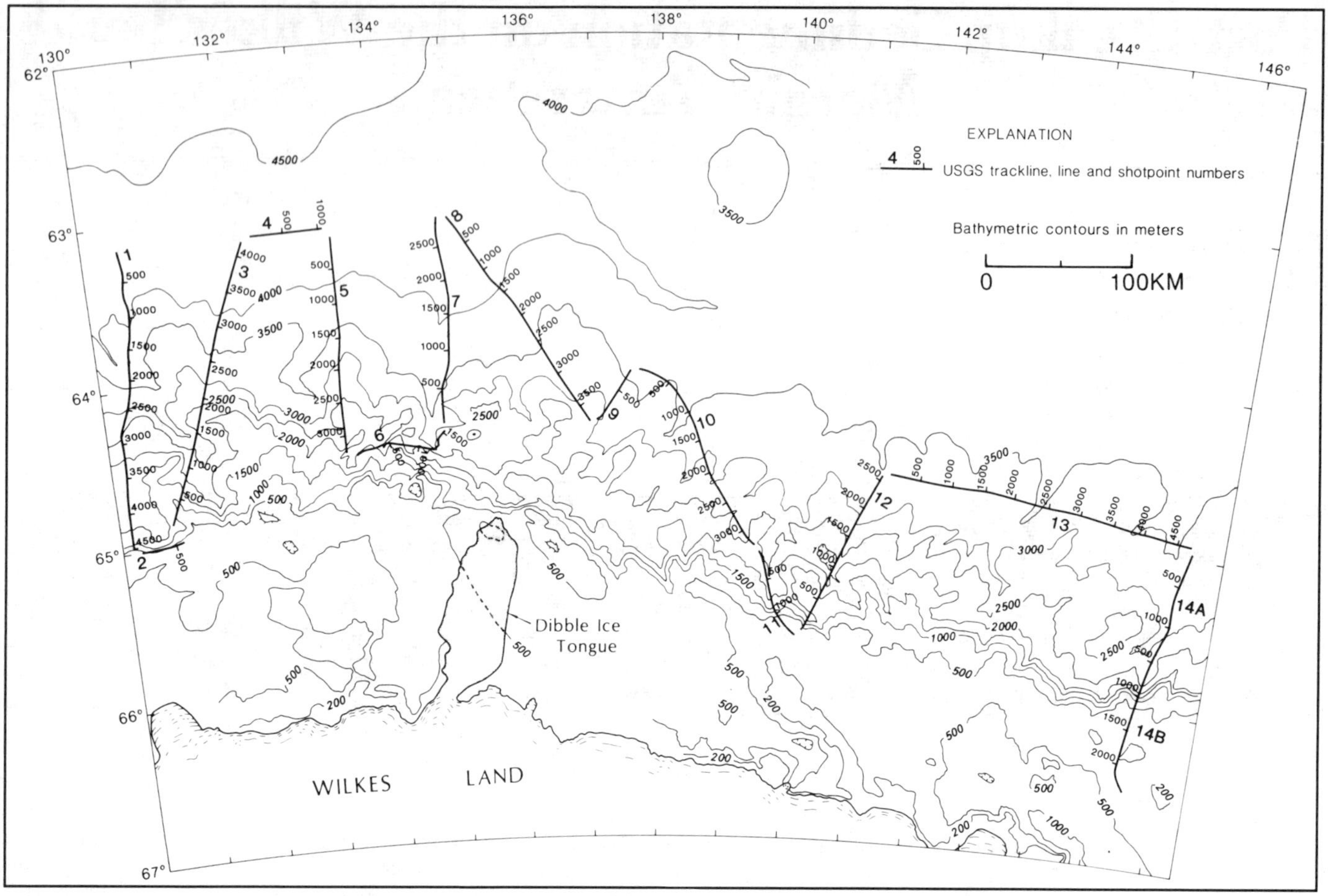

Figure 1—Tracklines of seismic-reflection profiles on the Wilkes Land continental margin. Trackline numbers and shotpoints are indicated on the lines.

center between Antarctica and Australia) between these times (Cande and Mutter, 1982). The spreading rate since 43 Ma is 22 mm/year. The rate of seafloor spreading determines the volume of spreading ridges, which in turn affects sea level (Pitman, 1978). The volume of ridges is relatively small during times of slow spreading, increasing the volume of the ocean basins and causing sea level to fall.

Multichannel seismic-reflection profiles collected in 1984 aboard the R/V *S.P. Lee* on the Wilkes Land continental margin between lat 63.0° and 66.5° S and long 130.0° and 146.5° E all display a feature inferred to be a breakup unconformity, and post-breakup depositional sequences (Figs. 1, 2; Eittreim and Smith, this volume). The objective of this paper is to reconstruct the post-breakup sedimentary evolution of the Wilkes Land continental margin by applying the principles of seismic sequence and facies analysis (Mitchum et al., 1977; Sangree and Widmier, 1977) to the multichannel profiles and interpreting the results in terms of the documented climatic and tectonic history of the region. High-resolution seismic-reflection profiles collected along the tracklines are similarly analyzed in order to deduce the most recent sedimentary regime.

The tracklines for the seismic-reflection profiles traverse the continental rise for the most part (Fig. 1). A few lines extend landward across the continental slope and onto the continental shelf. The continental rise consists of two segments: the steep upper rise (approximate water depth range 2000–3000 m) and the less steep lower rise (3000–4000 m). The morphology of the upper rise suggests that it consists of a series of submarine fans, traversed by deeply incised submarine canyons that extend as lower relief seavalleys across the lower rise (see Chase et al., this volume). The continental slope throughout most of the study area is steep (typically 8° but up to 16°) and narrow (typically 10 km) and extends from ~400 to 600 m at the shelf-slope break to ~2000 m water depth. Commonly, the decrease in seafloor declivity at the slope-rise boundary is sharp. At some places along the margin, the transition is not so abrupt, however, and the slope is less steep (4°) and relatively wide (25 km).

The multichannel seismic-reflection data were recorded 24-fold, using a 2.4 km long streamer and 1300-in.^{3} tuned array of five air guns as the sound source. Details of recording and processing parameters are given in Eittreim and Smith (this volume). High-resolution seismic-reflection profiles were recorded using 3.5- and 12-kHz tuned transducers.

SEISMIC STRATIGRAPHY

The stratigraphically lowest feature of concern in this paper is the breakup unconformity (K1 in Fig. 2). It was originally identified on the Wilkes Land margin and discussed by Eittreim et al. (1985). The correlative feature on the conjugate

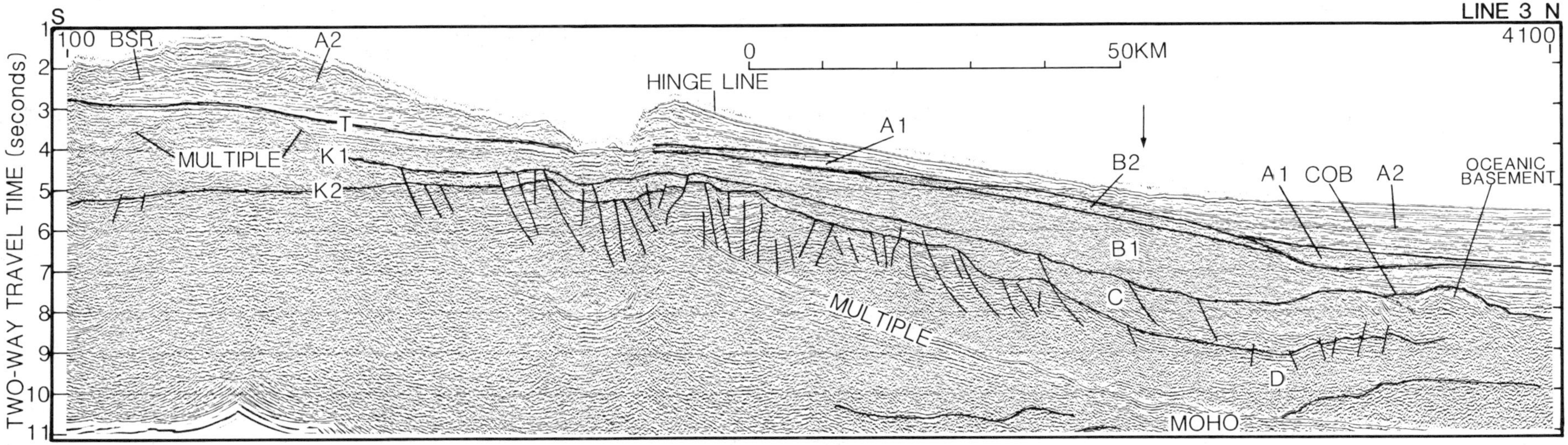

Figure 2—Seismic stratigraphic features identified in multichannel profiles. COB refers to the continent-ocean boundary, seismic-stratigraphic unit D is inferred to be block-faulted continental crust, K2 is the rift-onset unconformity (erosion of the subaerial continental surface that existed before rifting), unit C is the synrift sequence (deposited during rifting but before breakup), and K1 is the breakup unconformity (erosion in response to a crustal uplift event immediately prior to breakup; see Falvey, 1974). See Eittreim and Smith (this volume) for a discussion of these geologic features. Sequences A1, A2, B1, B2, and unconformity T are post-breakup features discussed in this paper. BSR is a bottom-simulating reflector (Kvenvolden et al., this volume). The transition between the upper and lower continental rise is approximately at the arrow. Vertical exaggeration = 7.

Australian margin was identified and discussed by Falvey (1974). The geology beneath the breakup unconformity is discussed by Eittreim and Smith (this volume). To summarize, block-faulted continental crust (D in Fig. 2) is truncated by a prerift unconformity (K2 in Fig. 2). The overlying synrift deposits (C in Fig. 2) extend seaward to the continent-ocean boundary (COB in Fig. 2) and oceanic basement. The breakup unconformity truncates sequence C and terminates at the COB.

Above the breakup unconformity on all profiles are two major seismic sequences A and B, each divided into two subsequences (A1, A2 and B1, B2) (Fig. 2). Sequences A and B are separated by a distinct unconformity (T in Fig. 2) that formed in response to a first-order change in water depth and sedimentary environment. Other seismic sequences and unconformities of limited extent appear on particular profiles.

Termination of the breakup unconformity at the COB is compatible with Falvey's (1974) suggestion that the unconformity forms during a final crustal uplift event prior to the onset of seafloor spreading. In some places, the unconformity is marked by a high-amplitude reflector, but typically it is defined only by the change in attitude, reflectivity, and configuration of strata between the underlying and overlying sequences. The unconformity generally deepens northward; commonly it is about 3–4 sec deep near the present shelf-slope break and about 8 sec deep at the northern end of profiles. Some normal faults appear to offset the unconformity (Fig. 2).

A seismic sequence (B1 in Figs. 2, 3A) of 2.1 sec (2.8 km) maximum thickness and 2.7 km/sec average interval velocity overlies the breakup unconformity and the oceanic crust north of it. Overall, the sequence has features of the oblique and sigmoid progradational seismic facies defined by Sangree and Widmier (1977). It exhibits distinct lateral changes in external form and reflector type from south to north. In the south, where the sequence is ~0.4–0.9 sec (0.5–1.2 km) thick, the reflectors are inclined gently seaward and tend to have high amplitude. The rough seafloor topography in this region partially obscures the subbottom reflectors, but they appear in general to have discontinuous and somewhat hummocky configuration. Gradationally northward the reflectors display greater continuity, smoothness, and parallelism, and they retain the high amplitude. North of a hingeline the reflectors become more steeply inclined and gradationally decrease in amplitude and continuity. Also, they diverge, and the sequence thickens appreciably. In the northernmost region where the sequence is relatively thin, the reflectors regain continuity and near-horizontal parallelism, although the amplitude is lower and the frequency is higher relative to the southern region.

Reflectors at the base of sequence B1 downlap on the breakup unconformity. The relation of reflectors to the upper boundary is varied. South of the hingeline, reflectors are nearly concordant to the upper boundary, but local discordance can be detected in some profiles; that is, the upper boundary is an unconformity that locally has low erosional relief. North of the hingeline, where the sequence thickens, beds become more discordant to the upper boundary (toplap), and the boundary itself steepens. Farthest north, where the sequence is thin, the upper boundary is conformable to slightly discordant.

On lines 1, 3, and 5 that traverse across the margin in the western part of the study area, a minor but distinct sequence (B2 in Figs. 2, 3B) occurs above the thickest part of sequence B1. It also exists on other profiles, but not as distinctly. The sequence is characterized by inclined internal reflectors generally of low frequency and amplitude, except at the southern extent of the sequence where the reflectors have high amplitude. Sequence B2 pinches out to the south on all profiles. It also pinches out to the north on some profiles, but on others the reflectors become relatively continuous and nearly horizontal, similar to the seaward part of sequence B1, which they overlie. Sequence B2 is thickest (~0.35 sec) in its central portion. The lower boundary of B2, and some internal reflections in B1, produce strong step-like reflector patterns suggestive of small-offset faults in some profiles. Reflector terminations at the upper and lower boundaries of sequence B2 generally are unclear, but downlap at the lower boundary and toplap at the upper boundary are locally evident.

A major unconformity (T in Fig. 2) occurs at the top of sequences B1 and B2. On the landward (southern) portion of profiles, it appears as a moderately continuous and smooth, strong, single or paired reflector. Seaward of the hingeline, the reflector becomes weaker to nonexistent where T is convex upward and has a relatively steep inclination to the north. At the northern end of profiles, T appears as a distinct reflector of varied amplitude.

Seismic sequence A overlies unconformity T and is divided into two parts (A1 and A2 in Figs. 2, 4A). Sequence A generally thickens to the east (Eittreim and Smith, this volume). The two subsequences have similar reflector types but different reflector configurations. The reflectors typically are strong and continuous, especially on the landward portions of profiles, but a seaward gradation to discontinuous and hummocky reflectors at certain stratigraphic levels is common.

Sequence A1 onlaps directly onto unconformity T (Fig. 5). Reflectors are strongly parallel except for local slight discordances. The sequence is 0–0.7 km thick in the south and thickens to about 1.2 km in the north. Average interval velocity is 2.4 km/sec. The upper boundary is difficult to define in many places, but reflectors are concordant or truncated at a low angle where the boundary can be located confidently.

The uppermost sequence (A2 in Fig. 2) has an average interval velocity of 2.0 km/sec. Reflector configurations within the sequence correspond to the morphology of the sea floor. In the south the sea floor is rough and deeply incised, and to the north it becomes smoother, although small-scale roughness is common. The seafloor morphology (Chase et al., this volume) suggests that the continental rise is composed of a series of deep-sea fans that on the upper rise are deeply incised by canyons that become wider and have less relief seaward. The seismic-reflection profiles demonstrate the aggradation of the fans, with subbottom reflectors of varied inclination in the south and smooth to moderately irregular reflectors in the north. Reflectors are abruptly truncated at high angles by the seafloor canyons in the south, and some of the canyons cut entirely through sequences A1 and A2, down into sequence B. Reflectors are concordant to the sea floor or truncated by it at low angles in the north. Reflectors terminate with onlap relations at the base of the sequence, although there are some local instances of downlap.

A strong bottom-simulating reflector (BSR) occurs typically within a subbottom depth of 0.8 sec in sequence A (Fig. 2). The strong reflectivity of the BSR causes a weak display of the underlying reflectors. The BSR may occur at the base of a frozen gas hydrate layer and is discussed further by Kvenvolden et al. (this volume).

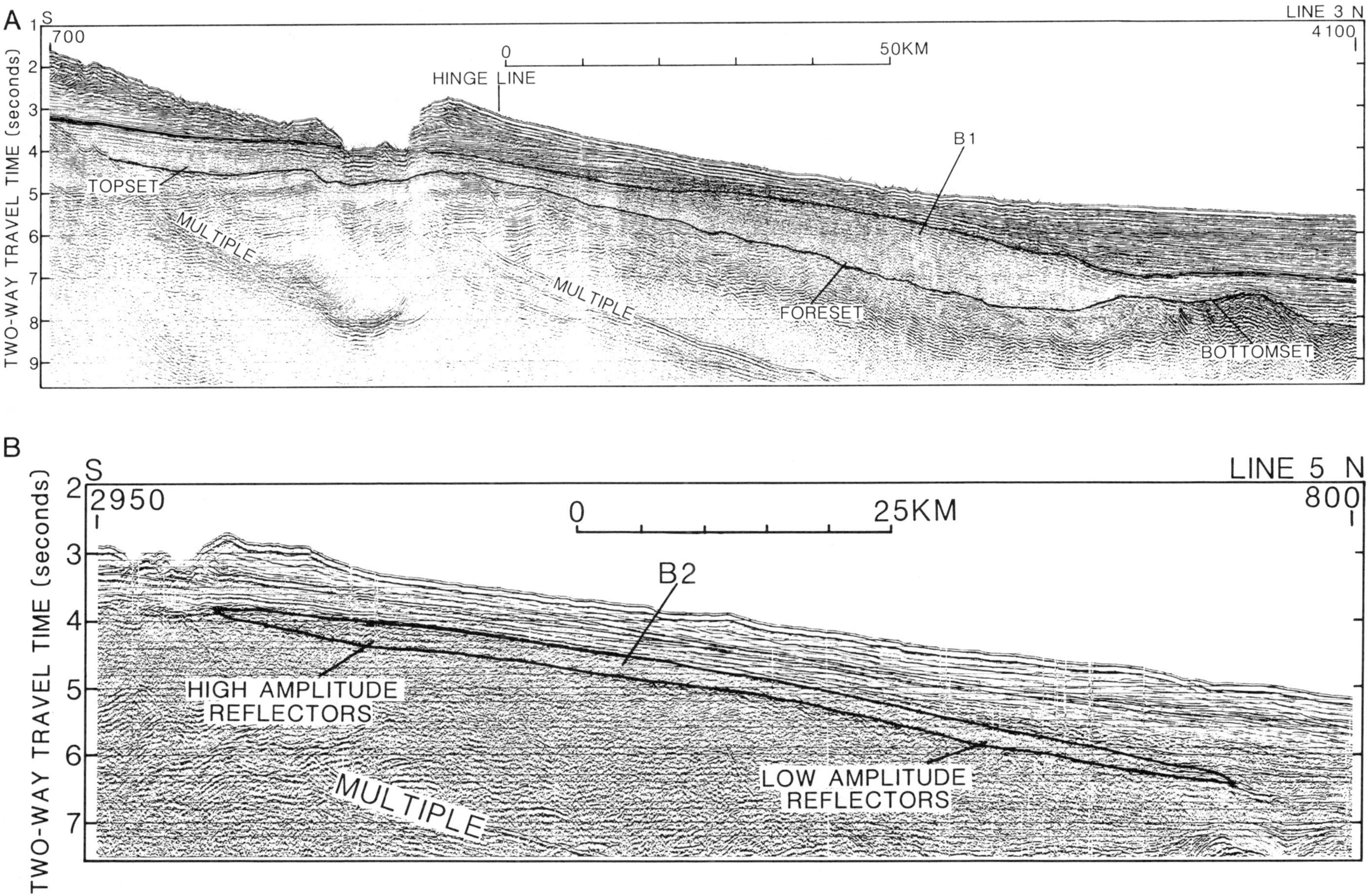

Figure 3—Multichannel profiles illustrating the features of the subunits of seismic sequence B. Locations of shotpoints at the ends of the profiles are shown in Figure 1. A. Unit B1; vertical exaggeration = 7. B. Unit B2; vertical exaggeration = 7.

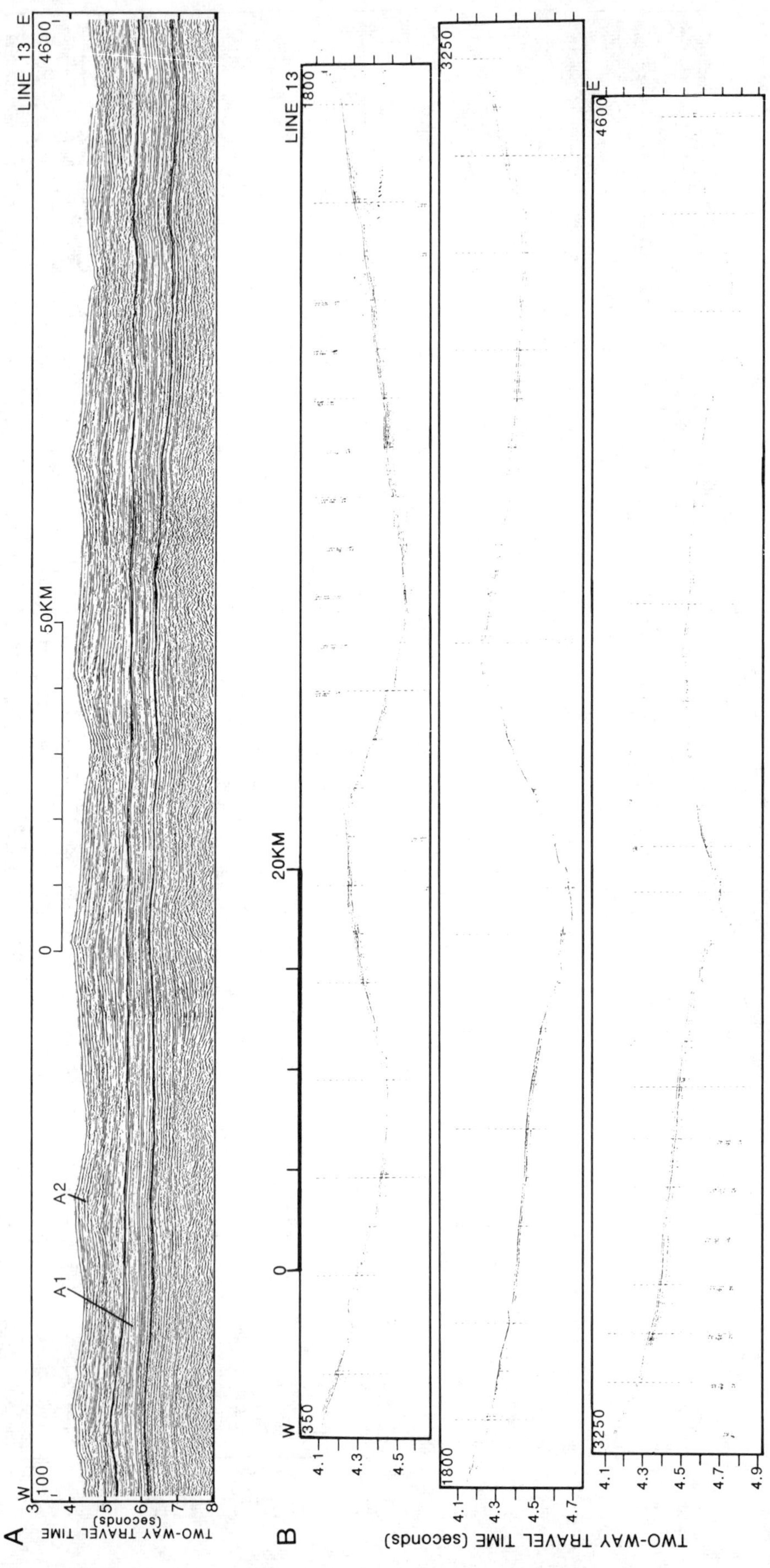

Figure 4—Seismic reflection profiles along line 13. Locations of shotpoints at the ends of the profiles are shown in Figure 1. A. Multichannel profile illustrating the features of seismic sequence A, divided into units A1 and A2. Vertical exaggeration = 7. B. 3.5-kHz profile along line 13 showing the shallow seismic stratigraphy across seavalleys on the lower continental rise. Vertical exaggeration = 13.

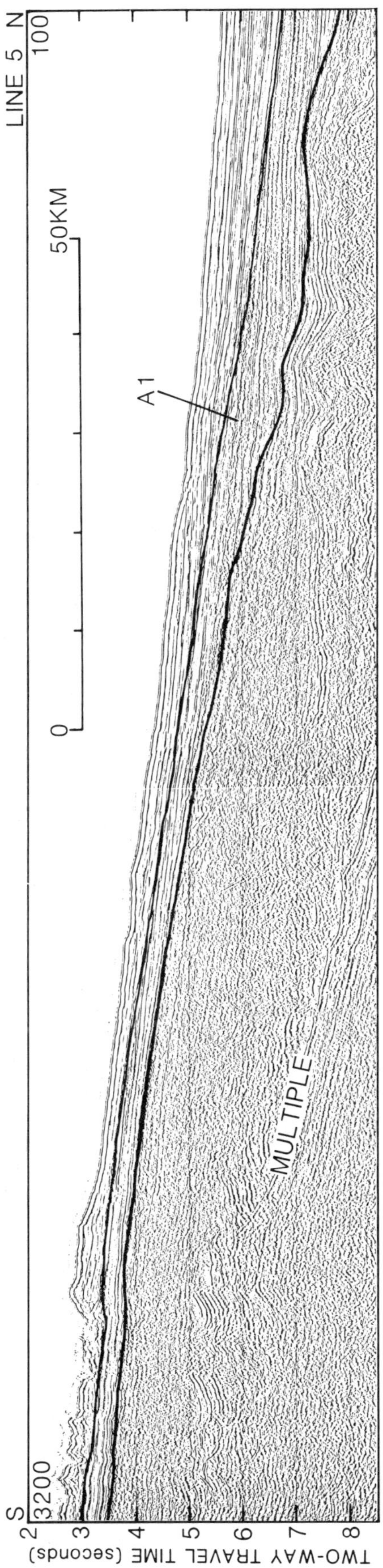

Figure 5—Multichannel profile illustrating the features of unit B1. Locations of shotpoints at the ends of the profiles are shown in Figure 1. Vertical exaggeration = 7.

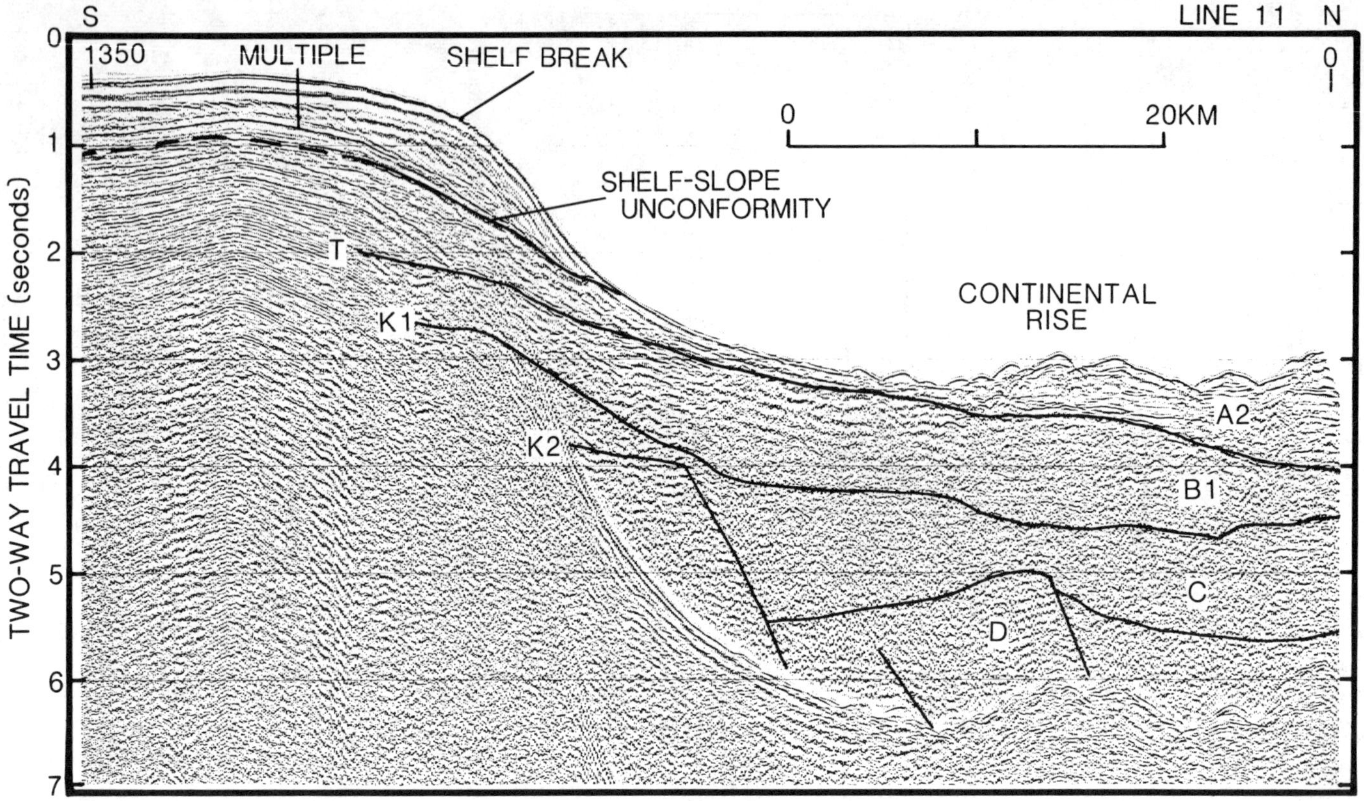

Figure 6—Multichannel profile across the continental shelf-slope break. Locations of shotpoints at the ends of the profile are shown in Figure 1. Vertical exaggeration = 7.

Line 13 trends latitudinally across the lower continental rise in the eastern part of the study area (Fig. 1). The seafloor morphology consists of shallow seavalleys and relatively sharp-crested intervalley fan buildups (Fig. 4A). The stratigraphy of sequence A2 consists of high-amplitude, inclined reflectors with discordances, and sharp hinges where the inclination abruptly changes. There are some local areas of hummocky reflectors. The sequence is similar in appearance to one displayed by Mitchum et al. (1977, Fig. 14) and attributed to deposition on migrating waves, but in our case the sequence was deposited by normal fan-building processes as attested to by the fact that the valleys in the seismic-reflection profile correspond to mapped north–south seavalleys (Chase et al., this volume).

Lines 1, 11, 12, and 14 traverse the steep continental slope and cross the shelf-slope break (Fig. 1). Although partly obscured by multiple reflections, seismic profiles show seaward-inclined sigmoid reflectors beneath the continental shelf and slope. Progradation of the slope over sequence A on the upper continental rise is clearly indicated in some profiles (the shelf-slope unconformity in Fig. 6).

The high-resolution 3.5-kHz profiles display two main types of reflectivity. The most prevalent type consists of a distinct, strong seafloor reflector with no subbottom reflectors that occurs over nearly all of the continental slope, most of the upper continental rise, and much of the lower rise. The other type is characterized by parallel, continuous reflectors that extend to 50 ms subbottom and occurs almost exclusively on valley walls and intervalley areas of the lower rise, although not on all such areas. The profile along line 13 nicely illustrates the two reflectivity types (Fig. 4B). A distinct seafloor reflector with no subbottom reflectors occurs on the floor of five of the six valleys on this line and continues up one of the gently sloping valley walls in four of them. The stratified sequence is present on the remaining valley floor and on the valley walls. The reflectors of the stratified sequence are truncated at the sea floor in some of the valleys.

Intermediate types of reflectivity, with vague, discontinuous subbottom reflectors, occur locally, especially on the upper continental rise. Types of reflectivity different from the two major types discussed above occur at the seaward end of lines 1 and 3. There is a series of steps on line 3, about 10 m high (Fig. 7A). The risers of some steps truncate faint reflectors; the multichannel profiles also indicate the truncation of reflectors at the sea floor. Landward of the steps, and also at the seaward end of line 1, the sea floor is hummocky and produces a weak, diffuse reflection, and there are some short, shallow subbottom reflectors (Fig. 7B). The 12-kHz profiles show closely spaced hyperbolae at the seaward end of line 1 (Fig. 7C).

Figure 7—High-resolution seismic-reflection profiles illustrating the types of bottom and subbottom returns. Locations of shotpoints at the ends of the profiles are shown in Figure 1. A. 3.5-kHz profile showing seafloor steps and truncated reflectors at the base of the continental rise. Vertical exaggeration = 13. B. 3.5-kHz showing weak, diffuse reflectors. Vertical exaggeration = 13. C. 12-kHz profile illustrating hyperbolic returns along the same portion of trackline as 6B. Vertical exaggeration = 13.

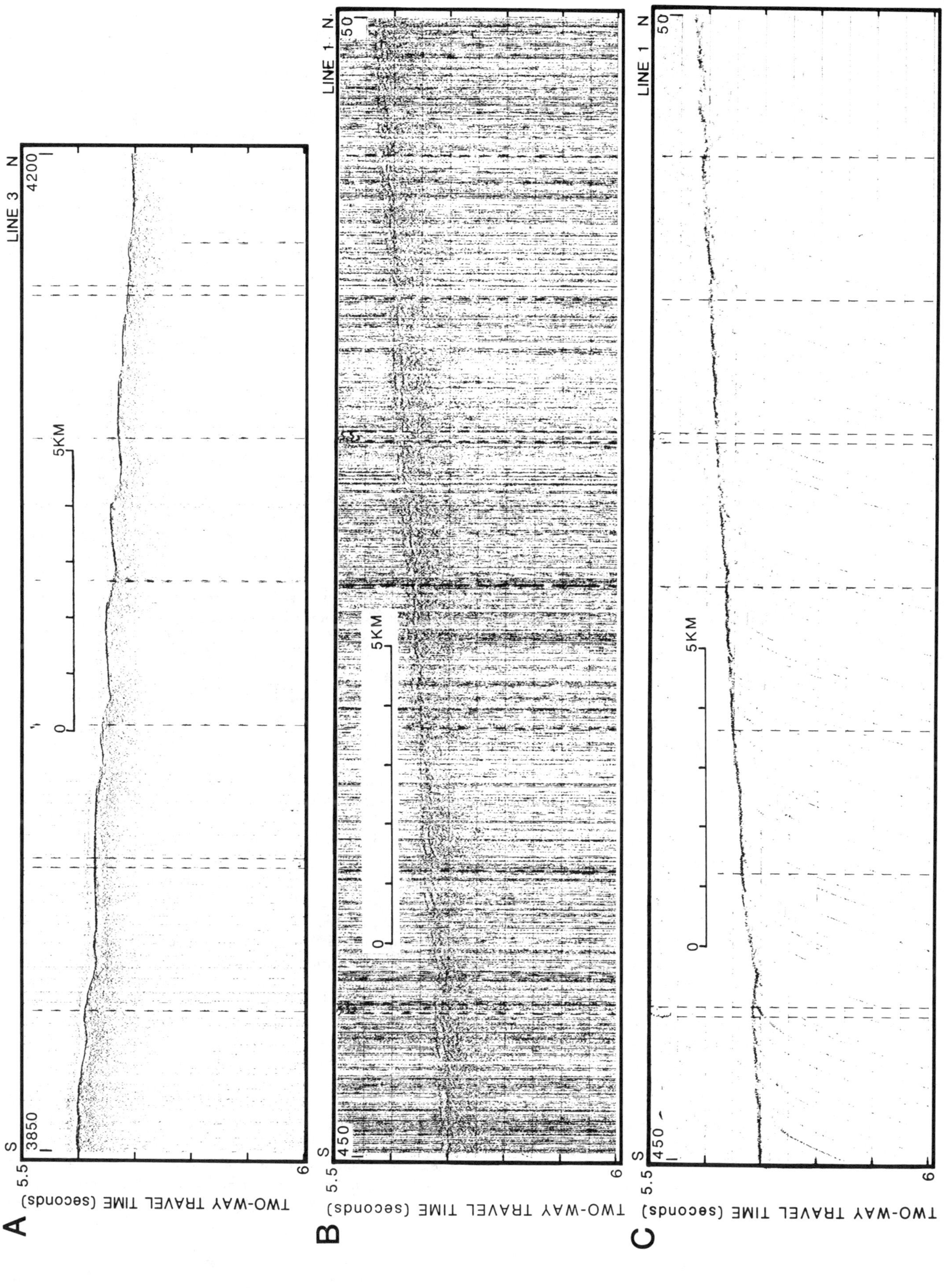
A
B
C
LINE 3
LINE 1
S
N
3850
4200
450
50
0
5KM
5.5
6
TWO-WAY TRAVEL TIME (seconds)

INTERPRETATION

Breakup of Antarctica and Australia was estimated on the basis of magnetic anomaly patterns by Weissel and Hayes (1972) at 55 Ma, but subsequent revision of the anomaly interpretations by Cande and Mutter (1982) place the event between 90 and 110 Ma and by Veevers (1986) specifically at 95 ± 5 Ma. Therefore, assuming that the breakup unconformity was eroded subaerially in response to a crustal-uplift event immediately prior to breakup (Falvey, 1974), this feature on the Wilkes Land margin probably is of mid-Cretaceous age but is perhaps as young as Paleocene. The morphology of the breakup unconformity may be due partly to subaerial erosion, but some of the largest relief features apparently are fault-controlled (Fig. 2), which indicates that normal-fault growth continued past the rifting stage.

We interpret sequence B1, because of the sigmoid to oblique reflection configuration pattern that downlaps onto the breakup unconformity and adjacent oceanic crust, to comprise nonmarine or continental shelf (topset), slope (foreset), and rise (bottomset) deposits that prograded across the young Wilkes Land margin as it subsided after breakup (Figs. 2, 3A). The hingeline controlled the initial position of the shelf-break (transition from topset to foreset) after breakup. The declivity of the foreset beds is rather gentle (~2°), about the same as the underlying breakup unconformity.

Early subsidence of the margin was rapid enough to allow aggradation of the nearly horizontal shelf deposits. However, the inclined, toplap termination of reflectors high in the continental slope section north of the hingeline, with no corresponding shelf deposits, implies that water depth eventually shallowed enough for sediment to by-pass the shelf while the slope continued to build seaward. The shallowing probably occurred because the rate of sediment accumulation exceeded the logarithmically decreasing subsidence rate (Sleep, 1971; Watts and Ryan, 1976), but this may have been abetted by the persistent drop of sea level after 86 Ma or by short-term sea level drops (Vail et al., 1977; Vail and Hardenbohl, 1979).

The discontinuous and somewhat hummocky reflector configurations at the south end of sequence B1 (e.g., near shotpoint 100 on Fig. 2) may signify a subaerial environment (Sangree and Widmier, 1977), and the position of the shoreline would be marked at the transition to more continuous reflectors northward. However, the poor display of the reflectors in the area of alleged subaerial deposits qualifies this inference; the sequence may be entirely marine. Evidence of subaerial exposure and weathering may be marked by the zone of high reflectivity along the southern part of unconformity T (Fig. 2).

Sequence B2 has the characteristics of shoreline and nonmarine deposits (high-amplitude, hard reflectors) transitional northward to shelfal marine shale deposits (low-amplitude reflectors; Fig. 3B) (Sangree and Widmier, 1977, Fig. 3). Evidently, deposition of sequence B1 terminated as the shoreline regressed, exposing the top of the shelf deposits to subaerial erosion that produced the low-relief upper surface that appears along the highly reflective southern portion of unconformity T in some profiles. Accumulation of sediment was areally and temporally limited after the regression, as evidenced by the restricted occurrence of sequence B2. The low-amplitude nature of the reflectors in the inferred shelf portion of this sequence suggests a more shale-prone lithology compared to sequence B1 (Sangree and Widmier, 1977).

We interpret the transition from the B-sequences to the A-sequences as caused by a major shoreline transgression and increase in water depth, leading from the subaerial and shallow-water environment during deposition of sequence B2 to the bathyal and abyssal environments that now exist in the study area. However, we do not see stratigraphic evidence of this transgression (for example in the form of onlapping sands) directly above unconformity T. An estimate of the amount of transgression can be made from the horizontal distance between the present-day shelf-break relative to the sequence B shelf-break, which is greatest (140 km) on profile 1. The modern shelf and slope are underlain by seaward-dipping, progradational deposits younger than unconformity T (Wannesson et al., 1985; Tanahashi et al., this volume), so the maximum landward shift of the shelf-break (and shoreline) was substantially greater.

Sequence A1 is an onlap-fill seismic facies (Sangree and Widmier, 1977), probably composed of contourite and hemipelagic deposits. We infer that contemporaneous submarine fans were being constructed landward of our profiles by sediment transported from the adjacent continent. The fans eventually prograded seaward over sequence A1 and appear on our profiles as sequence A2. Sequence A2 beneath the upper rise consists of channeled upper fan deposits, similar to those beneath the upper Laurentian Fan, for example (Piper and Normark, 1982). Sequence A2 grades seaward beneath the lower rise to nearly parallel-bedded lower fan deposits, which consist of contourites as well as distal turbidites and hemipelagic deposits. Contourites are an important constituent of the sediment at DSDP site 269 (located 230 km northeast of line 8), which probably is correlative to sequence A2. Furthermore, sediment cores collected on the lower rise contain contourites (Hampton et al., this volume), and a modern sediment-laden geostrophic contour current has been documented in the region (Eittreim et al., 1972). Distal sequence A2 is in many places indistinguishable from the underlying sequence A1 (Fig. 2), because of similarities in depositional processes.

The modern seafloor morphology on the upper rise is degradational, as indicated by the truncation of reflectors and deep incision of canyons (Fig. 2). Although in early post-T time the upper rise fans aggraded vigorously, they now have been cut off from their copious sediment supply, and succeeding current activity has cut deeply into the depositional surface. Sediment supply across the margin presently is small because of the dry-base glacial regime on land, and the onset of these conditions probably caused the observed change in sedimentary conditions. When wet-base conditions prevailed, glacial sediment supply fed the fans and caused the slope to prograde.

What was the timing of these events? Anderson (in press) reviewed evidence from other areas of offshore Antarctica indicating that during most of Neogene time the shelf prograded several tens of kilometers by fluvial-deltaic deposition in front of wet-base glaciers. Evidence of a prograding outer shelf is found all around the Antarctic continent (Anderson, in press; Haugland et al., 1985; Stagg, 1985). The prograded deposits beneath the Wilkes Land shelf probably are of similar age and origin (Fig. 6; Wannesson et al., 1985, Fig. 6; Tanahashi et al., this volume, Fig. 3E). The East and West Antarctic ice sheets expanded onto the continental shelf during Plio-Pleistocene time (Anderson, in press) and eroded the continental shelf to its present anomalously great depth. The glaciers eventually became dry-base, and significant growth of the Wilkes Land margin upper-rise fans ceased at this time.

The high-resolution profiles show areas of erosion and deposition on the modern continental rise (Fig. 4B). We interpret the parallel-bedded reflector pattern to represent sediment deposition mainly on canyon walls and intercanyon areas from the contour currents that presently sweep across the region (Eittreim et al., 1972). Sediment cores collected from this unit contain inferred contourite deposits (Hampton et al., this volume). Most other areas of the rise have a highly reflective sea floor (Fig. 4B), probably because of erosion or winnowing.

Whereas the contour current is depositional over much of the lower rise, it may be erosional on the lowermost rise, as implied by the stepped and rough topography at the seaward end of profiles 1 and 3 (Fig. 7). The hyperbolic reflections probably are from erosional furrows (Damuth, 1980) that are succeeded seaward by erosional steps that truncate subbottom reflectors.

The seavalleys in profile 13 (Fig. 4) display various states of activity. One valley is draped by contour-current deposits, whereas others are floored by a hard reflector, possibly from a coarse-grained lag deposit. Substantial downcutting is indicated for the two easternmost canyons, where the contour-current deposits are truncated, and lesser amounts for some of the others. The cause of the erosion or winnowing probably is turbidity currents generated in the canyons, either as a consequence of sediment that is transported along the outer shelf–upper slope and trapped in canyon heads, or of slumping related to headward erosion of the canyons, such as is common in eastern U.S. slope and rise canyons (Farre et al., 1983). Alternatively, the canyons may be eroded by saline density currents that occasionally form when heavy shelf water builds up sufficiently to surge through the dynamic barrier of the strong, contour-following current near the shelf-break (Jacobs et al., 1970).

DISCUSSION

We interpret the seismic stratigraphy above the breakup unconformity on the Wilkes Land margin to represent sediment that accumulated along the subsiding edge of the Antarctic continent after it separated from Australia. The deposits comprise the prograding planar nonmarine/shelf, slope, and rise facies typical of passive-margin sequences elsewhere, but the Wilkes Land post-breakup stratigraphy also has some aspects that make it distinctive.

The post-breakup deposits appear to be predominantly stratified siliciclastic sediment. In particular, we see no evidence of carbonate buildups, as dominate the upper post-breakup shelf deposits on the conjugate Australian margin (Boeuf and Doust, 1975), for example, or as line the reefed Jurassic paleo-shelf edge along Atlantic margins (Folger et al., 1979; Lehner and Ruiter, 1977). The carbonate deposits on the conjugate Australian margin are younger than middle Eocene (Boeuf and Doust, 1975) and probably correlate with sequences A1 and A2 on the Wilkes Land margin. The cooling climatic conditions that began at about this time (Wolfe, 1978) apparently prevented the development of a carbonate environment off Antarctica and commenced the dominant terrigenous-rich glacial sedimentation.

We propose that unconformity T is due to a relative sea-level drop that exposed the sequence-B1 shelf deposits. The seismic stratigraphy above unconformity T records a major southward displacement of the shelf-break, relative to its position beneath the unconformity, which we interpret to be associated with a shoreline transgression of eustatic, tectonic, or combined control. The transgression exceeded 140 km within the study area.

The ongoing thermal subsidence of the margin may be the principal mechanism that caused the transgression, but the abrupt facies change across unconformity T suggests a rapid transgression and therefore perhaps an increase in the subsidence rate or the added influence of a eustatic sea-level rise. A postrift subsidence history for the margin according to the McKenzie (1978) model suggests that the bulk of the subsidence of the margin occurred in the early few tens of million years after breakup. Therefore, an age of unconformity T is preferred that is within a few tens of million years of the mid-Cretaceous breakup (Eittreim and Smith, this volume).

The subsidence history based on the McKenzie (1978) model, constructed by Eittreim and Smith (this volume) at the location of shotpoint 2100 on line 1, is shown in the lower diagram of Figure 8. According to this diagram, unconformity T was at sea level about 78 Ma, but given the uncertainties regarding the age of the breakup unconformity relative to the initiation of subsidence (e.g., an extended "hot" period of upraised continental crust after the first generation of oceanic crust), the estimate may be on the order of 20 Ma younger. The upper diagram in Figure 8 shows the sea-level fluctuations postulated by Watts and Thorne (1984). Fluctuations of the correct magnitude and sense to have first exposed the sequence-B shelf, then rapidly submerged it, are possible at the times 65 Ma and 43 Ma (Fig. 7). The event at 65 Ma is preferred, based on the proximity to the time postulated when the top of sequence B subsided below wave base, based on the thickness (about 2.5 km) of sediment separating the rift-onset unconformity from unconformity T (Fig. 8).

An alternative explanation for the development of unconformity T is the major change in plate configuration, the increase in seafloor-spreading rate, and ridge position shift that occurred about 42 Ma (Cande and Mutter, 1982; Veevers, this volume). We cannot provide a quantitative test of this proposal nor suggest the specific mechanism, but we note that a rapid increase in subsidence rate and water depth along the conjugate Australian margin occurred at or near this time (upper Eocene) (Boeuf and Doust, 1975; Mutter et al., 1985).

ACKNOWLEDGMENTS

We thank Captain Alan McClenaghan and the crew of the *S.P. Lee* for their cooperation in achieving the goals of the study under the difficult operating conditions in Antarctic waters. Jon Childs guided the processing of multichannel data. Discussions with John Anderson improved our understanding of polar sedimentary processes. Reviews by Robert Bohannon, Art Grantz, and John Schlee are appreciated.

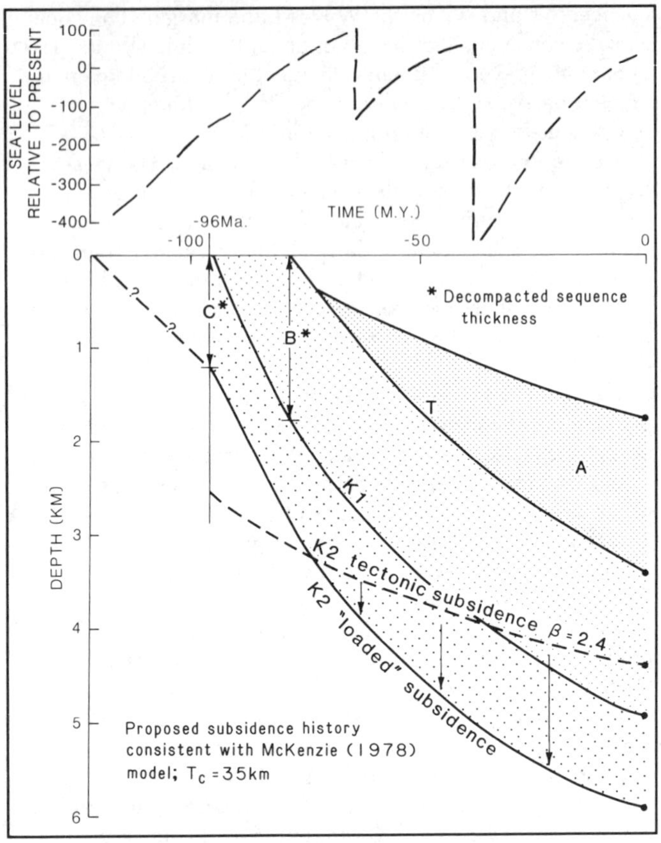

Figure 8—Proposed subsidence history at the location of shotpoint 2100 on line 1, based on the McKenzie (1978) crustal thinning and subsidence model with a crustal stretching factor $\beta = 2.4$, a crustal thickness (T_c) of 35 km, and assuming that sediment of sequences C and B were deposited at or near sea level (see Eittreim and Smith, this volume, for a more complete discussion). The sea-level curve in the top diagram is from Watts and Thorne (1984).

REFERENCES CITED

Anderson, J. B., in press, The Antarctic continental shelf: Results from recent marine geologic and geophysical investigations, *in* R. J. Tingey, ed., The geology of Antarctica: London, Oxford University Press.

Boeuf, M. G., and H. Doust, 1975, Structure and development of the southern margin of Australia: Australian Petroleum Exploration Association Journal, v. 15, p. 33-43.

Cande, S. C., and J. C. Mutter, 1982, A revised identification of the oldest sea-floor spreading anomalies between Australia and Antarctica: Earth and Planetary Science Letters, v. 58, p. 151-160.

Damuth, J. E., 1980, Use of high-frequency (3.5-12 kHz) echograms in the study of near-bottom sedimentation processes in the deep-sea: A review: Marine Geology, v. 38, p. 51-75.

Drewry, D. J., 1975, Initiation and growth of the East Antarctic ice sheet: Geological Society of London Journal, v. 131, p. 255-273.

Eittreim, S. A., P. M. Bruchhausen, and M. Ewing, 1972, Vertical distribution of turbidity in the south Indian and south Australian basins, *in* D. E. Hayes, ed., Antarctic oceanology. II. The Australian-New Zealand sector: American Geophysical Union, Antarctic Research Series, v. 19, p. 51-58.

Eittreim, S. A., M. A. Hampton, and J. R. Childs, 1985, Seismic-reflection signature of Cretaceous continental breakup on the Wilkes Land margin, Antarctica: Science, v. 229, p. 1082-1084.

Falvey, D. A., 1974, The development of continental margins in plate tectonic theory: Australian Petroleum Exploration Association Journal, v. 14, p. 95-106.

Farre, J. A., B. A. McGregor, W. B. F. Ryan, and J. M. Robb, 1983, Breaching the shelf break: Passage from youthful to mature phase in submarine canyon evolution: Society of Economic Paleontologists and Mineralogists Special Publication 33, p. 25-39.

Folger, D. W., W. P. Dillon, G. A. Grow, K. D. Klitgord, and J. S. Schlee, 1979, Evolution of the Atlantic continental margin of the United States, *in* M. Talwani, W. Hay, and W. B. F. Ryan, eds., Deep drilling results in the Atlantic Ocean: Continental margins and paleoenvironments: Washington, DC, American Geophysical Union, p. 87-108.

Haugland, K., Y. Kristoffersen, and A. Velde, 1985, Seismic investigations in the Weddell Sea Embayment: Tectonophysics, v. 114, p. 293-313.

Hayes, D. E., and L. A. Frakes, 1975, General synthesis Deep-Sea Drilling Project Leg 28, *in* D. E. Hayes and L. A. Frakes, eds., Initial reports of the Deep-Sea Drilling Project: Washington, DC, U. S. Government Printing Office, v. 29, p. 919.

Jacobs, S. J., A. F. Amos, and P. M. Bruchhausen, 1970, Ross Sea oceanography and Antarctic bottom water formation: Deep-Sea Research, v. 17, p. 935-962.

Lehner, P., and P. A. C. de Ruiter, 1977, Structural history of Atlantic margin of Africa: Bulletin of the American Association of Petroleum Geologists, v. 61, p. 961-981.

McKenzie, D. P., 1978, Some remarks on the development of sedimentary basins: Earth and Planetary Science Letters, v. 40, p. 25-32.

Mitchum, Jr., R. M., P. R. Vail, and J. B. Sangree, 1977, Seismic stratigraphy and global changes of sea level. Part 6. Stratigraphic interpretation of seismic reflection patterns in depositional sequences, *in* C. E. Payton, ed., Seismic stratigraphy—applications to hydrocarbon exploration: American Association of Petroleum Geologists Memoir 26, p. 117-133.

Mutter, J. C., K. A. Hegarty, S. C. Cande, and J. K. Weissel, 1985, Breakup between Australia and Antarctica: A brief review in the light of new data: Tectonophysics, v. 114, p. 255-279.

Piper, D. J. W., and W. R. Normark, 1982, Acoustic interpretation of Quaternary sedimentation and erosion on the channelled upper Laurentian Fan, Atlantic margin of Canada: Canadian Journal of Earth Sciences, v. 19, p. 1974-1984.

Pitman, III, W. C., 1978, Relationship between eustacy and stratigraphic sequences of passive margins: Geological Society of America Bulletin, v. 89, p. 1389-1403.

Sangree, J. B., and J. M. Widmier, 1977, Seismic stratigraphy and global changes of sea level. Part 9. Seismic interpretation of clastic depositional facies, *in* C. E. Payton, ed., Seismic stratigraphy—applications to hydrocarbon exploration: American Association of Petroleum Geologists Memoir 29, p. 165-184.

Shackleton, N. J., and J. P. Kennett, 1975a, Paleotemperature history of the Cenozoic and the initiation of Antarctic glaciation: Oxygen and carbon isotope analyses in DSDP sites 277, 279, and 281, *in* J. P. Kennett and R. E. Houtz, eds., Initial reports of the Deep-Sea Drilling Project: Washington, DC, U.S. Government Printing Office, v. 29, p. 743-755.

______, 1975b Late Cenozoic oxygen and carbon isotopic changes at Deep-Sea Drilling Project site 284: implications for glacial history of the northern hemisphere and Antarctica, *in* J. P. Kennett and R. E. Houtz, eds., Initial reports of the Deep-Sea Drilling

Project: Washington, DC, U.S. Government Printing Office, v. 29, p. 801–807.

Sleep, N. H., 1971, Thermal effects of the formation of Atlantic continental margins by continental breakup: Geophysical Journal of the Royal Astronomical Society, v. 24, p. 325–350.

Stagg, H. M. J., 1985, The structure and origin of Prydz Bay and MacRobertson Shelf, East Antarctica: Tectonophysics, v. 114, p. 315–340.

Vail, P. R., and J. Hardenbohl, 1979, Sea-level changes during the Tertiary: Oceanus, v. 22, p. 71–79.

Vail, P. R., R. M. Mitchum, Jr., and S. Thompson, III, 1977, Seismic stratigraphy and global changes of sea level. Part 4. Global cycles of relative changes of sea level, *in* C. E. Payton, ed., Seismic stratigraphy—applications to hydrocarbon exploration: American Association of Petroleum Geologists Memoir 26, p. 83–97.

Veevers, J. J., 1986, Breakup of Australia and Antarctica estimated as mid-Cretaceous (95 ± 5 Ma) from magnetic and seismic data at the continental margin: Earth and Planetary Science Letters, v. 77, p. 91–99.

Wannesson, J., M. Pelras, B. Petitperrin, M. Perret, and J. Segoufin, 1985, A geophysical transect of the Adelie margin, East Antarctica: Marine and Petroleum Geology, v. 2, p. 192–201.

Watts, A. B., and W. B. F. Ryan, 1976, Flexure of the lithosphere at continental margin basins: Tectonophysics, v. 36, p. 25–44.

Watts, A. B., and J. Thorne, 1984, Tectonics, global changes in sea level, and their relationships to stratigraphic sequences at the U. S. Atlantic continental margin: Marine and Petroleum Geology, v. 1, p. 319–339.

Weissel, J. K., and D. E. Hayes, 1972, Magnetic anomalies in the southeast Indian Ocean, *in* D. E. Hayes, ed., Antarctic oceanology. II. The Australian-New Zealand sector: American Geophysical Union, Antarctic Research Series, v. 19, p. 165–196.

Wolfe, J. A., 1978, A paleobotanical interpretation of Tertiary climates in the northern hemisphere: American Scientist, v. 66, p. 694–703.

Eittreim, S. L., and Hampton, M. A., 1987, *The Antarctic Continental Margin: Geology and Geophysics of Offshore Wilkes Land*, CPCEMR Earth Science Series, v. 5A: Houston, Texas, Circum-Pacific Council for Energy and Mineral Resources

Mechanism of Extension and Rifting at the Antarctica Continental Margin

G. W. Moore
S. L. Eittreim
U.S. Geological Survey
Menlo Park, California

A strong subbottom reflection from the crust-mantle interface extends from below oceanic crust to under the continental crust of Antarctica for more than 80 km. Based on interpretations of seismic-reflection profiles, we propose that during lithospheric extension hot mantle material (foliated peridotite) rose as a dike-like body to the base of the continental crust. As the newly emplaced mantle material widened, the continental crust was thinned by low-angle detachment faults that transported slabs of crustal material from both opposed plates directly onto the new mantle. The dike of mantle material continued to widen and to intrude to progressively shallower depths, and when the intrusion reached a critical subsurface depth, the weight of the remaining thin cover of continental crust at the toes of the slides was small enough to permit partial melting of the mantle peridotite to form basaltic magma. Subsequently, a rift broke through to the surface, and basaltic oceanic crust began to form. The new oceanic crust then extended to accommodate plate separation, and while spreading continued in the middle of the belt of oceanic crust, an essentially unfaulted layer of sediment was laid down across the continent-ocean boundary.

INTRODUCTION

Seismic-reflection profiles recorded aboard the U.S. Geological Survey research vessel *S.P. Lee* off Wilkes Land, Antarctica, display rarely seen reflections from the Mohorovicić discontinuity (Moho) at the continent-ocean boundary (Eittreim et al., 1985). Detection of these reflections is possible, in part, because only about 3 km of younger strata cover igneous rocks that mark the edge of the oceanic crust.

A problem at Wilkes Land and other areas of continental rifting is that elongation measured across bounding normal faults at the tops of tilt blocks seems to be less than what is needed to account for the observed crustal thinning by simple stretching. Measurements on migrated subbottom profiles of the normal faults recorded off Wilkes Land, which are near the seaward edge of the elongated continental area, indicate an extension of about 10%; values averaging 25% are observed over a wider area for other continental margins (Chenet et al., 1983) and for areas such as the U.S. Basin and Range province (Stewart, 1978). By contrast, if a continental block, at first consistently 35 km thick, deforms to a wedge that thins uniformly from 35 km thick toward the continent to 6 km thick toward the ocean, the extension would be 71%.

A second problem at Wilkes Land, and at other areas of continental rifting, is that a region of uplift producing an unconformity occurs within 100–150 km of the rift axis marked by the edge of the oceanic crust. Extension alone cannot elevate thinned continental crust to a position above its unthinned level, so as to produce an unconformity (Sleep, 1971; Steckler, 1985).

In this report, we endeavor to solve these problems by suggesting that the zone of surficial extension along which fault-displacement measurements are made is broad compared to the zone of principal crustal thinning, which is narrow (Bott, 1982). We also suggest that hot mantle material of asthenospheric temperature rises as a dike toward the crust (Royden et al., 1980). It stops directly below the low-density crust, which acts as a buoyancy cap. As the dike widens, upper crustal layers slide onto the new upper mantle creating a new segment of the Moho underlain by hot mantle, which elevates the thinned crust to sea level.

STYLE OF FAULTING AND CRUSTAL THICKNESS

A detailed analysis of the seismic-reflection sequences recorded on subbottom profiles that cross the continent-ocean boundary off Wilkes Land is presented and discussed in a separate report (Eittreim and Smith, this volume). Several of the profiles, on which digital processing corrected the depths and migrated sloping reflectors to their proper positions, provide evidence bearing on the mechanism of continental extension. For the present study, we have chosen examples from the profile that was recorded along *S.P. Lee* Line 1 (Figs. 1–4).

Seismic reflection (Fig. 4) and a sonobuoy seismic-refraction velocity exceeding 8 km/sec (Childs and Stagg, this volume) document the Moho below thinned continental crust near the edge of the oceanic crust. The Moho emerges from under the edge of the oceanic crust at a subbottom depth of about 9 km, with relatively little change in level (Fig. 4). Igneous rocks

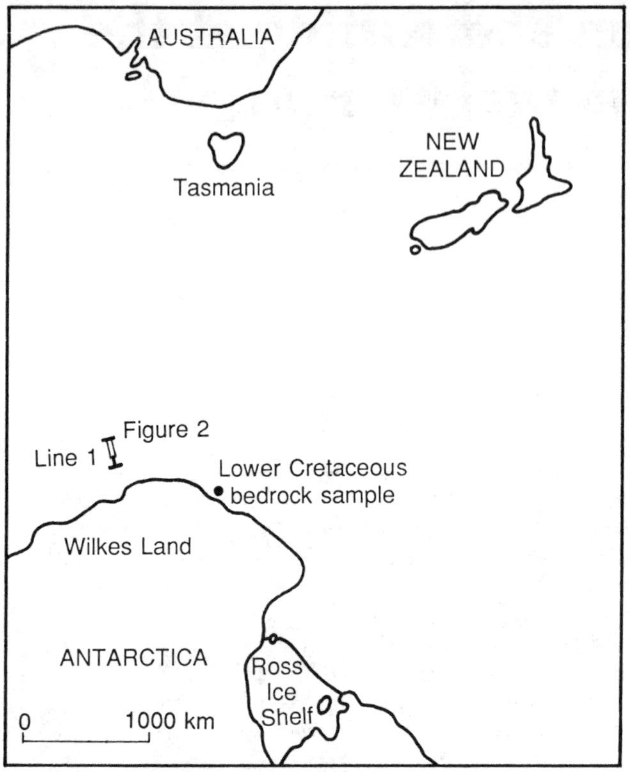

Figure 1—Index map to the Wilkes Land margin of Antarctica showing *S.P. Lee* Line 1, the segment of its subbottom profile drawn in Figure 2 (double line), and the location of an offshore Cretaceous bedrock sample.

associated with the oceanic crust near the continent–ocean boundary are 6 km thick. The adjacent continental crust, below the 3 km of postrift strata that lap across the boundary, consists of 2 km of strata at the top deposited during continental rifting, 3 km of little-deformed strata in tilted fault blocks, and a 1 km thick unit directly above the Moho that has few internal reflectors.

Near the continent–ocean boundary, the Moho closely underlies faulted but otherwise little deformed stratigraphic layers (Fig. 4). These layers, traced landward, constitute the uppermost units of the continental crust. As judged by an erosional unconformity along their upper surface (Fig. 3), the Moho lies directly below stratigraphically equivalent upper parts of the continental crust that were being subjected to erosion at the time the oceanic crust began to form.

Faults in stratified rocks so close above the Moho (and indeed inferred displacements on the Moho itself) indicate that these rocks were capable of sustaining brittle fracture.

Landward-dipping tilt blocks become especially conspicuous on the subbottom profiles about 50 km from the continent–ocean boundary (Figs. 2, 3). Typical dips of the strata are less than 10°, and the widths of the individual blocks average about 10 km. Vertical fault displacements are typically 400–600 m, with a maximum of about 1000 m. The tilt blocks are recorded to about 150 km landward from the edge of the continental crust, beyond which they are masked by multiple reflections and thicker sedimentary cover. But the data of Wannesson et al. (1985) show that they extend well under the continental shelf to near the shoreline, which is about 290 km from the continent–ocean boundary. They also may continue under the ice cover beyond that, because this part of Antarctica is occupied by the Wilkes subglacial basin, a large region of subsidence and sedimentary fill (Steed and Drewry, 1982). Half grabens such as those of the Newark Supergroup along the Atlantic coast of the United States, which formed in response to extension associated with continental breakup, in places are farther than 400 km from the edge of the oceanic crust (Hutchinson et al., 1983).

The upper parts of the faults that bound the tilt blocks off Wilkes Land have dips of about 45°, with some as shallow as 30°, and most of the faults show evidence of becoming less steep at depth. Several deep and nearly flat-lying reflectors that seem to cross-cut and pass beneath the strata may be major décollements (Figs. 2, 3).

A structurally high area in the form of a block-faulted anticline lies about 100 km landward from the continent–ocean boundary (Fig. 3). Truncation of beds on the anticline indicates that it was at sea level and subjected to erosion during the extension. The unconformity over the anticline has now sagged to a depth of 4.3 km, presumably because of later lithospheric cooling and sediment loading.

INITIATION OF RIFTING

Before the emplacement of oceanic crust during the Cretaceous (Cande and Mutter, 1982), the rifting that took place across the Australia–Antarctica supercontinent probably began by strike-slip shear caused by stresses emanating from active seafloor spreading that was then proceeding both northwest of Australia toward India and northeast of Australia–New Zealand toward the Pacific basin (Fig. 5). The smooth curve of the matching coasts and shelf-breaks of Australia and Antarctica (in contrast to the stairstep pattern of the present-day spreading axis) suggests an initial transform-dominated strike-slip fault zone along the new plate boundary, as does the idea that shear is the least energetic way to break through the lithosphere (Lachenbruch and Thompson, 1972).

The Ferrar Dolerite and Tasmanian Dolerite in this area, both of Jurassic age, were emplaced about 170 Ma, or approximately 70 million years before oceanic crust began to be emplaced at Wilkes Land. This earlier event seems to have occurred along a failed rift of the initial breakup of Gondwana (Ford and Kistler, 1980), and we assume that it did not bear a close relationship to the later rifting at Wilkes Land.

The Cretaceous breakup of the Australia–Antarctica supercontinent occurred along two plate boundaries, one between Australia and Antarctica, and one between Australia and New Zealand (Fig. 6). The same magnetic lineations are first recorded along both boundaries, indicating that both separations began at the same time.

The times of the first emplacement of oceanic crust and of the continental rifting that preceded it may be estimated by extrapolating the age and spacing of offshore magnetic lineations to the present edge of the continental crust and to its original position before continental extension. The spacing between magnetic anomaly 34 and the edge of the continental crust at *S.P. Lee* Line 1 is about 60 km. The ages of anomalies 24 and 34 are respectively 56 and 84 Ma (Palmer, 1983), and the spacing between them ranges from 150 to 210

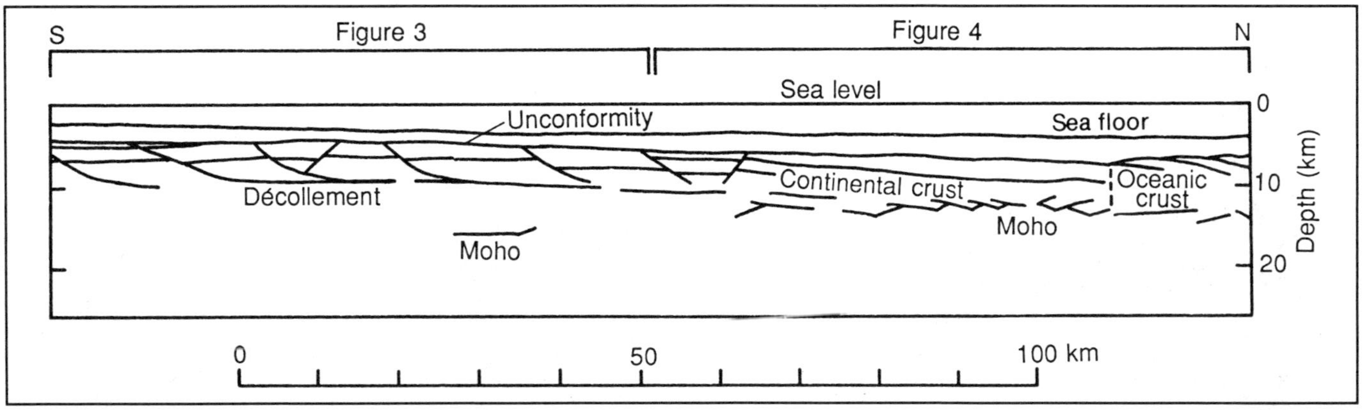

Figure 2—Line drawing of part of *S.P. Lee* Line 1 with no vertical exaggeration showing the locations of subbottom profiles reproduced in Figures 3 and 4. See Figure 1 for position of line.

km (Cande and Mutter, 1982). The above values contain uncertainties, but using the Palmer (1983) time scale and assuming a constant spreading rate from the initiation of spreading to 56 Ma, the Late Cretaceous (Cenomanian) emplacement of oceanic crust is estimated to have begun between 92 and 95 Ma (Veevers, 1986). Assuming the same rate, and continental extension at Antarctica of 100–150 km, the Early Cretaceous continental rifting is estimated to have begun between 105 and 123 Ma (Albian to Barremian). These values are consistent with the ages of cored Lower Cretaceous rocks (Domack et al., 1980) that were deposited in a synrift sag which formed at the time of initial rifting, and with fossil pollen of the same age (Truswell, 1983, and this volume) thought to have been eroded from these rocks during their later thermal uplift.

THERMAL AND ISOSTATIC CONDITIONS DURING THE RIFTING

The uplift that occurred during the Early Cretaceous before oceanic crust began to form off Wilkes Land, as indicated by the erosional unconformity on subbottom profiles and by the common occurrence of reworked fossil pollen from local Cretaceous rocks, is matched elsewhere in the world by other rifted margins and areas of active continental rifting (Morgan, 1983; Steckler, 1985). Crustal thinning during rifting to the extent seen on our profiles, considered by itself, would lead not to uplift but to isostatic subsidence. Hence, the general evidence for strong uplift points to nonuniform crustal extension and local heat anomalies, which produced a more complex regime than that of the simplified stretching model of McKenzie (1978). A minimum value for the magnitude of the thermal uplift at Wilkes Land can be estimated, after adjusting for sediment loading, by comparing the present 4.3 km minimum depth of the unconformity on the subbottom profiles at the anticline with an inferred original depth approximately equal to sea level.

Turning first to a comparison between the seafloor level of oceanic crust at an active spreading axis and that at an old seafloor area that has reached thermal equilibrium, lithospheric columns can be balanced to show the lithospheric density change resulting from cooling that causes the crest of a spreading axis to subside from an initial water depth of about 2.5 km to a final depth of about 6 km.

Under isostatic conditions, the sum of products of thickness times density for the various layers in a column of a lithospheric plate is equal to the sum for any other column. Therefore, if the thickness and density of layers in one column are known, an unknown thickness or density of a layer in another column can be determined by equating the sums for the two columns and solving for the unknown.

In the calculations to follow, a column height (depth of compensation) of 100 km has been selected. This is approximately the depth to the top of the asthenosphere below cold lithosphere, and the calculations assume identical density structure below that depth for all columns.

Assuming an oceanic crustal thickness of 6 km and densities at a spreading axis of 1.0, 2.7, and 3.3 g/cm^3 for the water, crust, and mantle, respectively, we find that the observed thermal subsidence from an active spreading ridge to an area of thermal equilibrium can be accounted for by an increase in upper-mantle density of 0.09 g/cm^3 (rounded to 0.1 g/cm^3 in Fig. 7).

Cretaceous temperature and density conditions at Wilkes Land can be estimated using the balanced-column method by comparing conditions at an area of active continental rifting with the change in upper-mantle density determined above for oceanic rifts. If the assumption is made that the temperature of the mantle material under an actively extending continent is the same as that under an oceanic spreading axis, the balanced-column method shows that thinned continental crust can be elevated to sea level and subjected to erosion. Under such conditions, the crustal thickness at the shoreline would be 15.6 km (Fig. 7). With the later cooling to thermal equilibrium, crust 15.6 km thick would sink to a depth of 3.5 km. At Wilkes Land, the eroded rocks over the anticline were depressed still farther by 1.8 km of postrift sediment loading, and they now lie 4.3 km below sea level. The balanced-column method suggests that the density of this postrift sediment is 2.4 g/cm^3.

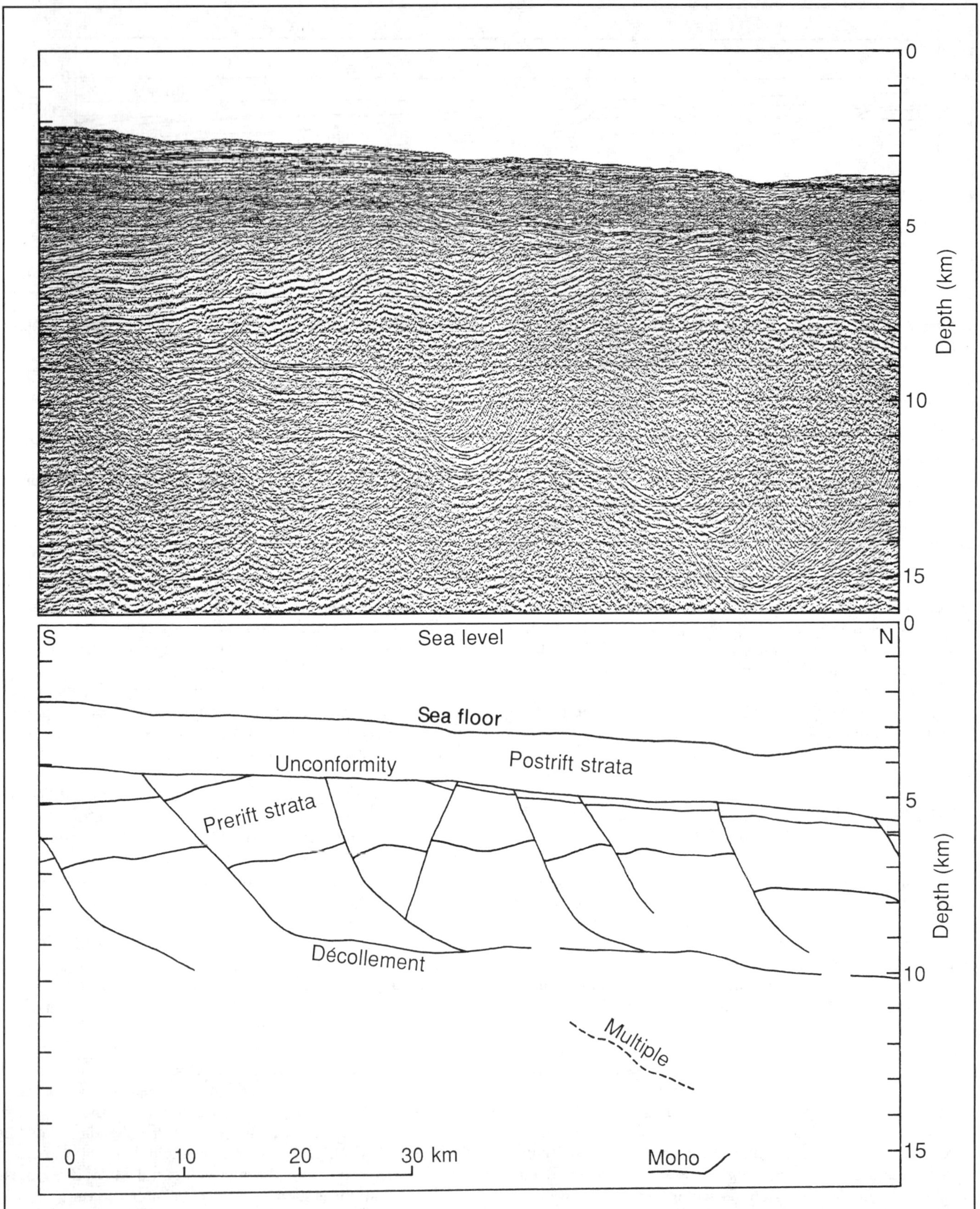

Figure 3—Subbottom profile and line drawing showing tilt blocks and uplift unconformity about 100 km landward from the continent-ocean boundary on *S.P. Lee* Line 1. Joins with Figure 4; see Figure 2 for location; vertical exaggeration is 2.9:1.

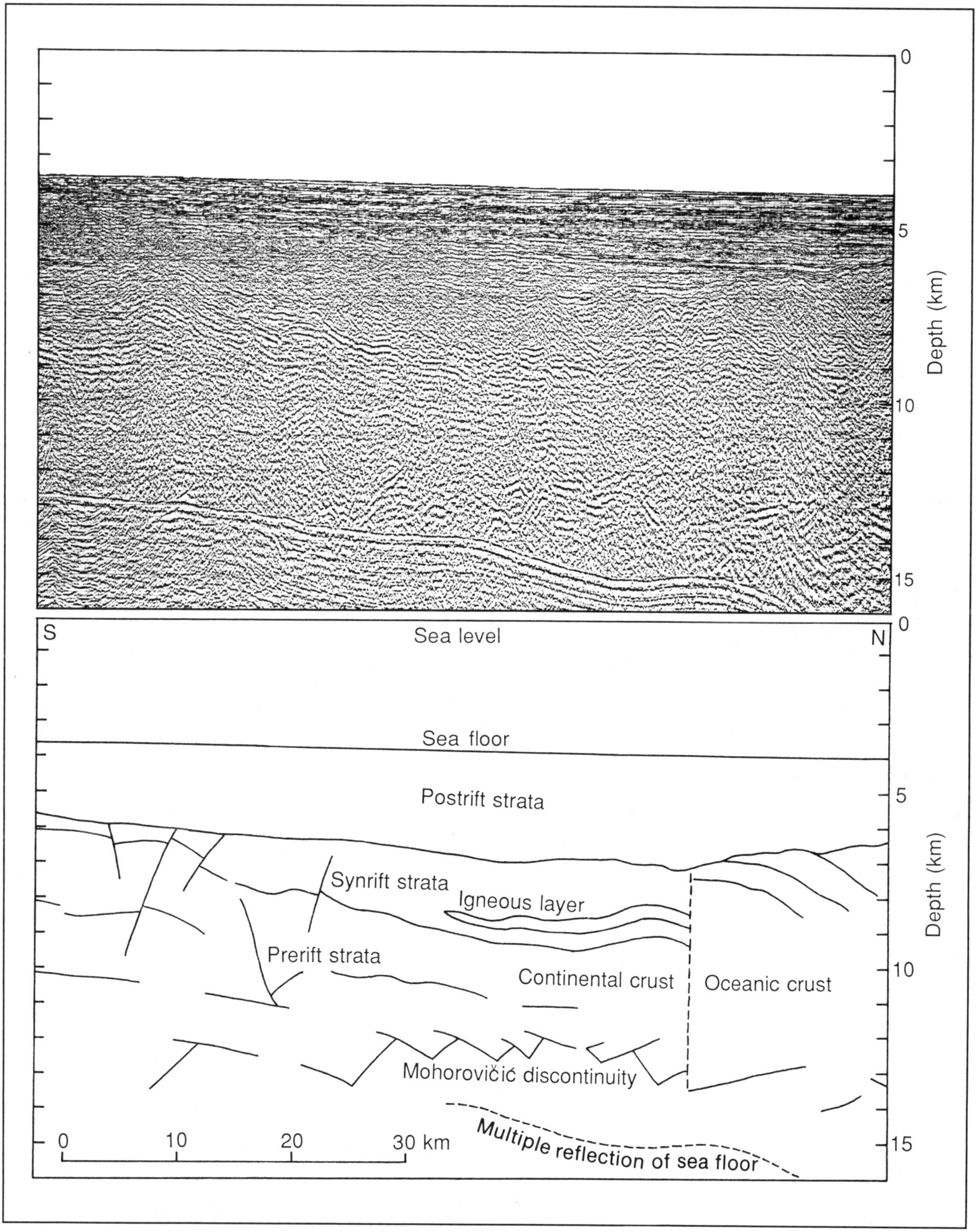

Figure 4—Subbottom profile and line drawing showing the continuation of the Moho from beneath oceanic crust to beneath continental crust on *S.P. Lee* Line 1. Joins with Figure 3; see Figure 2 for location; vertical exaggeration is 2.9:1.

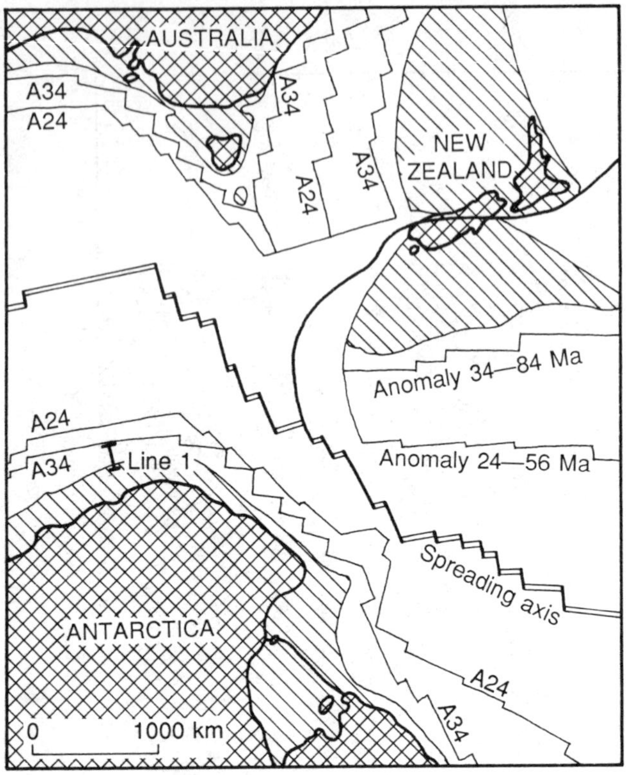

Figure 5—Selected magnetic anomalies showing how initial Australia-Antarctica and Australia-New Zealand spreading was succeeded by only the present Australia-Antarctica spreading; single-line pattern represents continental shelf; adapted from Circum-Pacific Map Project (1983).

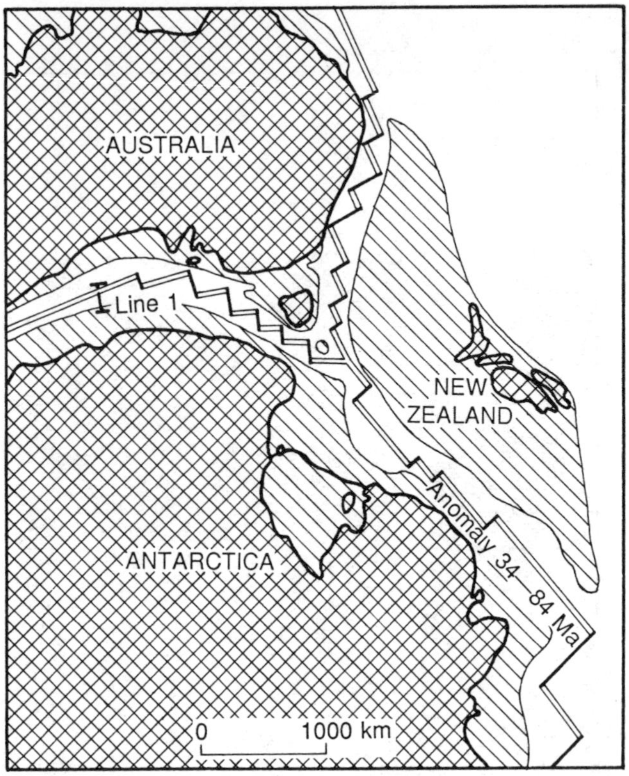

Figure 6—Configuration of Australia, Antarctica, and New Zealand during the Late Cretaceous at the time of anomaly 34 (84 Ma), soon after the breakup of the supercontinent.

MODEL OF EXTENSION

The following facts need to be explained by a mechanism for the extension and rifting at the Antarctica continental margin: (1) The Moho passes reasonably smoothly from beneath the oceanic crust to beneath the continental crust; (2) continental half grabens mainly tilt away from the oceanic crust; (3) deep reflectors that cut across and beneath the strata of the continental crust slope at a low angle toward the continent–ocean boundary; (4) as the boundary is approached, progressively younger stratigraphic levels are affected by tilting and faulting; (5) the entire lower part of the crust seems to be missing near the boundary; and (6) an eroded anticline, now at a depth of 4.3 km, lies about 100 km from the continent–ocean boundary, indicating that the strata there, despite relatively thin crust, once were uplifted above sea level, presumably because of the thermally expanded state of underlying materials.

To explain the evidence from Wilkes Land, we suggest a process in which crustal material at rifted continents and continental margins slides down along deep detachment faults onto newly emplaced mantle material (Moore and Eittreim, 1985). Laboratory studies of olivine deformation suggest that the upper 16 km of the mantle is in the brittle-fracture regime (Sawyer, 1985). An initial shearing and a slight separation of the lithospheric mantle created by world-wide plate reorganization caused a narrow dike of solid deep-mantle material to intrude toward the surface (Fig. 8). This hot mantle material, initially from the lower lithosphere, later from the asthenosphere, rose to the level of the base of the crust, where its upward movement stopped because of the low-density crustal material above. The 35-km crust, previously broken along a smooth shear fracture, tended to separate in response to the separation of the lower lithosphere. But such a thick layer of crust cannot support itself laterally (Hubbert, 1945), and a crustal belt 400–500 km wide on each side of the deep-mantle intrusion began to move relatively toward the break.

As the lithosphere continued to extend, mantle material continued to intrude upward, and the crust extended by a series of detachments that rooted at progressively shallower levels as the mantle intrusion came closer to the surface.

Soon after the underlying intrusion and extension began, they became self-generating as parts of a locally driven ridge-push process (Hales, 1969; Moore, 1975). Owing to the balance between the new ridge-push force and the forces at all the other boundaries of the newly separated lithospheric plates, the direction of relative motion at this boundary became different from that during the initial shear break. The smoothly curved spreading axis of the first stage evolved into a stairstep boundary with transform faults, apparently to minimize the length of the spreading-axis segments of the boundary (Lachenbruch and Thompson, 1972).

The initial break cut through the whole original thickness of the crust, and the first detachment faults to form descended deepest and extended farthest from the line of intrusion at the base of the crust. The upwardly curved fault surfaces resulted in rotation of half grabens (rollover anticlines) away from the

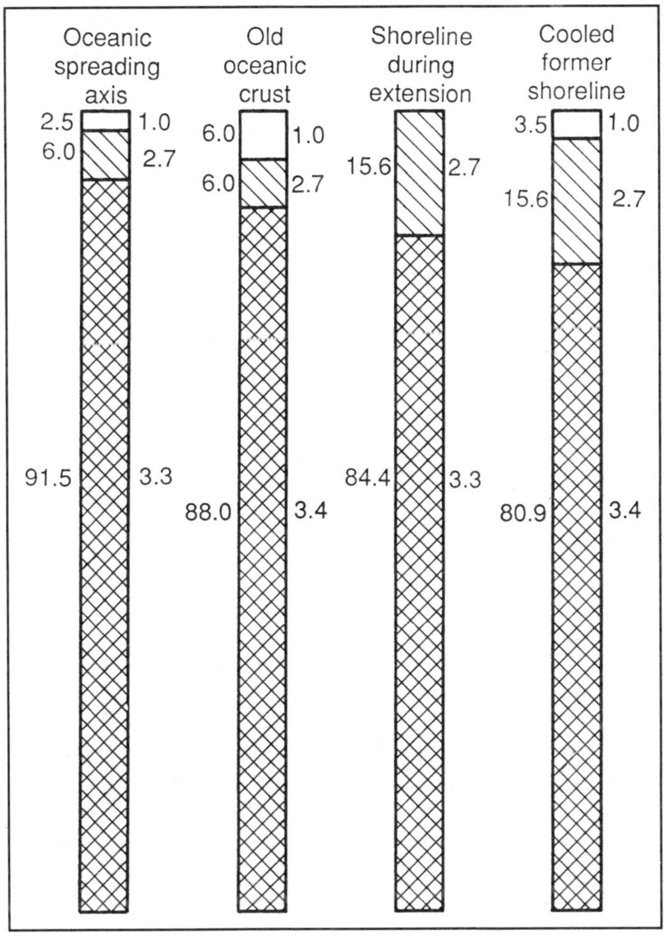

Figure 7—Thickness of water, crust, and mantle in isostatically balanced 100-km columns at oceanic and continental-margin areas during active extension and at thermal equilibrium. Thicknesses (km) are to left of columns and densities (g/cm^3) to right.

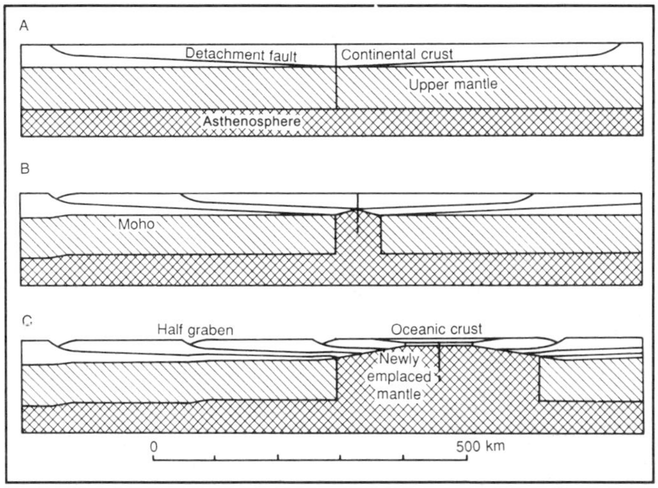

Figure 8—Model of continental breakup by intrusion of hot mantle material to the base of the crust, followed by sliding of upper crustal layers onto the newly emplaced mantle. A. Vertical strike-slip (transform) fault divides the lithosphere, and the crust becomes laterally unsupported because of plate separation. B. Low-angle detachment faults cut across the crust, and the toes of the detachment blocks slide across mantle material intruded to progressively shallower depth. C. Basalt separates from the mantle material when the crust is thinned to about 6 km, and the resulting oceanic crust laterally supports the continental crust and stops its sliding.

break (Fig. 8B), and the average slope of the detachments from their surface outcrop to the base of the crust is about 6°.

The leading edges of the continental slabs from each flank abutted at the axis of rifting, while the slabs slid off their respective continents and across the new mantle material. The local top of the Moho served as the baselevel for the slides. As the level of mantle intrusion rose gradually, the baselevel also rose, and the geostatic head at the crustal detachments was reduced. The 450–500 km slides that formed first were deactivated when the baselevel became higher. New slides formed, both thinner and in a narrower belt along the axis, scaled to the reduced depth to the top of the mantle intrusion (Fig. 8C).

When the mantle material approached to about 6 km from the top of the crust, partial melting occurred in a zone in the upper part of the mantle because of its decompression (Green and Ringwood, 1963). Basalt then formed sills within the crust and erupted to the surface. At this stage, the active upper detached slabs extended only about 100 km from the center of symmetry and were composed of only the upper part of the continental crust. As seen on the subbottom profiles, stratified upper crustal layers, cut by brittle fracturing, are superimposed almost directly onto the Moho.

Then a continuous belt of oceanic crust began to form, and this greatly changed the geostatic conditions near the rift. The column of gabbro and basalt of the oceanic crust raised the baselevel for the slides all the way to the surface and provided lateral support for the continental crust. This overcame the tendency for continental extension and stopped further sliding. Finally, a continuous layer of sediment was laid down over both types of crust at the continent-ocean boundary.

Near the surface, the major décollements are spaced rather far apart. The highest décollement on *S.P. Lee* Line 1 is 115 km long and 5 km thick (Fig. 2). But steps equal to the vertical spacing of the major décollements do not occur at the Moho. The irregularities seen on the profiles at the Moho are of the order of 1 to 2 km (Fig. 4). This implies that as the Moho progressively becomes shallower between each major décollement, the toes of the slides become sheared in small steps 1–2 km thick. This brittle deformation under very thin crust differs from the several planar reflections that mark the Moho under the thicker crust of the U.S. Basin and Range province (Klemperer et al., 1986).

Most of the crustal thinning occurs in belts 100–150 km wide on each side of the central rift, and a column of hot material from deep below in the mantle is emplaced directly below the thinned continental crust. This seems to offer a good explanation for the uplift and unconformity that occurs 100–150 km from the continent-ocean boundary (Fig. 9). Closer to the boundary, where the continental crust is very thin and therefore stood below sea level even at maximum lithospheric temperature, thick Upper Cretaceous sedimentary beds prograde toward the edge of the oceanic crust.

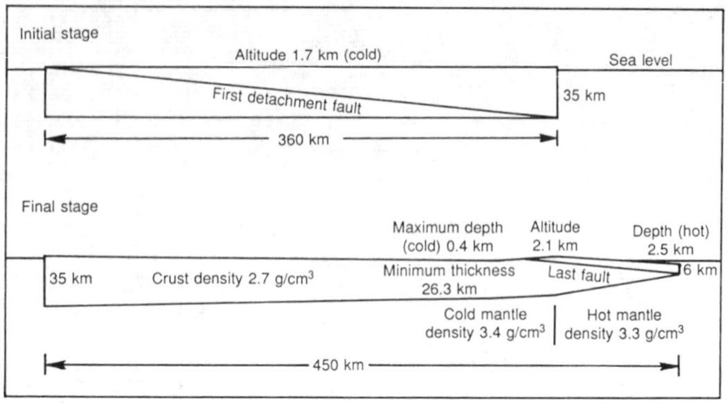

Figure 9—Diagram to illustrate the depression during crustal thinning caused by detachment faulting and the uplift during lithospheric heating caused by mantle intrusion, when the lithosphere is extended by 25%. Depth of compensation of 100 km used for the calculations.

CONCLUSIONS

Evidence on subbottom profiles for thin continental crust adjacent to the continent–ocean boundary off Wilkes Land is compatible with lithospheric extension associated with the solid intrusion of deep-mantle material up to the level of the base of the crust, accompanied by sliding of detached slabs of upper crustal material away from the continent. The crustal material is inferred to have slid initially onto the newly emplaced and hot mantle material at levels near the prerifting crustal thickness, then to gradually shallower levels as the mantle material intruded progressively higher. When the intrusion reached a subbottom depth approaching 6 km, oceanic crustal material began to melt out of the mantle material, and plate separation by ordinary seafloor spreading commenced. This caused the sliding and extension of the adjacent continental crust to cease abruptly, as seen by the near absence of faults in younger strata that lap across the boundary between the continental and oceanic crust.

The compositional interface that formed when crustal material slid onto newly emplaced mantle material constitutes an entirely new segment of the Moho. Were such a structure to be found in outcrop, it would consist of a shear-zone transition between thermally metamorphosed crustal material and foliated peridotite of the mantle.

Other passive continental margins, such as at the Triassic rift valleys of the Atlantic margin of the United States, show an area of continental rifting more than 400 km wide, and we suggest that a belt of principal crustal thinning at depth, with a half width of 100–150 km, is accompanied at the former surface by a belt of extension 400–500 km wide. This relationship explains a roughly threefold disparity between surficial extension and crustal thinning that has puzzled investigators of other areas of rifting (Chenet et al., 1983).

Although broad areas of continental crust adjacent to new oceanic crust have extended by up to 25%, phrases like "crustal stretching" and "crustal necking" convey false impressions of the process. We believe that multiple slabs separated by detachment surfaces, similar to that interpreted at upper crustal levels in the Basin and Range province of the western United States (Wernicke, 1985), and akin to a slow-moving coherent landslide, provides a more accurate analogy. But an important difference between Wernicke's (1985) model and ours is that the descending shear surface does not root deeply in the mantle. It stops near the Moho, and it lengthens and also constitutes a newly formed reach of the Moho, as the base of the crust comes into contact with mobilized mantle material along the shear surface. The detached slab moves onto the top of the newly emplaced mantle much like a track-laying railroad car moves onto its newly laid track.

During the later stages of extension, mainly upper crustal layers are involved in the multiple-sliding process. The initial detachment surface, which extended through the full thickness of the crust, reached up to the land surface farthest from the eventual location of oceanic crust. As the mantle intrusion became progressively shallower, crustal masses that were thinner and less extensive became involved in the sliding. Immediately before the oceanic crust began to be emplaced, the Wilkes Land profiles show that stratified upper crustal materials were the ones that were being superimposed directly onto mantle material. The entire lower crust is missing, so that rocks believed to be of Paleozoic and Mesozoic age are faulted directly onto mantle material.

The extreme shallowness of the Moho under the edge of the continental crust, as seen on the Wilkes Land profiles, greatly constrains the acceptable hypotheses for continental thinning. The thermal effects leading to the observed uplift and unconformity might be permitted by a delamination model in which less dense asthenospheric material engulfs slightly cooler and therefore more dense lithospheric mantle material (Brown and Girdler, 1980), but this mechanism cannot cause the thinning of continental crust, because no way seems available to increase the density of the continental crust sufficiently for it to sink into the mantle.

The crustal-sliding model accounts for most of the phenomena associated with the continuum represented by continental rift valleys, narrow spreading seas, and rifted passive continental margins. The 100–150 km wide region of uplift and unconformity that flanks both rift valleys on continents and oceanic crust at passive continental margins is inferred to result from low lithospheric density as a result of the high temperature under the toes of the sliding crustal material. Similarly, the hills of peridotite that crop out at some rifted margins (Nicholls et al., 1981; Boillot et al., 1986; Bonatti et al., 1986) can be explained more easily by continental slides that bulldoze up mantle material intruded to shallow depth than by mechanisms that involve only crustal stretching and necking.

ACKNOWLEDGMENTS

We are indebted to Captain Alan J. McClenaghan and the ship's crew for the success of this cruise. Gregory L. Smith processed the digital subbottom profiles, and Jonathan R. Childs and Norman H. Sleep gave us useful suggestions. The manuscript was improved as a result of critical reviews by Michael S. Marlow and Tracy L. Vallier.

REFERENCES CITED

Boillot, G., E. L. Winterer, et al., 1986, ODP Leg 103 drills into rift structures: Geotimes, v. 31, n. 1, p. 15–17.

Bonatti, E., G. Ottonello, and P. R. Hamlyn, 1986, Peridotites from the island of Zabargad (St. John), Red Sea: Petrology and geochemistry: Journal of Geophysical Research, v. 91, p. 599-631.

Bott, M. H. P., 1982, Stress based tectonic mechanisms at passive continental margins: American Geophysical Union, Geodynamics Series, v. 6, p. 147-153.

Brown, C., and R. W. Girdler, 1980, Interpretation of Africa gravity and its implication for the breakup of the continents: Journal of Geophysical Research, v. 85, p. 6443-6455.

Cande, S. C., and J. C. Mutter, 1982, A revised identification of the oldest sea floor spreading anomalies between Australia and Antarctica: Earth and Planetary Science Letters, v. 58, p. 151-161.

Chenet, P., L. Montadert, H. Gairaud, and D. Roberts, 1983, Extension ratio measurements on the Galicia, Portugal, and northern Biscay continental margins: Implications for evolutionary models of passive continental margins: American Association of Petroleum Geologists Memoir 34, p. 703-715.

Circum-Pacific Map Project, 1983, Plate-tectonic map of the Circum-Pacific region, Antarctica sheet: American Association of Petroleum Geologists, ed. 2. Scale 1:10,000,000.

Domack, E. W., W. W. Fairchild, and J. B. Anderson, 1980, Lower Cretaceous sediment from the East Antarctic continental shelf: Nature, v. 287, p. 625-626.

Eittreim, S. L., M. A. Hampton, and J. R. Childs, 1985, Seismic-reflection signature of Cretaceous continental breakup on the Wilkes Land margin, Antarctica: Science, v. 229, p. 1082-1084.

Ford, A. B., and R. W. Kistler, 1980, K-Ar age, composition, and origin of Mesozoic mafic rocks related to Ferrar Group, Pensacola Mountains, Antarctica: New Zealand Journal of Geology and Geophysics, v. 23, p. 371-390.

Green, D. H., and A. E. Ringwood, 1963, Mineral assemblages in a model mantle composition: Journal of Geophysical Research, v. 68, p. 937-945.

Hales, A. L., 1969, Gravitational sliding and continental drift: Earth and Planetary Science Letters, v. 6, p. 31-34.

Hubbert, M. K., 1945, Strength of the Earth: Bulletin of the American Association of Petroleum Geologists, v. 29, p. 1630-1653.

Hutchinson, D. R., J. A. Grow, K. D. Klitgord, and B. A. Swift, 1983, Deep structure and evolution of the Carolina trough: American Association of Petroleum Geologists Memoir 34, p. 129-152.

Klemperer, S. L., T. A. Hauge, E. C. Hauser, J. E. Oliver, and C. J. Potter, 1986, The moho in the northern Basin and Range province, Nevada, along the Cocorp 40°N seismic-reflection transect: Geological Society of America Bulletin, v. 97, p. 603-618.

Lachenbruch, A. H., and G. A. Thompson, 1972, Oceanic ridges and transform faults; their intersection angles and resistance to plate motion: Earth and Planetary Science Letters, v. 15, p. 116-122.

McKenzie, D., 1978, Some remarks on the development of sedimentary basins: Earth and Planetary Science Letters, v. 40, p. 25-32.

Moore, G. W., 1975, Pump-like tidal driving of tectonic plates: Bulletin of the American Association of Petroleum Geologists, v. 59, p. 1020-1021.

Moore, G. W., and S. L. Eittreim, 1985, Mechanism of continental rifting—Wilkes Land evidence: Eos, v. 66, p. 1098-1099.

Morgan, P., 1983, Constraints on rift thermal processes for heat flow and uplift: Tectonophysics, v. 94, p. 277-298.

Nicholls, I. A., J. Ferguson, H. Jones, G. P. Marks, and J. C. Mutter, 1981, Ultramafic blocks from the ocean floor southwest of Australia: Earth and Planetary Science Letters, v. 56, p. 362-374.

Palmer, A. R., 1983, The Decade of North American Geology 1983 geologic time scale: Geology, v. 11, p. 503-504.

Royden, L., J. G. Sclater, and R. P. Von Hertzen, 1980, Continental margin subsidence and heat flow: Important parameters in formation of petroleum hydrocarbons: Bulletin of the American Association of Petroleum Geologists, v. 64, p. 173-187.

Sawyer, D. S., 1985, Brittle failure in the upper mantle during extension of continental lithosphere: Journal of Geophysical Research, v. 90, p. 3021-3025.

Sleep, N. H., 1971, Thermal effects of the formation of Atlantic continental margins by continental break up; Royal Astronomical Society Geophysical Journal, v. 45, p. 325-350.

Steckler, M. S., 1985, Uplift and extension at the Gulf of Suez: Indications of induced mantle convection: Nature, v. 317, p. 135-139.

Steed, R. H. N., and D. J. Drewry, 1982, Radio echo sounding investigations of Wilkes Land, Antarctica, *in* C. Craddock, ed., Antarctic geoscience: Madison, University of Wisconsin Press, p. 969-976.

Stewart, J. H., 1978, Basin and Range structure in western North America—a review: Geological Society of America Memoir 152, p. 1-31.

Truswell, E. M., 1983, Geological implications of recycled palynomorphs in continental shelf sediments around Antarctica, *in* R. L. Oliver, P. R. James, and J. B. Jago, eds., Antarctic earth science: Canberra, Australian Academy of Science, p. 394-399.

Veevers, J. J., 1986, Breakup of Australia and Antarctica estimated as mid-Cretaceous (95 ± 5 Ma) from magnetic and seismic data at the continental margin: Earth and Planetary Science Letters, v. 77, p. 91-99.

Wannesson, J., M. Pelras, B. Petitperrin, M. Perret, and J. Segoufin, 1985, A geophysical transect of the Adelie margin, East Antarctica: Marine and Petroleum Geology, v. 2, p. 192-201.

Wernicke, B., 1985, Uniform-sense normal shear of the continental lithosphere: Canadian Journal of Earth Sciences, v. 22, p. 108-125.

Eittreim, S. L., and Hampton, M. A., 1987, *The Antarctic Continental Margin: Geology and Geophysics of Offshore Wilkes Land*, CPCEMR Earth Science Series, v. 5A: Houston, Texas, Circum-Pacific Council for Energy and Mineral Resources

The Deep Crustal Structure of the Wilkes Land Continental Margin

J. R. Childs
U.S. Geological Survey
Menlo Park, California

H. M. J. Stagg
Bureau of Mineral Resources
Canberra, Australia

Marine refraction, gravity, and magnetic data from the Wilkes Land margin of Antarctica are used with multichannel seismic profiles to model the crustal structure over the continent–ocean boundary (COB) of this passive margin. The crustal types encountered in the study include normal oceanic crust, marginal rift-basin crust, and thinned continental crust. The latter two crustal types comprise the crustal transitional zone, which extends approximately 35 km landward of the COB and is characterized by heterogeneous velocity structure. Gravity models suggest that the transitional zone is underlain by a body that is slightly less dense (3.15 g/cm^3) than typical for mantle material. Inconsistencies between the sonobuoy data from the Wilkes Land margin and similar data from the corresponding conjugate margin of southern Australia are noted.

INTRODUCTION

The 1984 geophysical survey of the continental margin of Wilkes Land, Antarctica, conducted aboard the R/V *S.P. Lee* (Figs. 1, 2) yielded considerable quantities of marine refraction, gravity, and magnetics data. These data provide valuable insight into the nature and origin of the crust underlying this largely unknown geologic region. In this paper, we will examine these data in detail and then propose appropriate crustal models.

Although the margin of East Antarctica has been investigated sporadically during the past 15 years, the surveys have suffered seriously from a combination of factors, including low-resolution and low-power geophysical apparatus, operational restrictions caused by ice conditions, poor navigation, or inadequate survey time. While this survey similarly suffered to some extent from time constraints and adverse ice conditions, the quality of the data is high and allows the definition of the deep crustal structure at the continent–ocean boundary (COB). Furthermore, the extent of the survey is great enough to allow some generalizations to be made about the margin-wide structure.

The primary geophysical data collected during the survey were 24-fold seismic profiles, described elsewhere in this volume (Eittreim and Smith). The multichannel lines were used extensively to constrain the geometry of the gravity models. Other data include:

1. Seismic-refraction and wide-angle reflection data: A total of 30 sonobuoy stations were shot, extending in range between 7 and 53 km. The acoustic source for these profiles consisted of the 1300-in.^{3} multichannel seismic air gun array; receivers were the U.S. Navy SSQ-41 sonobuoy. Solutions for sediment and crustal horizons are presented and are used for control of geometry and densities in gravity modeling.
2. Magnetic data were recorded continuously with a Geometrics proton-precession marine gradiometer. The survey was conducted in a magnetically quiet zone, and the magnetic profiles for the most part reflect this.
3. Gravity data were recorded continuously with a LaCoste and Romberg sea gravimeter on a two-axis stabilized platform. A free-air gravity map of the Wilkes Land margin, incorporating previous data, is presented, as well as crustal models derived from gravity and sonobuoy data.

Previous Investigations

The Wilkes Land continental margin has been studied by Houtz and Markl (1972), Cande and Mutter (1982), Veevers (1982), and Tsumuraya et al. (1985). However, the most detailed examination of the crustal structure offshore Wilkes Land was done by König (1980, this volume), based primarily on data acquired by the U.S.N.S. *Eltanin* in the late 1960s and early 1970s. König's principal conclusions can be summarized as follows:

1. The patterns of available geophysical data on the Wilkes Land margin are highly correlative to the corresponding patterns along the southern margin of Australia. This close correspondence suggests that the present margins of the

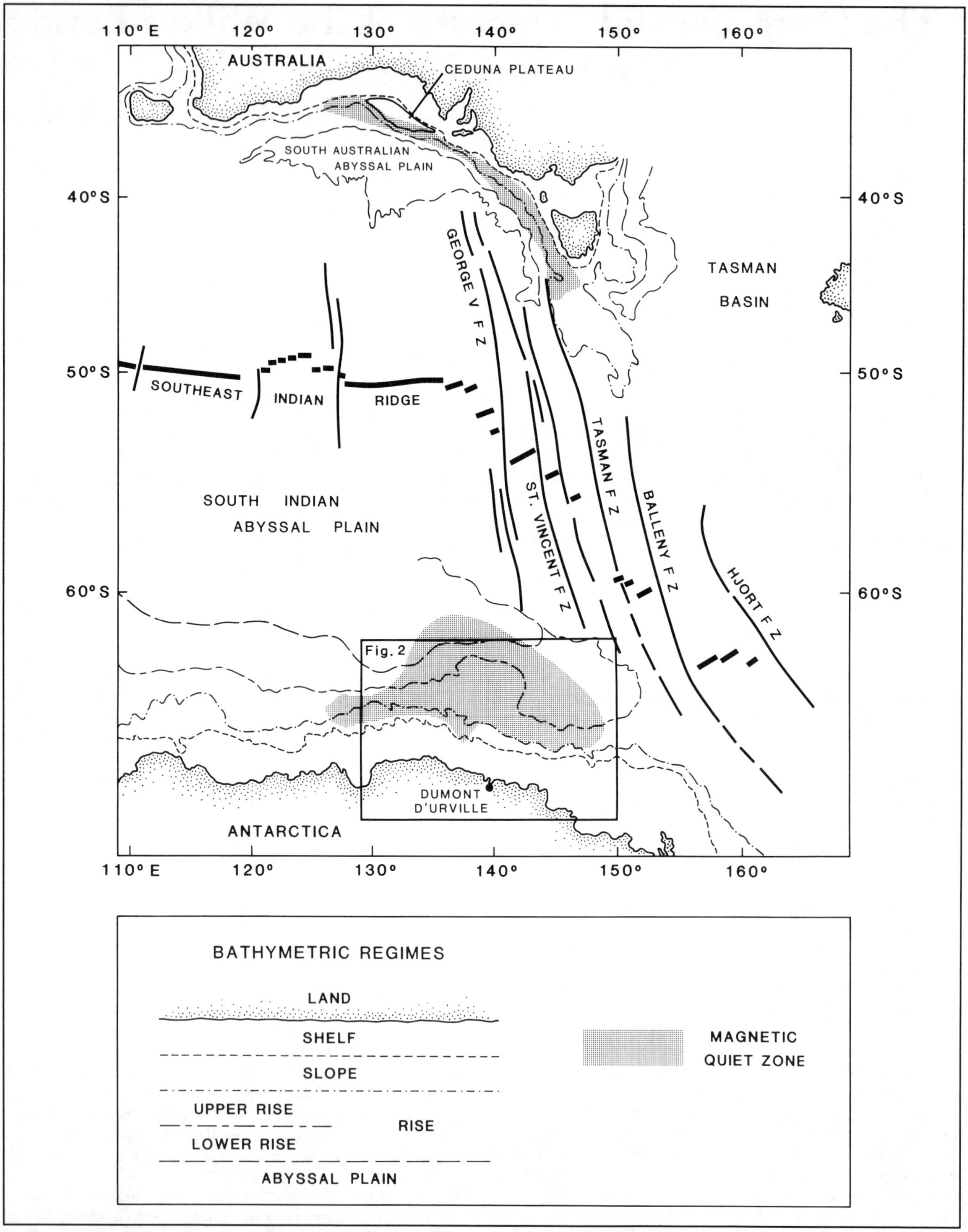

Figure 1—Geographic and physiographic map of the Southeast Indian Ocean (from Hayes, 1972). © 1972 American Geophysical Union. Outline of study area is indicated in the south-central portion of the map. Mercator projection.

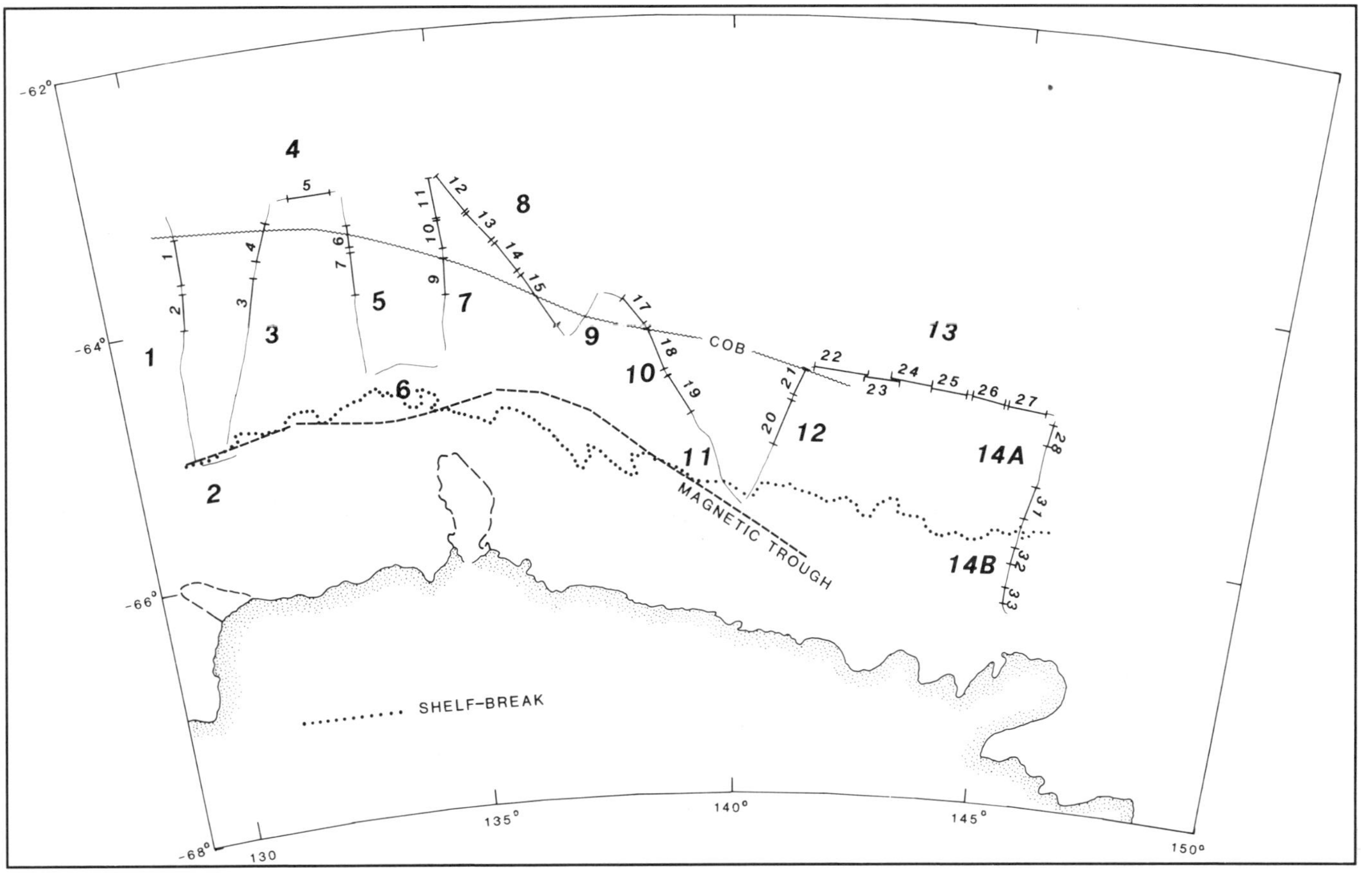

Figure 2—Map of the study area showing multichannel seismic tracklines (larger numbers; Eittreim and Smith, this volume), and locations of sonobuoy stations (smaller numbers). Polar stereographic projection.

Southeast Indian Ocean are formed by the fossil remnants of a once continuous and very broad (250–300 km) intracontinental rift system. The southern Australia and Wilkes Land margins are a conjugate pair.

2. Oceanic crust does not extend south of the seaward border of the magnetic quiet zone (MQZ) on the Wilkes Land margin.
3. The entire extent of the wide Australia–Antarctica paleorift appears to be a primary structural feature dating from the inception of the rift and not a result of subsequent crustal extension.

VELOCITY DATA

Of the 33 refraction stations attempted during the survey, 30 were successful in recording refractions from sedimentary and/or crustal layers. The locations of these stations are shown in Figure 2. Recording and reduction of the data were done according to methods outlined by Childs and Cooper (1978). The inversion of the refraction data was by conventional ray-tracing calculations (Dobrin, 1960; Chapter 5) from travel times picked from the analog records. The reflection data were inverted using the method suggested by Le Pichon et al. (1968). Both refraction and reflection results were corrected for dipping interfaces, calculated from time differences measured on the seismic-reflection records that were converted to depth using the approximate velocity solutions calculated without the dip correction. The results from the wide-angle reflections are presented in Table 1, and results from refractions in Table 2 and Figures 3, 4, and 5.

Refractions from the upper crustal layer (velocity of 5.0–5.8 km/sec) were recorded at 9 stations, while refractions from the primary crustal layer (velocity of 5.8–6.9 km/sec) were recorded at 14 stations; both crustal layers were recorded at only two stations. There is overlap in the velocities from these two layers, and where only one layer was recorded it is not always possible to conclude with certainty to which of these two horizons the refractor should be assigned. Possible mantle arrivals were recorded at stations 2 (8.55 km/sec) and 15 (8.22 km/sec), and possibly at station 18 (7.54 km/sec). In general, however, the maximum ranges of sonobuoy reception were less than the critical range of mantle refractions (see Cooper et al., in press).

The COB is clearly defined in the reflection seismic data for lines 1, 3, 5, 7, and possibly 8 (Eittreim and Smith, this volume). Relying on the seismic reflection profiles, Eittreim and Smith interpreted the crust as consisting of three types, from north to south: normal oceanic; marginal rift basin; and subsided continental. The marginal rift basin and subsided continental crust are together considered to comprise the transition between oceanic crust and normal thickness continental crust and will therefore be referred to as "transitional crust." The transitional crust extends from the COB to the magnetic trough, approximately 100–120 km south of the COB. Based on the seismic-reflection interpretation, the type of crust that underlies a particular sonobuoy can be assumed with some

Table 1—Sonobuoy wide-angle reflection velocity solutions from the Wilkes Land margin.[1]

SONO	LAT	LONG	LINE	CDP RANGE	WATER DEPTH	V1 (H1)	V2 (H2)	V3 (H3)	V4 (H4)	V5 (H5)
1	-63 19.7	130 25.4	1	439 -1258	4.120	1.65 (0.410)	2.06 (0.467)	2.81 (1.191)		
2	-63 44.0	130 23.8	1	1419 -2161	3.734	1.52 (0.435)	2.28 (0.423)	2.69 (3.296)		
3	-63 03.8	131 25.0	3	2128 -2927	2.855	1.49 (0.244)	1.74 (0.323)	2.69 (1.502)	3.88 (1.610)	
4	-63 35.8	131 43.8	3	3235 -3888	3.849	1.44 (0.528)				
5	-63 09.2	132 24.7	4	237 - 992	4.230	1.47 (0.420)	2.47 (1.047)			
6	-63 14.8	133 18.6	5	541 - 936	4.153	1.93 (0.860)	2.31 (1.076)			
7	-63 37.5	133 20.6	5	1039 -1754	3.666	1.65 (0.577)	2.08 (0.729)			
9	-64 01.5	134 54.2	7	835 -1457	3.257					
10	-63 40.7	134 56.4	7	1602 -2106	3.911	1.97 (0.462)	2.72 (1.095)			
11	-63 25.7	134 52.4	7	2188 -2890	4.069	1.72 (0.332)	1.85 (0.250)	2.54 (0.479)		
12	-63 06.7	134 55.3	8	15 - 842	4.121	1.68 (0.317)	2.01 (0.475)	2.37 (0.371)		
13	-63 25.4	135 23.5	8	900 -1624	3.991	1.71 (0.430)	1.87 (0.308)	2.77 (0.839)	4.02 (2.854)	
14	-63 39.6	135 49.8	8	1725 -2410	3.547	1.51 (0.263)	1.60 (0.591)			
15	-63 55.2	136 14.6	8	2528 -3515	3.374	1.75 (0.629)	2.03 (0.313)	2.43 (0.440)	2.69 (0.635)	
16	-64 20.8	136 51.5	8	3581 - 543	2.379	1.67 (0.507)	2.43 (1.181)	3.06 (1.416)		
17	-64 08.4	138 00.6	10	377 - 978	3.388	1.53 (0.455)	2.11 (0.475)	2.42 (0.514)	3.22 (0.830)	
18	-64 22.3	138 25.3	10	1065 -1901	3.346	1.74 (0.667)	2.28 (0.598)	3.30 (0.927)	2.87 (0.817)	
19	-64 44.0	138 45.8	10	2014 -2842	2.141	1.97 (0.903)	2.94 (0.616)	3.52 (0.893)		
20	-65 15.5	140 44.5	12	1087 -1944	2.134	1.85 (0.681)	2.10 (0.654)			
21	-64 53.6	141 05.9	12	2038 -2625	3.084	1.89 (0.723)	2.79 (0.688)			
22	-64 40.4	141 29.2	13	84 -1105	3.285	1.73 (0.503)	2.45 (0.615)	3.22 (0.810)	3.75 (2.368)	
23	-64 43.2	142 24.9	13	1061 -1767	3.078	1.68 (0.601)	2.82 (0.835)	3.18 (2.965)		
24	-64 44.0	142 53.8	13	1556 -2368	3.204	1.68 (0.301)	1.85 (0.338)	2.71 (0.939)	3.80 (0.728)	4.00 (1.232)
25	-64 46.8	143 40.9	13	2394 -3043	3.357	1.55 (0.489)	2.78 (1.005)	3.55 (1.228)		
26	-64 49.2	144 26.8	13	3155 -3792	3.108	1.85 (0.896)	2.67 (0.734)	3.30 (0.838)	2.99 (0.764)	
27	-64 52.4	145 09.3	13	3863 -4548	3.362	2.34 (1.557)	2.98 (0.833)	5.12 (2.320)		
28	-64 59.0	146 01.2	14A	107 - 445	3.156					
31	-65 29.3	145 48.5	14B	93 - 655	2.790	1.72 (0.580)	2.98 (1.435)	3.15 (0.655)	3.58 (0.538)	3.65 (1.460)
32	-65 58.3	145 31.6	14B	1183 -1428	0.434					
33	-66 16.4	145 24.4	14B	1864 -2159	0.189					

[1]Depths and thicknesses in km; velocities in km/sec. Groups are, from top, western, central, and eastern subsets shown in Figure 4.

Table 2—Sonobuoy refraction velocity solutions from the Wilkes Land margin.[1]

				SEAFLOOR	SEDIMENTS				CRUST	
SONO	RANGE	TYPE CRUST[2]	WATER DEPTH	VSF (H1)	V2 (H2)	V3 (H3)	V4 (H4)	V5 (H5)	V6 (H6)	V7
1	39.0	M	4.120	2.02 (0.571)		2.94 (0.968)	3.80 (2.433)		6.31	
2	31.5	M	3.733	2.05 (1.500)			3.79 (2.391)	5.28 (0.880)	6.57 (9.141)	8.55
3	39.5	M	2.855	1.8* (0.644)	2.06 (0.452)	2.97 (0.637)	3.57 (2.374)	5.75		
4	33.5	M	3.849	1.8* (0.541)	2.34 (1.194)	3.33 (1.8*1)	4.01 (3.151)			7.68
5	37.0	O	4.231	1.8* (0.357)	2.51 (0.548)	3.13 (2.957)		5.09 (1.533)		7.93
6	37.0	M	4.156	1.8* (0.239)	2.18 (1.433)	3.29 (1.717)			6.00	
7	35.0	M	3.665	1.8* (0.271)	2.17 (0.847)	2.74 (1.021)	3.74 (2.452)		6.49	
9	31.5	M	3.257	1.83 (1.324)		3.00 (0.831)	3.46 (1.553)		6.42	
10	24.0	O	3.901	1.8* (0.457)	2.26 (0.789)	2.90 (0.400)	3.44 (0.984)		5.79	
11	35.5	O	4.069	1.8* (0.401)	2.31 (0.777)	2.66 (0.792)	3.40 (1.970)	5.02		
12	39.0	O	4.114	1.8* (0.240)	2.27 (1.684)		3.82 (2.119)	5.22 (0.892)	6.87	
13	30.0	O	3.925	1.88 (0.739)		2.58 (1.161)	3.65 (2.810)		6.87	
14	30.0	O	3.626	1.8* (0.564)	2.45 (0.724)	3.05				
15	52.5	M	3.372	1.8* (0.560)	2.37 (0.797)	2.90 (1.144)	3.79 (2.569)		6.12 (7.160)	8.22
16	7.0	M	2.379	1.8* (1.178)		3.24				
17	28.5	O	3.390	1.8* (0.607)	2.34 (0.829)	2.93 (0.298)	3.55 (3.789)			7.87
18	38.0	M	3.329	1.83 (0.479)	2.48 (0.904)	2.65 (0.621)	3.56 (3.907)		6.09 (5.131)	7.54
19	38.0	M	2.754	1.8* (0.806)	2.47 (0.705)	2.69 (0.863)	3.84 (4.538)		6.07	
20	39.5	M	2.103	1.8* (0.922)	2.35 (0.730)	3.12 (1.165)	3.83 (4.436)	5.69		
21	26.0	M	3.080	1.8* (0.625)	2.66 (2.032)	3.88 (4.238)			6.74	
22	46.5	M	3.289	1.8* (0.298)	2.20 (0.791)		3.68 (4.531)	5.63		
23	31.5	M	3.077	1.8* (0.615)	2.18 (0.601)	3.52 (1.465)	4.06 (2.969)	5.29		
24	36.0	M	3.207	1.8* (0.234)	2.23 (0.860)	3.83 (1.388)	4.40 (3.566)		6.51	
25	31.5	M	3.339	1.78 (0.321)	2.25 (0.705)	3.84 (5.599)				7.71
26	31.0	M	3.109	1.8* (0.201)	2.27 (1.089)	3.97 (3.455)		4.61		
27	33.5	M	3.366	1.8* (0.655)	2.94 (0.208)	3.34 (3.681)		5.00		
28	19.5	M	3.157	1.8* (0.544)	2.47 (0.717)		4.16 (5.262)		6.39	
31	29.0	C	2.791	1.8* (0.629)	2.40 (0.654)	3.64 (1.675)	4.09			
32	12.5	C	0.462	2.50 (0.877)	3.43 (0.696)	3.69				
33	14.5	C	0.187	1.8* (0.388)	2.71 (1.489)	3.62 (1.070)	4.36			

[1]Depths and thicknesses in km; velocities in km/sec. Groups are, from top, western, central, and eastern subsets shown in Figure 4.
[2]M, marginal rift basin; O, oceanic; C, thinned continental. Designation of crustal types based on the seismic-reflection data (Eittreim and Smith, this volume).
*Assumed seafloor refraction velocity.

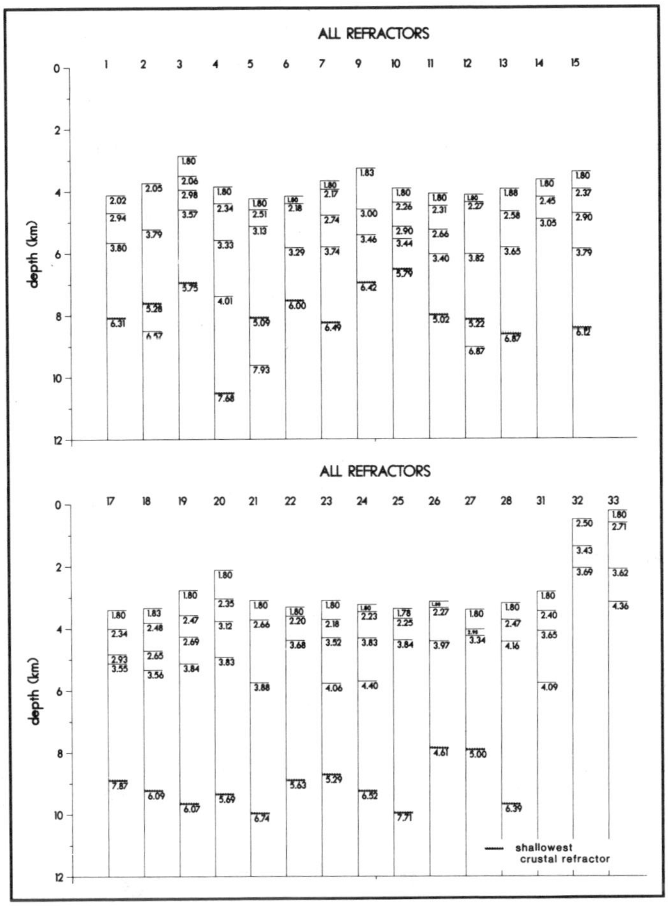

Figure 3—Refraction solutions for all sonobuoy stations. Solutions at depths greater than 12 km at stations 2, 15, and 18 are not shown.

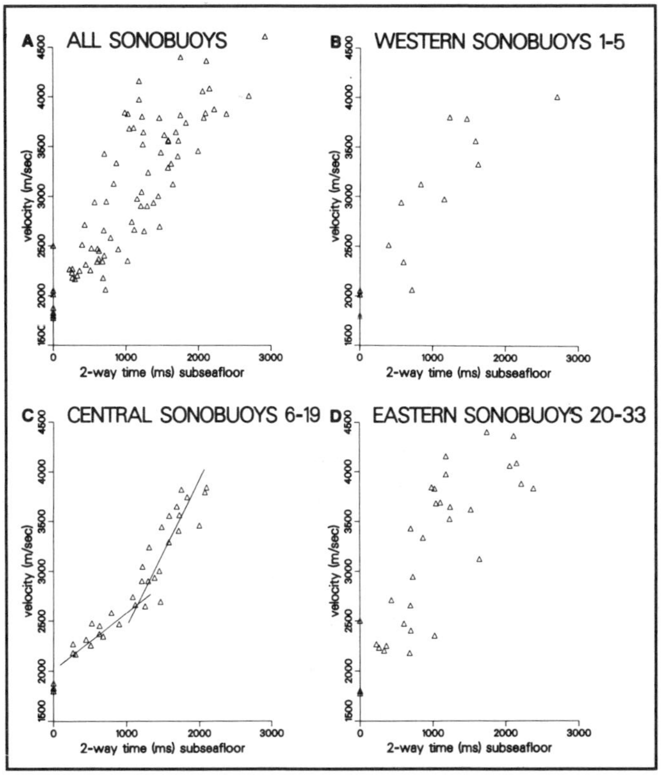

Figure 4—A. Plot of velocity against two-way reflection time below sea floor for all sonobuoys. B. Plot of velocity against two-way reflection time below sea floor for sonobuoys 1–5. C. Plot of velocity against two-way reflection time below sea floor for sonobuoys 6–19. D. Plot of velocity against two-way reflection time below sea floor for sonobuoys 20–33.

confidence. Using this criteria, we observe that 7 stations were recorded over oceanic crust, 23 were astride the COB or over the marginal rift basin crust south of the COB, and the remaining 7 were over subsided continental crust. Table 2 identifies the crust underlying each sonobuoy as oceanic (O), marginal rift basin (M), or subsided continental (C). Despite the benefit of good crustal definition, the crustal refraction velocities cannot be divided by value into subsets consistent with the crustal type.

Sedimentary Section

On the Wilkes Land margin, the sediment thickness is on the order of 4–5 km, approximately 2.5 km pre-breakup sediment and 1.5 km post-breakup. In this respect, the Wilkes Land margin differs substantially from the conjugate southern Australian margin, where the total sedimentary column is as much as 12 km thick beneath the Ceduna Plateau (Fig. 1), of which approximately 7 km is pre-breakup and 5 km is post-breakup (Fraser and Tilbury, 1979). Because of the larger sediment thickness beneath the Australian margin in both the pre- and post-breakup sections, the differences cannot be attributed solely to dissimilar sedimentation regimes since the Cretaceous breakup. Furthermore, the sediments on the Wilkes Land margin underlie the continental slope and rise and do not form a marginal plateau similar to the Ceduna Plateau.

Distribution patterns of the sediment from multichannel seismic analysis are shown in isopach maps by Eittreim and Smith (this volume) and in the sonobuoy results in Figures 3 and 5. The thickest sediments (greater than 7 km) were encountered in the eastern portion of the survey area. A basement high of indeterminate trend seen on line 7 appears to mark the western boundary of this eastern basin. Farther west the sediments again thicken, but to depths shallower than in the east.

Interpretation of seismic profiles reveal that several prominent sedimentary sequences extend throughout the survey area. The most prominent of these is the basal sedimentary layer, labeled D (K1–basement) by Eittreim and Smith (this volume), which has a refraction velocity of 3.4–4.2 km/sec and a maximum thickness on the upper continental rise of more than 5 km. The velocity contrast between D and the overlying layer is relatively large, averaging 0.8 km/sec, indicating that the boundary probably represents a major discontinuity.

The next prominent horizon, with refraction velocity of 2–2.4 km/sec, is at the top of a sequence rarely more than 1.5 km thick. This sequence is thickest on the mid to upper rise, thins dramatically under the shelf edge, and perhaps thickens again under the shelf. This layer probably consists of a Late Cretaceous and Tertiary veneer of post-breakup marine sedi-

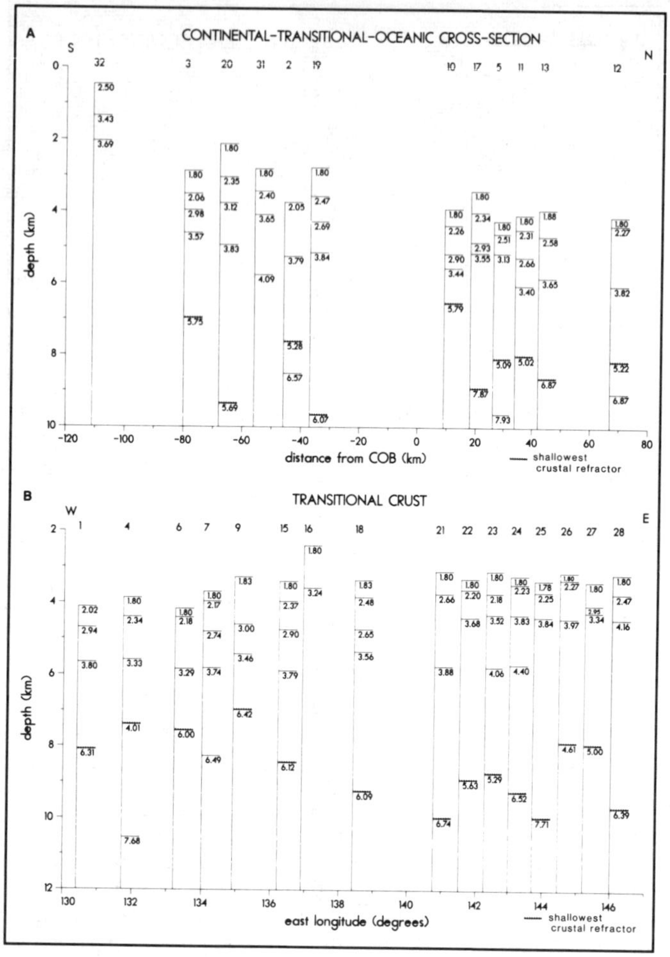

Figure 5—A. Sonobuoy solutions plotted as a function of distance from the continent-ocean boundary. (For locations, see Fig. 2.) B. Sonobuoy solutions from transitional crust, plotted as a function of east longitude. (For locations, see Fig. 2.)

ments beneath the slope, and Tertiary sediments over oceanic crust.

The shallowest sequence, with a velocity of 1.8–2.1 km/sec, forms a widespread veneer except on the continental shelf, where we assume that it has been eroded by ice.

Velocity-Depth Profiles

Refraction velocities for all sonobuoys have been plotted against sub-seafloor travel-time in Figure 4A. The scatter observed with all solutions is considerable. However, the solutions do suggest three subsets: western (Stations 1–5; Fig. 4B), central (Stations 6–19; Fig. 4C), and eastern (Stations 20–33; Fig. 4D). The central subset shows a tight relation between velocity and travel-time, with very few points being scattered from the main distribution. Drawing from a worldwide sample of sonobuoy solutions, Houtz (1974) observed that there is little statistical justification for a polynomial relation between velocity and travel-time (meaning that a linear fit to the time data should be sufficient). However, Figure 4C indicates a definite break in slope at about 1.1 sec two-way travel-time and a velocity of 2.5 km/sec. One may either fit the data by second-order polynomial or by two straight lines with different regression coefficients. The second option provides a statistically better fit and partially satisfies Houtz' conclusions concerning the linear relationship of velocity with travel time. The change in velocity gradient probably corresponds to a significant and permanent change in sedimentation rate or type and correlates roughly with the early Tertiary horizon (T) as interpreted by Eittreim and Smith (this volume).

The western and eastern subsets (Figs. 4B, 4D) differ substantially from the central subset. The scatter, particularly in the eastern data, precludes the calculation of credible regression coefficients. This scatter probably arises from large differences in the thickness of the younger section caused by erosion or differential deposition.

Crustal Section

Most of the refraction stations were shot on the middle and upper continental rise over transitional crust, as determined from seismic-reflection profiles (Eittreim and Smith, this volume). Only one station (19) displays a velocity structure that could be considered typical of continental crust. Two other stations (32 and 33), presumably located on continental crust, display refractions from the upper crust only, which is not definitive. At Station 2, the primary crustal layer has a velocity higher than normal for continental crust.

Seven stations over oceanic crust are shown in Figure 5A. Although the total section of sedimentary and crustal rock would have here been expected to be the thinnest recorded in the survey, the profiles do not support this. A sedimentary layer 2.5–5 km thick overlies what is presumably oceanic layer 2 (5.1–5.8 km/sec), beneath which is observed oceanic layer 3 (6.4–6.9 km/sec).

The crustal velocities recorded at stations 6 and 15 (6.0 and 6.1 km/sec) lie midway between typical velocities for oceanic layers 2 and 3, but, considering the velocity variations reported for layer 2 and its subdivisions 2a, 2b, and 2c (see, for example, Houtz and Ewing, 1976), both of these stations probably recorded layer 2c. Mantle refractions were recorded at station 15.

Twelve stations over the marginal rift basin crust (the band 30–40 km wide south of the COB) recorded crustal refractions (Fig. 5B). Of these stations, nine recorded refractions from the primary crustal layer. If any feature stands out in these refractions, it is their great variability—velocities range from 6.1 to 7.4 km/sec, and there is little regularity in their distribution. The implication is that this transitional regime is laterally and vertically inhomogeneous.

MAGNETICS

The magnetic field at high latitudes is very susceptible to large temporal variations (storm activity and diurnal changes) that render the interpretation of magnetic anomalies highly problematic. In an effort to filter the temporal variations from the spatial variations of interest, the magnetic field was measured with a gradiometer system, which consists of two proton-precession magnetometers separated horizontally by 150 m. The near sensor was towed 300 m aft of the ship, and both sensors were towed at a depth of approximately 15 m. Data were measured at a sampling interval of 4 sec, with a sensitivity of 1 nanotesla (nT). The method used to reconstruct the magnetic profile from the gradiometer data has been described

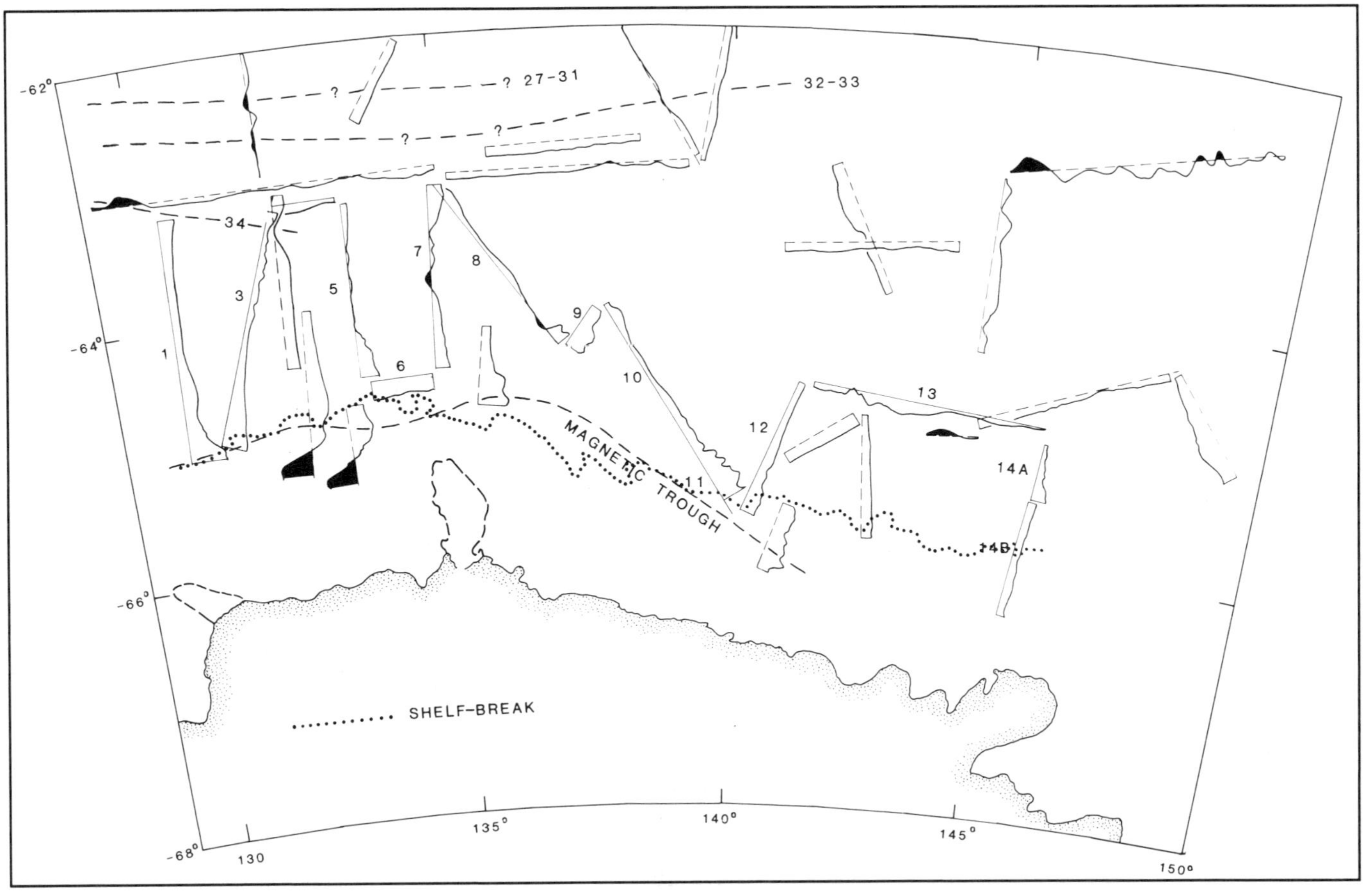

Figure 6—Residual magnetic profiles, after removal of IGRF 1980 reference field, plotted perpendicular to the tracklines. Positive free-air anomalies are filled; tracklines represent free-air anomaly of zero. *Lee* tracklines are solid with trackline numbers. Other profiles from earlier *Eltanin*, Japanese, and Australian cruises are shown for reference with dashed tracklines; for identification refer to Veevers (1986). Seafloor-spreading anomalies identified by Cande and Mutter (1982) are shown as dashed lines, as is the magnetic trough (T) identified by König (1980).

by Hansen (1984, 1985), and the advantages and limitations of the method are further described in Hansen and Childs (in press). Because the temporal variations are indistinguishable from the spatial variations of interest, gradiometer reconstruction of the magnetic field is considered essential for accurate interpretation. The reconstructed profiles are displayed in Figure 6, and both single-sensor and gradiometer profiles are shown for comparison in the appendix. Base station records from Dumont d'Urville (Fig. 1) were examined but did not prove useful in verifying the validity of the reconstruction method, probably because of the great distance between the survey area and base station location. However, comparison of the magnetometer base station records from Scott base with the temporal variations determined from gradiometer reconstruction of the subsequent Ross Sea data show a high degree of coherence (Hansen and Childs, in press; Behrendt et al., in press).

Total-field measurements have been reduced to residual or anomalous magnetic values by removal of the IGRF 1980 reference field (IAGA, 1976). Figure 6 shows in map view the reconstructed profiles along the major survey lines. For reference, the seafloor-spreading anomalies of König (1980), revised by Cande and Mutter (1982), are also shown. Since the survey was carried out largely in the magnetic quiet zone (MQZ; Fig. 1), there is little information in the magnetic anomaly profiles, with the exception of a 300-nT high in the middle of line 7, which corresponds to a basement ridge clearly discernable on the multichannel seismic record. (For additional discussion of the MQZ and the comparable quiet zone on the Australian margin, see Weissel and Hayes, 1972; Mutter, 1978; Talwani et al., 1979; and Veevers, 1986.) The southern ends of lines 1, 3, and 11 just reach the magnetic trough, a broad −500-nT low. Veevers (1986) has described the magnetic data in the Southern Indian Ocean resulting from the breakup of Australia and Antarctica. He attributes the positive anomaly identified as A34 by Cande and Mutter (1982) as the edge effect of the COB and dates the Australian-Antarctic breakup as mid-Cretaceous (95 ± 5 Ma). The *Lee* magnetic tracklines extended as far as this anomaly only at the northern extent of Line 3 and serve primarily to confirm the extent of the MQZ and to locate the survey within the context of the regional magnetic data.

GRAVITY

Gravity data were originally collected at a sampling interval of 20 sec, and desampled to 2 min or greater for plotting and modeling. The data have been Eötvös corrected and reduced to free-air anomalies. Cross-coupling analysis and correction

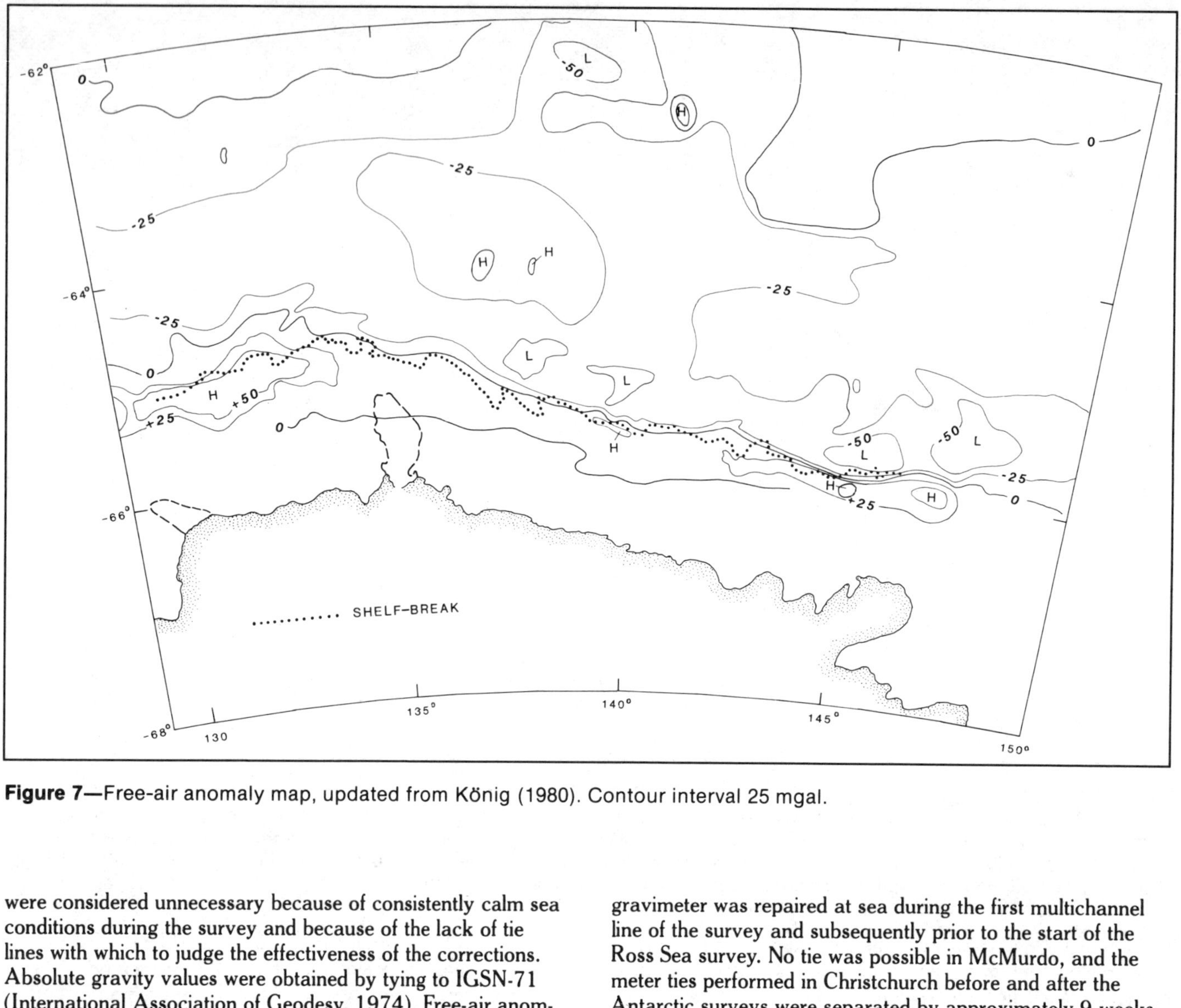

Figure 7—Free-air anomaly map, updated from König (1980). Contour interval 25 mgal.

were considered unnecessary because of consistently calm sea conditions during the survey and because of the lack of tie lines with which to judge the effectiveness of the corrections. Absolute gravity values were obtained by tying to IGSN-71 (International Association of Geodesy, 1974). Free-air anomalies were calculated by removal of the 1967 reference field (International Association of Geodesy, 1971). The data were edited to remove spurious values at turns, and to remove erroneous Eötvös corrections introduced by navigational errors. The edited data were then filtered to remove wavelengths of less than about 20 min (approximately 3 km). Since much of the survey was conducted off the continental shelf, this high-pass limit is not expected to have removed any geological information.

A free-air gravity anomaly map (approximately 1:5,500,000 scale) of this region was published by König (1980) based upon *Eltanin* data collected by Lamont-Doherty Geological Observatory between 1968 and 1972. Data were recovered from a copy of this map by digitizing trackline intersections of contours and were combined with the present data to create the updated free-air anomaly map shown in Figure 7. Comparison of the *Lee* data with the *Eltanin* contours indicated a consistent offset of 21 to 25 mgal throughout the data set. The source of this offset is unresolved but could arise from any one of several causes. The most likely source is an incorrect tie of the *Lee* data to international gravity network. The gravimeter was repaired at sea during the first multichannel line of the survey and subsequently prior to the start of the Ross Sea survey. No tie was possible in McMurdo, and the meter ties performed in Christchurch before and after the Antarctic surveys were separated by approximately 9 weeks. Assuming the *Lee* datum to be the source of the difference, a constant of 23 mgal was added to the *Lee* data prior to combination with the *Eltanin* data. The resulting data set is remarkably consistent. Comparison of proximate data values shows agreement between the two data sets that is well within the 6.5 mgal mean crossover error (standard deviation 4.91 mgal) calculated by König (1980) for the *Eltanin* data.

The resulting map remains very similar to König's within this region. All prominent features described by König remain with minor modifications. A broad 25-mgal regional low in the central portion of the map corresponds very closely with the MQZ, bounded to the north by anomaly 34, and widening to the east of long 133° E, where that anomaly terminates. The 75-mgal gradient across the northwest–southeast high-low pair in the north-central area of the map is probably related to the George V Fracture Zone, which terminates just north of the map at approximately long 145° E (Fig. 1). South of the MQZ is a steep gradient associated with the continental slope and rise, with positive anomalies as high as 75-mgal that coincide with the magnetic trough. The relatively complex anomaly patterns over the slope in the vicinity of long 129–134° E and long 144–148° E reflect the complex bathymetry resulting from extensive canyon formation (see Chase et al., this volume).

Gravity Models

The edge effect anomaly is frequently the dominant feature of the gravity field at a continental margin. This edge effect, featuring a high at the top of the continental slope and a low at the foot, is the result of differences in the gradients of the two major opposing gravitational influences at a seawards crossing of a continental margin—an increase in water depth (decreasing gravity) and the thinning of the crust (increasing gravity). The *Lee* survey lines occupy the region of greatest change, extending from just seaward of the edge-effect high to the corresponding low farther offshore. This location increases the uncertainty in the degree of isostatic compensation, a problem partly alleviated by the good seismic control.

The analysis of stacking velocities along line 3 indicates significant seismic discontinuities at horizons T, K1, and K2 (Eittreim and Smith, this volume). These horizons can be traced reliably on all lines modeled (see appendix to this paper). The boundary between the sedimentary section and igneous/crystalline basement is rather uncertain; we have located it on the basis of a vague discontinuity between stratified and reflection-free portions of the seismic section in conjunction with the refraction velocities. This boundary is the principal area of adjustment for the fitting of the shorter wavelength anomalies beneath the continental rise. Moho is clearly visible on several lines, and an initial estimate of its depth under oceanic crust has been taken from the migrated multichannel depth section of line 3. Initially, Moho was extrapolated southward from deep water assuming isostatic equilibrium; minor adjustments were then made to match the longer wavelength gravity features. The dense lower crustal bodies incorporated in the final models will be discussed in the following sections.

Densities of the layers T–K1, K1–K2, and K2–basement (Table 3) were determined from the refraction and stacking velocities using the Nafe–Drake velocity-density curves (Press, 1966). The density of the material above horizon T was determined from the magnitude of the gravity anomaly across the midslope canyon on line 3. The resulting density (2.1 g/cm^3) is somewhat higher than that predicted from the Nafe–Drake curves. Finally, the densities used to model crust and mantle are 2.85 and 3.3 g/cm^3, respectively. The latter may be somewhat low for lithosphere of age 90–100 Ma; substitution of a higher density can be accommodated with only minor adjustments to the Moho.

Gravity profiles for all lines are shown in the appendix. Gravity models for lines 3, 5, and 7 are shown in Figures 8, 9, and 10. The free-air anomalies have been modeled in each case, using an adaptation of the standard Talwani two-dimensional potential-field algorithm (Saltus and Blakely, 1983). The seismic-reflection data provide excellent control of the sedimentary section and of Moho in deep water, while the refraction data give a reasonable first estimate of basement depths on the continental rise. Such control allows a reliable model of shallow, midcrustal, and deep crustal horizons. These three lines were chosen for having the most distinct crossings of the COB and the best velocity control. (The gravity meter was, unfortunately, out of service for most of line 1.)

Table 3—Sediment and crustal densities used in gravity modeling.

Layer	Density	Layer description
1	1.03	seawater
2	2.10	post-horizon T (? post-Eocene)
3	2.20	K1-T (? Upper Cretaceous - Eocene)
4	2.45	K1-K2 (? synrift)
5	2.55	K2-basement (? pre-rift)
6	2.85	main crustal layer
7	3.15	? amphibolite-grade metamorphics
8	3.30	mantle

Line 3 provides the best illustration of the sedimentary section, extending from abyssal plain to upper slope. The sedimentary layer K2–basement (density 2.55 g/cm^3) shows a radical variation in thickness from about 3 km under the central rise to zero at 30–50 km from the COB and then to about 1 km under the upper slope. Thinning at the COB is attributed to erosion and stretching at rift onset. Fitting the gravity data at the southern extremity of the profile, near the shelf-break, required that the deep sedimentary interfaces and the basement surface shoal quite abruptly in this vicinity, which suggests that the shelf edge is controlled by a major upper crustal feature such as a basement fault or monoclinal hinge. This conclusion is supported by the refraction data at stations 28–33 (Figs. 2, 3), which show the deep sedimentary refractor on the shelf about 3 km shallower than on the adjacent slope.

Sequence K1–K2 likewise shows radical variation in thickness, but in a sense opposite to that of the underlying section. The sequence thins to almost zero over the midrise crest of K2 and forms a trough 2 km thick under the upper slope, thickening to 3 km under the lower rise where it abuts oceanic basement. The remaining sedimentary layers, T–K1 and seabed–T, form a post-breakup blanket of relatively uniform thickness over both continental and oceanic crust.

There are difficulties in devising a geologically reasonable crustal model that fits the gravity and seismic data. Initially, line 3 was modeled with separate layers for continental and oceanic crust. However, the density variation at the COB produced a gravity step that could not be removed by any logical adjustment of the shallow interfaces. Therefore, a single-crust model developed that could be fit to the observed gravity by simple adjustments of the Moho (Fig. 8A).

This model, however, conflicts with the reflection seismic data. The gravity model in Figure 8A incorporates thinned crust (Moho at 12–13 km depth) over a distance of about 35 km landward of the COB. The Moho interpreted from the seismic sections (Eittreim and Smith, this volume) dips landward from the COB to a depth of 17–18 km about 30 km from the COB. To reconcile this difference, a dense body must be introduced into the lower crust. Justification for such a body may be found at refraction station 4, immediately to the south of the COB (Fig. 5B), which contains a 7.7 km/sec refractor at a depth of about 10.5 km, and from the reflection data that show a flat-lying deep crustal reflector at about 11–12 km in

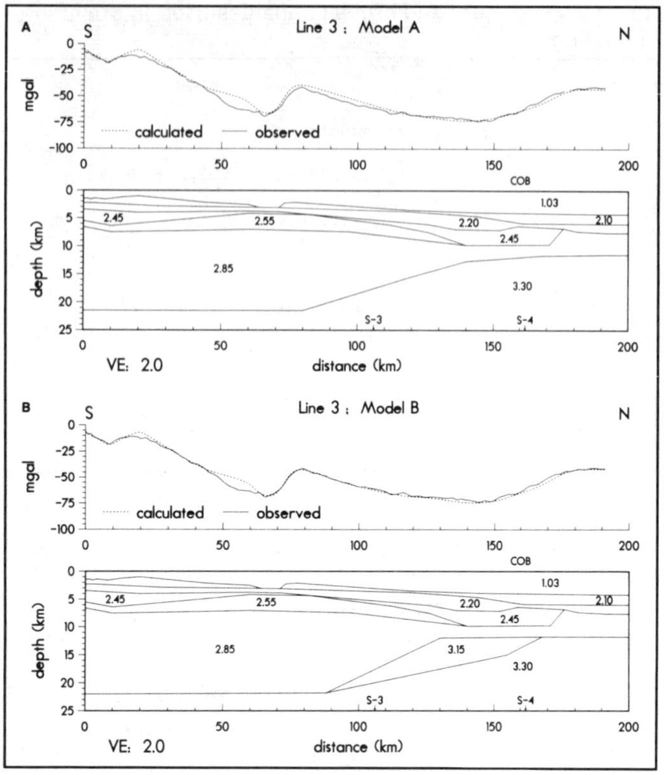

Figure 8—A. Gravity model along line 3. Densities are in g/cm^3. This model assumes a single-density crust without a dense, lower crustal body. Locations of sonobuoy stations for control are shown by small arrows at the bottom of the model. Average crustal density of 2.85 g/cm^3; 3.3 g/cm^3 body extends to a depth of 32.6 km. B. Gravity model along line 3. Densities are in g/cm ^{3}. Identical to the model in Figure 7, except for the addition of a high-density lower crustal body (3.15 g/cm^3) and consequent adjustment of the Moho to fit the observed gravity data. Average crustal density of 2.85 g/cm^3; 3.3 g/cm^3 body extends to a depth of 32.6 km.

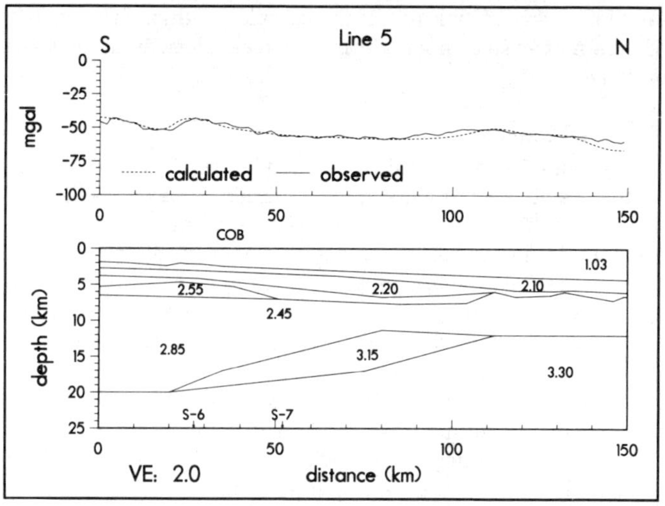

Figure 9—Gravity model along line 5. Densities in g/cm^3. Locations of sonobuoy stations for control are shown by small arrows at the bottom of the model. Average crustal density of 2.85 g/cm^3; 3.3 g/cm^3 body extends to a depth of 30.2 km.

the same area. This body may be equivalent to one proposed beneath the Australian margin by Falvey (1974) and Falvey and Middleton (1981). The top of this dense body was placed at 12 km, and the bottom was then adjusted to fit the observed data (Fig. 8B). The density used for this thin body, 3.15 g/cm^3, was derived from the refraction velocity and the Nafe–Drake curves; densities within the range 3.1–3.2 g/cm^3 can be accommodated by the model. Similar bodies were incorporated in the models for lines 5 and 7.

The crustal structure landward of the COB is remarkably uniform on lines 3, 5, and 7. Moho lies at about 20–22 km under the upper slope and midrise, with the thickness of the main crustal layer being 13–15 km. Over the marginal rift basin crust and north of the COB, the Moho shallows to about 12 km depth over a distance of some 80 km. This implies that crust thins to as little as 2 km (line 3) adjacent to the COB. On the oceanic side of the COB, the Moho is fairly constant at about 11.5–12 km depth on lines 3 and 5. On line 7, however, there is a prominent gravity high of approximately 30 mgal over oceanic crust, which can be accommodated by either of two simple modifications to the model. The first requires an increase in crustal density over a distance of some 45 km, holding the crustal thickness constant (Fig. 10A). However, the resulting density (3.15 g/cm^3) is unrealistically high considering the uniform oceanic crust immediately to the west. The second, preferred model (Fig. 10B) requires the Moho to shallow sharply to a depth of 8.5–9 km, leaving a total sediment/crust thickness of 4–5 km at the seaward extent of the line.

DISCUSSION

Although geophysical coverage of the margin of East Antarctica opposite Australia is sparse, we are fortunate in that the southern margin of Australia has been extensively studied in the last decade (Boeuf and Doust, 1975; König and Talwani, 1977; Mutter, 1978; Willcox, 1978; Fraser and Tilbury, 1979; Talwani et al., 1979; König, 1980; Houtz, 1981). These studies have focused primarily on the sedimentary structure and history from seismic-reflection data and on the crustal structure from gravity, magnetic, and seismic-refraction data. Talwani et al. (1979) interpreted an extensive set of sonobuoy data, principally from the western half of the Great Australian Bight (GAB). Although the number of stations they analyzed was about the same as the number we have analyzed here, the stations extended over a much greater area, and hence provide a broad background to this study. They found that the sonobuoy solutions could be divided into three sets (Groups I, II, and III), in addition to the standard oceanic crust. Group I had a velocity structure similar to oceanic crust, except for a total crustal thickness (13–18 km) higher than normal for oceanic crust (approximately 11 km). The Group III profiles were similar to continental crust, while Group II had an unusual crustal structure, including a main crustal layer with a velocity of 7.2–7.5 km/sec. Depth to Moho was not established for Group III stations, and the only Group II station that recorded Moho had a total crustal thickness of 13.5 km. They further found that Group II stations were concentrated in the Australian MQZ west of the GAB, while the Group I and III stations were intermingled throughout the GAB. The few sonobuoys that recorded crustal refrac-

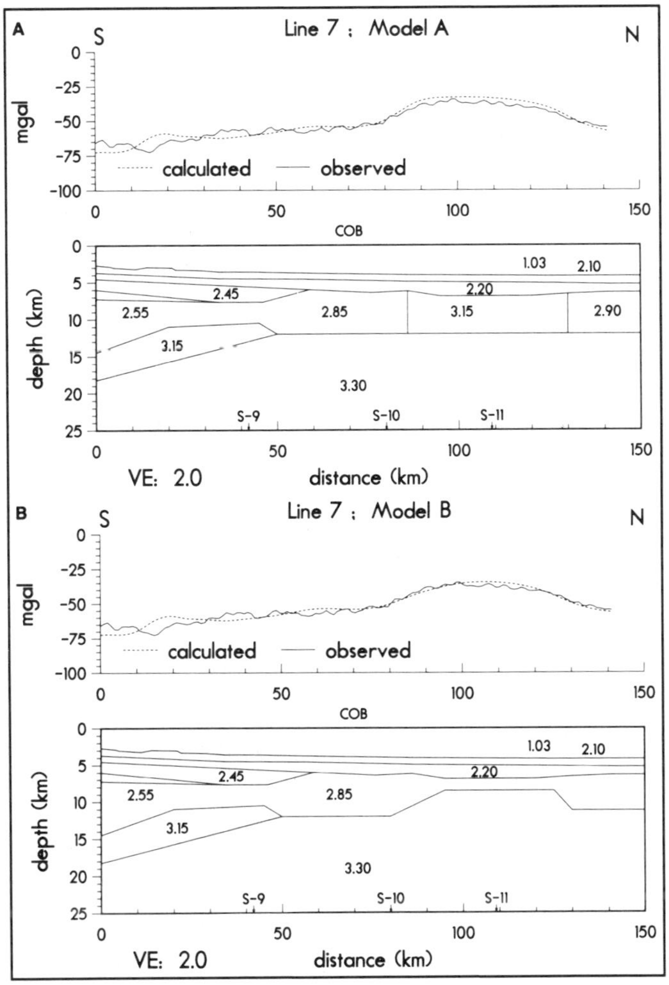

Figure 10—A. Gravity model along line 7. Densities in g/cm^3. This model assumes that the positive FAA at the northern end of the line is due to a crustal body of high density. Locations of sonobuoy stations for control are shown by small arrows at the bottom of the model. Average crustal density of 2.85 g/cm^3; 3.3 g/cm^3 body extends to a depth of 29 km. B. Gravity model along line 7. Densities in g/cm^3. This model assumes that the positive FAA at the northern end of the line is due to the shallowing Moho. Average crustal density of 2.85 g/cm^3; 3.3 g/cm^3 body extends to a depth of 29 km.

tions from the Australian margin opposite Wilkes Land were primarily of the Group I type.

Talwani et al. (1979) proposed that the Australian MQZ is characterized by a variable velocity structure, being neither typically oceanic nor typically continental. They suggested that the quiet zone crust was formed in a 150–200 km wide rift valley and that, by the time of breakup, the crust had evolved to a unique form that they referred to as "rift valley crust." The occurrence of a corresponding magnetic quiet zone on the Wilkes Land margin certainly suggests a similar underlying crustal structure (König, this volume).

The sonobuoy data here do not closely resemble the data from the Australian margin, but the precise placement on the COB from seismic-reflection data allows the location of a sonobuoy station relative to the COB with reasonable certainty. The results from the oceanic sonobuoy stations do not represent typical oceanic sections. The velocities from the main crustal layer are generally lower than would be expected (6.0–6.9 km/sec), though it has already been suggested that the velocities of 6.0 and 6.1 km/sec may originate from layer 2c rather than from layer 3.

The majority of our sonobuoys are from the marginal rift-basin crust immediately landward of the COB. The seismic-reflection data indicate that the stations shown in Figure 5B all lie in this complex zone. The variety of crustal velocities reinforces this apparent complexity and points to a nonuniform composition for the marginal rift basin crust. At stations 9 and 24 the structure is apparently continental; at stations 1, 7, and 21 it resembles oceanic crust; while stations 4, 18, and 25 show an unusual structure with the main crustal refractor being about 7.3–7.4 km/sec. At station 18, this refractor underlies what may be interpreted as a standard continental layer of velocity 6.0 km/sec, while at the other two stations it appears to directly underlie the sediments.

We propose that the variation in crustal structure observed in the MQZ on the Australian margin in Talwani et al. (1979) is also seen in this data at a smaller scale although the details of the two structures differ.

Figure 11 is a composite crustal section compiled from refraction profiles of several lines combined with the three gravity models. This profile is comparable to similar profiles from the conjugate Australian margin (König and Talwani, 1977, figure 8c; Fraser and Tilbury, 1979, figure 20). The width of the main sedimentary accumulation is similar in all cases, and on both margins it is bounded landward by a probable major fault or hingeline near the present-day shelf edge and oceanward by a basement ridge near the COB. However, the Australian margin has the appearance in places (particularly the Ceduna Plateau) of a classical sediment-rich Atlantic-type margin, particularly prior to breakup, whereas the Wilkes Land margin appears to have been relatively starved for sediment.

Falvey and Mutter (1981) provided a summary of the possible driving mechanisms for passive-margin subsidence, with particular reference to the margins of Australia. They presented a generalized cross section of an Australian-type passive continental margin. It is reasonable to extend such a model to the conjugate margin of Antarctica. Falvey and Mutter's (1981) favored model was based primarily on an earlier model by Falvey (1974), refined by Falvey and Middleton (1981). The model proposes that subsidence is driven by an increase in density, or contraction of the lower part of the crust, that results from large-scale metamorphism of greenschist-grade rocks to amphibolite grade or higher. This metamorphism is generated by heat from below the lithosphere moving into the lower crust. They further surmised that crustal extension was a significant subsidence-inducing factor only in the region close to the COB.

We believe that our gravity models, in combination with the seismic refraction data and the deep reflection data, lend strong support to this model. The model of line 3, with a single, thinned crustal layer extending 40 km landward of the COB, is compatible with crustal extension. It may also be argued by proponents of crustal metamorphism that the metamorphosed zone proposed here is unreasonably confined to a local zone near the COB. The gravity models could also accommodate a dense, deep crustal layer that extended landward of the southern end of the modeled profiles and that fit

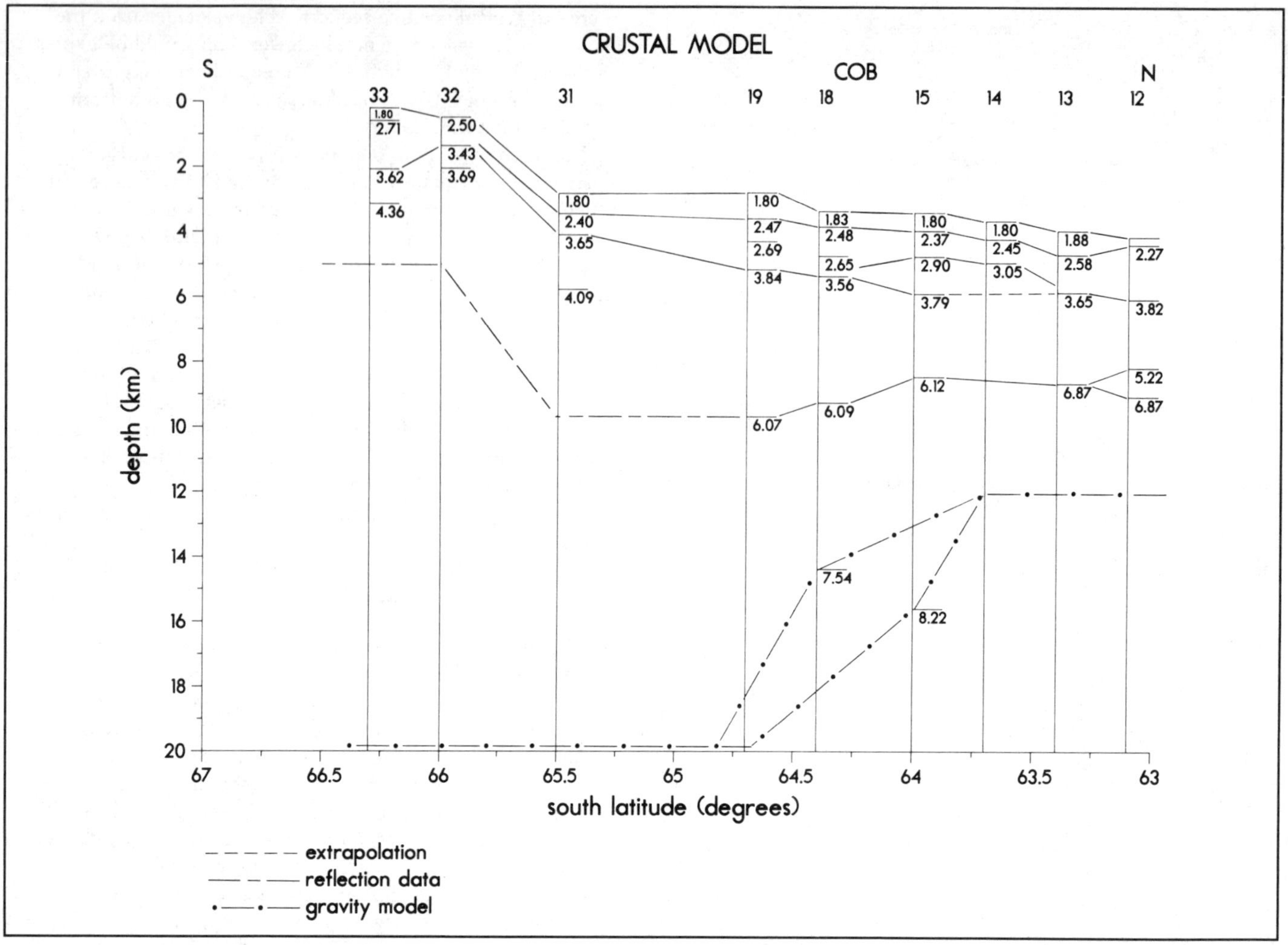

Figure 11—Composite structural profile of the Wilkes Land margin based primarily on selected sonobuoy stations (solid lines) and gravity models, supplemented by seismic reflection profiles.

the observed data equally well. Our only reason for not so doing is the lack of direct evidence in our data for such a deep layer.

CONCLUSIONS

Seismic-reflection, refraction, and gravity data from the Wilkes Land margin provide the basis for models of the crustal structure over the continent–ocean boundary at the passive margin of Antarctica. The model thus derived is characterized by extreme thinning of the crust for approximately 35 km landward of the COB, underlain by a body with density (3.15 g/cm^3) intermediate between crust and mantle rock. The marginal rift basin crust is also characterized by variable refraction velocities, indicating considerable inhomogeneity. The reflection and refraction data vary considerably from corresponding data along the southern margin of Australia, but these differences may arise from small and inconsistent sampling, because in other respects the two margins are similar.

APPENDIX

Geophysical data from selected multichannel seismic lines. Each line is represented by three plots:

1. Top: Sonobuoy refraction data plotted over line drawings of sea floor and horizons T, K1, K2, V, and M interpreted from multichannel seismic sections (Eittreim and Smith, this volume). Sonobuoy station numbers are identified at the top. MT and COB represent locations of the magnetic trough and continent–ocean boundary (Fig. 2). Refraction velocity solutions from thin layers have been deleted for legibility.
2. Middle: Gravity profiles with free-air gravity anomaly as darker line, and Eötvös correction as lighter line.
3. Bottom: Magnetic profiles with reconstructed gradiometer data as darker line, and single sensor magnetometer as lighter line.

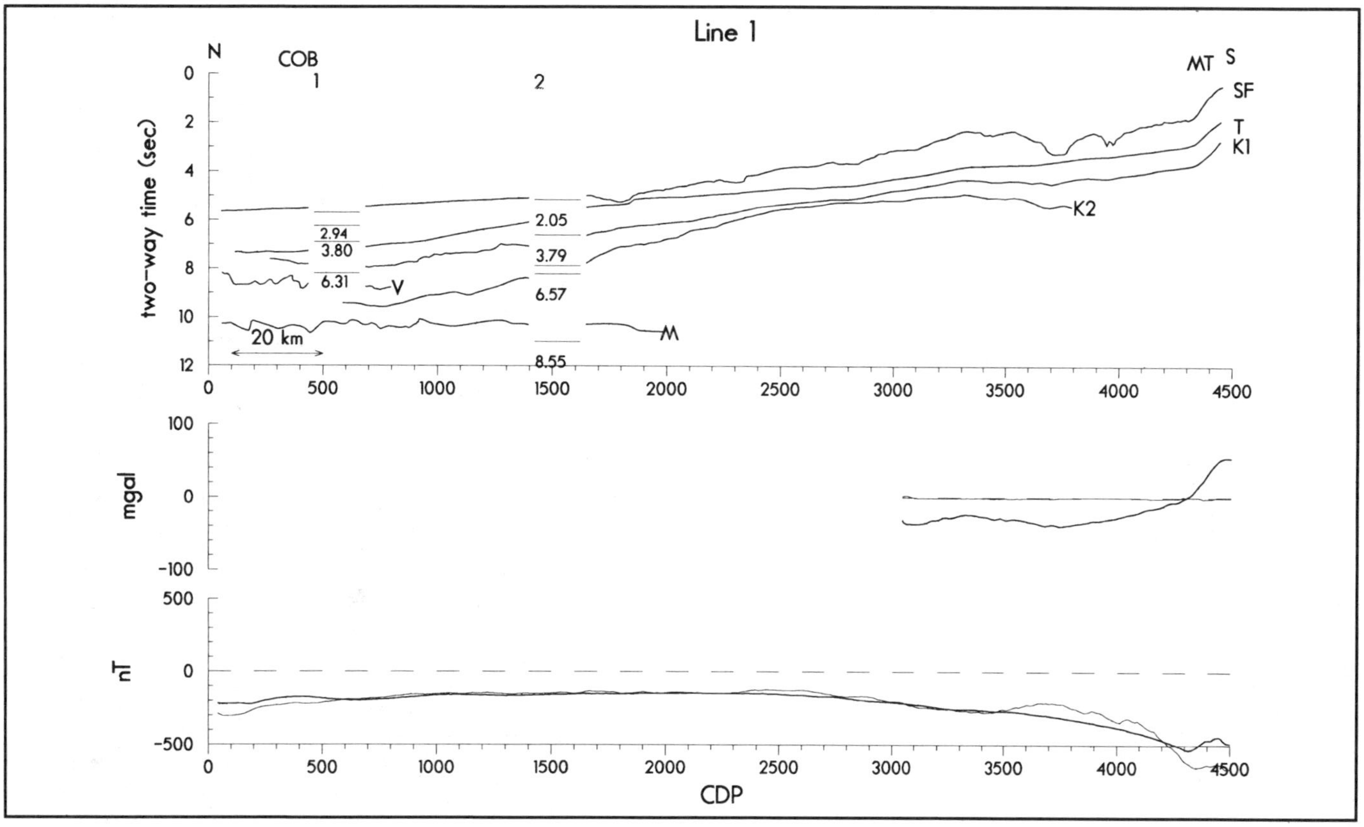

Appendix 1

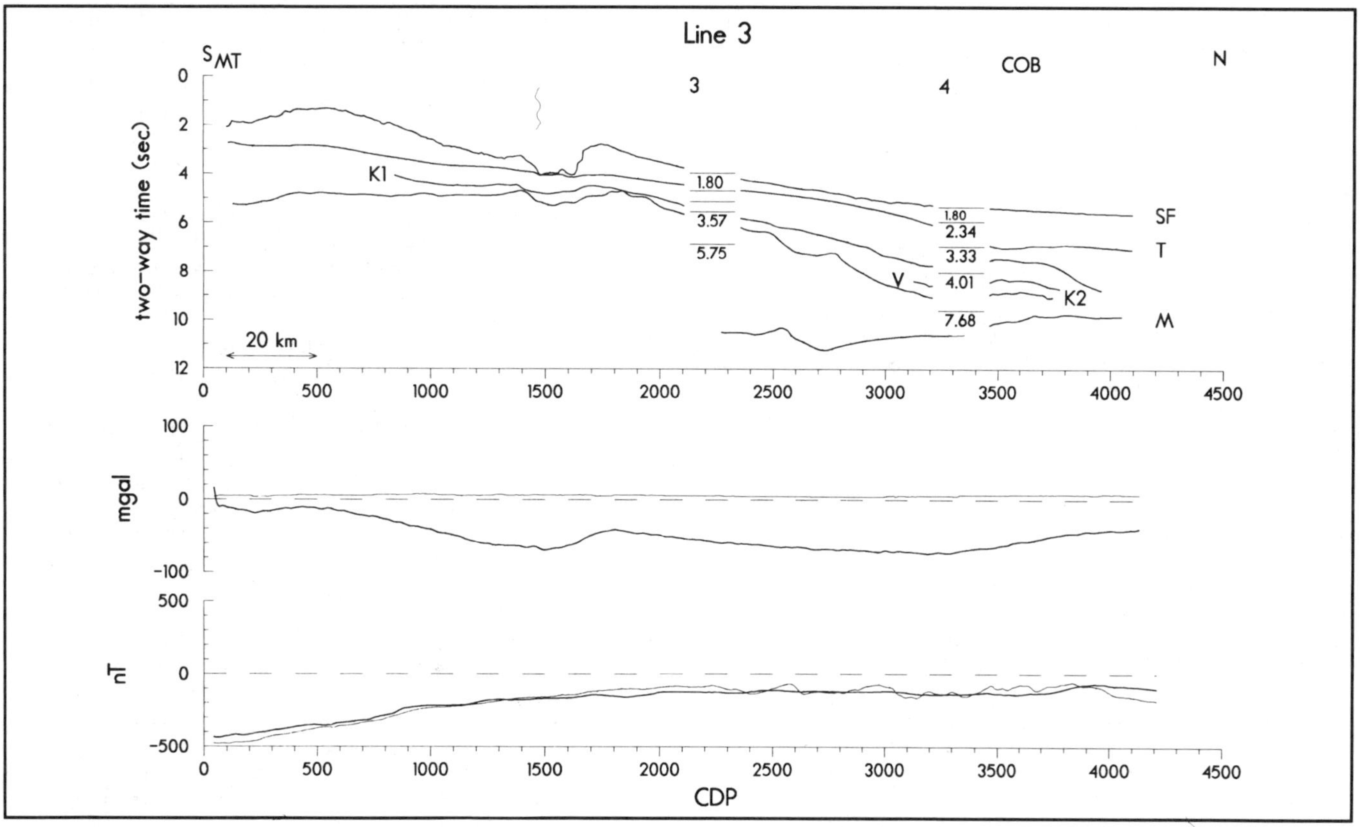

Appendix 2

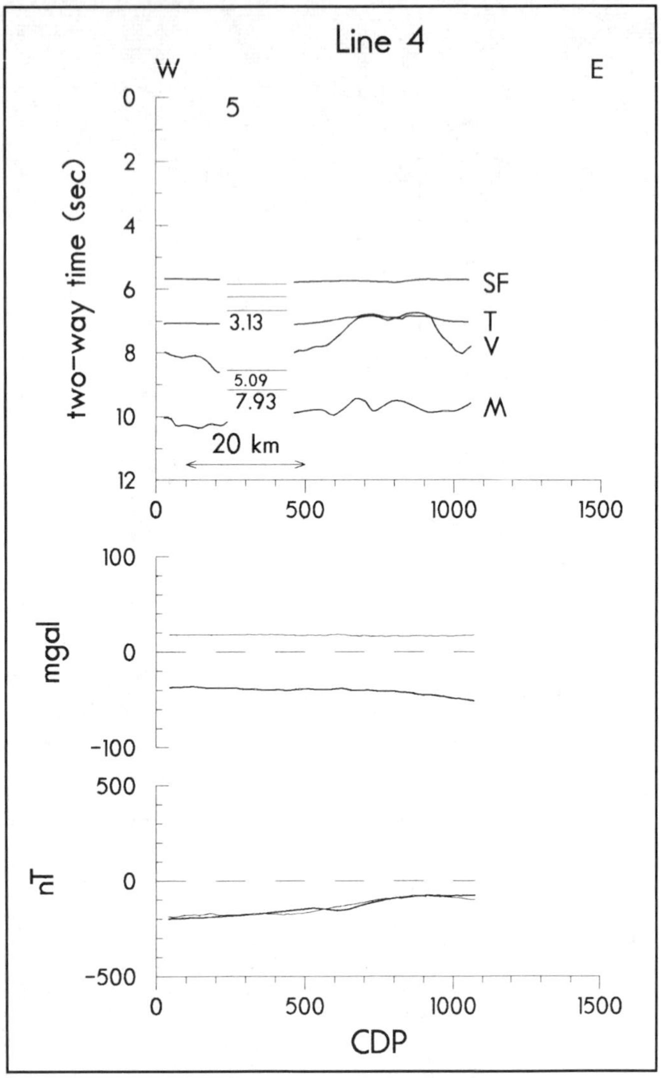

Appendix 3

ACKNOWLEDGMENTS

We would like to thank Steve Eittreim for his collaboration in the interpretation and comparison of the seismic reflection records; Guy Cochrane and Steve May of the U.S. Geological Survey; and Michael König and Robert Houtz for their critical reviews of the manuscript.

REFERENCES CITED

Behrendt, J. C., A. K. Cooper, and A. Yuan, in press, Interpretation of marine magnetic gradiometer and multichannel seismic-reflection observations over the western Ross Sea shelf, Antarctica, *in* A. K. Cooper and F. J. Davey, eds., The Antarctic continental margin: Geology and geophysics of the Western Ross Sea: Circum-Pacific Council for Energy and Mineral Resources Earth Science Series, v. 5B.

Boeuf, M. G., and H. Doust, 1975, Structure and development of the southern margin of Australia: Australian Petroleum Exploration Association Journal, v. 15, p. 33–43.

Cande, S. C., and J. C. Mutter, 1982, A revised identification of the oldest sea-floor spreading anomalies between Australia and Antarctica: Earth and Planetary Science Letters, v. 58, p. 151–160.

Childs, J. C., and A. K. Cooper, 1978, Collection, reduction, and interpretation of marine seismic sonobuoy data: U.S. Geological Survey Open File Report 78-442.

Cooper, A. K., F. J. Davey, and G. R. Cochrane, in press, Structure of extensionally rifted crust beneath the western Ross Sea and Iselin Bank, Antarctica, from sonobuoy seismic data, *in* A. K. Cooper and F. J. Davey, eds., The Antarctic continental margin: Geology and geophysics of the western Ross Sea: Circum-Pacific Council for Energy and Mineral Resources Earth Science Series, v. 5B.

Dobrin, M. B., 1960, Introduction to geophysical prospecting: New York, McGraw-Hill, 446 p.

Falvey, D. A., 1974, Continental margin development and plate tectonic theory: Australian Petroleum Exploration Association Journal, v. 14, p. 95–106.

Falvey, D. A., and M. F. Middleton, 1981, Passive continental margins: Evidence for a prebreakup deep crustal metamorphic subsidence mechanism: Oceanologia Acta, v. 3, p. 103–114.

Falvey, D. A., and J. C. Mutter, 1981, Regional plate tectonics and the evolution of Australia's passive continental margins: Australian Bureau of Mineral Resources Journal, v. 6, p. 1–29.

Fraser, A. R., and L. A. Tilbury, 1979, Structure and stratigraphy of the Ceduna Terrace region, Great Australian Bight: Australian Petroleum Exploration Association Journal, v. 19, p. 53–65.

Hansen, R. O., 1984, Two approaches to total field reconstruction from gradiometer data: Society of Exploration Geophysicists expanded abstracts for 1984 technical program: Tulsa, Society of Exploration Geophysicists, p. 245–247.

———, 1985, Reconstruction of time-variation free total fields from marine gradiometer data: Sunnyvale, CA, EGG Geometrics Technical Report 29.

Hansen, R. O., and J. R. Childs, in press, The Antarctic continental margin magnetic gradiometer data: Suppression of time variations, *in* A. K. Cooper and F. J. Davey, eds., The Antarctic continental margin: Geology and geophysics of the western Ross Sea: Circum-Pacific Council for Energy and Mineral Resources Earth Science Series, v. 5B.

Hayes, D. E., 1972, Marine geophysics of the Southeast Indian Ocean, *in* D. E. Hayes, ed., Antarctic oceanology. II. The Australian–New Zealand sector: American Geophysical Union, Antarctic Research Series, v. 19, p. 165–196.

Houtz, R. E., 1974, Preliminary study of global sound velocities from sonobuoy data, *in* L. Hampton, ed., Physics of sound in marine sediments: New York, Plenum Press, p. 519–535.

———, 1981, Comparison of sediment sound-velocity functions from conjugate margins: Geological Society of America Bulletin, v. 92, p. 262–267.

Houtz, R. E., and J. Ewing, 1976, Upper crustal structure as a function of plate age: Journal of Geophysical Research, v. 81, p. 2490–2498.

Houtz, R. E., and R. G. Markl, 1972, Seismic profiler data between Antarctica and Australia, *in* D. E. Hayes, ed., Antarctic oceanology. II. The Australian–New Zealand sector: American Geophysical Union, Antarctic Research Series, v. 19, p. 147–164.

International Association of Geodesy, 1971, Geodetic reference system 1967: Special Publication 3, 116 p.

———, 1974, The international gravity standardization net 1971 (IGSN-71): Special Publication 4, 194 p.

IAGA Division 1 Study Group, 1976, International geomagnetic reference field: Transactions, American Geophysical Union, v. 57, p. 120–121.

König, M., 1980, Geophysical investigations of the southern continental margin of Australia and the conjugate sector of East Antarctica: Ph.D. Thesis, Columbia University, New York, 337 p.

König, M., and M. Talwani, 1977, A geophysical study of the southern continental margin of Australia; Great Australian Bight and western sectors: Geological Society of America Bulletin, v. 88, p. 1000–1014.

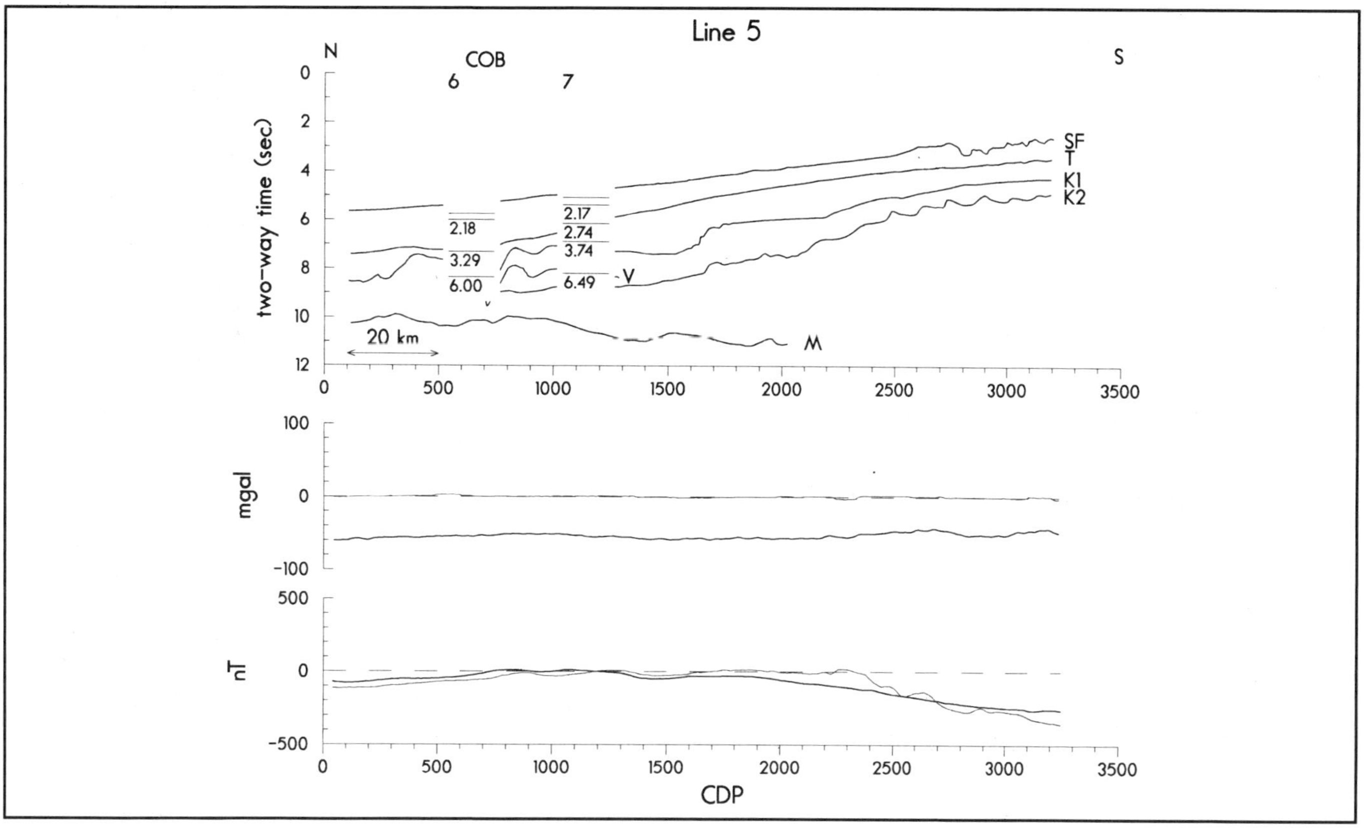

Appendix 4

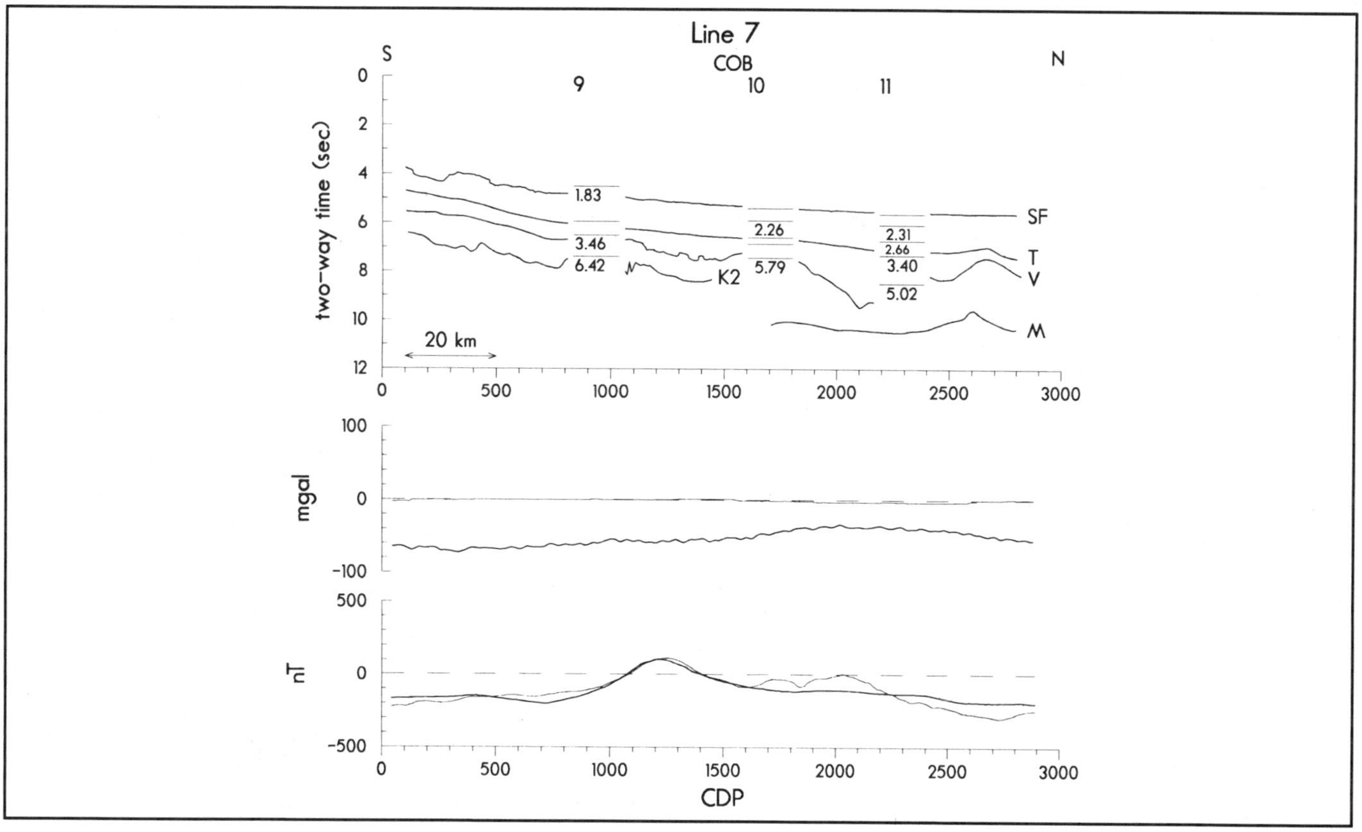

Appendix 5

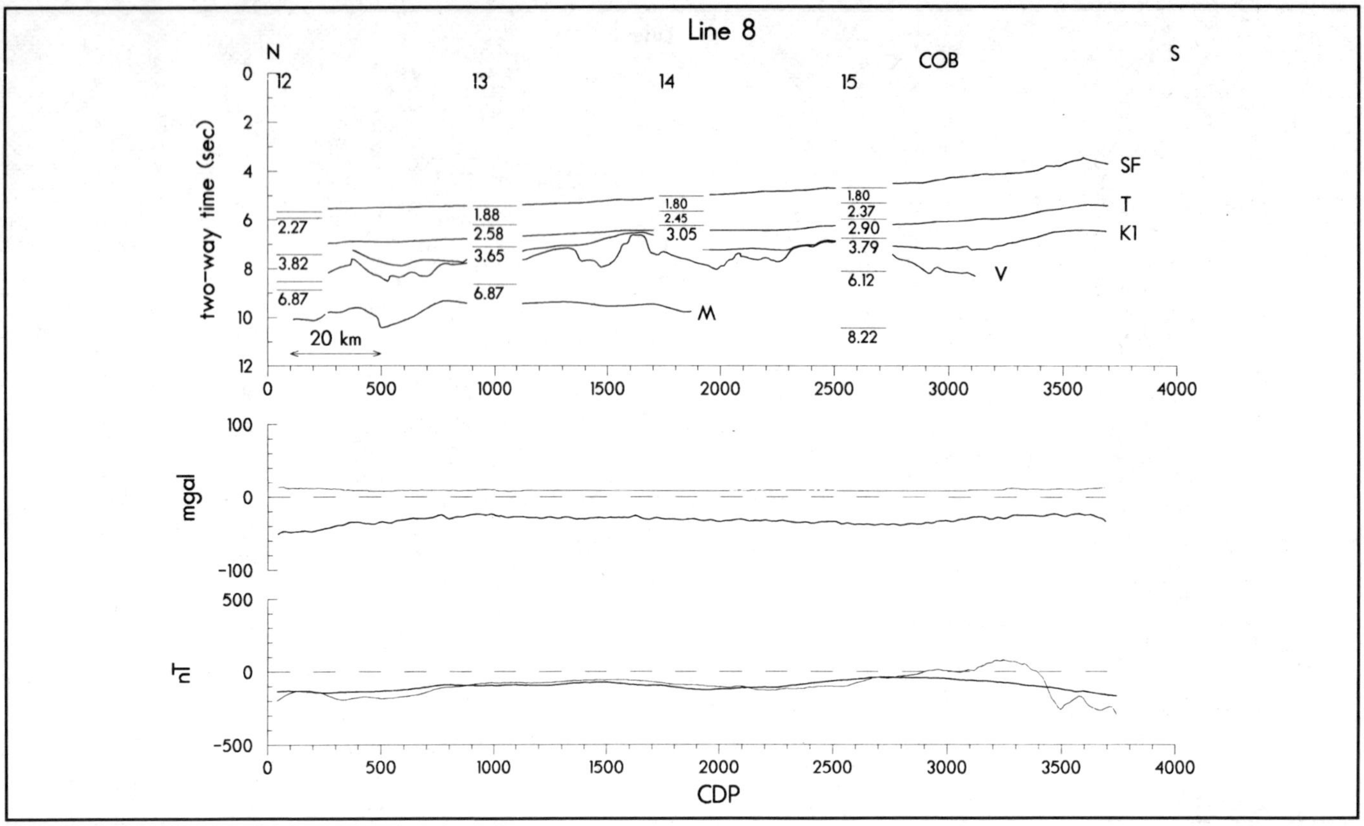

Appendix 6

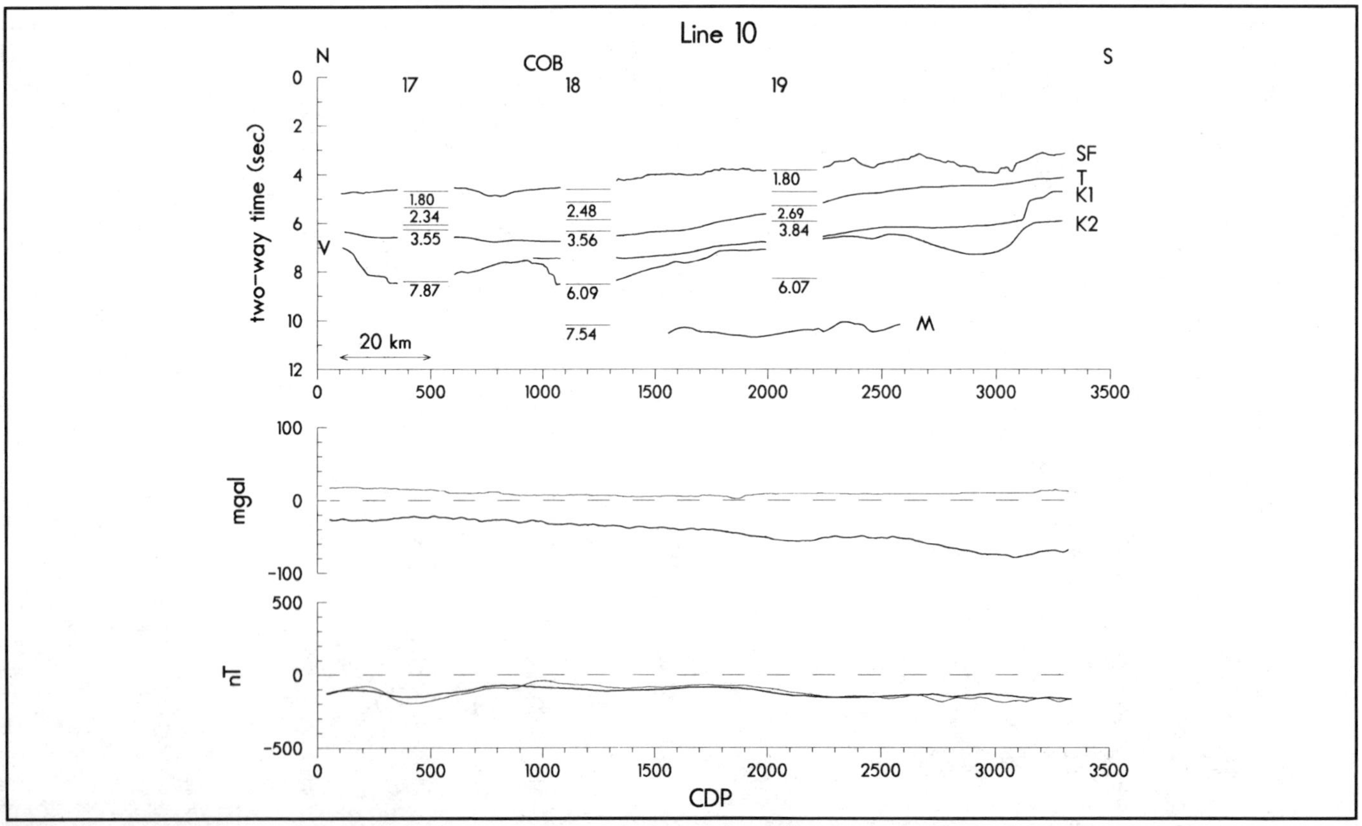

Appendix 7

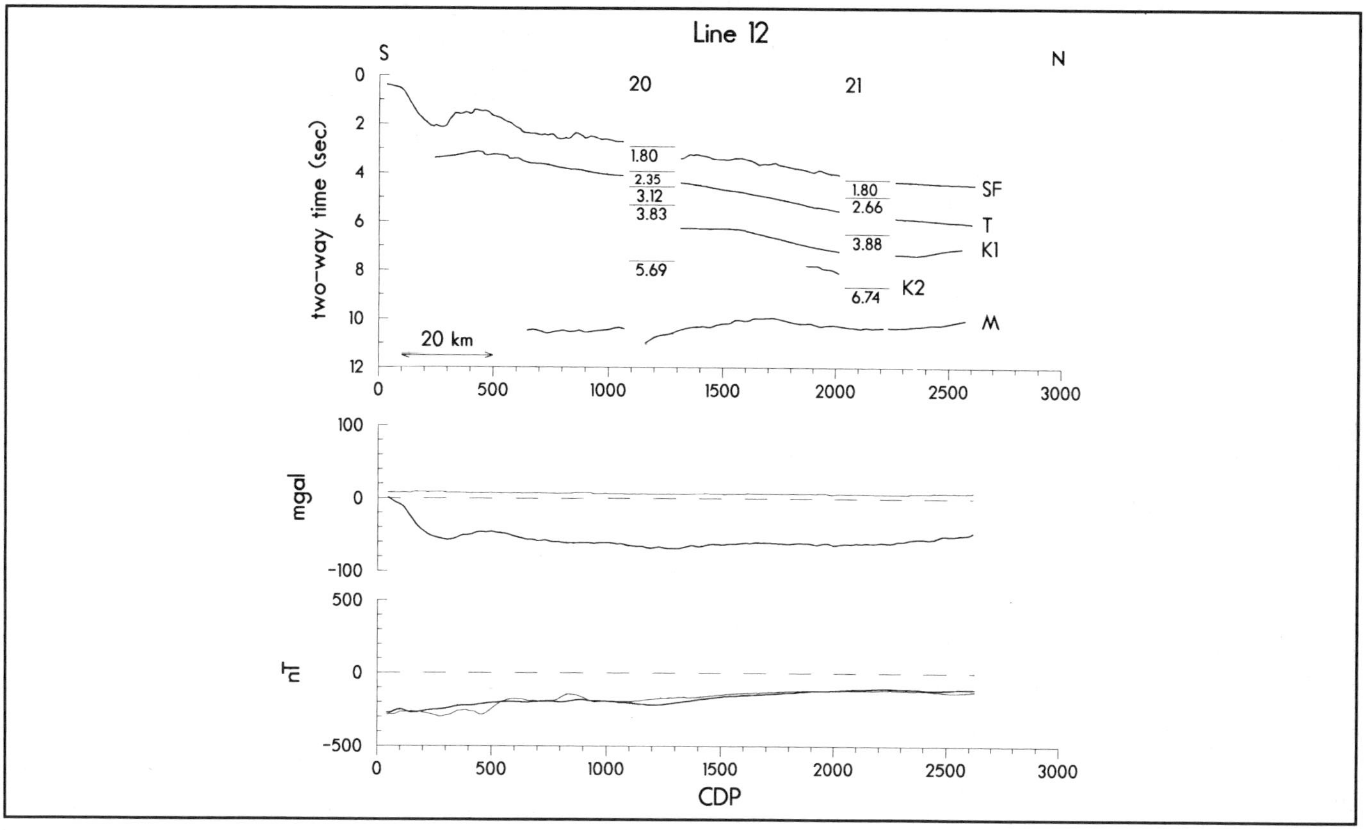

Appendix 8

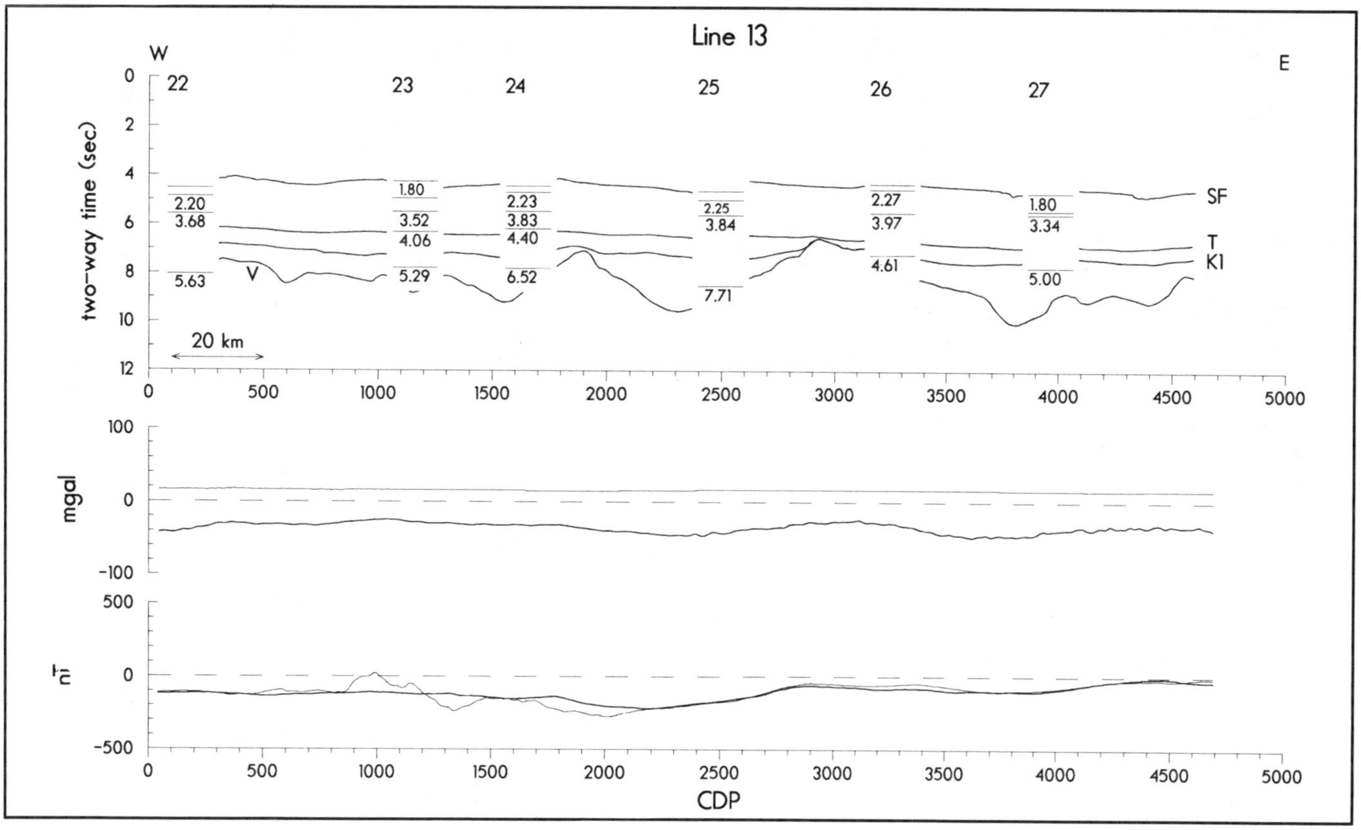

Appendix 9

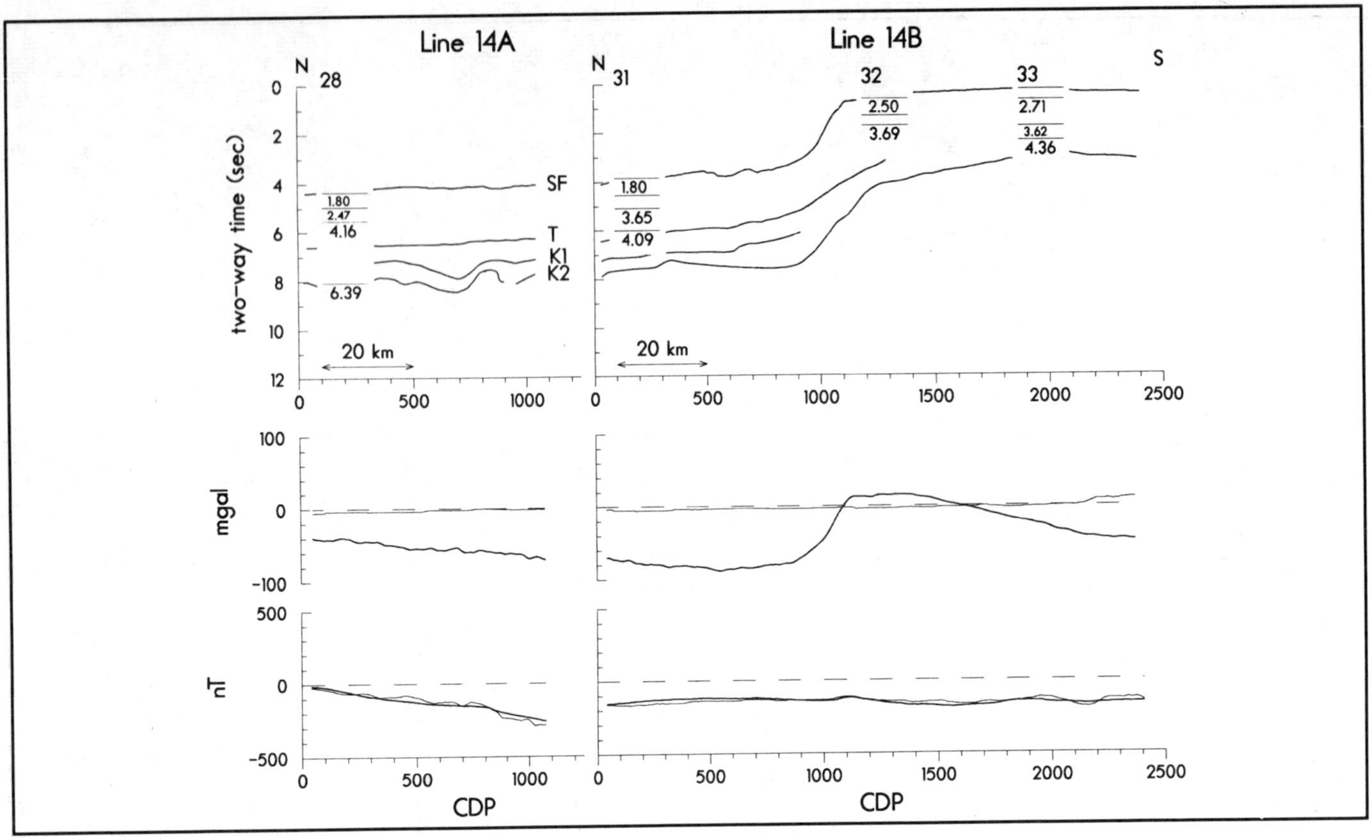

Appendix 10

Le Pichon, X., J. Ewing, and R. E. Houtz, 1968, Deep-sea sediment velocity determination made while reflection profiling: Journal of Geophysical Research, v. 73, p. 2597–2614.

Mutter, J. C., 1978, An analysis of the Mesozoic-early Cenozoic rifting history of Australia and Antarctica: M. Sc. Thesis, University of Sydney, Sydney, Australia, 223 p.

Press, F., 1966, Seismic velocities, *in* S. P. Clark, ed., Handbook of physical constants: Geological Society of America Memoir 97, p. 195–218.

Saltus, R. W., and R. J. Blakely, 1983, Hypermag: An interactive two-dimensional gravity and magnetic modelling program: U.S. Geological Survey Open-File Report 83-241, 28 p.

Talwani, M., J. C. Mutter, R. E. Houtz, and M. König, 1979, The crustal structure and evolution of the area underlying the magnetic quiet zone on the margin south of Australia, *in* J. S. Watkins, L. Montadert, and P. W. Dickerson, eds., Geological and geophysical investigations of continental margins: American Association of Petroleum Geologists Memoir 29, p. 151–175.

Tsumuraya, Y., M. Tanahashi, T. Saki, T. Machihara, and N. Asakura, 1985, Preliminary report of the marine geophysical and geological surveys off Wilkes Land, Antarctica in 1983–1984: Japan National Institute of Polar Research Memoir 37, p. 48–62.

Veevers, J. J., 1982, Australian–Antarctic depression from the mid-ocean ridge to adjacent continents: Nature, v. 295, p. 315–317.

______, 1986, Breakup of Australia and Antarctica estimated as mid-Cretaceous (95 ± 5 Ma) from magnetic and seismic data at the continental margin: Earth and Planetary Science Letters, v. 77, p. 91–99.

Weissel, J. K., and D. E. Hayes, 1972, Magnetic anomalies in the Southeast Indian Ocean, *in* D. E. Hayes, ed., Antarctic oceanology. II. The Australian–New Zealand sector: American Geophysical Union, Antarctic Research Series, v. 19, p. 165–196.

Willcox, J. B., 1978, The Great Australian Bight: A regional interpretation of gravity, magnetic, and seismic data from the continental margin survey: Australian Bureau of Mineral Resources Report 201.

Eittreim, S. L., and Hampton, M. A., 1987, *The Antarctic Continental Margin: Geology and Geophysics of Offshore Wilkes Land*, CPCEMR Earth Science Series, v. 5A: Houston, Texas, Circum-Pacific Council for Energy and Mineral Resources

Geophysical Data from the Continental Margin off Wilkes Land, Antarctica—Implications for Breakup and Dispersal of Australia-Antarctica

M. König
Columbia University
New York, New York

Geophysical data collected in the Southeast Indian Ocean by the U.S. R/V *Eltanin* are presented and used, together with the results from *Glomar Challenger* leg 28, in a study of the continental margin off Wilkes Land, Antarctica.

A distinctive pattern of geophysical anomalies delimits and characterizes a magnetic quiet zone situated over the continental slope and rise. The data document that the quiet zone is fundamentally different from the ocean basin to the north and the continent to the south.

The pattern of geophysical anomalies off Wilkes Land, Antarctica, shows a remarkable degree of similarity to the pattern of gravity, magnetics, and seismic data observed on the southern continental margin of Australia. This similarity is the basis for extending the interpretation of the Australian quiet zone as overlying the remnant of a fossil rift system to the conjugate quiet zone off Wilkes Land.

On each of the margins of the Southeast Indian Ocean, the oldest seafloor-spreading magnetic lineation is anomaly 34, originally identified as anomaly 22. This isochron lies close to the seaward edge of the magnetic quiet zone and is parallel to it. The lineation allows a new pole of rotation to be determined for the predrift fit of Australia and Antarctica; this pole is located at lat 1.5° N, long 37.0° E, with a rotation angle for the closing of the Southeast Indian Ocean of 28.0°.

The new reconstruction is less tight than those previously proposed, for it includes the 250 km wide rift zone as it existed just prior to continental breakup. The fit further avoids any overlap of continental material and fully satisfies the geologic constraints governing the reassembly of Australia and Antarctica.

Finite-rotation parameters were determined for each of the younger isochrons and used to trace the first 38 Ma of northward drift of Australia following the onset of seafloor spreading. The magnetic data indicate the diachronous opening of the Southeast Indian Ocean from west to east. The geologic evidence pertaining to the evolution of the southern continental margin of Australia supports the history of seafloor spreading as deduced from the marine magnetic lineations.

INTRODUCTION

The earliest proposed continental reconstructions of Gondwanaland, e.g., those of Wegener (1929) and du Toit (1937), showed southern Australia juxtaposed against East Antarctica (for location see Fig. 1), in a predrift configuration that appears intuitively acceptable and geologically feasible. More recent computer methods of fitting selected isobaths, such as employed by Sproll and Dietz (1969), generally support the previously proposed rejoining of Australia and Antarctica, as does Smith and Hallam's (1970) reassembly of Gondwanaland, which is primarily based on geological data. Using seafloor-spreading magnetic lineations, Weissel and Hayes (1971, 1972) presented evidence for the early Eocene separation of Australia and Antarctica about 52 Ma (the age of anomaly 22 on the Kent and Gradstein [1986] polarity-reversal time scale), which was subsequently revised by Cande and Mutter (1982) to a Late Cretaceous age of, most likely, approximately 95 Ma (Veevers, 1986), with an estimated uncertainty of ±5 Ma. Weissel and Hayes (1971, 1972) also reported the existence of magnetic quiet zones off both southern Australia and the conjugate sector of Antarctica. Such quiet zones are commonly observed at rifted continental margins (Poehls et al., 1973). Together with the observed marine magnetic lineations and the evident lack of crustal subduction along either the southern Australian or the conjugate Antarctic margin, they corroborate the much earlier expressed and then more speculative notion of predrift contiguity of Australia and Antarctica.

While recent studies of the southern continental margin of Australia (Fig. 1), notably those of Boeuf and Doust (1975), Deighton et al. (1976), König and Talwani (1977), Talwani et al. (1979), and Hegarty (1985) have shed light on the structure and pre-seafloor-spreading evolution of the area occupied by the magnetic quiet zone, very little comparable work has been done on the corresponding region off the Antarctic conti-

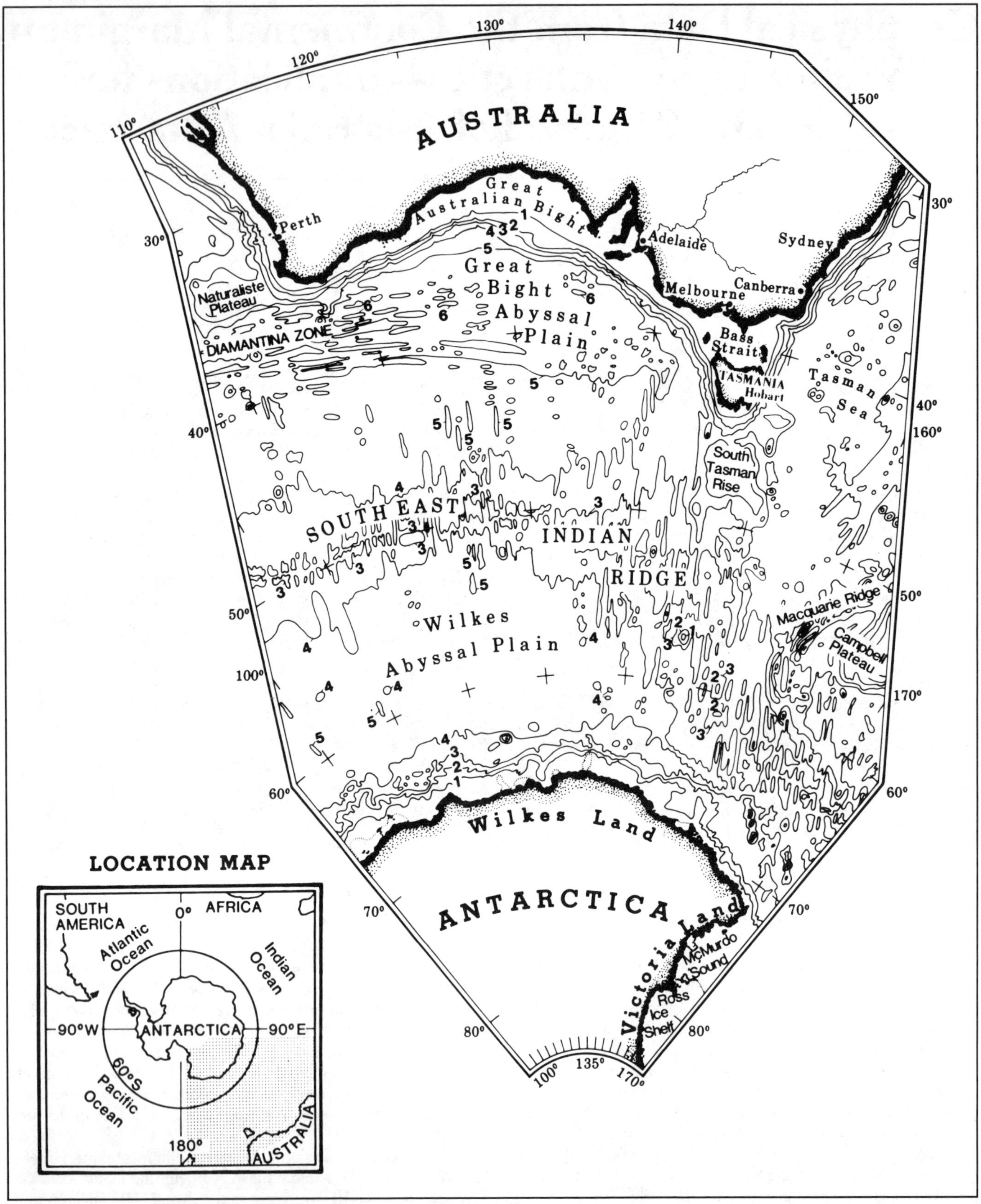

Figure 1—Map showing bathymetry of the Southeast Indian Ocean and geography of its margins (polar stereographic projection). Depth contours are labeled in units of kilometers; prominent offshore physiographic features are identified, as are the major cities of Australia and the main geographic regions of the sector of Antarctica shown. The Diamantina Zone off southwestern Australia is an area of very rugged oceanfloor topography.

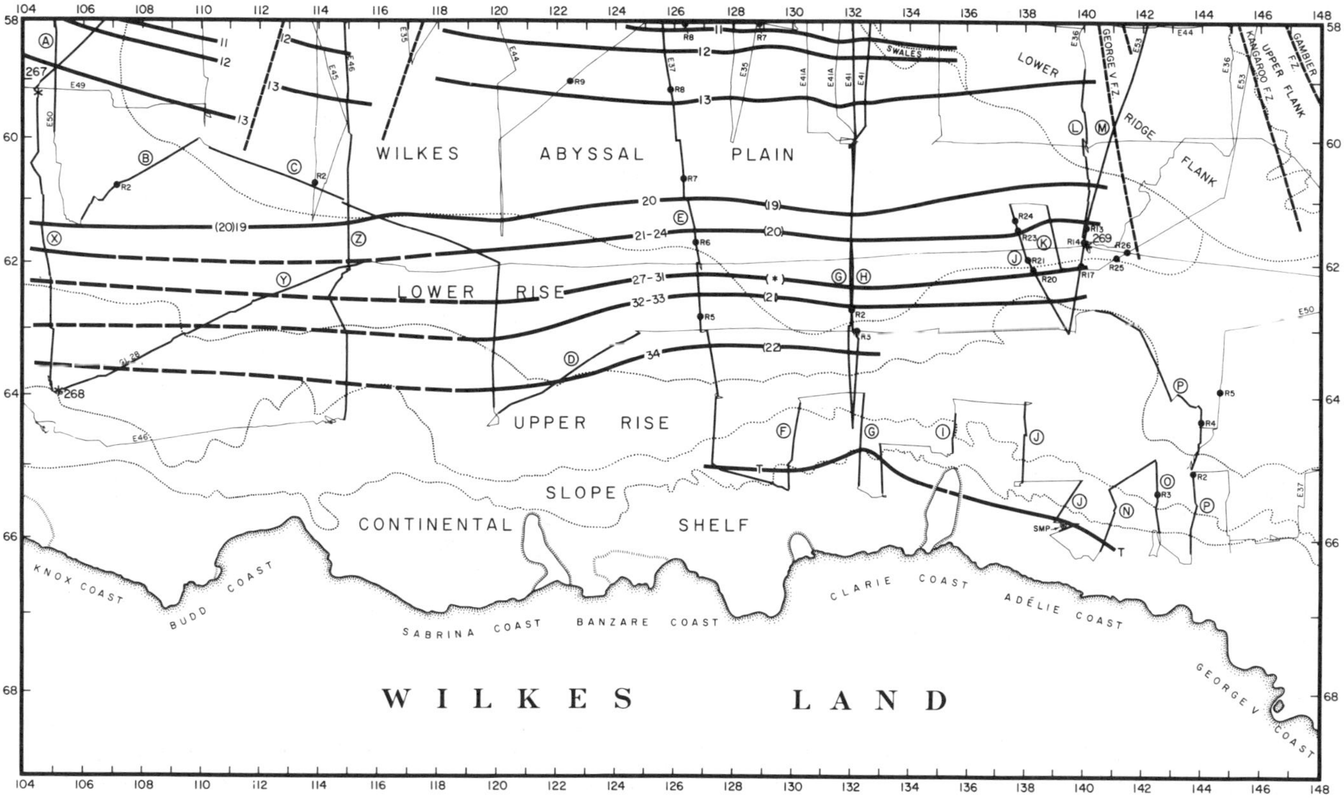

Figure 2—Track chart of the continental margin off Wilkes Land, Antarctica (Mercator projection). The numbered solid lines, dashed where correlations are tentative, represent marine magnetic anomalies identified according to the Kent and Gradstein (1986) time scale of geomagnetic reversals. Anomaly identifications are as proposed by Cande and Mutter (1982) with the original designations of Weissel and Hayes (1972), extended by König (1980), enclosed in parentheses. The heavy line labeled "T" indicates the magnetic trough on the continental shelf and slope. Ships' tracks are shown as light and heavy solid lines, the lettered heavy portions identifying profiles shown in Figures 3, 4, and 5. Sonobuoy stations are indicated by large dots and identified by numbers preceded by "R." Large, numbered asterisks represent deep-sea drilling sites (*Glomar Challenger* leg 28), and the small encircled asterisk designated "SMP" marks the location of the South Magnetic Pole (epoch 1975.0) as taken from the geomagnetic-field chart published by the Defense Mapping Agency Hydrographic Center (1975). Ship identifications refer to U.S.N.S. *Eltanin* (E) and D. V. *Glomar Challenger* (GL). Small dots indicate the boundaries of physiographic provinces according to Hayes and Conolly (1972) and Hayes and Wall (1975). Ice or glacier tongues, ice shelves, and the coastline of Antarctica are shown as represented on the Geologic Map of Antarctica (Craddock, 1972a).

nent (Fig. 1), mostly because of its remoteness and hostile climate. During the approximately 4½ year long operation of the research vessel *Eltanin* in the Southeast Indian Ocean, cruises in high southern latitudes were undertaken that also ventured onto the continental margin off Wilkes Land, Antarctica. These cruises and *Glomar Challenger* leg 28 are shown in Figure 2, together with the boundaries of the physiographic provinces of the area (taken from Hayes and Conolly, 1972, and Hayes and Wall, 1975), marine magnetic lineations, DSDP sites, and sonobuoy-station locations indicated by numbers preceded by "R." Several types of geophysical data were obtained on all tracks (except for the *Eltanin* 46 cruise for which only bathymetry is available) and form the basis of this study. Some of these data have already been presented by Houtz and Markl (1972), Houtz (1974), Talwani et al. (1979), Cande and Mutter (1982), and Veevers (1986). The results of the latest geophysical survey in this region, the research cruise of R/V *S.P. Lee* in 1984 (Eittreim et al., 1984), are extensively discussed in the remaining papers of this volume.

PRESENTATION OF DATA

The data presented in Figures 3, 4, and 5 consist of magnetics, gravity, and bathymetry profiles, respectively, projected at right angles to the margin. The magnetic anomalies are total-intensity data with either the International Geomagnetic Reference Field 1965.0 (IAGA, 1969) or the one given by Cain et al. (1968) removed. The gravity data consist of free-air and "two-dimensional" Airy-type isostatic anomalies; details of computation and significance to continental-margin studies of these isostatic anomalies were discussed by König and Talwani (1977). In addition to these profiles, seismic sections representing the sonobuoy solutions from the stations indicated in Figure 2 are shown in Figure 6.

The magnetic lineations as originally identified by Weissel and Hayes (1972)—complemented by those data to which their identifications were later extended by König (1980)—and since then revised by Cande and Mutter (1982), are labeled and correlated by solid lines in Figure 2. The dashed lines represent tentative correlations, and the anomaly identification

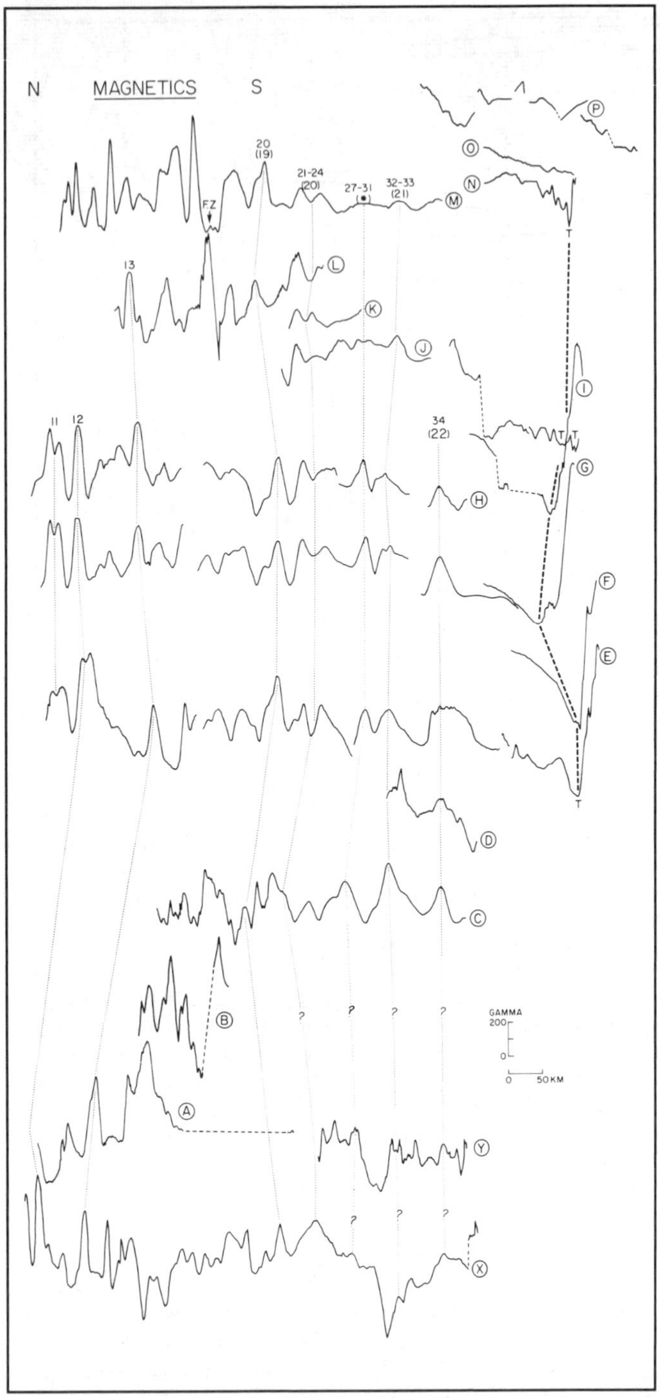

Figure 3—Projected magnetic-anomaly profiles from the Wilkes Land sector of the East Antarctic continental margin. For location of profiles see Figure 2. (Magnetic measurements were not made on *Eltanin* cruise 46.) The azimuth of projection is 180°. The profiles are aligned with respect to anomaly 34(22), as indicated, or lineation 32–33(21), where anomaly 34(22) is not observed. The magnetic-anomaly identifications are the same as in Figure 2; the lineations are shown correlated by dotted lines, with queries indicating tentative correlations. The heavy dashed line correlates the magnetic trough (T) along the margin; "F. Z." denotes the occurrence of a fracture zone on profile M. The magnetic-quiet zone lies north of the magnetic trough and south of isochron 34(22) or, where it is not present, anomaly 32–33(21).

by Weissel and Hayes (1972), with the addition made by König (1980), is shown in parentheses. Both Weissel and Hayes' (1972) and Cande and Mutter's (1982) identifications have not been verified by independent means, such as drilling, but are based on the matching of synthetic and observed magnetic-anomaly profiles. The evidence as discussed by König (1980) and Cande and Mutter (1982) favors the latest proposed identification of the oldest portion of the anomaly sequence. Both anomaly designations are given in this study, however, with the original identifications enclosed in brackets or parentheses, to reflect the uncertainty regarding the age of the earliest oceanic crust in the Southeast Indian Ocean.

Anomaly 34[22] is the oldest identified magnetic lineation (Fig. 2), and although not very well defined because of the small number of cruises sampling it, this isochron does appear to vanish somewhere east of long 132°30′ E, probably near long 135° E based on the latest magnetic data collected on the *S.P. Lee* cruise (Childs and Stagg, this volume). It is not observed on *Eltanin* track P. To the west, lineation 34[22] has only been tentatively recognized on the two *Glomar Challenger* track segments crossing it; Hayes and Wall's (1975) magnetic-anomaly identification on profile X does not extend beyond lineations older than anomaly 21–24[20]. However, DSDP site 268 probably lies in a magnetic quiet zone (Hayes et al., 1975a). Landward of anomaly 34[22] (Fig. 3) lies an area where the magnetic-anomaly field is either greatly subdued or entirely featureless—the magnetic quiet zone. A distinctive trough-shaped magnetic low forms the southern boundary of the quiet zone. The high-amplitude, short-wavelength anomalies sampled landward of the magnetic trough (Fig. 3, profiles E, F, I) are characteristic of shallow continental basement.

Anomaly 21–24[20] and younger isochrons are better surveyed than anomaly 34[22] as shown in Figure 2. In the east, all of them appear to abut against a major north-northwest–south-southeast-trending fracture zone, which is expressed both magnetically and topographically on profile M (Figs. 3, 5) and known to extend northward to the area northwest of Tasmania (Weissel and Hayes, 1972). Magnetic anomalies 34[22] to 11 off Antarctica generally trend east–west parallel to the continental margin to the south and the Southeast Indian Ridge farther north. However, noticeable deviations from an east–west orientation, involving anomalies 32–33[21] and 34[22] in particular, occur near long 125° E. This observation will be reconsidered later in connection with the derivation of a new pole of rotation for the Australia–Antarctica reconstruction.

The greatest variations in the free-air and isostatic gravity fields occur over the magnetic quiet zone and near its boundaries (Fig. 4). A pronounced isostatic gravity low, reaching −30 mgal or more negative values on several profiles, coincides with the quiet zone. Near the seaward boundary, i.e., anomaly 34[22] or lineation 32–33[21], a fairly steep gradient forms the transition from the quiet-zone low to the generally slightly positive gravity anomalies associated with oceanic crust. Local undulations, for the most part unrelated to topography, represent the only variations of the otherwise uniform gravity-anomaly field in this area. The three long profiles E, G, and H of Figure 4 illustrate particularly well the contrast between the generally smooth gravity field in the oceanic region from anomaly 11 south to and including anomaly 32–33[21], and the large variations in the gravity field that

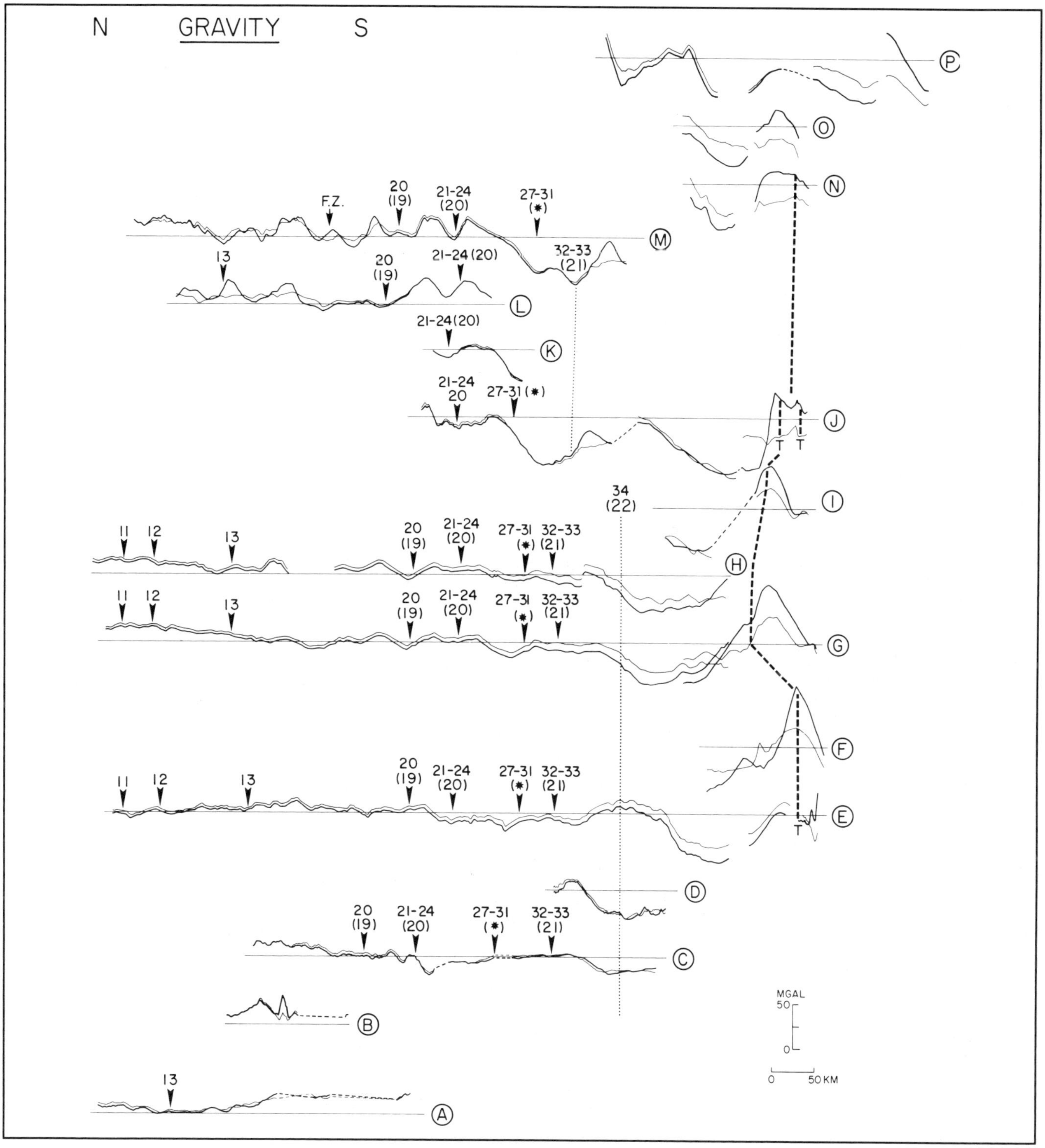

Figure 4—Projected gravity-anomaly profiles from the Wilkes Land sector of the East Antarctic continental margin. For location of profiles see Figure 2. (Gravity measurements were not made on the *Glomar Challenger* cruise or on *Eltanin* leg 46.) The azimuth of projection is 180°. The profiles are aligned with respect to anomaly 34(22) or anomaly 32–33(21) as indicated by dotted lines. Numbered arrows on each profile mark the locations of the magnetic lineations on the corresponding tracks of Figure 3. The heavy dashed line represents the correlation of the magnetic trough (T) along the margin; "F.Z." denotes the occurrence of a fracture zone on profile M. The two gravity anomaly curves shown for each profile represent the observed free-air (heavy trace) and computed two-dimensional isostatic (light trace) gravity anomalies, the latter being based on the Airy-Heiskanen scheme of isostatic compensation. The association of the magnetic trough with a pronounced free-air and isostatic gravity high, the very low gravity field over the magnetic quiet zone, and, particularly where anomaly 34(22) forms the seaward quiet-zone boundary, the existence of a steep gravity gradient (or local high) near the oldest recognized isochron are readily apparent.

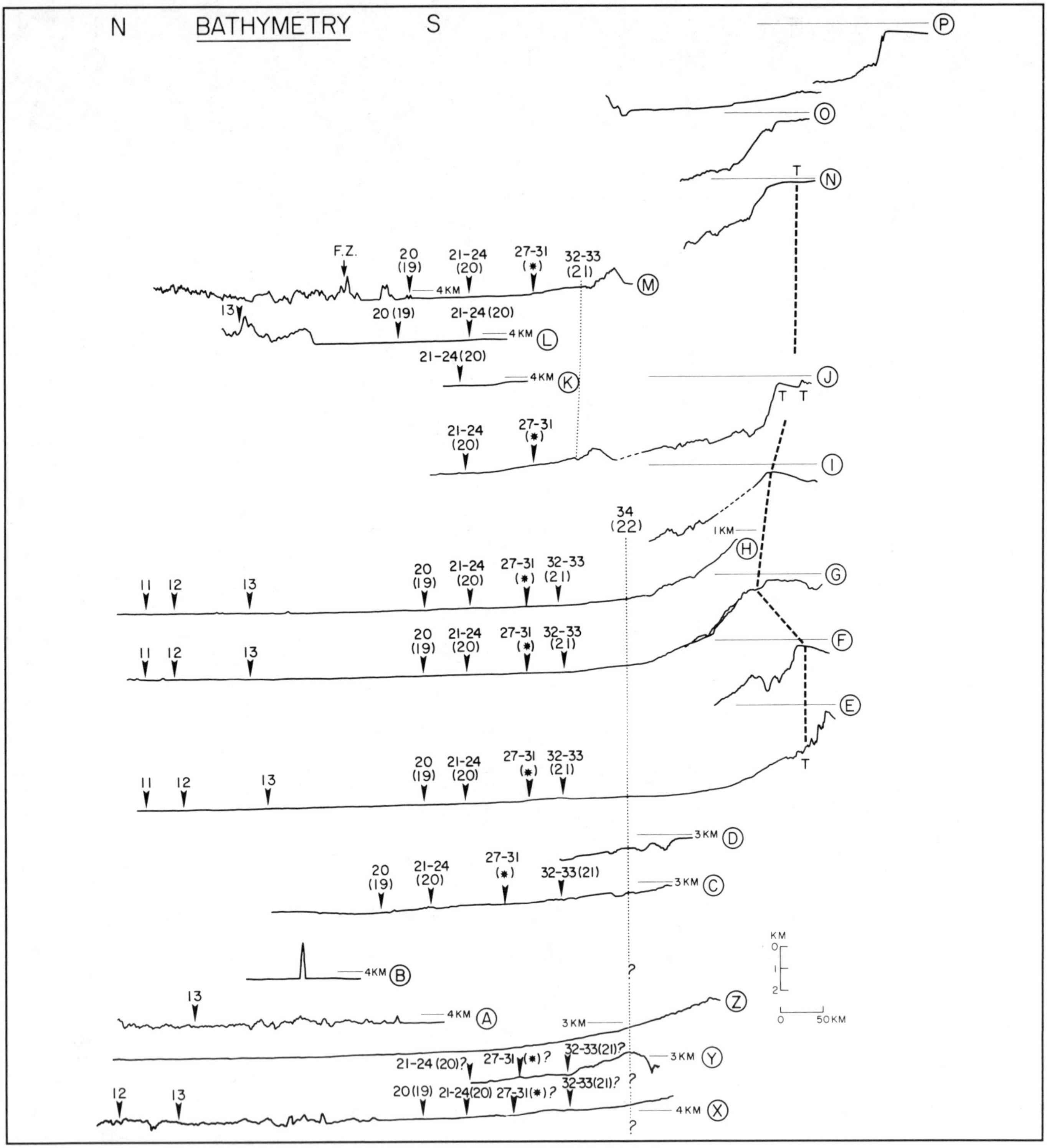

Figure 5—Projected bathymetry profiles from the Wilkes Land sector of the East Antarctic continental margin. For location of profiles see Figure 2. The azimuth of projection is 180°. The profiles are aligned with respect to anomaly 34(22) or anomaly 32-33(21) as indicated by dotted lines. Numbered arrows on each profile mark the locations of the magnetic lineations on the corresponding tracks of Figure 3, queries denoting tentative identifications or correlations. The dashed line represents the correlation of the magnetic trough along the margin; "F.Z." indicates a fracture zone traversed by profile M. Horizontal lines above profiles E, F, G, I, J, N, O, and P denote sea level, and the short line next to each of the remaining profiles is marked as to water depth represented. The shift in position of the magnetic trough from continental shelf to slope towards the west, the lack of diagnostic features associated with the quiet-zone boundaries, except on profiles J and M, and the extremely smooth bathymetry over most of the ocean basin as well as its comparatively shallow depth, both indicative of a high influx of sediments, are easily recognizable.

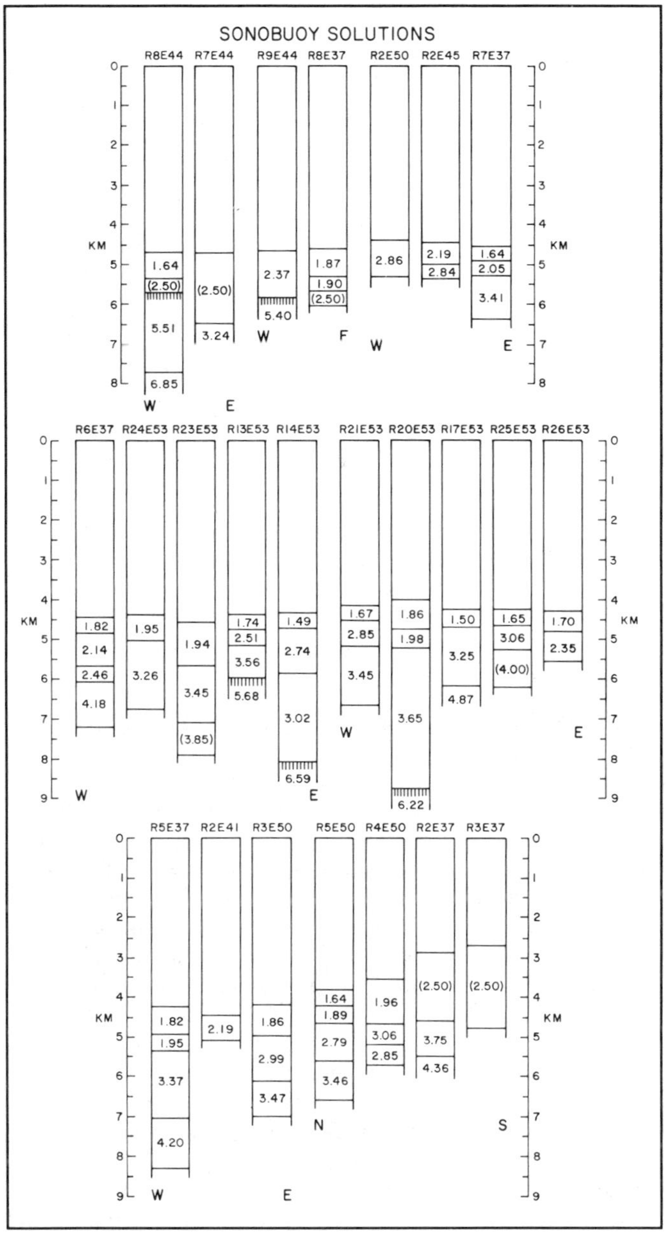

Figure 6—Sonobuoy solutions for the continental-margin stations off Wilkes Land, Antarctica. (See also Table 1.) The locations of the sonobuoy stations are shown in Figure 2. The seismic stations are arranged in three groups according to their position with respect to the observed magnetic lineations off Wilkes Land. Stations north of anomaly 20(19) are shown at the top, those situated south of lineation 20(19) but north of anomaly 32-33(21) are in the center, and the sonobuoy stations landward of this isochron are shown at the bottom. Within each group, subgroups are arranged from north (at the left) to south (at the right), and individual stations of each subgroup are aligned mostly west to east as indicated. Seismic-velocity values are given in units of km/sec; assumed interval velocities are enclosed in parentheses. Hatching denotes the top of basement, and the blank portion of each column represents water. It is apparent that basement was not reached south of anomaly 32-33(21) and that depth to basement and sediment thickness both show a general increase from the ocean basin toward the continent.

begin just south of this isochron and encompass the magnetic quiet zone and its boundaries. The corresponding profiles of Figure 5 in addition clearly demonstrate that both the large gravity variations and the majority of the smaller ones are unrelated to topography.

Figures 2 and 4 show that the gravity gradient at the seaward quiet-zone boundary is situated south of isochron 32–33[21] in the west up to long 132°30′ E but shifts northward to a position between anomalies 32–33[21] and 21–24[20] east of that meridian. Thus, the eastward widening of the magnetic quiet zone, caused by the disappearance of anomaly 34[22], is mirrored in the gravity field. The landward quiet-zone boundary is reflected in the gravity field as a free-air and isostatic gravity high located in the vicinity of the magnetic trough; it is best developed on those profiles (E, F, G, I) that also show the clearest expression of the trough. The isostatic high is generally considerably reduced in amplitude in comparison with the free-air anomaly high thus pointing to the gravity "edge effect" as a significant factor contributing to the large positive gravity anomaly (maximally close to +75 mgal) near the shelf edge.

The bathymetry profiles presented in Figure 5 show no abnormal features associated with either anomaly 34[22] or the magnetic trough. Anomaly 34[22] is located on the continental rise, and the magnetic trough, for the most part, is situated close to the shelf-break and only in one instance on the slope (profile E). Thus, physiographically the magnetic quiet zone comprises a narrow strip of the continental shelf (east of long 133°E), the entire slope, and a sizable portion of the continental rise (Fig. 2). Bathymetrically, the seaward-adjoining area is virtually featureless owing to the extensive sedimentation that has rendered the Wilkes Abyssal Plain about 1 km shallower than the Great Australian Bight Abyssal Plain north of the ridge (Hayes and Conolly, 1972). A gentle mound-like feature that probably represents a sediment ridge occurs on three profiles (J, M, P; Fig. 5) in the area where anomaly 34[22] is not observed. The ridge is not present on profiles farther west and thus appears to be restricted to the area characterized by a northward salient of the upper-rise province (Fig. 2). It is perhaps no coincidence that this widening of the upper rise almost parallels the widening of the magnetic quiet zone east of the meridian where anomaly 34[22] is last observed.

The available seismic-refraction (sonobuoy) data are summarized in Figure 6 and Table 1. Some of these results have already been discussed by Houtz and Markl (1972) and Houtz (1974) together with the information gained from seismic-reflection profiling. Most of the sonobuoy stations indicated in Figure 2 are located over oceanic crust, and none exist landward of anomaly 34[22] in the magnetic quiet zone, but four stations are situated just north of the magnetic trough off Adélie coast. Even in the area underlain by oceanic crust, the sediment cover is generally quite thick and reaches maxima of 1.8 km or greater (at station R7E44) north of anomaly 20[19], 4.8 km (at station R20E53) near isochron 27–31[*], and 4.1 km (at station R5E37) between lineations 32–33[21] and 34[22] (Fig. 2). Basement was reached at a few stations but not at any of those situated south of anomaly 32–33[21]; the velocities encountered are typical of oceanic layers 2 and 3. On the seismic-profiler records, basement is generally not visible beneath the sedimentary cover from the continental

Table 1—Sonobuoy Solutions from the continental margin off Wilkes Land, Antarctica.

Sonobuoy station	Water depth (km)	v_1 (km/sec)	σ_1 (km/sec)	h_1 (km)	v_2 (km/sec)	σ_2 (km/sec)	h_2 (km)	v_3 (km/sec)	σ_3 (km/sec)	h_3 (km)	v_4 (km/sec)	σ_4 (km/sec)	h_4 (km)	Latitude (S)	Longitude (E)
R2E37†	2.87	(2.50)	—	1.72	3.75	—	0.89	4.36	—	—	—	—	—	65°05′	143°47′
R3E37	2.70	(2.50)	—	2.08	—	—	—	—	—	—	—	—	—	65°22′	142°33′
R5E37†	4.23	1.82	0.09	0.70	1.95	0.38	0.41	3.37	0.10	1.69	4.20	0.57	1.27	62°43′	126°54′
R6E37†	4.43	1.82	0.30	0.40	2.14	0.05	0.83	2.46	0.16	0.41	4.18	—	1.15*	61°39′	126°43′
R7E37†	4.54	1.64	0.09	0.36	2.05	0.11	0.39	3.41	0.12	1.08	—	—	—	60°38′	126°19′
R8E37†	4.62	1.87	0.12	0.68	1.90	0.29	0.36	(2.50)	—	0.37*	—	—	—	59°10′	125°53′
R2E41†	4.45	2.19	0.11	0.64	—	—	—	—	—	—	—	—	—	62°42′	132°01′
R7E44	4.71	(2.50)	—	1.76	3.24	—	—	—	—	—	—	—	—	58°01′	128°50′
R8E44†	4.69	1.64	0.09	0.65	(2.50)	—	0.37	5.51	—	2.00	6.85	—	—	58°01′	126°21′
R9E44†	4.65	2.37	0.10	1.17	5.40	—	—	—	—	—	—	—	—	59°02′	122°23′
R2E45	4.44	2.19	0.15	0.56	2.84	0.41	0.37	—	—	—	—	—	—	60°44′	113°51′
R2E50	4.39	2.86	0.13	0.92	—	—	—	—	—	—	—	—	—	60°46′	107°12′
R3E50	4.19	1.86	0.07	0.78	2.99	0.10	1.12	3.47	0.42	0.89	—	—	—	63°01′	132°06′
R4E50	3.54	1.96	0.07	1.13	3.06	0.10	0.51	2.85	0.11	0.51	—	—	—	64°21′	144°01′
R5E50	3.80	1.64	0.05	0.41	1.89	0.22	0.44	2.79	0.10	0.95	3.46	0.08	0.99	63°55′	144°39′
R13E53	4.37	1.74	0.19	0.36	2.51	0.40	0.43	3.56	0.30	0.80	5.68	—	—	61°25′	140°04′
R14E53	4.32	1.49	0.11	0.40	2.74	0.08	1.14	3.02	—	2.22	6.59	—	—	61°38′	140°01′
R17E53	4.24	1.50	0.09	0.46	3.25	0.08	1.48	4.87	—	—	—	—	—	62°02′	139°54′
R20E53	4.00	1.86	0.08	0.73	1.98	0.29	0.48	3.65	—	3.55	6.22	—	—	62°04′	138°16′
R21E53	4.15	1.67	0.30	0.37	2.85	0.50	0.66	3.45	0.16	1.46	—	—	—	61°54′	138°04′
R23E53	4.57	1.94	0.10	1.08	3.45	0.33	1.44	(3.85)	—	0.82*	—	—	—	61°27′	137°43′
R24E53	4.38	1.95	0.08	0.65	3.26	—	1.72*	—	—	—	—	—	—	61°18′	137°38′
R25E53	4.23	1.65	0.17	0.41	3.06	0.08	0.62	(4.00)	—	0.95*	—	—	—	61°54′	141°03′
R26E53	4.29	1.70	0.06	0.52	2.35	0.10	0.74	—	—	—	—	—	—	61°49′	141°23′

Notes: Sonobuoy stations identified by daggers are from Houtz and Markl (1972). Except for assumed interval velocities (enclosed in parentheses), seismic velocities not followed by standard deviations were obtained from refraction arrivals. Bottom-layer thicknesses denoted by asterisks were determined from observed last reflection arrivals. v, seismic velocity; o, standard deviation of interval velocity; h, layer thickness.

slope seaward until well into the abyssal plain (Houtz and Markl, 1972).

Deep-Sea Drilling Project site 268 most likely lies inside the magnetic quiet zone close to its seaward border (Fig. 2). The section consists of clayey sediments ranging in age from Quaternary to middle Oligocene and includes ice-rafted detritus of early Miocene or younger age (Hayes et al., 1975a). The total sediment cover is thought to be at least 1.5 km thick, but only the top 474.5 m were penetrated.

At site 267 (Fig. 2), basement was reached and consists of fairly typical tholeiitic oceanfloor basalt (Ford, 1975). Oligocene to Miocene and Miocene to Pleistocene ooze amounting to a combined total thickness of about 207 m overlies the basalt (Hayes et al., 1975b). The contribution of ice rafting to the sediment supply since at least late Miocene time is apparent at this site. The age of the basal sediment is compatible with the late Eocene age of the basalt suggested by the marine magnetic lineations.

The 958 m thick section cored at site 269 (Fig. 2) consists of predominantly silts and clays of late Oligocene to Quaternary age (Hayes et al., 1975c); the total sedimentary section is more than 1.3 km thick. The age of the oldest sediments is considerably less than that of the basement (late Paleocene to early Eocene [mid-Eocene]) as deduced from magnetic lineations. The influence of terrigenous sediment is less pronounced at this site than at DSDP site 268, where continent-derived detritus dominates the section, and is less still at site 267 (Kemp et al., 1975).

Thus, the sonobuoy, seismic-profiler, and deep-sea drilling data combined indicate a rather thick sedimentary cover overlying typical oceanic basement; as the continental margin is approached from the north, the thickness of the sedimentary section and the basement depth increase. The *Eltanin* and *Glomar Challenger* seismic data are quite limited in that they give no clue as to basement depth or nature of the crust beneath the quiet zone; the recent *S.P. Lee* cruise, however, has contributed much to alleviate this lack of information (Childs and Stagg, this volume).

SYNTHESIS OF ANTARCTIC AND AUSTRALIAN DATA

A high degree of similarity exists between the geophysical-anomaly patterns of the southern Australian and conjugate Antarctic continental margins (Fig. 7). Magnetic quiet zones are located off both continents and extend from the shelf or slope seaward well into the continental rise. A prominent trough-like magnetic anomaly, occurring with great consistency along the entire southern Australian margin, and a corresponding magnetic low, less well defined in extent but nonetheless observed near the shelf-break or on the slope off Wilkes Land, Antarctica, in each case forms the landward boundary of the magnetic quiet zone. Occupying the area between the magnetic trough and the continent are high-amplitude, short-wavelength magnetic anomalies of continental origin; they are extensively observed on the southern Australian shelf but less well documented for the area immediately offshore from Wilkes Land. On its seaward side, the magnetic quiet zone terminates against the oldest marine magnetic lineation, anomaly 34[22]. This isochron is known with certainty not to exist east of long 131°30′ E on the southern Australian margin (König, 1980) and similarly seems to disappear somewhere east of long 135° E off Antarctica (Childs and Stagg, this volume). To the east of these meridians, the magnetic quiet zone widens seaward with anomaly 32–33[21] forming its boundary both off South Australia and off Clarie and Adélie coasts, Antarctica. These relations are shown in Figure 7 where the magnetic quiet zones are indicated by ruling and key magnetic lineations near the continental margins are represented by dotted lines.

In general, the free-air gravity field displays similar characteristics on the opposing continental margins of southern Australia and East Antarctica. A regional gravity low, considerably deeper off Australia than off Antarctica, overlies the magnetic quiet zone; landward of it an isostatic (and free-air) gravity high is associated with the magnetic trough but varies somewhat in position relative to the trough and in amplitude on either margin, and the seaward quiet-zone boundary is

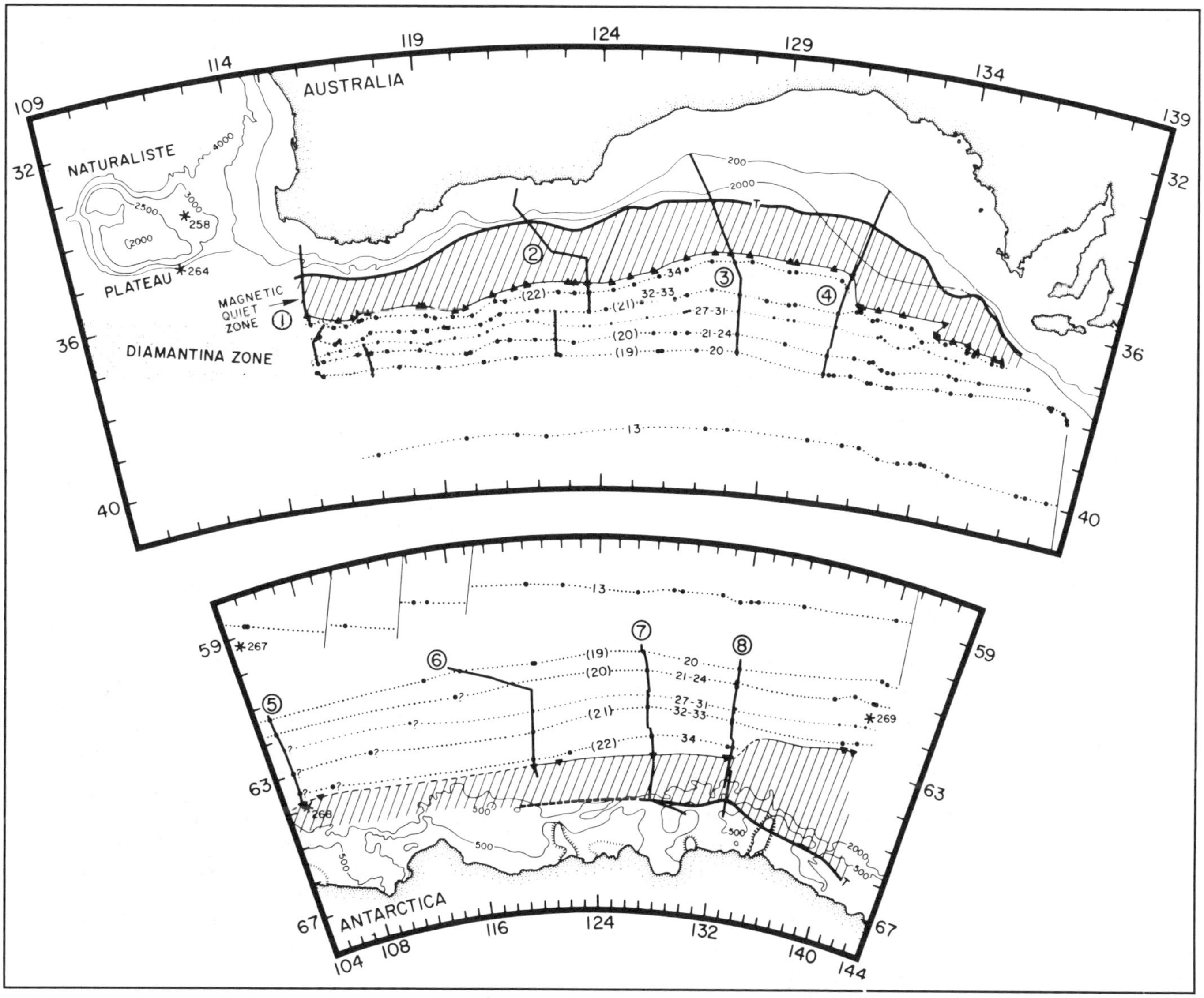

Figure 7—Marginal quiet zones and key magnetic lineations off southern Australia (top) and Wilkes Land, Antarctica (bottom) (polar stereographic projections). The magnetic quiet zone on each continental margin is indicated by ruling extending from the magnetic trough (T), shown as a solid trace, dashed where inferred, to the seaward quiet-zone boundary represented by triangles (with apices pointing landward) connected by solid lines, dashed where conjectural. The marine magnetic lineations are shown as heavy dots joined by light dotted lines; they are labeled according to the Kent and Gradstein (in press) sequence of geomagnetic-polarity reversals. Dots and triangles represent actual data points, and the queries next to the magnetic anomalies denote their identification as tentative. Faint north–south lines are major fracture zones. Deep-sea drilling sites in the map areas are shown as numbered asterisks. The heavy solid lines labeled 1 through 8 are segments of ships' tracks shown as projected profiles in Figure 8A. The coastlines (including ice or glacier tongues and ice shelves on the Antarctic margin) and isobaths of the Australian and Antarctic continents are taken from the Tectonic Map of Australia and New Guinea (Geological Society of Australia, 1971) and the Geologic Map of Antarctica (Craddock, 1972a) in conjunction with the topographic map published by the American Geographical Society (1970), respectively. The magnetic lineations are numbered following Cande and Mutter (1982), with the original identifications of Weissel and Hayes (1972), modified by König (1980), enclosed in parentheses. The depth contours delineating the Naturaliste Plateau are based on Markl's (1974) bathymetric map of the Indian Ocean.

expressed in the gravity field as a pronounced gradient. In the Great Australian Bight and locally off Banzare coast, Antarctica, this gradient is replaced by an isostatic (and free-air) gravity high. At both continental margins, anomaly 34[22] is typically located on either the seaward gradient or this gravity high where extant.

In Figure 8A, four profiles from each continental margin are shown that demonstrate the close correspondence of the geophysical-anomaly patterns. Both bathymetry and gravity profiles in addition clearly display the observed regional differences in seafloor depth and level of the free-air gravity field, respectively. The bathymetric bulge over the continental slope and upper rise of track 4 is the Ceduna Plateau; no obvious corresponding feature is apparent on profile 8.

The seismic data show that the magnetic quiet zones are associated with larger-than-normal thicknesses of sediment. No seismic information beyond this general increase in sediment thickness toward the continent can be extracted from the *Eltanin* and *Glomar Challenger* data; considerably more detailed seismic results pertaining to structure and sediment distribution of the quiet-zone area off southern Australia are available, however (Talwani et al., 1979). There, deeply subsided basement buried by a thick section of terrigenous sediments and a much thinner marine sequence underlies the area occupied by the magnetic quiet zone; the depth to basement is about 6–7 km but locally varies from less than 5 km in the west to a maximum of more than 11 km in the east (Boeuf and Doust, 1975), the sediment thickness being typically 2–4 km, but locally in excess of 10 km. Landward of the magnetic trough, the basement is shallow and overlain by a sedimentary section that is thin, except for the considerably greater thicknesses that are found in the shelf and slope sedimentary basins (Hegarty, 1985). Near the seaward quiet-zone boundary, oceanic basement peaks are typically seen on seismic-reflection profiles in the vicinity of anomaly 34[22], either beneath a thin veneer of sediment or, in the Diamantina Zone in the western part of the margin south of Australia (Fig. 7), with no sediment cover at all. The basement high observed near magnetic lineation 34[22] merges seaward with rough oceanic basement that rises gently toward the Southeast Indian Ridge. Seismic-refraction data from the quiet zone off southern Australia indicate large variations in seismic velocity representative of a high degree of heterogeneity with neither continental nor oceanic velocities predominating (Talwani et al., 1979); where measured beneath the crustal section, the mantle velocities display normal values. The seismic data were interpreted by these authors as indicating three distinct crustal types for the Australian margin: continental crust landward of the magnetic trough, "quiet-zone crust" underlying the thick sedimentary section in the magnetic quiet zone, and oceanic crust seaward of the quiet zone.

König and Talwani (1977) modeled the magnetic-trough anomaly as the effect of a normal fault in continental basement and the regional gravity low over the magnetic quiet zone as arising from a mass-deficient structure characterized by deeply subsided basement overlain by a thick sediment deposit. Talwani et al. (1979), extending this interpretation in light of much more extensive seismic data, suggested a taphrogenic origin for the region of the magnetic quiet zone related to rifting of East Gondwanaland and subsequent rupture into Australia and Antarctica. In this tectonic-evolutionary scheme, the magnetic trough delineates the boundary fault of the rift valley, the sediment-filled basin beneath the quiet zone represents the remnant of the original rift-graben trough containing a thick sedimentary deposit (corresponding to the rift-valley sequence of Falvey, 1974, overlain by a thin veneer of postrift marine sediments), and the seaward quiet-zone boundary marks the location where the rift graben broke apart and oceanic crust was injected. Thus, the occurrence of magnetic quiet zones off both southern Australia and the conjugate sector of East Antarctica, the similarity of the geophysical anomalies associated with them, and the almost identical nature of this association imply the existence of paired fossil remnants of a once-continuous, active rift graben. These rift fragments are now separated by ocean floor generated at the evolving Australian–Antarctic plate boundary in the Southeast Indian Ocean.

Figure 8B shows the profiles of Figure 8A joined at the seaward quiet-zone boundaries, i.e., with the ocean floor eliminated. The combined profiles represent the extent of the original rift graben just before breakup and its geophysical-anomaly signature except for the effect of postrift modifications. On profiles 4/8 and 3/7 (Fig. 8B) the rift zone is approximately 250 km wide, the site of rupture being located just about at its center.

AUSTRALIA-ANTARCTICA RECONSTRUCTION AND THE EARLY HISTORY OF SEAFLOOR SPREADING IN THE SOUTHEAST INDIAN OCEAN

Determination of New Finite-Rotation Parameters

Examination of the geophysical data from the southern Australian and conjugate Antarctic continental margins has resulted in a clearer demarcation of oceanic and continental structural provinces than the selection of particular isobaths, which has commonly been employed in continental reassemblies, is able to provide. As pointed out by Hallam (1976), the tightness of fit is the most poorly determined parameter in such reconstructions; a priori it would seem unlikely that the continent–ocean boundary is always located on the continental slope. Off southern Australia and Wilkes Land, Antarctica, the region of typical oceanic structure lies seaward of the magnetic quiet zone, and unquestionably continental areas extend at least as far seaward as the magnetic-trough anomalies. Part of the region underlying the magnetic quiet zone in the Great Australian Bight displays continental characteristics (Boeuf and Doust, 1975; Talwani et al., 1979), and the interpretation of the marginal quiet zones in the Southeast Indian Ocean as fossil rift-graben fragments implies that these areas were once continental in nature. As a first approximation and logical choice, the seaward edge of the oceanic crust on each of the two corresponding margins was used for the reconstruction of Australia–Antarctica. Off either continent, magnetic lineation 34[22] is both more readily recognized than the exact location of the seaward quiet-zone boundary and located in close proximity to it (Fig. 7); hence this isochron was used in the derivation of a finite-rotation pole for the reconstruction of Australia–Antarctica. Two distinctive characteristics of anomaly 34[22] were found to serve as constraints in the determination of the best-fitting pole; the rapid and well-documented disappearance of this isochron on the southern Australian margin east of long 131°30′ E and its characteristic southward bend near long 127° E (Fig. 7), mirrored by an identical change in trend of lineation 34[22] offshore Banzare coast near long 126° E (Fig. 2). In Figure 9, the Antarctic lineation (open circles) is shown stationary with anomaly 34[22] located off southern Australia rotated onto it (closed circles). The solid line passing through the two anomaly sets represents the location of the Southeast Indian Ridge at anomaly-34[22] time (about 84[52] Ma ago on the Kent and Gradstein [1986] time scale). The quality of the fit is excellent, especially in view of the mismatch that results when previously proposed poles are used to rotate the Australian isochron onto its Antarctic counterpart. The pole locations and rotation angles of these and all other rotations performed in this study are listed in Table 2.

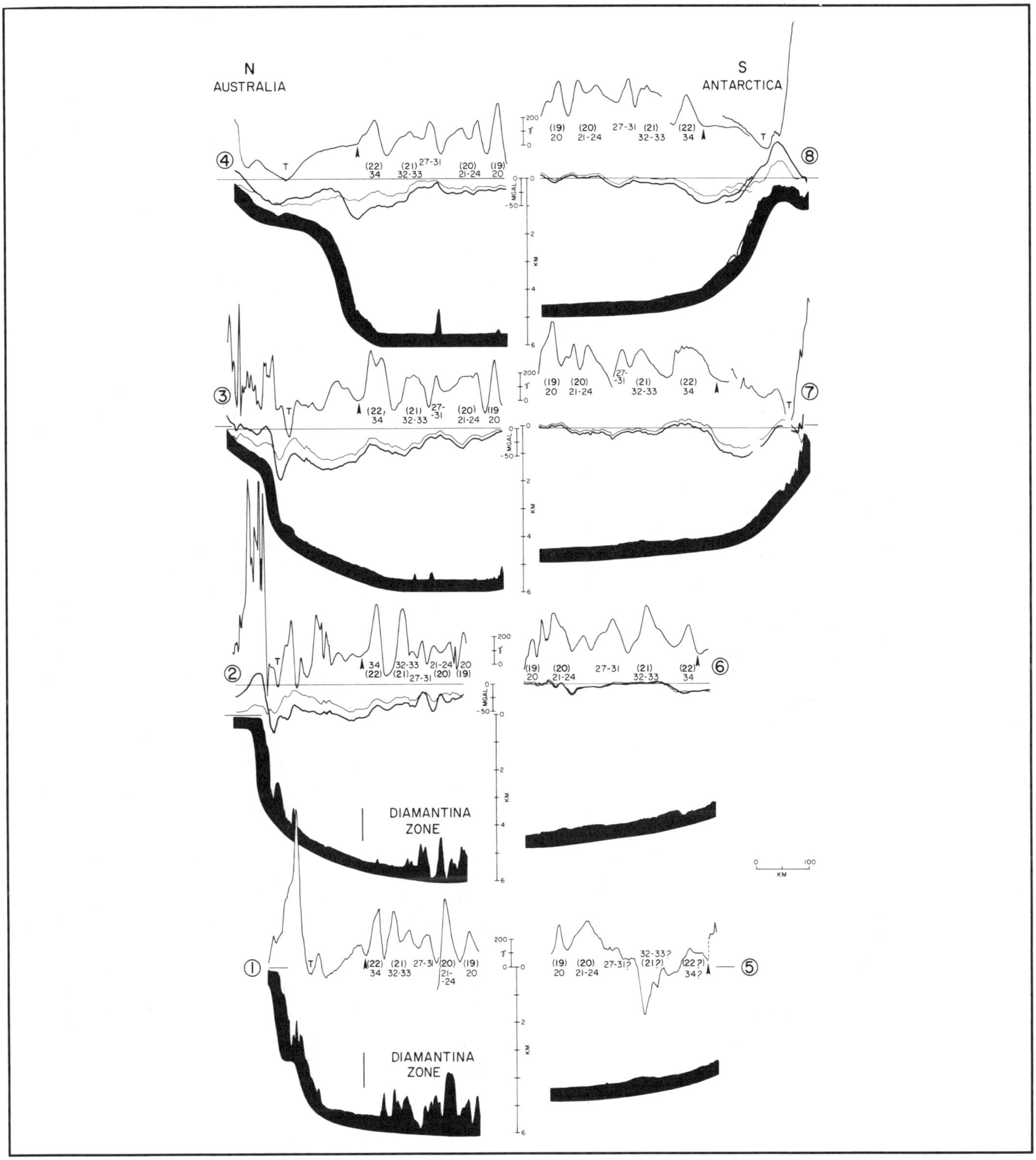

Figure 8A—Comparison of projected profiles from the conjugate rifted margins of the Southeast Indian Ocean. For location of profiles see Figure 7. The azimuth of projection is 180°. The profiles are aligned with respect to magnetic lineation 34(22). For each profile, magnetic anomalies, free-air (heavy trace) and isostatic (light trace) gravity anomalies, and bathymetry are given and identified by the corresponding scales; gravity data are not available for tracks 1 and 5. On the magnetic profiles, "T" denotes the magnetic trough, and the arrow indicates the seaward quiet-zone boundary. Anomaly identifications are identical to those of Figure 7. Ticks at the landward ends of the bathymetry profiles denote sea level. The high degree of symmetry of the profiles from opposing margins, particularly the existence of magnetic quiet zones and the expression of their boundaries in the geophysical-anomaly signature, is striking. Also apparent is the difference in seafloor depth and regional level of gravity of the two basins in the Southeast Indian Ocean.

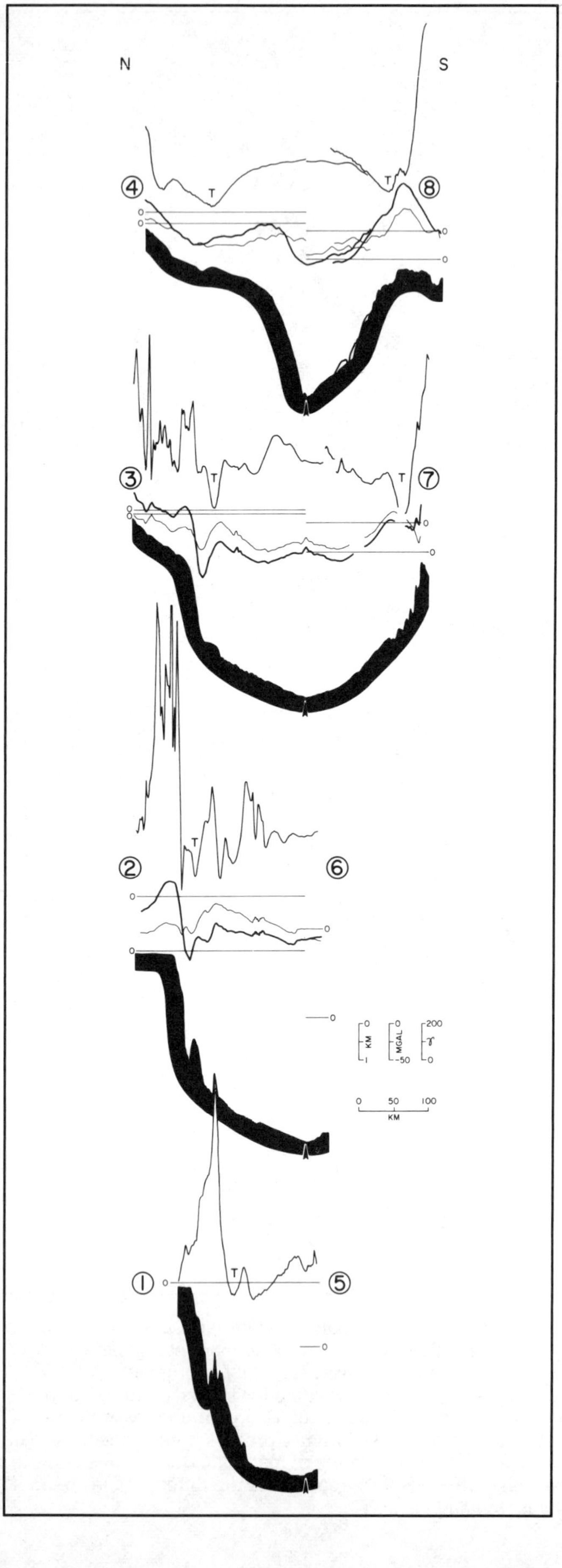

In order to determine the error margin in the geographic location of the new pole of rotation, the area of mismatch as a function of pole position of the 84[52]-Ma isochrons from both margins was examined. Lines were drawn through the discrete magnetic-anomaly data points as in Figure 7, and trial pole positions were used to generate values for the areal mismatch in the same fashion as done by McKenzie and Sclater (1971). The pole minimizing the misfit is the new pole of rotation located at lat 1.5° N, long 37.0° E; the location is accurate to within ±1.5° in latitude and ±1.2° in longitude.

With the pole locations given by Griffiths (1974), Norton and Molnar (1977), and Weissel et al. (1977) for the Australian–Antarctic reconstruction kept unchanged, the angles of rotation were varied in order to produce the best fits shown in Figure 9. The discrepancies were minimized by eye in the area east of long 126° E at the expense of the region to the west, since it was found the three poles cannot generate a close fit along the entire length of the isochron off Antarctica. In addition to the obvious mismatch of rotated and stationary lineations west of about long 120° E for all three poles, the eastern terminus of the Australian lineation in each case falls considerably west of the easternmost observed occurrence of anomaly 34[22] off Clarie coast. This violates the most crucial constraint governing the Australian–Antarctic reconstruction; for as Figure 3 demonstrates, lineation 34[22] is very well defined on profiles G and H (see Fig. 2 for location), yielding the two data points near long 132° E (Fig. 9). These control points therefore constrain any match involving anomaly 34[22]. Thus, the new finite-rotation parameters for the reconstruction of Australia–Antarctica result in a better fit of the oldest magnetic lineations than the previously proposed ones are able to provide.

New Australia-Antarctica Fit

The new finite-rotation pole for the closing of the Southeast Indian Ocean at anomaly-34[22] time was used to rotate Australia more closely toward Antarctica, which was held fixed, so as to superimpose the seaward quiet-zone boundaries of the two continental margins (Figs. 10, 13A). In their study of the southern margin of Australia, König and Talwani (1977) determined the seaward quiet-zone boundary to be situated on the landward side of, and very close to, the distinctive peak marking the oldest lineation. Veevers (1986) found a similar relationship to hold for both the southern Australian and the Wilkes Land, Antarctica, margins. The fit of the quiet-zone boundaries is very good as illustrated in Figure 13A; the

Figure 8B—Juxtaposition of representative projected profiles from the conjugate rifted margins of the Southeast Indian Ocean with postrift oceanic crust removed. The profiles are identical to those of Figure 8A except for the eliminated ocean floor and associated magnetic and gravity anomalies; the arrow on each combined bathymetry curve indicates the seaward quiet zone boundaries of the original profiles. Each set of anomaly profiles is shown joined irrespective of the absolute values of the anomalies, which results in smooth curves extending across the entire rift zone. The rift is roughly 250 km wide, with the locus of eventual breakup at its center.

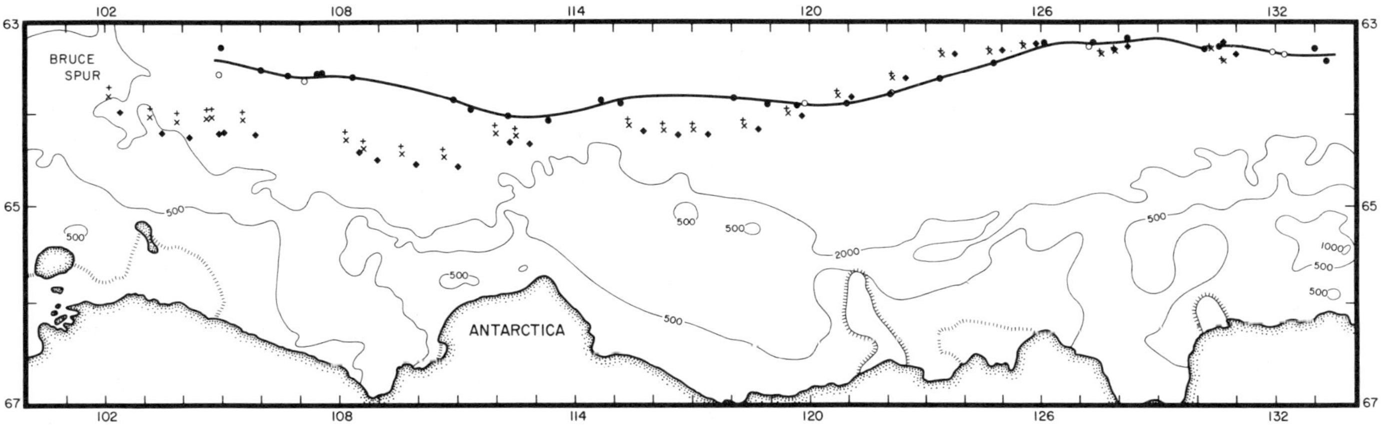

Figure 9—Rotation of anomaly 34(22) from the southern Australian continental margin onto lineation 34(22) situated off Wilkes Land, Antarctica (Mercator projection). Antarctica is held fixed in its present position; the data source for coastlines and isobaths shown is as given for Figure 7, and the magnetic-lineation data points presented are likewise those indicated in Figure 7. Stationary observed-anomaly locations on the Antarctic margin are denoted by open circles; all other symbols represent rotated anomaly data points from the southern Australian continental margin. Different rotation parameters are denoted by different symbols; solid dots constitute the best fit, and the solid line joining them with the open circles marks the inferred position of the Southeast Indian Ridge at anomaly-34(22) time, about 84(52) Ma before present. Crosses, Xs, and diamonds represent attempted closest fits of the Australian and Antarctic sets of magnetic-anomaly data points based on the finite-rotation pole positions given by Weissel et al. (1977), Griffiths (1974), and Norton and Molnar (1977), respectively. The best-fitting angles of rotation were determined by minimization, through inspection, of the mismatch of conjugate sets of points (cf. Table 2). The good fit of open and closed circles and the significant deviation of the data points shown as crosses, Xs, and diamonds from the ridge axis, particularly in the west, are readily apparent.

Table 2—Parameters for finite rotations of Australia to Antarctica.

		New rotation parameters				Old rotation parameters					
		Pole location		Rotation angle		Pole location		Rotation Angle			
Anomaly no.	Time (Ma B.P.)	(lat)	(long)	(°)	Illustration of rotation	(lat)	(long)	(°)	Illustration of rotation	Reference	Comments
Reconstruction	>94 (>54)	1.5° N	37.0° E	[30.05]	Fig. 12	3.4° S	38.8° E	31.20	—	Sproll & Dietz (1969)	Angle in brackets only used for comparison purposes; old rotation parameters not considered in this study
		1.5° N	37.0° E	28.00	Figs. 10, 11, 13A	11.9° N	30.8° E	30.90	Fig. 12	Norton & Molnar (1977)	—
						10.8° N	32.2° E	30.60	Figs. 11, 12	Griffiths (1974)	Best fit to 2-km isobath determined by eye; no angle given by Griffiths
						10.3° N	32.7° E	30.00	Fig. 12	Weissel et al. (1977)	—
34 (22)	84 (~52)	1.5° N	37.0° E	27.58	Figs. 9, 13B	11.9° N	30.8° E	28.40	Fig. 9	Norton & Molnar (1977)	Best-fitting rotation angles determined by eye; rotations not previously investigated
						10.8° N	32.2° E	28.25	Fig. 9	Griffiths (1974)	
						10.3° N	32.7° E	28.15	Fig. 9	Weissel et al. (1977)	
Pre-final Separation	>80 (>50.5)	5.0° N	34.5° E	26.60	Fig. 13C	—	—	—	—	—	
32-33 (21)	~76 (49.5)	5.0° N	34.5° E	26.37	Fig. 13D	—	—	—	—	—	These four rotations have not previously been investigated
27-31 (*)	~66 (46.5-48.5)	5.0° N	34.5° E	25.55	Fig. 13E	—	—	—	—	—	
21-24 (20)	~52 (45.5)	9.0° N	32.3° E	24.68	Fig. 13F	—	—	—	—	—	
20 (19)	~45.5 (43.8)	13.0° N	31.5° E	24.10	Fig. 13G	14.0° N	28.8° E	24.80	—	Sclater et al. (1981)	Old rotation parameters not considered in this study
13	~35.5	12.0° N	34.0° E	20.20	—	11.9° N	34.4° E	20.50	—	Weissel et al. (1977)	Rotation not considered in this study

The estimated uncertainty in position of the new finite-rotation poles is ±1.5° in latitude and ±1.2° in longitude.
Numbers in parentheses in first two columns pertain to anomaly identifications of Weissel and Hayes (1972) and König (1980).

landward edges of the oceanic crust are represented by triangles with apices pointing to the margin whose seaward boundary they represent. Also shown are the adjoining magnetic quiet zones and the two continents in their predrift configuration. Depicted in Figure 10 is a more comprehensive reconstruction that combines geological information and geophysical data; as before, Antarctica is held fixed in its present position. Table 3 provides the locations of the seismic stations shown in Figure 10. As before, Antarctica is held fixed in its present position. The same parameters of rotation as employed in Figures 10 and 13A were used in the detailed comparison of key geologic features from Tasmania and northern Victoria Land that is represented by Figure 11. Also indicated in this figure (by dashes) are the outlines of Tasmania and southeastern Australia rotated about Griffiths' (1974) reconstruction pole.

In comparison with previously proposed reassemblies of Australia and Antarctica (e.g., Sproll and Dietz, 1969; Smith and Hallam, 1970; McKenzie and Sclater, 1971; Griffiths, 1974; Norton and Molnar, 1977; Weissel et al., 1977; Barron et al., 1978), the fit shown in Figure 10 is less tight, leaving a

Table 3—Seismic stations shown in Figure 10.

Station no.	Station identification	Type of crust[1]	Rotated position (°S lat)	Rotated position (°E long)	Original location (°S lat)	Original location (°E long)	Reference
1	V16-35	III	61.849 62.018	105.881 106.521	34.855 34.961	115.624 116.025	Hawkins et al. (1965b)
2	V16-36	III	62.251 62.025	106.501 106.600	35.191 34.961	116.069 116.071	Hawkins et al. (1965b)
3	V16-37	I	62.949 63.616	106.343 106.120	35.889 36.563	116.153 116.119	Hawkins et al. (1965a)
4	E53-58	III	62.511	107.948	35.318	116.937	König & Talwani (1977)
5	V33-2	II	63.417	110.929	35.980	118.770	Talwani et al. (1979)
6	V33-3	II	63.735	110.834	36.300	118.780	Talwani et al. (1979)
7	V33-29	II	63.932	116.087	36.188	121.658	Talwani et al. (1979)
8	V33-28	II	63.136	116.492	35.378	121.778	Talwani et al. (1979)
9	V33-27	II	63.674	116.608	35.908	121.908	Talwani et al. (1979)
10	V33-5	II	63.060	117.219	35.270	122.170	Talwani et al. (1979)
11	V33-6	II	63.338	117.889	35.520	122.570	Talwani et al. (1979)
12	E53-55	(III)	62.198	120.737	34.288	124.053	König & Talwani (1977)
13	V33-30	III	63.708	120.638	35.798	124.108	Talwani et al. (1979)
14	V33-13	II	62.470	123.676	34.500	125.720	Talwani et al. (1979)
15	V33-33	III	62.894	124.132	34.918	125.988	Talwani et al. (1979)
16	V33-32	III	62.840	124.728	34.858	126.318	Talwani et al. (1979)
17	V33-12	I	63.253	124.804	35.270	126.370	Talwani et al. (1979)
18	V33-34	I	62.794	125.233	34.808	126.598	Talwani et al. (1979)
19	E53-51	(III)	62.104	125.498	34.117	126.737	König & Talwani (1977)
20	V33-14	III	62.769	125.954	34.780	127.000	Talwani et al. (1979)
21	V33-11	I	63.150	126.401	35.160	127.250	Talwani et al. (1979)
22	V33-15	I	62.938	127.501	34.950	127.860	Talwani et al. (1979)
23	V33-37	III	62.318	128.731	34.338	128.558	Talwani et al. (1979)
24	V33-36	II	63.329	130.232	35.368	129.358	Talwani et al. (1979)
25	V33-18	III	62.906	130.508	34.950	129.530	Talwani et al. (1979)
26	E53-48	III	61.096	130.617	33.143	129.677	König & Talwani (1977)
27	V33-39	(III)	62.927	132.976	35.028	130.898	Talwani et al. (1979)
28	V33-38	(III)	62.560	133.176	34.668	131.038	Talwani et al. (1979)
29	V33-40	I	63.982	136.269	36.198	132.598	Talwani et al. (1979)
30	V33-21	I	63.994	139.828	36.390	134.520	Talwani et al. (1979)
31	V16-40	I	64.029 64.135	139.911 143.141	36.430 36.746	134.559 136.277	Hawkins et al. (1965a)
32	V16-41	III	63.579 63.833	142.640 143.207	36.162 36.453	136.120 136.375	Hawkins et al. (1965a)
33	V33-42	III	65.202	149.343	38.318	139.288	Talwani et al. (1979)
34	E37-16	(III)	65.711	154.797	39.397	141.928	Houtz & Markl (1972)

[1]Crustal-type designations (enclosed in parentheses where inferred) according to criteria given by Talwani et al. (1979).

considerable gap between the 2000-m isobaths of the two margins west of long 148° E but at the same time avoiding any overlap of reportedly continental regions in contrast to all of the above fits. This, of course, is due to the different criteria used in the definition of the edge of oceanic crust. While commonly assumed to abut against continental crust on the slope (Carey, 1958) at various water depths such as 2000 m, 1000 fms, or 500 fms, typical oceanic crust is known to extend no farther landward than the seaward boundaries of the magnetic quiet zones off southern Australia (Veevers, 1986) and the Wilkes Land sector of Antarctica (Eittreim et al., 1984). These boundaries typically lie at depths of 3.5–4 km off East Antarctica and 5–5.5 km off southern Australia (Fig. 8A). The magnetic trough south of Australia and that portion of its Antarctic counterpart which is situated off the western region of Clarie coast roughly follow the 2000-m depth contour on each margin and, in the case of Australia, indicates the southern edge of the continent (König and Talwani, 1977). Almost all of the previously proposed reconstructions are based on the match of the 2000-m isobaths of Australia and Antarctica. Therefore, they imply the oceanic nature of the crust underlying the magnetic quiet zones on both margins. The available seismic information, however, does not support an oceanic origin for the crust beneath the marginal quiet zones (Talwani et al., 1979; Childs and Stagg, this volume), but rather supports the juxtaposition of the seaward quiet-zone boundaries in the new predrift fit of Australia and Antarctica.

Other bathymetric features previously considered in relation to the reassembly of Australia and Antarctica are the Naturaliste and Bruce plateaus off southwestern Australia and seaward of the Knox coast–Queen Mary coast region of East Antarctica, respectively (Sproll and Dietz, 1969; Markl, 1978). In Figure 10 they are shown to fit closely in the vicinity of the north–south segment of the Antarctic 4000-m isobath; again there is no overlap of presumed continental structures. Also indicated in Figure 10 by the additional 1000-m and 3000-m isobaths are the Ceduna Plateau in the eastern part of the Great Australian Bight and the Eyre Plateau farther west. Both features consist of thick sediment deposits overlying subsided continental basement (Boeuf and Doust, 1975; Talwani et al., 1979). On the corresponding margin of Antarctica, the depth contours show no bathymetric features of comparable dimension. The asymmetry in submarine topography of the two margins is also well displayed by profiles 4 and 8 in Figure 8A.

Main Geologic Elements of Southern Australia and East Antarctica

The principal tectonic units of the Australian continent are indicated in Figure 10. Offshore, the main structural features are the Bremer Basin, part of the Eucla Basin, the Polda Trough, and the Duntroon and Otway basins. These sedimentary basins contain sequences typically ranging in age from Early Cretaceous to late Tertiary; they are commonly fault-bounded on the north and usually reach maximum basement depths of several kilometers below sea level. Vast expanses of Precambrian terrain are exposed onshore and constitute the Australian shield. Large areas are also occupied by shallow sedimentary basins such as the Eucla and Murray basins. In marked contrast to the thin, undeformed sequences of predominantly Mesozoic–Cenozoic age occupying these basins, thick, deformed sections of geosynclinal sedimentary rock are contained in the two mobile belts occurring in the area shown in Figure 10, the Adelaide Orogen, flanked by the Kanmantoo Trough, and the Lachlan Orogen as part of the Tasman Orogenic Zone (Solomon and Griffiths, 1972). Both mobile belts were subjected to major phases of diastrophism during the early Paleozoic.

All the rock exposures along the coast of Antarctica west of about long 148° E (Fig. 10), though widely scattered, are Precambrian in age and consist of dominantly high-grade metamorphic gneisses and schists along with charnockitic rocks. Similar rocks occur outside the map area elsewhere in East Antarctica (Craddock, 1972a) and are representative of a basement complex of mainly early Precambrian age that underlies almost all of East Antarctica and constitutes the East Antarctic shield (Hamilton, 1967; Grikurov et al., 1972; Elliot, 1975). Overlying this basement complex are upper Precambrian to Cambrian detrital sedimentary rocks (exposed on Knox and Adèlie coasts) and some volcanic rocks representing typically flat-lying platform deposits. The easternmost exposures occurring in the area shown in Figure 10 belong to the Ross Orogen, the structural backbone of the Transantarctic Mountains, which extend from northern Victoria Land toward the South Pole and beyond. The thick sedimentary sequence was deformed, metamorphosed, and intruded by granitic plutons during the Ordovician Ross Orogeny. Superimposed on the platform deposits of the shield region are the essentially undisturbed sedimentary beds of the Beacon Supergroup (Barrett et al., 1972), the Antarctic equivalent of the Gondwana sequence on the other southern continents. The Beacon Supergroup strata range in age from Devonian or older to Jurassic. The largest exposure in the area of Figure 10 is the occurrence on George V coast. Basaltic flows of Jurassic age overlie the Beacon Supergroup rocks in the Ross Orogen. Cenozoic tectonic activity appears to be responsible for uplift and formation of the modern Transantarctic Mountains. The Antarctic ice sheet may have formed as early as 25 Ma ago (Hayes and Frakes, 1975) but has certainly been in existence for the past 7 Ma (Craddock, 1970; Calkin and Nichols, 1972).

Geologic Constraints on Australia–Antarctica Fit

Several lines of geologic and geophysical evidence exist for the reconstruction of Australia and Antarctica, and much of this evidence has been summarized by Smith and Hallam (1970). It includes similarities in large-scale crustal structure of Australia and East Antarctica (along with significant differences between West and East Antarctica) (Doyle and Everingham, 1964; Adams, 1972; Bentley, 1973; Mathur, 1974; Drewry, 1976); structural trends in and composition of the Precambrian crystalline basement rocks (Craddock, 1970; Ravich, 1972); and similarities of the Paleozoic and Mesozoic fossil records on both continents (Plumstead, 1964; Colbert, 1972). The continuation of the Ross Orogen into the Adelaide mobile belt in South Australia appears to be well established on geological grounds (Hamilton, 1976; Sproll and Dietz, 1969; Frakes and Crowell, 1970; Smith and Hallam, 1970; Craddock, 1972b), but the most useful geologic data in terms of constraining the reassembly of the two continents have been obtained in the eastern regions of both Australia and East Antarctica. Geologic mapping in northern Victoria Land, in particular (Nathan and Skinner, 1972; Dow and Neall, 1972), has enabled such constraints to be established (Laird et al., 1977).

In Figure 11, Tasmania is shown fitted against northern Victoria Land; the rotation parameters used are the same as those for Figure 10. The faint dashed outline of Tasmania shown is based on Griffiths' (1974) finite rotation for his Australia–Antarctica reconstruction. The rocks exposed in northern Victoria Land belong to the Ross Orogen (Wilson Group) in the west and the Borchgrevink Orogen (Bowers Group, Robertson Bay Group) east of it (Fig. 11). The high-grade schists and gneisses of the Wilson Group (4 of legend) are adjoined on the east by a narrow fault-bounded strip of Bowers Group rocks (7 of legend) and a more extensive area occupied by the Robertson Bay Group (5 of legend). Both of these rock units consist of thick sequences of clastic sediment of late Precambrian or Cambrian age that were weakly metamorphosed and strongly deformed during the Silurian Borchgrevink Orogeny (Nathan and Skinner, 1972). Two areas of Precambrian rocks are exposed on Tasmania, one in the northwest corner of the island, the other farther east and extending southward to the coast (Spry and Banks, 1962). The Dundas Trough ("D.T." and 7 of legend) lies between these two terrains; it underwent a complex history of sedimentation, deformation, and volcanism that culminated with the intrusion of granitic plutons in the Devonian.

Griffiths (1974) suggested the correlation of the Dundas Trough sequence with the strata of the Bowers Group, and Laird et al. (1977) presented new geological evidence in support of this correlation. The similarities in lithology, strati-

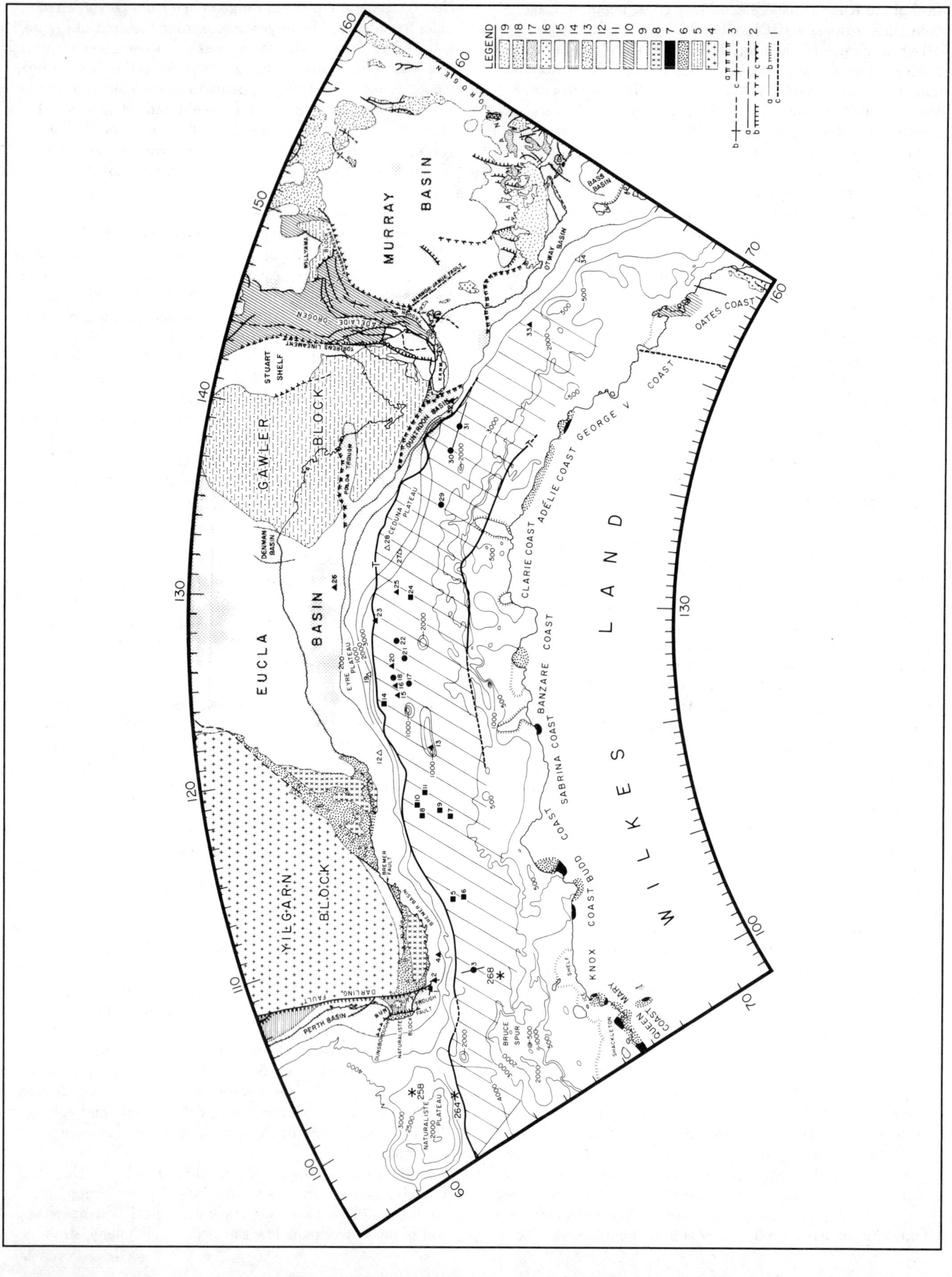
LEGEND
MURRAY BASIN
GAWLER BLOCK
STUART SHELF
ADELAIDE OROGEN
TORRENS LINEAMENT
WILLYAMA BLOCK
BASS BASIN
OTWAY BASIN
DUNTROON BASIN
DENMAN BASIN
EUCLA BASIN
EYRE PLATEAU
CEDUNA PLATEAU
YILGARN BLOCK
BREMER FAULT
BREMER BASIN
DARLING FAULT
PERTH BASIN
NATURALISTE BLOCK
NATURALISTE PLATEAU
BRUCE SPUR
WILKES LAND
OATES COAST
GEORGE V COAST
ADÉLIE COAST
CLARIE COAST
BANZARE COAST
SABRINA COAST
BUDD COAST
KNOX COAST
QUEEN MARY COAST
SHACKLETON

graphic succession, inferred depositional environment, as well as timing and nature of tectonic activity led them to postulate contiguity of the central portions of the Cambrian and Early Ordovician depositional troughs in Tasmania and northern Victoria Land. Both the Bowers Group and Dundas Trough occurrences are rather narrow (Fig. 11); they are distinctly different in character from other Cambrian geosynclinal rocks in southern Australia and northern Victoria Land (Laird et al., 1977) and thus possess the characteristics necessary to serve as firm constraint in the reassembly of southeastern Australia/Tasmania and northern Victoria Land. Laird et al.'s (1977) fit is close to Griffiths' (1974) reconstruction, shown in Figure 11 by a dashed line, but differs from it in detail by a slight counterclockwise rotation. This difference is considered negligible, however.

The reconstruction of Australia and Antarctica shown in Figures 10 and 11 results in a new position of Tasmania that is in close proximity with those in the reassemblies of Griffiths (1974) and Laird et al. (1977). As with the Griffiths reconstruction, the trends of the Bowers Group and Dundas Trough are not exactly aligned (Fig. 11), but since the new fit is less tight than those previously proposed and in view of the observed changes in strike of the Dundas Trough, the mismatch in trend is considered insignificant and not detrimental to the validity of the reconstruction. On the contrary, the proposed fit satisfies all known geologic constraints very well considering it was independently derived from marine magnetic lineations.

Comparison of New Fit with Previous Reconstructions

A comparison of three previously proposed reconstructions (Griffiths, 1974; Norton and Molnar, 1977; Weissel et al., 1977) with the new fit (Figs. 10, 11) is shown in Figure 12. Represented by different combinations of dots and dashes are the four outlines of the Australian continent and the corresponding 2000-m isobaths, with the Antarctic coastline and 2000-m depth contour shown by solid traces. The positions of both Griffiths' (1974) and the new pole of rotation were left unchanged (Table 2), but the angle of rotation was varied so as to determine the closest match, judged by eye, of the Australian and Antarctic 2000-m isobaths. With respect to the reconstruction of Figure 10, the fit represented by dashes in Figure 12 is considerably tighter and more closely resembles the other three reconstructions shown. This facilitates the comparison of the different positions of the Australian continent with each other and with that of Antarctica. Also indicated are the two magnetic-trough anomalies; the rotation of the Australian trough was done with the same parameters as used for the dashed continental outline.

Figure 10—(opposite) Reconstruction of the predrift configuration of Australia–Antarctica (polar stereographic projection). Antarctica is held fixed in its present position. The data sources for continental outlines and bathymetry shown are as given for Figure 7. Deep-sea drilling sites in the area are identified by numbered asterisks, and seismic-station locations are indicated by numbered solid circles, squares, and triangles (open where crustal type is inferred) representing crustal types I, II, and III of Talwani et al. (1979), respectively (see Table 3). The Australian-Antarctic rift zone is shown as the shaded region between the magnetic-trough anomalies indicated by solid (dashed where projected) lines labeled "T"; where not shown abutting against the magnetic trough, the rift zone in its landward extent is conjectural. The major geologic-tectonic features represented are taken from the Tectonic Map of Australia and New Guinea (Geological Society of Australia, 1971) and the Geologic map of Antarctica (Craddock, 1972a). Legend: 1. Geologic boundaries—a, contact between geologic-tectonic units; 1000-m depth-to-basement contour outlining Bass Basin; b, approximate limit of preserved Upper Cretaceous to Holocene sediments; c, inferred western boundary of Ross Orogen according to Craddock (1972b). 2. Faults—a, type unknown; long dashes where approximate, short dashes where concealed by younger platform cover; b, high-angle fault, normal or reverse; ticks denote downthrown block; dashed where approximate or concealed by younger platform cover; c, thrust fault; teeth point to upper plate. 3. Flexures and folds—a, flexure, concealed by younger platform cover; teeth denote downwarped side; b, anticline, dashed where concealed by younger platform cover; c, syncline, dashed were concealed by younger platform cover. 4. Undifferentiated granite and gneiss and metasedimentary rocks (>2.2 Ga in age). 5. Precambrian metamorphic complexes (phases of deformation 1.9–1.6 Ga B.P.) and mainly acid extrusives (1.6–1.4 Ga), Gawler and Willyama Blocks (Australia); Precambrian schists and quartzites, greenschist facies (East Antarctica). 6. Precambrian amphibolites (1.9–1.6 Ga) and granulites (1.4–1.2 Ga), Albany–Fraser Province; late Precambrian granulites (670 Ma), Naturaliste Block (southwestern Australia); Precambrian gneisses and schists, granulite facies (East Antarctica). 7. Precambrian charnockite bodies. 8. Precambrian granite (1.2–1.0 Ga), Albany–Fraser Province (Western Australia); Precambrian granite, adamellite, and syenite intrusives, George V coast (East Antarctica). 9. Precambrian metasedimentary and volcanic rocks (? deformation 1 Ga to 700 Ma), Tyenna–Rocky Cape orogenic domain (King Island and western Tasmania); older Precambrian medium- to high-grade metasedimentary rocks (Wilson Group), Transantarctic Mountains (northern Victoria Land, Antarctica). 10. Precambrian to Ordovician (deformation Middle Cambrian to Ordovician) geosynclinal sedimentary and limited volcanic rocks, Adelaide Orogen (South Australia/New South Wales); upper Precambrian metasedimentary rocks, moderately to strongly deformed (East Antarctica). 11. Upper Precambrian to Middle Cambrian sedimentary platform cover (South Australia); upper Precambrian sedimentary and volcanic rocks, undeformed (East Antarctica). 12. Mid-Cambrian to mid-Ordovician (Kanmantoo Trough) and mid-Ordovician to Devonian (Lachlan Orogen) geosynclinal sedimentary and volcanic rocks. 13. Mid-Devonian to mid-Carboniferous sediments, acid volcanics and granites. 14. Mesozoic platform cover (Permian to Triassic; Jurassic to mid-Cretaceous). 15. Mesozoic–Cenozoic platform cover (Permian and Permian to Triassic locally; Jurassic to mid-Cretaceous; mid-Cretaceous to Miocene; Miocene to Holocene), Australia; Beacon Supergroup, East Antarctica (beds locally as old as Devonian, as young as Jurassic). 16. Mid-Cambrian to mid-Ordovician granite (eastern Australia); Cambrian to Ordovician mainly felsic intrusives, Transantarctic Mountains (northern Victoria Land, Antarctica). 17. Mid-Devonian to mid-Carboniferous granite and acid extrusives. 18. Cretaceous flood basalt. 19. Cenozoic alkali-basalt spreads.

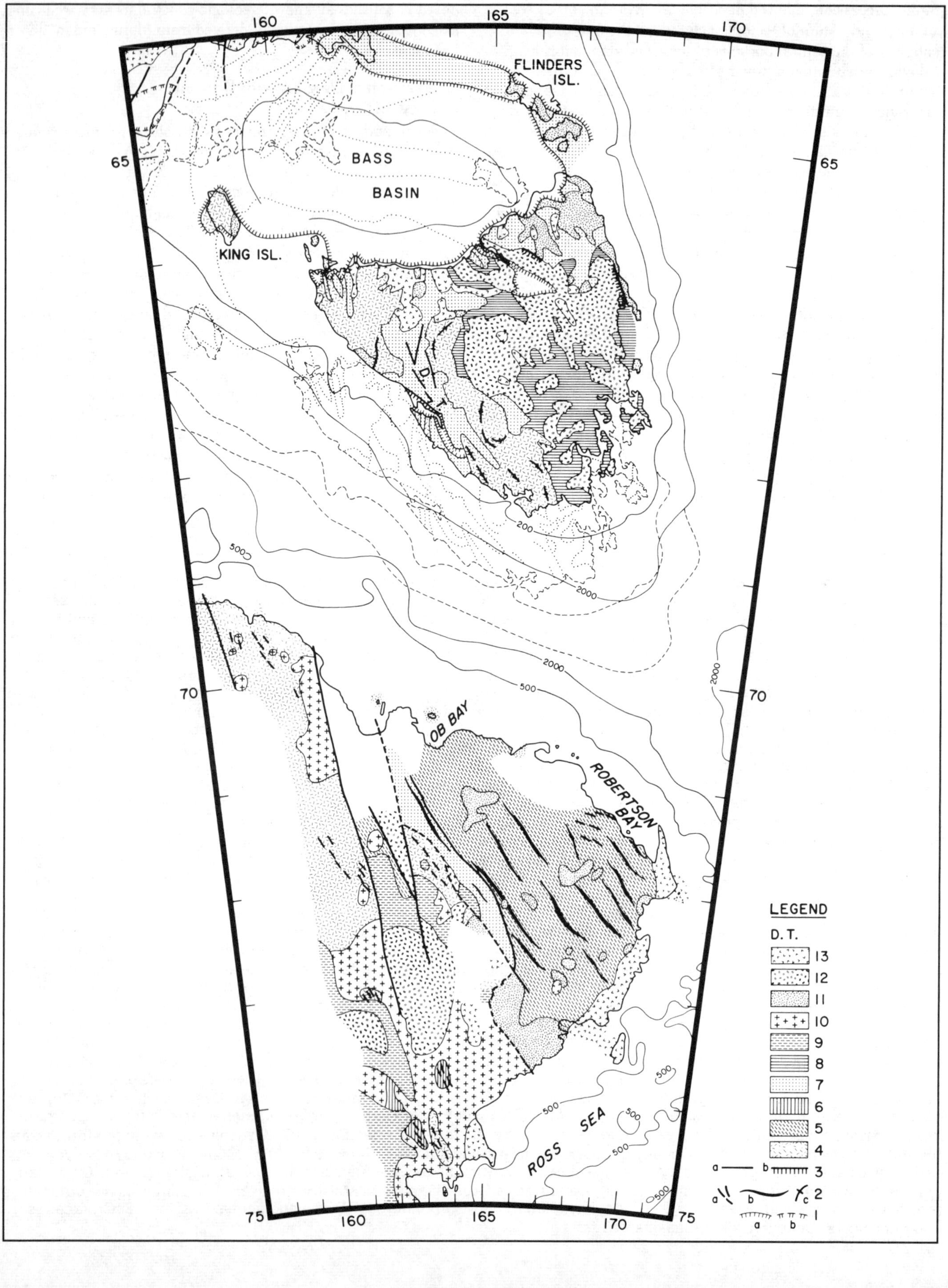
FLINDERS ISL.
BASS
BASIN
KING ISL.
D.T.
OB BAY
ROBERTSON BAY
ROSS SEA
200
500
2000
LEGEND
D. T.
13
12
11
10
9
8
7
6
5
4
a b 3
a b c 2
a b 1
160
165
170
65
70
75

All three previously proposed fits considered here, but especially those of Griffiths (1974) and Norton and Molnar (1977), result in rotated positions of the Australian continent that are very close to each other. There are gaps between the conjugate 2000-m isobaths in all of these fits, notably in the Great Australian Bight (off Banzare coast), as well as regions of significant overlap in three of the reconstructions, such as southeast of Kangaroo Island (off eastern George V coast) and around the southwestern tips of the three continental outlines of Australia. In keeping with Sproll and Dietz's (1969) view, Bruce Plateau off western Knox coast is considered continental in nature. The fourth rotated position of Australia lies considerably to the east of the other three; this relative eastward shift amounts to about 163 km on the average and also results in the Australian continent clearing Bruce Spur as defined by the 2000-m isobath on the Antarctic side.

A comparison of the dashed outline of the Australian continent in Figure 12 with the coastline in Figure 10 reveals that toward the eastern boundary of the map area, Australia in the looser fit (Fig. 10) lies considerably west of the position the continent assumes when rotated against Antarctica to fit more tightly at the 2000-m isobath (Fig. 12). This is neatly illustrated by the location of King Island (northwest of Tasmania); it lies well within the map area of Figure 10 but falls almost entirely outside the same area in Figure 12. The good agreement in the positions of Tasmania relative to each other that is shown in Figure 11 would therefore not exist if the rotation angle for the tighter fit of Figure 12 were used, for Tasmania would be located substantially farther east in such a reconstruction.

Thus, the new pole for the reassembly of Australia and Antarctica satisfies the established geologic constraints only for the smaller angle of rotation (Figs. 10, 11). A substantially tighter fit based on the new pole, such as represented by Figure 12, violates these constraints. Thus, a different pole of rotation would be necessary if Australia and Antarctica were to be matched at their 2000-m isobaths.

Early History of Seafloor Spreading in the Southeast Indian Ocean

Previous investigations of the evolution of the Southeast Indian Ocean were aimed at the younger history of seafloor spreading from anomaly-13 time (~35 Ma; Kent and Gradstein, 1986) to the present (Weissel and Hayes, 1971, 1972); dealt with its history in large time frames by-passing the evolution before anomaly-18 time (Weissel et al., 1977); or examined only the onset of oceanfloor generation (Cande and Mutter, 1982; Veevers, 1986). In Figures 13A through 13G the first 48.5[10] Ma of seafloor spreading and northward movement of Australia, from predrift to anomaly-20[19] time (~45.5[43.8] Ma), are depicted at intervals of several million years. The magnetic anomalies shown in Figure 7 were used to derive finite-rotation parameters for the five oldest isochrons as well as for two additional time markers, the quiet-zone boundaries adjacent to lineation 34[22] and, in areas where anomaly 34[22] is not observed, next to anomaly 32–33[21]. Trend and extent of the magnetic anomalies and their eastward abuttal against the George V Fracture Zone (Fig. 2) were found to constitute constraints that allowed both the derivation of the new finite-rotation parameters and an estimate of the error margin in the pole positions to be made. The accuracy in the location of the fracture zone was estimated to be ±0.75° of longitude; for each isochron, the mismatch was determined in the manner employed for lineation 34[22], yielding an accuracy in pole location of ±1.5° in latitude and ±1.2° in longitude. The new finite-rotation parameters are listed in Table 2. Instantaneous poles of rotation were calculated from the finite-rotation poles and found to fall in the vicinity of these poles, and all of the new pole locations are more tightly grouped than the finite-rotation and stage poles given

---

Figure 11—(opposite) Reconstruction of the predrift configuration of Tasmania–northern Victoria Land, Antarctica (polar stereographic projection). Antarctica is held fixed in its present position. The data sources for coastlines and isobaths are the same as those for Figure 7, and the sources for the geologic–tectonic features are as given for Figure 10. The dashed outlines of southeastern Australia and Tasmania and the corresponding 200-m and 2000-m isobaths represent the predrift configuration based on Griffiths' (1974) finite-rotation parameters (cf. Table 2); geologic features in this fit are sketched by dotted lines. Dashes and dots are only shown where not overlapping the solid coastlines or isobaths. Legend: "D.T.," Dundas Trough. 1. Limits of preserved sediments—a, limit of Upper Cretaceous to Holocene sediments; b, limit of Jurassic to Lower Cretaceous sediments concealed by younger platform cover. 2. Structural trends—a, trend of structural grain from foliation or fold axes; b, trend of fold axis; c, structural trend in medium- to high-grade metamorphic rocks. 3. Faults—a, major fault (Antarctica); fault of unknown type, dashed where approximate (Australia and Tasmania); b, high-angle fault, normal or reverse; ticks denote downthrown block. 4. Precambrian metasedimentary and volcanic rocks (? deformation 1 Ga to 700 Ma), Tyenna–Rocky Cape orogenic domain (western Tasmania and King Island); older Precambrian medium-to high-grade metasedimentary rocks (Wilson Group), Transantarctic Mountains (northern Victoria Land, Antarctica). 5. Upper Precambrian metasedimentary rocks, moderately to strongly deformed (Robertson Bay Group). 6. Lower Paleozoic or upper Precambrian metasedimentary rocks, commonly folded (Ross Supergroup strata). 7. Cambrian to mid-Devonian geosynclinal sedimentary and volcanic deposits, Lachlan Orogen (mostly Dundas Trough), Tasmania; Cambrian to Ordovician folded metasedimentary rocks (Bowers Group), Transantarctic Mountains, northern Victoria Land. 8. Permian to Triassic platform cover. 9. Beacon Supergroup; beds locally as old as Devonian, as young as Jurassic. 10. Cambrian to Ordovician intrusives, mainly felsic in composition. 11. Mid-Devonian to mid-Carboniferous intrusive granite (King and Cape Barren Islands, Tasmania); Devonian to (?) Carboniferous intrusives, felsic in composition (northern Victoria Land, Antarctica). 12. Jurassic dolerite (Tasmania); Jurassic basaltic volcanic rocks, Transantarctic Mountains (northern Victoria Land, Antarctica). 13. Cenozoic alkali–basalt spreads (southeastern Australia and Tasmania); Cenozoic volcanic rocks, mainly basalt (northern Victoria Land, Antarctica). The heavy dashed line extending southeast from the northern coast of Victoria Land near long 162° E, forming the approximate western boundary of the Bowers Group rocks, marks the boundary between the Ross and Borchgrevink Orogens as given by Craddock (1972b).

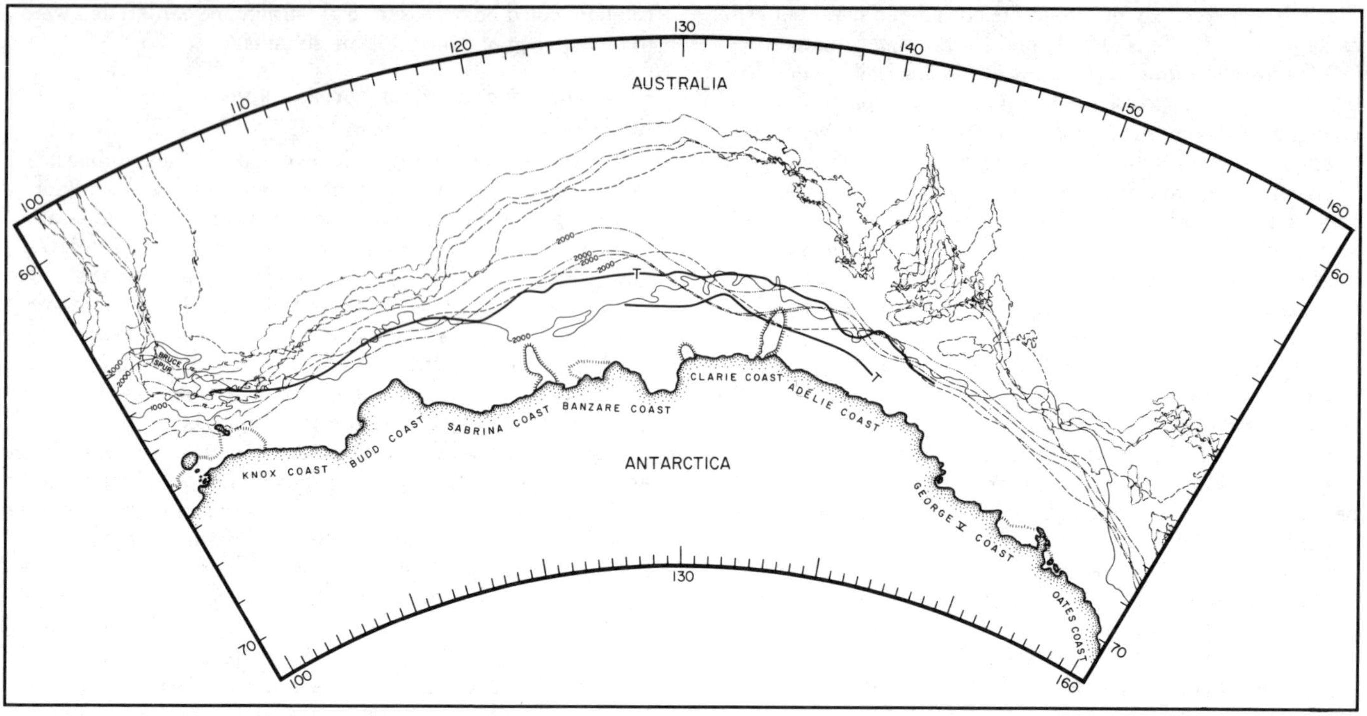

Figure 12—Comparison of different reconstructions of Australia–Antarctica (polar stereographic projection). Antarctica is held fixed in its present position. The data sources for bathymetric contours and coastlines shown are as given for Figure 7. The different positions of the Australian continent and corresponding 2000-m isobaths are represented by various combinations of dots and dashes depending on the finite-rotation parameters used. In addition to the stationary 2000-m depth contour off Antarctica, the 1000-m and 3000-m isobaths delineating Bruce Plateau off westernmost Knox Coast are shown. The location of the magnetic trough (T) off southern Australia is the rotated position of this anomaly based on the new finite-rotation pole. All reconstructions are attempts to match the 2000-m isobaths off southern Australia and the Wilkes Land sector of Antarctica; this match was determined by eye for the fit represented by dashes and the one based on the rotation-pole location of Griffiths (1974). The key to the four reconstructions shown is (see also Table 2): —·—·—·—, Norton and Molnar (1977); —··—··—, Griffiths (1974); —···—···—, Weissel et al. (1977); ————————, this study. There is overlap of southeastern Australia and the region off George V coast, Antarctica, present in all reconstructions, as well as sizable mismatch between three of the Australian coastlines and Bruce Spur. Also noteworthy is the significant eastward shift of Australia in the new fit relative to the positions of the continent in the old reconstructions.

by Weissel and Hayes (1972). The rotation parameters for anomaly 18 and younger lineations are well documented and were presented by Weissel et al. (1977, Table III).

The reconstruction of Australia and Antarctica back to the rift stage just prior to breakup is shown in Figure 13A. The quality of the fit can be judged from the alignment of the symbols representing the quiet zone boundaries; the positions of Australia and Antarctica and of the magnetic quiet zone (indicated by ruling) are the same as in Figure 10. Shortly after breakup, at anomaly-34[22] time, the two continents were in the configuration shown in Figure 13B; in the interpretation of the data presented here, a narrow tongue of ocean floor extended from the west and abutted against still-continuous rift crust east of about long 133°30′ E. This implies that while sea floor was being generated in the narrow ocean basin in the west, extension was taking place to the east of it. However, this is not the only possible interpretation although it is believed to be the most likely one; a different explanation (Veevers, 1986) will be discussed later. Figure 13C shows the two continents drifted still farther apart but not yet fully separated; at this time maximum extension east of the young ocean basin was achieved. As depicted in Figure 13D, the break had propagated east through the rift zone at anomaly-32–33[21] time; the area of oceanic crust is extremely narrow east of about long 137°30′ E. Figures 13E, 13F, and 13G represent Australia, Antarctica, and the intervening ocean basin at anomaly-27–31[*], anomaly-21–24[20], and anomaly-20[19] time, respectively. Despite the sizable width of the young ocean in this area, farther to the east, south of Tasmania, Australia and Antarctica were still connected. It was not until much later, in the mid- to late Oligocene (about anomaly-10 time and later), when the South Tasman Rise cleared the northeastern tip of Victoria Land, that a regime of active deep-sea circulation was established in the region (Kennett et al., 1972, 1974).

DISCUSSION

The principal new element in the predrift reconstruction of Australia and Antarctica (Figs. 10, 13A) is the inclusion of the marginal magnetic quiet zones located off both southern Australia and Wilkes Land, Antarctica. The basic premise for joining the quiet zones at their seaward boundaries is the notion that the crust underlying these zones is not oceanic in origin. The seismic evidence from the southern Australian con-

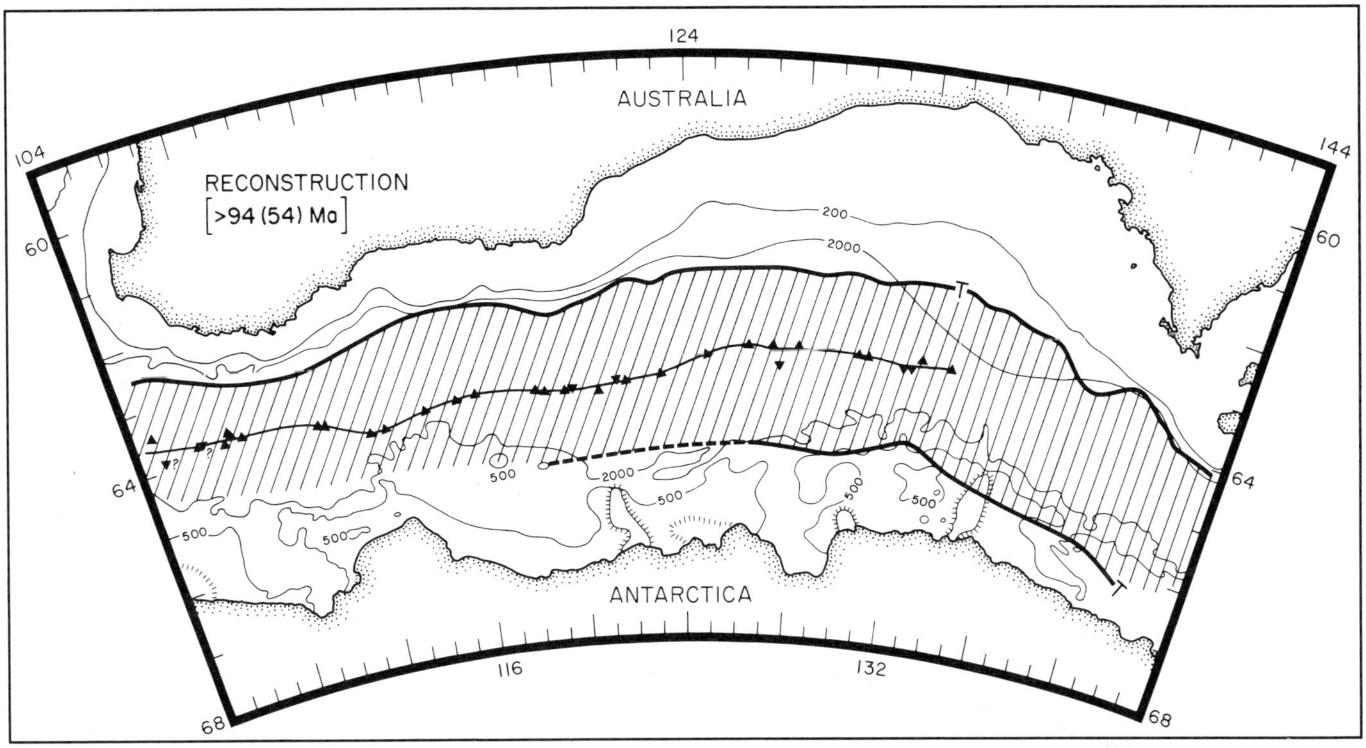

Figure 13A—Predrift reconstruction of Australia–Antarctica (polar stereographic projection). Antarctica is held fixed in its present position. Data sources for continental outlines and bathymetry are as given for Figure 7. The heavy lines marked "T" (broken where extrapolated) represent the magnetic-trough anomalies located on the continental margins off southern Australia and off Wilkes Land, Antarctica. Ruling denotes the extent of the juxtaposed marginal magnetic quiet zones constituting the Australian–Antarctic rift zone. Where not shown abutting against the magnetic trough, the quiet zone off Antarctica is conjectural in its landward extent. Triangles define the seaward boundaries of the magnetic quiet zones and point with their apices to the interiors of the quiet zones whose boundaries they demarcate. Queries indicate tentative identifications. The triangles are joined by a line representing the location of eventual continental breakup. The good quality of the fit as well as the near-symmetrical location of the line of rupture are noteworthy.

tinental margin (Talwani et al., 1979) firmly supports this premise, as do the latest data from the Wilkes Land margin of Antarctica (Childs and Stagg, this volume). The Australian data also indicate that the crust beneath the magnetic quiet zone is thinner than, and in other respects different from, typical continental crust. This led Talwani et al. (1979) to propose that the magnetic quiet zone is floored by "rift crust."

Weissel and Hayes' (1972) initial identification of the oldest seafloor-spreading lineation in the Southeast Indian Ocean as anomaly 22 implied the onset of ocean-crust generation roughly 53 Ma ago, the approximate age of anomaly 22 (Kent and Gradstein, 1986), as the sequence of geomagnetic-polarity reversals extends back in time well beyond 53 Ma. The recent reidentification of this isochron as anomaly 34, proposed by Cande and Mutter (1982), opened up the possibility that oceanic crust of uniform normal polarity underlies the magnetic quiet zone, for anomaly 34 corresponds to the younger boundary of the long period of uniform geomagnetic polarity in the mid- to Late Cretaceous (84–118 Ma; Kent and Gradstein, 1986).

One unsatisfactory aspect, however, of Cande and Mutter's (1982) proposed magnetic model is the rather poor fit of their anomaly 34 to the observed pronounced positive peak located at the landward end of the oceanic magnetic-lineation sequence. This problem has recently been addressed by Veevers (1986), who devised a model for the oldest portion of the oceanic basement that produces a good fit of calculated and observed magnetic anomalies. The essential feature of this new model is the termination of oceanic crust about 45 km landward of the positive anomaly peak at the quiet-zone boundary. This distance corresponds to a time interval of 10 Ma (Fig. 14), based on the spreading rate of 4.5 mm/a reported by Cande and Mutter (1982). (The uncertainty in the determination of the spreading rate results in an estimated error margin of ±5 Ma for the age of the oldest oceanic crust.) Seismic-reflection data from the Australian and Antarctic margins (Boeuf and Doust, 1975; Eittreim et al., 1984; Eittreim and Smith, this volume) consistently show a basement peak in the vicinity of the oldest magnetic isochron. Landward of this basement high, in contrast to the area seaward of it, the observed basement does not exhibit the hyperbolic reflection pattern typically associated with oceanic crust. These latest results thus provide conclusive evidence that, off both southern Australia and Wilkes Land, Antarctica, oceanic crust terminates very close to the seaward boundary of the magnetic quiet zone as determined from the observed seafloor-spreading lineations.

Although the new predrift fit of Australia and Antarctica (Figs. 10, 13A) is superior to the previously proposed reconstructions, it cannot accurately describe the positions of the

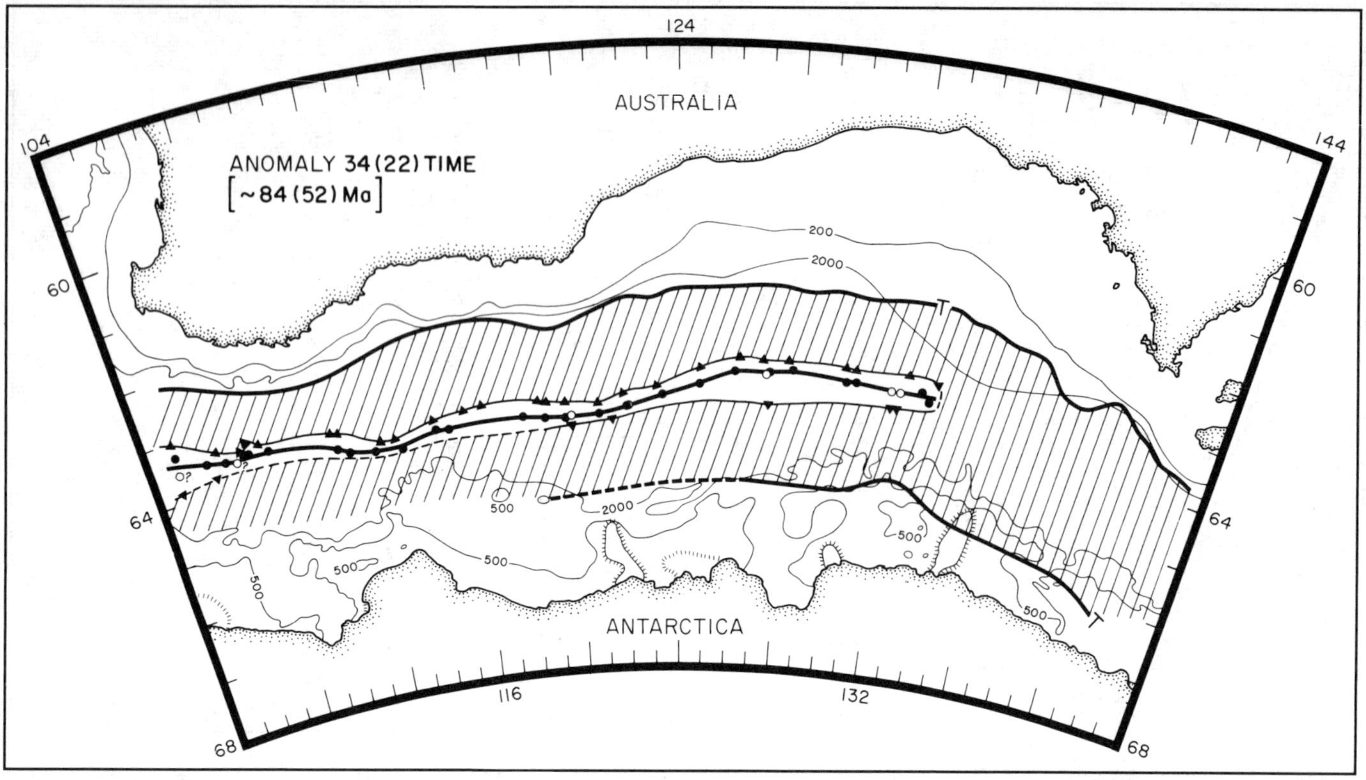

Figure 13B—Reconstruction of the Southeast Indian Ocean and of Australia–Antarctica for anomaly-34(22) time (polar stereographic projection). Antarctica is held fixed in its present position. Data sources for continental outlines, bathymetry, and magnetic lineations are as given for Figure 7. The magnetic trough is represented in the same way as in Figure 13A. Triangles define the seaward boundaries of the marginal magnetic quiet zones and are connected by solid or dashed (where tentative) lines. Ruling represents the area occupied by the magnetic quiet zones or, where continuous between the bounding magnetic troughs, by the remaining intact portion of the rift zone. Where not shown abutting against the magnetic trough, the quiet zone off Antarctica is conjectural in its landward extent. The central anomaly is indicated by large open and closed circles; open symbols represent the observed locations of the stationary Antarctic lineation (queried where identification is tentative), and solid circles denote the rotated positions of the Australian magnetic-anomaly data points. The solid line passing through open and closed circles marks the inferred position of the mid-ocean ridge. The blank space lying between the magnetic quiet zones constitutes oceanic crust generated by seafloor spreading. The narrow tongue of ocean floor extending between the severed portions of the rift and abutting eastward against the still-continuous rift zone is a conspicuous feature of this reconstruction.

continents before the onset of rifting; rather, it depicts the two fragments of Gondwanaland as they were situated at the end of the rift phase, immediately prior to the injection of oceanic crust. Knowledge of the amount of crustal extension for the basement underlying the marginal magnetic quiet zones is required for the accurate prerift reconstruction of Australia and Antarctica. The available seismic evidence documents that the crust beneath the quiet zone south of Australia was subjected to extension. The seismic-refraction data attest to significantly reduced crustal thicknesses in the quiet-zone area (Talwani et al., 1979), and the seismic-reflection data (Boeuf and Doust, 1975; Falvey and Mutter, 1981) show tilted fault blocks successively downdropped away from the major basement fault bounding the quiet zone on its landward side. Quantitative estimates of crustal extension for the Australian margin have yielded inconsistent results, however. Based on the observed amount of total subsidence, Mutter and Cande (1983) and Hegarty (1985) estimated 350–400% of stretching for the quiet zone crust. The estimate of extension obtained from the fault-block geometry as displayed on marine seismic-reflection records yielded an average value of 1.3 for the stretching factor β (Hegarty, 1985), corresponding to 30% of crustal extension. Hegarty's detailed analysis of the stratigraphic record at a number of exploration-well sites on the continental shelf of southern Australia likewise gave values ranging from 1.1 to 1.5. Hegarty (1985) concluded that crustal extension at the margin on the whole was moderate, between 10% and 50%.

These estimates of extension do not reveal a consistent picture for the Australian continental margin; they are based on different methods and were derived from diverse geologic settings. As such, they are likely to reflect local effects and may, in addition, indicate crustal extension that varies across the margin. At this time, therefore, the estimates on balance cannot be considered conclusive evidence as regards magnitude or spatial variation of crustal extension; they do, however, point to stretching as one of the key processes responsible for creating the unusual characteristics of the present-day "rift crust" beneath the magnetic quiet zones off southern Australia and Wilkes Land, Antarctica.

The eastward disappearance of anomaly 34[22] on the Australian margin is well documented (König, 1980). Similarly, this isochron appears to vanish in the vicinity of long

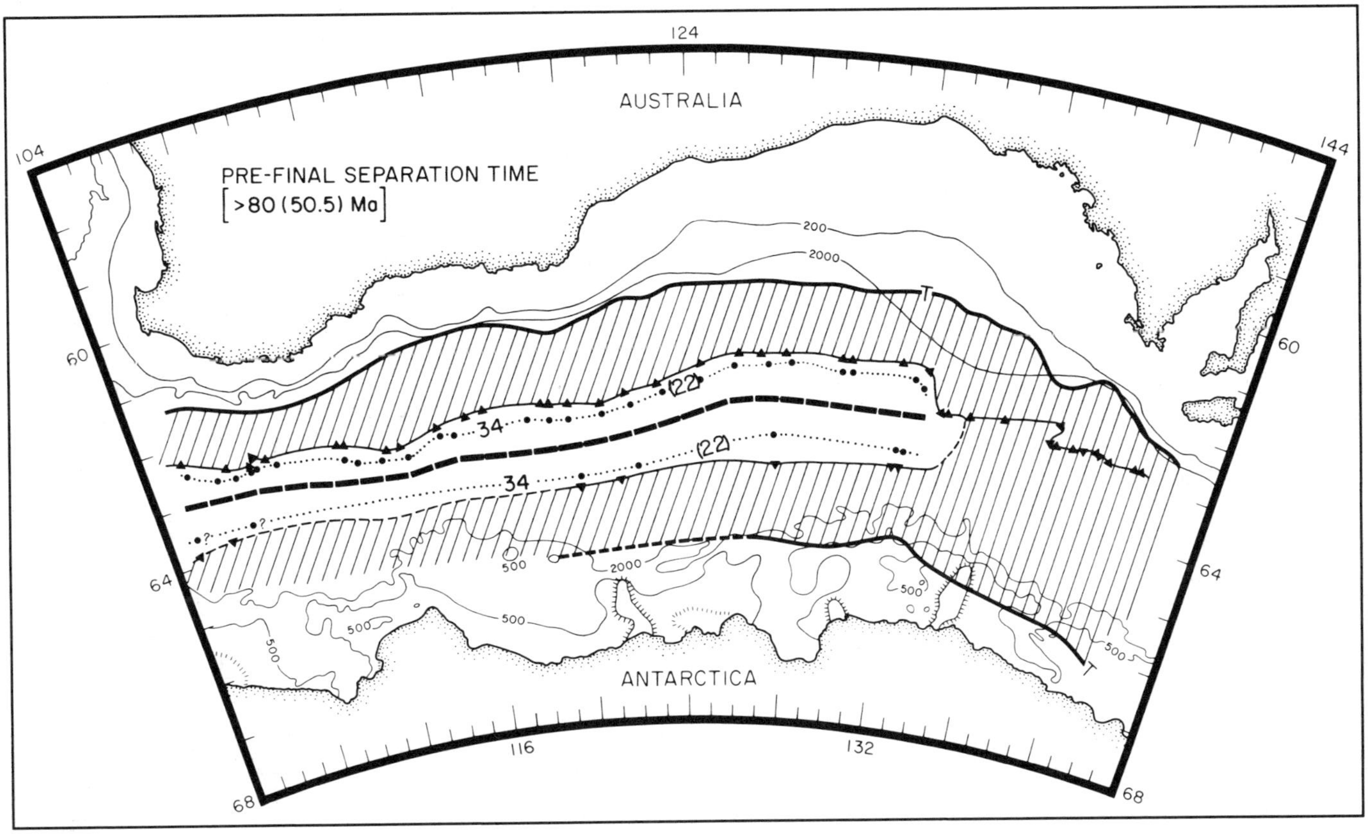

Figure 13C—Reconstruction of the Southeast Indian Ocean and of Australia–Antarctica for the time just prior to final continental breakup (polar stereographic projection). Antarctica is held fixed in its present position. Data sources are as given for Figure 7; the magnetic trough is represented in the same way as in Figure 13A. Triangles connected by solid and dashed lines and situated at the edge of the shaded area and the ruling denote the same geophysical-anomaly features as in Figure 13B; triangles joined by a line inside the shaded region are analogous to the identical symbols of Figure 13A. Observed magnetic lineations are indicated by large dots (queried where recognition is uncertain) and identified by numbered small dots joining the larger symbols. The magnetic-anomaly designation follows that of Cande and Mutter (1982) with the original identification of Weissel and Hayes (1972), extended by König (1980), indicated in parentheses. The mid-ocean ridge is shown schematically by a heavy dashed line near the center of the mostly blank space lying between the magnetic quiet zones and representing oceanic crust generated by seafloor spreading. The fairly wide region of ocean floor abutting eastward against the still-continuous rift zone, extended to maximum width just before rupturing, is a conspicuous feature of this reconstruction.

133° E off Wilkes Land, Antarctica (Figs. 2, 7). On both margins, the younger magnetic-anomaly sequence is observed seaward of the area where anomaly 34[22] does not exist. These observations imply the time-transgressive initiation of seafloor spreading in the Southeast Indian Ocean as depicted in Figures 13B to 13D.

Recently, Veevers (1986) argued for the synchronous opening of the Southeast Indian Ocean along the entire length of the Australian–Antarctic rift zone. He explained the apparent time-transgressive onset of seafloor spreading, observed off southern Australia, by a series of ridge-axis jumps from south to north, leaving the ocean floor generated prior to each ridge jump behind on the Antarctic margin. The evidence crucial to the documentation of these events lies in a poorly explored sector off the coast of Wilkes Land east of long 133° E (Fig. 2). The *Eltanin* tracks in this area consist of only short north–south segments on the continental slope and rise. The magnetic measurements from these tracks indicate small, nonlineated anomalies (Fig. 3) interpreted as the quiet-zone signature. The latest data obtained by R/V *S.P. Lee* (Childs and Stagg, this volume) enable the oldest isochron, anomaly 34[22], to be delineated eastward to about long 132°, possibly 135° E. East of this meridian, the new magnetic profiles are virtually featureless, displaying the typical characteristics of a magnetic quiet zone. The seismic data corroborate these magnetic observations; the identification of the edge of the oceanic crust, the so-called continent-ocean boundary (COB), is easily carried out up to long 135° E, but becomes tentative east of this meridian (Eittreim and Smith, this volume). The well-developed magnetic anomaly on *Lee* profile 7 and its complete absence on track 8 just to the east is a particularly striking observation and a virtual mirror image of the conditions near long 131°30′ E on the conjugate Australian margin (König, 1980). Thus, the evidence that exists today on balance favors the eastward vanishing of the oldest marine magnetic lineation off Wilkes Land, Antarctica, and implies the diachronous opening of the Southeast Indian Ocean.

Information gained from the stratigraphic record to a number of petroleum-exploration wells on the southern continental margin of Australia constitutes additional independent

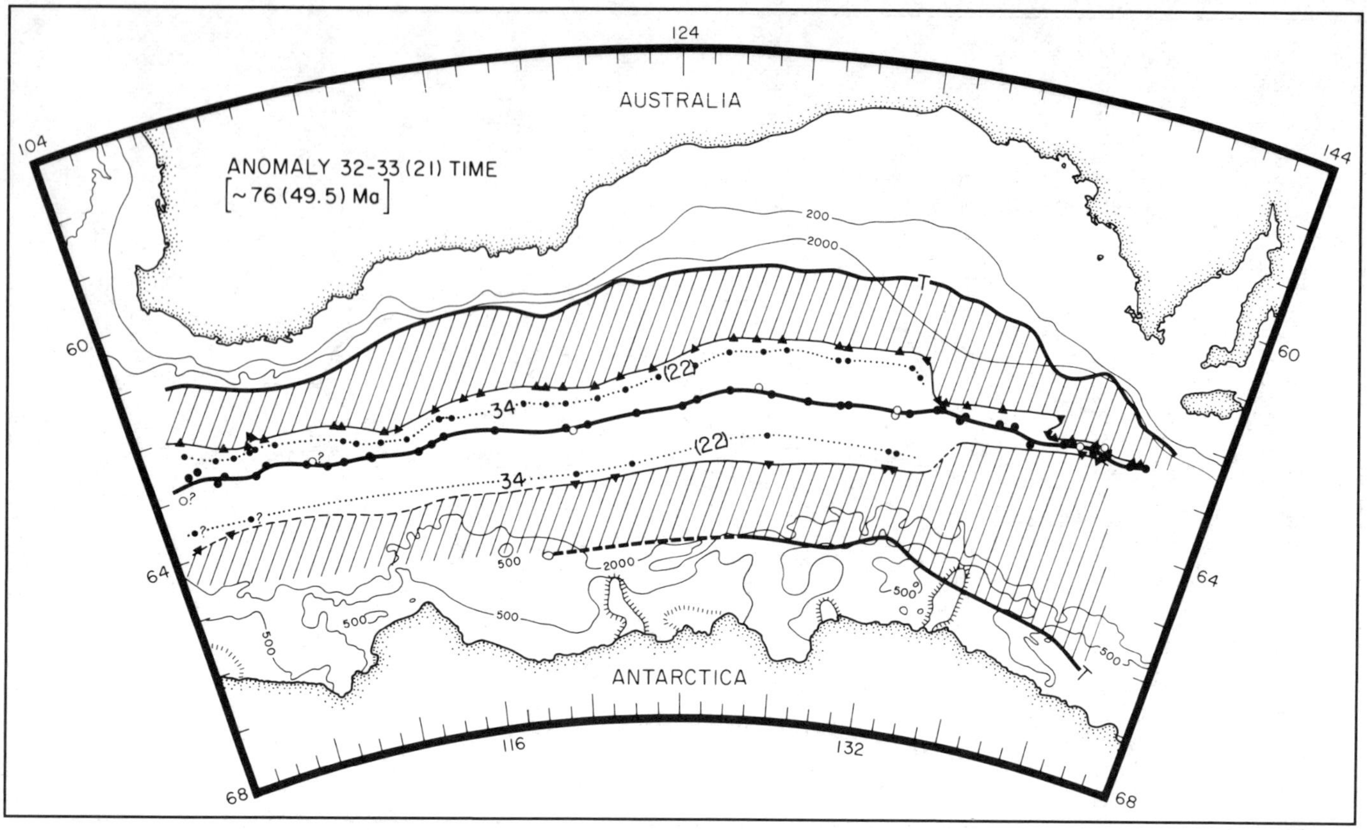

Figure 13D—Reconstruction of the Southeast Indian Ocean for anomaly-32–33(21) time (polar stereographic projection). Antarctica is held fixed in its present position. Data sources are as given for Figure 7; geophysical-anomaly features are represented in the same way as in Figures 13B and 13C. The eastward break of oceanic crust through the rift zone at this time is apparent.

evidence pertaining to the evolution of the margin. Griffiths (1971) interpreted the geologic data to indicate the eastward propagation of rifting and the occurrence, in the Late Cretaceous, of a major marine transgression from west to east. Contrasting environments of Cretaceous sedimentation for the western and eastern parts of the present-day Australian margin, i.e., marginally marine and continental, respectively, were noted by Deighton et al. (1976). These authors also postulated that in Late Cretaceous time a shallow gulf extended from the Naturaliste Plateau (Fig. 7) as far east as the Otway Basin.

Recently, Hegarty (1985) derived the tectonic-subsidence history from the stratigraphic sequence as sampled by a number of wells on the Australian margin. Two distinct phases of subsidence were found to have occurred; one is a period of very rapid subsidence along the margin believed to reflect the rift phase of margin evolution, and the second is a phase of slow subsidence characteristic of lithospheric cooling initiated with the onset of seafloor spreading. The transition from rapid to slow subsidence is very well defined and geologically nearly instantaneous. Mutter et al. (1985), using these results, demonstrated the diachronous nature of this change in subsidence rate along the Australian margin. The transition from fast to slow subsidence occurred earliest in the west and took place at progressively later times in the east (Fig. 14). These authors further showed that the timing of the distinct change in subsidence rate conforms well to the onset of seafloor spreading in the Late Cretaceous as deduced from the pattern of marine magnetic lineations.

Taken as a whole, then, the geologic evidence from the southern Australian margin strongly suggests diachronous opening of the Southeast Indian Ocean, beginning in the west and propagating eastward with time. This breakup history is compatible with the observed pattern of seafloor-spreading magnetic lineations as described here but appears difficult to explain with the model of early seafloor evolution recently proposed by Veevers (1986).

The diachronous separation of a landmass into separate continents is not a phenomenon that was restricted to the evolution of Australia and Antarctica. In fact, it has been argued that this is the rule rather than the exception when a plate ruptures (Courtillot, 1982; Vink, 1982; Courtillot and Vink, 1983). The separation of Arabia from Africa, which has created the Gulf of Aden and the Red Sea, is a currently active example of such continental breakup. The Gulf of Aden owes its shape to the westward propagation towards Afar of the rift system (Cochran, 1981; Courtillot and Vink, 1983), and the Red Sea is a young ocean that has been spreading in its southern part for the past 5 Ma but is still in the final rift stage in the north (Roeser, 1975; Cochran, 1983). Vink (1982) discussed other examples, notably the Gulf of California and the Mesozoic breakup of Africa and South America.

As for Australia–Antarctica, it is intriguing to note that during the time when rifting propagated eastward, continental

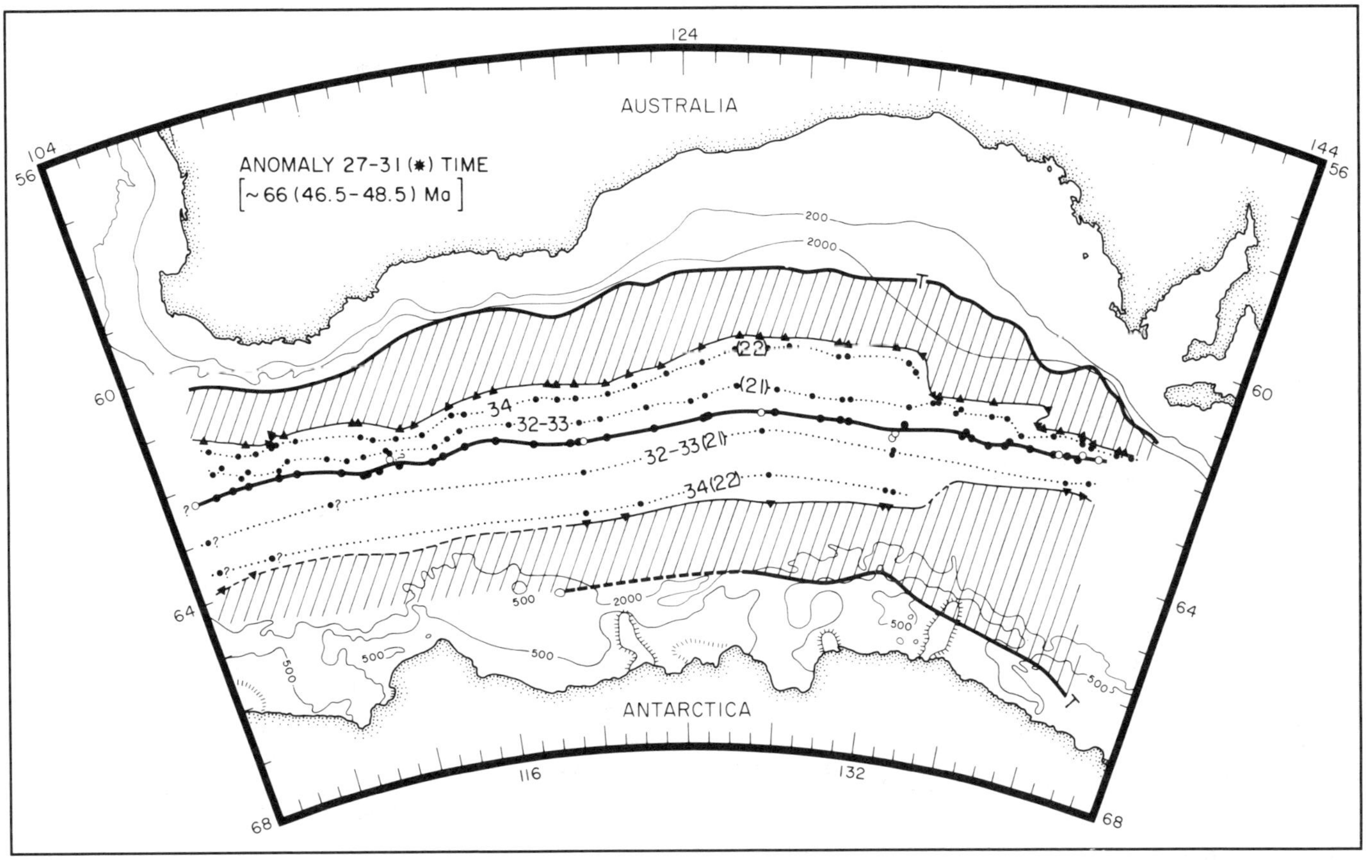

Figure 13E—Reconstruction of the Southeast Indian Ocean for anomaly-27–31(*) time (polar stereographic projection). Antarctica is held fixed in its present position. Data sources are as given for Figure 7; geophysical-anomaly features are represented in the same way as in Figures 13B and 13C. The good fit of stationary and rotated central-anomaly positions, as in the case of lineations 32–33(21) and 34(22) shown in Figures 13D and 13B, respectively, is apparent.

separation was very slow (Fig. 14), slower, in fact, than has been observed elsewhere in the world's oceans. This period of very slow (4.5 mm/a) spreading lasted for some 45 Ma, based on the onset of seafloor generation about 94 Ma ago, and in turn was followed by a brief 4-Ma interval of accelerated (9 mm/a) spreading before the final (22-mm/a) half-rate of separation of Australia from Antarctica was established slightly less than 45 Ma ago (Fig. 14). This sequence of events can perhaps be most readily explained, in a qualitative fashion, in terms of great resistive forces offered to crustal extension and progressive tearing by the still-contiguous East Gondwana continental remnant. With continental separation nearing completion, the rate of motion increased sharply, until it reached a steady-state equilibrium once the final severing of the connection between the two new continents had occurred and Australia was free to move northward independently of Antarctica.

CONCLUSIONS

1. The geophysical data collected at the continental margin off Wilkes Land, Antarctica, document the existence of a magnetic quiet zone on the continental slope and rise. The quiet zone is characterized by a distinct pattern of geophysical anomalies that document its fundamental difference with respect to the ocean basin to the north and the continent to the south.
2. The magnetic quiet zone off Wilkes Land, Antarctica, is a virtual mirror image of the quiet zone that is situated on the conjugate southern continental margin of Australia. There it has been interpreted as overlying the remnant of a continental-rift system that was active in the past and preceded the breakup of Australia and Antarctica in the Late Cretaceous. The close similarity of the Antarctic marginal quiet zone to the southern Australian one strongly suggests the existence of a corresponding partial fossil rift system beneath the continental slope and rise off Wilkes Land, Antarctica.
3. On each of the conjugate margins of the Southeast Indian Ocean, the oldest seafloor-spreading lineation, anomaly 34 (22 according to the initial identification of Weissel and Hayes, 1972), lies close to and parallels the seaward edge of the magnetic quiet zone for most of its length. This condition has been used as the basis for the derivation of a new pole of rotation for the predrift fit of Australia and Antarctica; the new pole is located at lat 1.5° N, long 37.0° E, the angle of rotation for the closing of the Southeast Indian Ocean being 28.0°. Likewise, new finite-rotation parameters were determined for the younger magnetic isochrons and used to trace the 38-Ma period of the evolution of the Southeast Indian Ocean that followed the initial opening.
4. The new continental reconstruction is less tight than the previously proposed ones since it includes the approxi-

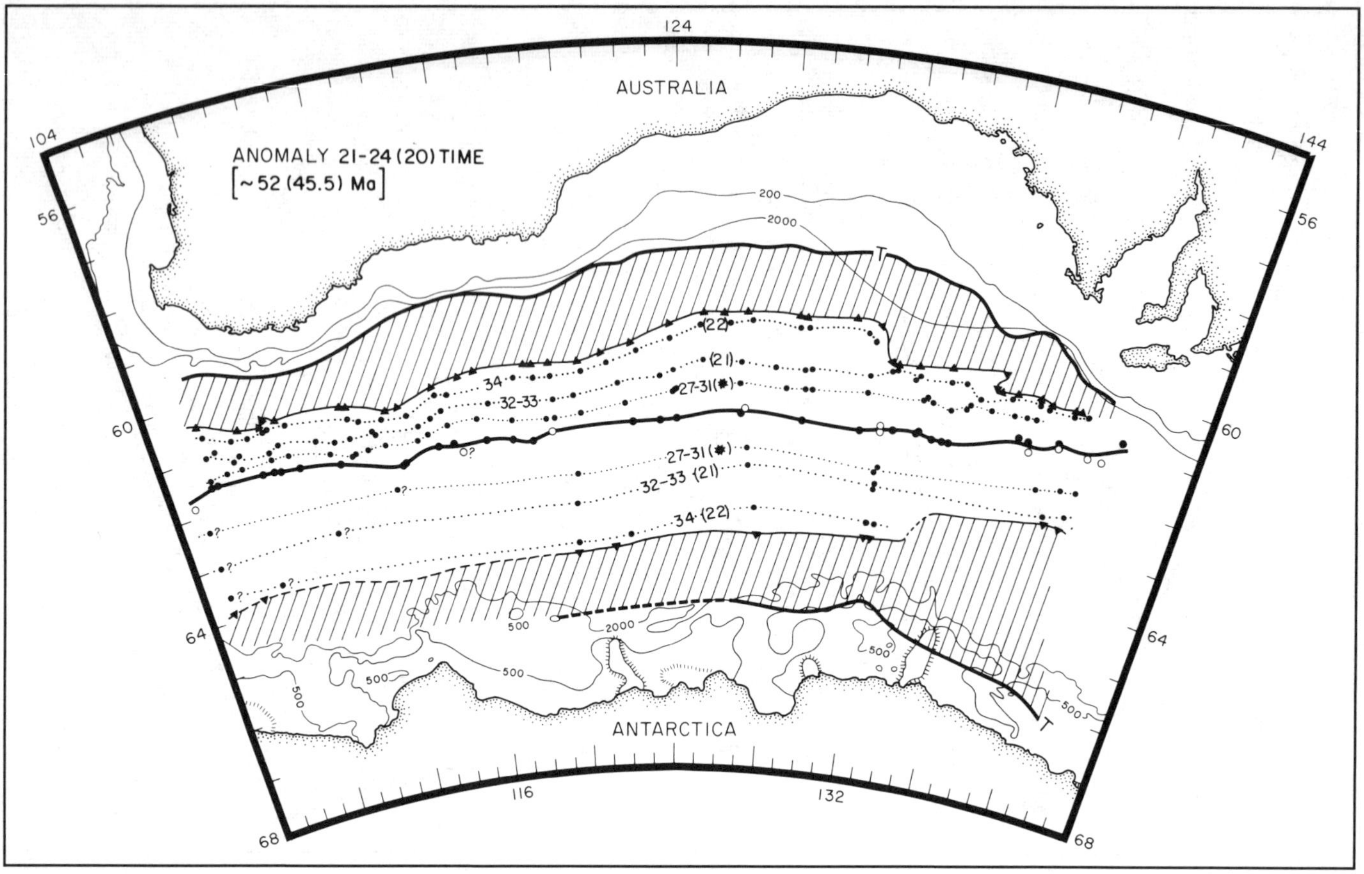

Figure 13F—Reconstruction of the Southeast Indian Ocean for anomaly-21–24(20) time (polar stereographic projection). Antarctica is held fixed in its present position. Data sources are as given for Figure 7; geophysical-anomaly features are represented in the same way as in Figures 13B and 13C.

mately 250 km wide rift zone as it existed just prior to breakup. The new fit avoids any overlap of continental material and at the same time fully satisfies the independently established geologic constraints governing the reassembly of Australia and Antarctica.

5. The eastward disappearance of anomaly 34[22] on both margins of the Southeast Indian Ocean together with the existence and uninterrupted extent of the younger isochrons along the margins implies the progressive opening from west to east of the Southeast Indian Ocean. While very slow seafloor spreading occurred in the west, extension and progressive tearing of the rift crust took place in the east for a period of probably 45 Ma. During the last 4 Ma before full separation was achieved, seafloor spreading accelerated to more than twice its previous rate and roughly doubled again with the final breakup of the East Gondwana continent into Australia and Antarctica. The existing geologic information on the evolution of the southern Australian margin is fully compatible with the diachronous opening of the Southeast Indian Ocean as deduced from the pattern of marine magnetic lineations.
6. Two major unresolved problems remain. The age of the oldest oceanic crust in the Southeast Indian Ocean has not yet been determined by independent geophysical methods. Direct sampling by drilling appears indispensable in order for the onset of seafloor spreading to be firmly established. The amount and variation of extension of the crust underlying the marginal magnetic quiet zones require further study. This is necessary for the proper reassembly of Australia and Antarctica to their prerift Gondwana positions.

ACKNOWLEDGMENTS

The great majority of the work was done while the writer was affiliated with Lamont-Doherty Geological Observatory of Columbia University, and constitutes part of his Ph. D. thesis. The data acquisition, reduction, and analysis were supported by the National Science Foundation through grants GA-894, GA-1543, GA-1121, GV-12825, GV-23334, GV-27472, GV-40896, and OPP74-02238. The relentless efforts, despite often adverse conditions, of the officers, crews, and scientific personnel aboard U.S.N.S. *Eltanin*, which made the successful acquisition of the geophysical data on the Antarctic margin off Wilkes Land possible, is gratefully acknowledged.

While at the Aldridge Laboratory of Applied Geophysics, the writer has been supported by Project MIDAS, a consortium of ten oil and geophysical companies, and through grants TO-82K0782 (ONR) and EAR-84-12207 (NSF).

Most of the figures were drafted, in expert fashion, by Virginia Rippon. The manuscript was critically reviewed by S. C. Cande and J. C. Mutter of Lamont-Doherty Geological Observatory, and by two anonymous reviewers. Their constructive criticism served to improve this paper and is gratefully acknowledged.

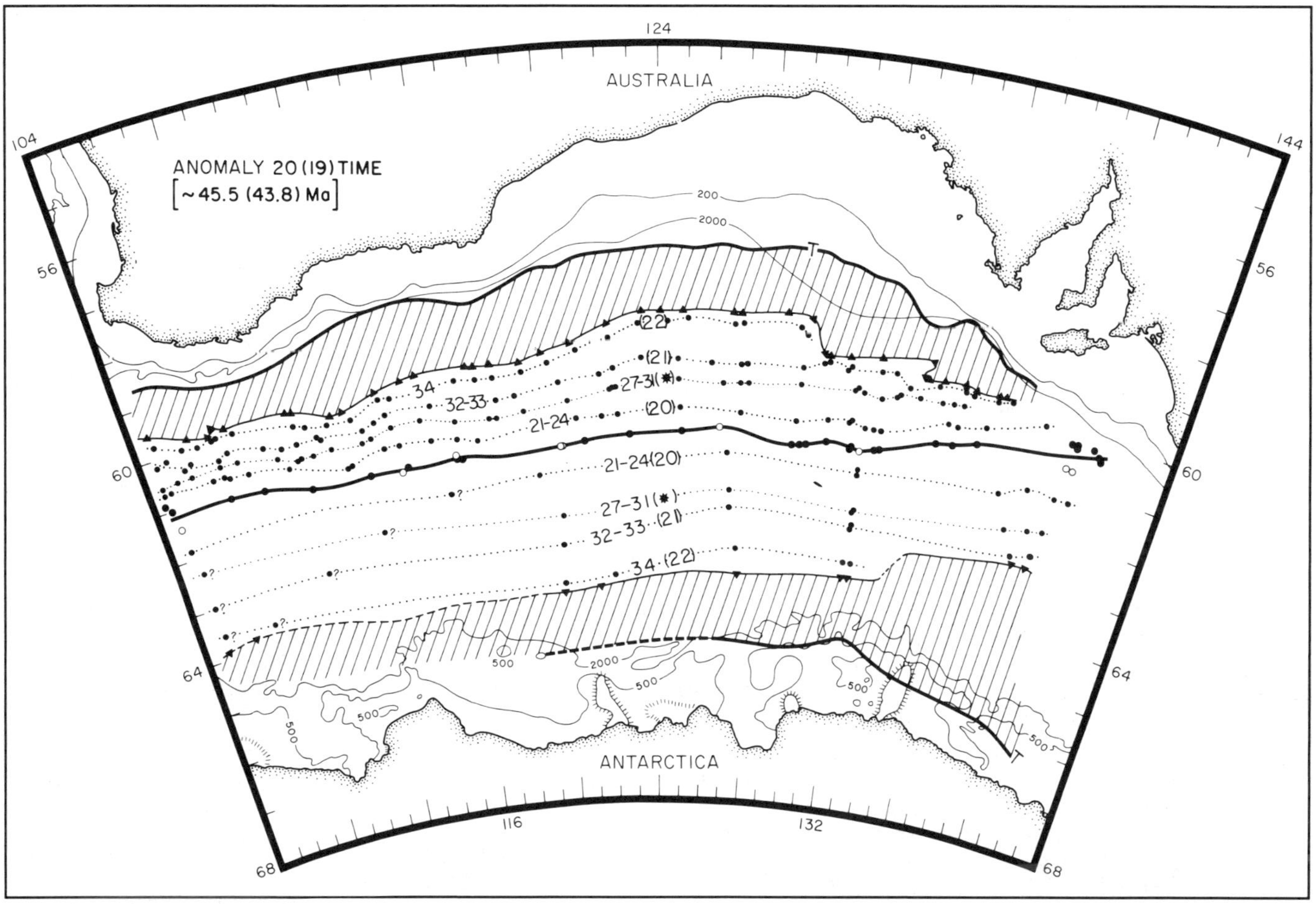

Figure 13G—Reconstruction of the Southeast Indian Ocean for anomaly-20(19) time (polar stereographic projection). Antarctica is held fixed in its present position. Data sources are as given for Figure 7; geophysical-anomaly features are represented in the same way as in Figures 13B and 13C.

REFERENCES CITED

Adams, R. D., 1972, Dispersion-wave studies in Antarctica, *in* R. J. Adie, ed., Antarctic geology and geophysics: Oslo, Universitetsforlaget, p. 473–480.

American Geographical Society, 1970, Antarctica: New York, American Geographical Society Map Series of the World, sheet 13. Scale 1:5,000,000.

Barrett, P. J., G. W. Grindley, and P. N. Webb, 1972, The Beacon Supergroup of East Antarctica, *in* R. J. Adie, ed., Antarctic geology and geophysics: Oslo, Universitetsforlaget, p. 319–332.

Barron, E. J., C. G. A. Harrison, and W. W. Hay, 1978, A revised reconstruction of the southern continents: Transactions, American Geophysical Union, v. 59, p. 436–449.

Bentley, C. R., 1973, Crustal structure of Antarctica, *in* S. Mueller, ed., The structure of the earth's crust based on seismic data: Tectonophysics, v. 20, p. 229–240.

Boeuf, M. G., and H. Doust, 1975, Structure and development of the southern margin of Australia: Australian Petroleum Exploration Association Journal, v. 15, p. 33–43.

Cain, J. C., S. J. Hendricks, W. E. Daniels, and D. C. Jensen, 1968, Computation of the main geomagnetic field from spherical harmonic expansions: Greenbelt, MD, National Space Sciences Data Center, Data User's Note NSSDC 68-11.

Calkin, P. E., and R. L. Nichols, 1972, Quaternary studies in Antarctica, *in* R. J. Adie, ed., Antarctic geology and geophysics: Oslo, Universitetsforlaget, p. 625–643.

Cande, S. C., and J. C. Mutter, 1982, A revised identification of the oldest sea-floor spreading anomalies between Australia and Antarctica: Earth and Planetary Science Letters, v. 58, p. 151–160.

Carey, S. W., 1958, A tectonic approach to continental drift, *in* S. W. Carey, ed., Continental drift—a symposium: Hobart, Tasmania, University of Tasmania, p. 177–355.

Cochran, J. R., 1981, The Gulf of Aden: Structure and evolution of a young ocean basin and continental margin: Journal of Geophysical Research, v. 86, p. 263–287.

______, 1983, A model for development of Red Sea: Bulletin of the American Association of Petroleum Geologists, v. 67, p. 41–69.

Colbert, E. H., 1972, Antarctic Triassic tetrapods, *in* R. J. Adie, ed., Antarctic geology and geophysics: Oslo, Universitetsforlaget, p. 393–401.

Courtillot, V., 1982, Propagating rifts and continental breakup: Tectonics, v. 1, p. 239–250.

Courtillot, V., and G. E. Vink, 1983, How continents break up: Scientific American, v. 249, n. 1, p. 42–49.

Craddock, C., 1970, Antarctic geology and Gondwanaland: Antarctic Journal, v. 5, p. 53–57.

______, comp., 1972a, Geologic map of Antarctica: New York, American Geographical Society. Scale 1:5,000,000.

______, 1972b, Antarctic tectonics, *in* R. J. Adie, ed., Antarctic geology and geophysics: Oslo, Universitetsforlaget, p. 449–455.

Defense Mapping Agency Hydrographic Center, 1975, Charts of the earth's magnetic field, epoch 1975.0: Washington, DC. Scale 1:39,000,000.

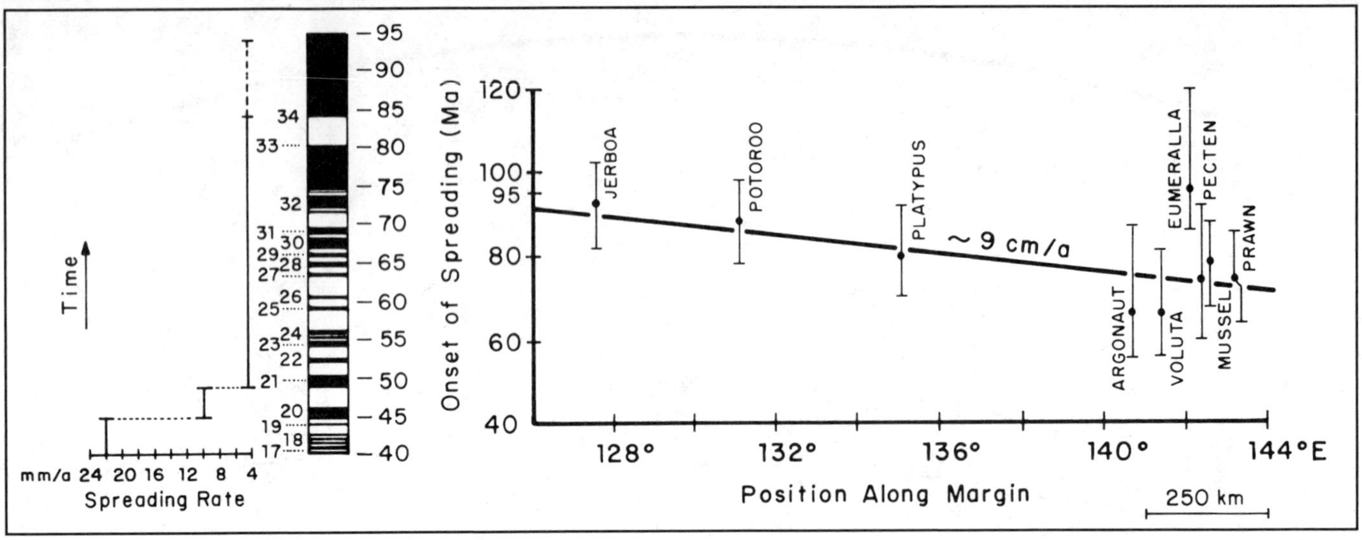

Figure 14—Plot of the time of breakup vs. location along the margin as deduced from petroleum-exploration well records at the southern continental margin of Australia (right-hand side), and of the relationship between spreading rate and magnetic-polarity reversals during the first 55 Ma of seafloor evolution of the Southeast Indian Ocean (left-hand side). The plot on the right is modified from Mutter et al. (1985); the magnetic polarity-reversal time scale is that of Kent and Gradstein (1986), except for the length of the anomaly-19 interval, which is from Cande and Mutter (1982), as are the spreading rates indicated. The initiation of seafloor spreading at 95 Ma B.P. is in accordance with the model proposed by Veevers (1986).

Deighton, I., D. A. Falvey, and D. J. Taylor, 1976, Depositional environments and geotectonic framework: Southern Australian continental margin: Australian Petroleum Exploration Association Journal, v. 16, p. 25-36.

Dow, J. A. S., and V. E. Neall, 1972, Summary of the geology of lower Rennick Glacier, northern Victoria Land, *in* R. J. Adie, ed., Antarctic geology and geophysics: Oslo, Universitetsforlaget, p. 339-344.

Doyle, H. A., and I. B. Everingham, 1964, Seismic velocities and crustal structure in southern Australia: Geological Society of Australia Journal, v. 11, p. 141-150.

Drewry, D. J., 1976, Sedimentary basins of the East Antarctic craton from geophysical evidence, *in* M. H. P. Bott, ed., Sedimentary basins of continental margins and cratons: Tectonophysics, v. 36, p. 301-314.

Eittreim, S. L., A. K. Cooper, et al., 1984, Marine geological and geophysical investigations of the Antarctic continental margin, 1984: U.S. Geological Survey Circular 935, 12 p.

Elliot, D. H., 1975, Tectonics of Antarctica: A review: American Journal of Science, v. 275-A, p. 45-106.

Falvey, D. A., 1974, The development of continental margins in plate tectonic theory: Australian Petroleum Exploration Association Journal, v. 14, p. 95-106.

Falvey, D. A., and J. C. Mutter, 1981, Regional plate tectonics and the evolution of Australia's passive continental margins: Bureau of Mineral Resources Journal of Australian Geology and Geophysics, v. 6, p. 1-29.

Ford, A. B., 1975, Volcanic rocks of Naturaliste Plateau, eastern Indian Ocean, site 264, DSDP leg 28, *in* D. E. Hayes, L. A. Frakes, et al., Initial reports of the Deep-Sea Drilling Project: Washington, DC, U.S. Government Printing Office, v. 28, p. 821-833.

Frakes, L. A., and J. C. Crowell, 1970, Geologic evidence for the place of Antarctica in Gondwanaland: Antarctic Journal, v. 5, p. 67-69.

Geological Society of Australia, 1971, Tectonic map of Australia and New Guinea: Sydney, Australia. Scale 1:5,000,000.

Griffiths, J. R., 1971, Continental-margin tectonics and the evolution of southeast Australia: Australian Petroleum Exploration Association Journal, v. 11, p. 75-79.

______, 1974, Revised continental fit of Australia and Antarctica: Nature, v. 249, p. 336-338.

Grikurov, G. E., M. G. Ravich, and D. S. Soloviev, 1972, Tectonics of Antarctica, *in* R. J. Adie, ed., Antarctic geology and geophysics: Oslo, Universitetsforlaget, p. 457-468.

Hallam, A., 1976, How closely did the continents fit together?: Nature, v. 262, p. 94-95.

Hamilton, W., 1967, Tectonics of Antarctica: Tectonophysics, v. 4, p. 555-568.

Hawkins, L. V., J. F. Hennion, J. E. Nafe, and H. A. Doyle, 1965a, Marine seismic-refraction studies on the continental margin to the south of Australia: Deep-Sea Research, v. 12, p. 479-495.

Hawkins, L. V., J. F. Hennion, J. E. Nafe, and R. F. Thyer, 1965b, Geophysical investigations in the area of the Perth Basin, Western Australia: Geophysics, v. 30, p. 1026-1052.

Hayes, D. E., and J. R. Conolly, 1972, Morphology of the Southeast Indian Ocean, *in* D. E. Hayes, ed., Antarctic oceanology. II. The Australian-New Zealand sector: American Geophysical Union, Antarctic Research Series, v. 19, p. 125-145.

Hayes, D. E., and L. A. Frakes, 1975, General synthesis, Deep-Sea Drilling Project leg 28, *in* D. E. Hayes, L. A. Frakes, et al., Initial reports of the Deep-Sea Drilling Project: Washington, DC, U.S. Government Printing Office, v. 28, p. 919-942.

Hayes, D. E., and R. E. Wall, 1975, Bathymetric, magnetic, and seismic-reflection data: Glomar Challenger, leg 28, *in* D. E. Hayes, L. A. Frakes, et al., Initial reports of the Deep-Sea Drilling Project: Washington, DC, U.S. Government Printing Office, v. 28, p. 945-975.

Hayes, D. E., L. A. Frakes, P. J. Barrett, D. A. Burns, P. -H. Chen, A. B. Ford, A. G. Kaneps, E. M. Kemp, D. W. McCollum, D. J. W. Piper, R. E. Wall, and P. N. Webb, 1975a, Site 268, *in* D. E. Hayes, L. A. Frakes, et al., Initial reports of the Deep-Sea Drilling Project: Washington, DC, U.S. Government Printing Office, v. 28, p. 153-177.

———, 1975b, Site 267, *in* D. E. Hayes, L. A. Frakes et al., Initial reports of the Deep-Sea Drilling Project: Washington, DC, U.S. Government Printing Office, v. 28, p. 121–151.

———, 1975c, Site 269, *in* D. E. Hayes, L. A. Frakes, et al., Initial reports of the Deep-Sea Drilling Project: Washington, DC, U.S. Government Printing Office, v. 28, p. 179–210.

Hegarty, K. A., 1985, Origin and evolution of selected plate boundaries: Ph.D. Thesis, Columbia University, New York.

Houtz, R. E., 1974, Continental margin of Antarctica: Pacific–Indian sectors, *in* C. A. Burk, and C. L. Drake, eds., The geology of continental margins: New York, Springer-Verlag, p. 655–658.

Houtz, R. E., and R. G. Markl, 1972, Seismic-profiler data between Antarctica and Australia, *in* D. E. Hayes, ed., Antarctic oceanology. II. The Australian–New Zealand sector: American Geophysical Union, Antarctic Research Series, v. 19, p. 147–164.

IAGA Commission 2 Working Group 4, 1969, International geomagnetic reference field 1965.0: Journal of Geophysical Research, v. 74, p. 4407–4408.

Kemp, E. M., L. A. Frakes, and D. E. Hayes, 1975, Paleoclimatic significance of diachronous biogenic facies, leg 28, Deep-Sea Drilling Project, *in* D. E. Hayes, L. A. Frakes, et al., Initial reports of the Deep-Sea Drilling Project: Washington, DC, U.S. Government Printing Office, v. 28, p. 909–917.

Kennett, J. P., R. E. Burns, J. E. Andrews, M. Churkin, Jr., T. A. Davies, P. Dumitrica, A. R. Edwards, J. S. Galehouse, G. H. Packham, and G. J. van der Lingen, 1972, Australian–Antarctic continental drift, paleocirculation changes and Oligocene deep-sea erosion: Nature Physical Science, v. 239, p. 51–55.

Kennett, J. P., R. E. Houtz, P. B. Andrews, A. R. Edwards, V. A. Gostin, M. Hajós, M. A. Hampton, D. G. Jenkins, S. V. Margolis, A. T. Ovenshine, and K. Perch-Nielsen, 1974, Development of the Circum-Antarctic Current: Science, v. 186, p. 144–147.

Kent, D. V., and F. M. Gradstein, 1986, A Jurassic to Recent chronology, *in* B. E. Tucholke, and P. R. Vogt, eds., The western north Atlantic region: Boulder, Geological Society of America, The geology of North America, v. M, p. 45–50.

König, M., 1980, Geophysical investigations of the southern continental margin of Australia and the conjugate sector of East Antarctica: Ph.D. Thesis, Columbia University, New York.

König, M., and M. Talwani, 1977, A geophysical study of the southern continental margin of Australia: Great Australian Bight and western sections: Geological Society of America Bulletin, v. 88, p. 1000–1014.

Laird, M. G., R. A. Cooper, and J. B. Jago, 1977, New data on the lower Palaeozoic sequence of northern Victoria Land, Antarctica, and its significance for Australian–Antarctic relations in the Palaeozoic: Nature, v. 265, p. 107–110.

Markl, R. G., 1974, Bathymetric map of the eastern Indian Ocean, *in* T. A. Davies, B. P. Luyendyk, et al., Initial reports of the Deep-Sea Drilling Project: Washington, DC, U.S. Government Printing Office, v. 26, p. 967–968.

———, 1978, Basement morphology and rift geometry near the former junction of India, Australia, and Antarctica: Earth and Planetary Science Letters, v. 39, p. 211–225.

Mathur, S. P., 1974, Crustal structure in southwestern Australia from seismic and gravity data: Tectonophysics, v. 24, p. 151–182.

McKenzie, D., and J. G. Sclater, 1971, The evolution of the Indian Ocean since the Late Cretaceous: Geophysical Journal of the Royal Astronomical Society, v. 25, p. 437–528.

Mutter, J. C., and S. C. Cande, 1983, The early opening between Broken Ridge and Kerguelan Plateau: Earth and Planetary Science Letters, v. 65, p. 369–376.

Mutter, J. C., K. A. Hegarty, S. C. Cande, and J. K. Weissel, 1985, Breakup between Australia and Antarctica: A brief review in the light of new data: Tectonophysics, v. 114, p. 255–279.

Nathan, S., and D. N. B. Skinner, 1972, Recent advances in the pre-Permian geology of northern Victoria Land, *in* R. J. Adie, ed., Antarctic geology and geophysics: Oslo, Universitetsforlaget, p. 333–337.

Norton, I. O., and P. Molnar, 1977, Implications of a revised fit between Australia and Antarctica for the evolution of the eastern Indian Ocean: Nature, v. 267, p. 338–340.

Plumstead, E. P., 1964, Palaeobotany of Antarctica, *in* R. J. Adie, ed., Antarctic geology: Amsterdam, North-Holland Publishing Company, p. 637–654.

Poehls, K. A., B. P. Luyendyk, and J. R. Heirtzler, 1973, Magnetic smooth zones in the world's oceans: Journal of Geophysical Research, v. 78, p. 6985–6997.

Ravich, M. G., 1972, Regional metamorphism of the Antarctic platform crystalline basement, *in* R. J. Adie, ed., Antarctic geology and geophysics: Oslo, Universitetsforlaget, p. 505–515.

Roeser, H.-A., 1975, A detailed magnetic survey of the southern Red Sea: Geologisches Jahrbuch, v. 13, p. 131–153.

Sclater, J. G., R. L. Fisher, P. Patriat, C. R. Tapscott, and B. Parsons, 1981, Eocene to Recent development of the Southwest Indian Ridge, a consequence of the evolution of the Indian Ocean triple junction: Geophysical Journal of the Royal Astronomical Society, v. 64, p. 587–604.

Smith, A. G., and A. Hallam, 1970, The fit of the southern continents: Nature, v. 225, p. 139–144.

Solomon, M., and J. R. Griffiths, 1972, Tectonic evolution of the Tasman Orogenic Zone, eastern Australia: Nature Physical Science, v. 237, p. 3–6.

Sproll, W. P., and R. S. Dietz, 1969, Morphological continental-drift fit of Australia and Antarctica: Nature, v. 222, p. 345–348.

Spry, A., and M. R. Banks, eds., 1962, The geology of Tasmania: Geological Society of Australia Journal, v. 9, p. 107–362.

Talwani, M., J. C. Mutter, R. E. Houtz, and M. König, 1979, The crustal structure and evolution of the area underlying the magnetic quiet zone on the margin south of Australia, *in* J. S. Watkins, L. Montadert, and P. W. Dickerson, eds., Geological and geophysical investigations of continental margins: American Association of Petroleum Geologists Memoir 29, p. 151–175.

Toit, A. L. du, 1937, Our wandering continents: Edinburgh, Scotland, Oliver and Boyd, 366 p.

Veevers, J. J., 1986, Breakup of Australia and Antarctica estimated as mid-Cretaceous (95 ±5 Ma) from magnetic and seismic data at the continental margin: Earth and Planetary Science Letters, v. 77, p. 91–99.

Vink, G. E., 1982, Continental rifting and the implications for plate tectonic reconstructions: Journal of Geophysical Research, v. 87, p. 10677–10688.

Wegener, A., 1929, Die Entstehung der Kontinente und Ozeane: Braunschweig, Germany, Verlag Friedrich Vieweg und Sohn, fourth edition, 231 p.

Weissel, J. K., and D. E. Hayes, 1971, Asymmetric sea-floor spreading south of Australia: Nature, v. 231, p. 518–522.

———, 1972, Magnetic anomalies in the Southeast Indian Ocean, *in* D. E. Hayes, ed., Antarctic oceanology. II. The Australian–New Zealand sector: American Geophysical Union, Antarctic Research Series, v. 19, p. 165–196.

Weissel, J. K., D. E. Hayes, and E. M. Herron, 1977, Plate tectonics synthesis: The displacements between Australia, New Zealand, and Antarctica since the Late Cretaceous: Marine Geology, v. 25, p. 231–277.

Eittreim, S. L., and Hampton, M. A., 1987, *The Antarctic Continental Margin: Geology and Geophysics of Offshore Wilkes Land*, CPCEMR Earth Science Series, v. 5A: Houston, Texas, Circum-Pacific Council for Energy and Mineral Resources

Marine Topography of Offshore Antarctica

T. E. Chase
B. A. Seekins
J. D. Young
S. L. Eittreim
U.S. Geological Survey
Menlo Park, California

Two maps of the offshore topography off Wilkes Land, Antarctica, are presented here with the seafloor contours on different projections. Both polar stereographic and Mercator projections are presented to give the reader an option of either for future investigations. Less distortion makes the polar stereographic projection favorable for high latitudes, while Mercator projection is useful for hand plotting of data while on shipboard and in the laboratory.

These maps are the result of previous surveys in this region that have resulted in several previous maps (among them Vanney and Johnson, 1979; Anonymous, 1980), upgraded by data collected on the U.S. Geological Survey ship *S.P. Lee* on cruise L1-84-AA. Additional information about procedures used in preparing these maps and others in this area can be obtained by contacting the authors.

Perhaps the most striking features revealed are the dendritic, deeply cut canyon systems of the upper continental rise. While these canyons begin at various upper levels on the rise, their deeply incised portions end at about 3500 m. The shelf-slope break ranges from 400 to 600 m depth. At midlevel on the upper continental rise (about 2500 m), isolated highs occur that are presumably caused by headward erosion and encirclement of portions of the upper rise fan sediment.

REFERENCES CITED

Anonymous, 1980, General bathymetric chart of the oceans (GEBCO), Antarctic sheet, scale 1:6 million: Hydrographic Chart Distribution Office, Department of Fisheries and Oceans, Ottawa.

Vanney, J. R., and G. L. Johnson, 1979, The seafloor morphology seaward of Terre Adelie (Antarctica): Deutsches Hydrologisches Zeitschrift, v. 32, p. 77–87.

Eittreim, S. L., and Hampton, M. A., 1987, *The Antarctic Continental Margin: Geology and Geophysics of Offshore Wilkes Land*,
CPCEMR Earth Science Series, v. 5A: Houston, Texas, Circum-Pacific Council for Energy and Mineral Resources

Geology of Sediment Cores from the George V Continental Margin, Antarctica

M. A. Hampton
U.S. Geological Survey
Menlo Park, California

J. H. Kravitz
Office of Naval Research
Arlington, Virginia

G. Luepke
U.S. Geological Survey
Menlo Park, California

Sediment cores collected from the continental shelf off George V Land, Antarctica, contain Holocene diatom mud and Pleistocene glacial-marine diamicton. Cores from the continental slope contain sandy to gravelly sediment that has been reworked by strong geostrophic currents. Cores from the continental rise consist of alternating massive and laminated units that were deposited by hemipelagic and fluctuating traction-current processes, respectively. Between-core and down-core mineralogical variations observed in the sediment probably were caused by hydraulic sorting processes; there is no compelling reason to invoke multiple source terrains. The gravel-sized clasts, although lithologically heterogeneous, are represented compositionally by rocks that crop out along the George V and adjacent coasts. Most quartz sand grains have a glacially imparted morphology, with various degrees of superimposed rounding acquired in fluvial and submarine environments. A subordinate but distinct group of eolian grains probably was derived from a polar desert that once existed in front of the advancing ice cap.

The physical properties of these polar sediment types have values mostly within ranges reported for marine sediment collected world-wide. Data from laboratory consolidation tests confirm that the shelf diamictons are glacial-marine deposits rather than lodgement or basal till, because no effects of previous ice-loading could be detected in the tests. Measurements of shear strength indicate that sediment underlying slopes of the magnitude depicted on bathymetric charts are stable under gravitational loads.

INTRODUCTION

Seven sediment cores were collected on the George V continental margin, during the January 1984 cruise of the USGS research vessel *S.P. Lee*, to extend the existing knowledge of the sedimentary process in this area and to measure the physical properties of a variety of polar sediment types (Fig. 1). Of particular interest are the continental slope and rise, because these areas have been studied less intensively than the continental shelf (Piper and Brisco, 1975; Anderson et al., 1979; Domack, 1980, 1982; Domack and Anderson, 1983). Sampling time was limited during the cruise because the main objective was to collect multichannel seismic-reflection data, but the seven cores discussed in this report provide reconnaissance-type information that is largely consistent with, and extends, the story constructed from previous studies.

As with most polar and subpolar marine regions, the sedimentary environment of the George V continental margin has varied drastically during Cenozoic time. Ice advances onto the continental shelf during the Pleistocene sculpted the sea floor into a series of flat banks and deep depressions (Vanney and Johnson, 1979; Fig. 1), and lodgement till and glacial-marine diamicton were emplaced during at least two episodes of deposition (Domack, 1982). Holocene continental ice extends only to the George V coast, except for the floating ice tongues of the Mertz and Ninnis glaciers. Holocene sedimentation on the shelf includes accumulation of diatomaceous mud at depths greater than 500 m (in the depressions) and reworking of glacial sediment by currents and ice at shallower depths (on the banks). Transport of sediment across the shelf is notably minor because no rivers or wet-base glaciers deliver sediment to the coastline and because the net tilt of the shelf is toward the continent. Strong geostrophic currents near the shelf-break are believed to rework glacial sediment there, leaving a winnowed lag at the surface (Domack and Anderson, 1983). Sediment gravity flows are active on steep parts of the shelf, for instance, locally on the flanks of the depressions (Domack, 1982; Wright et al., 1983). Deposition of ice-rafted sediment

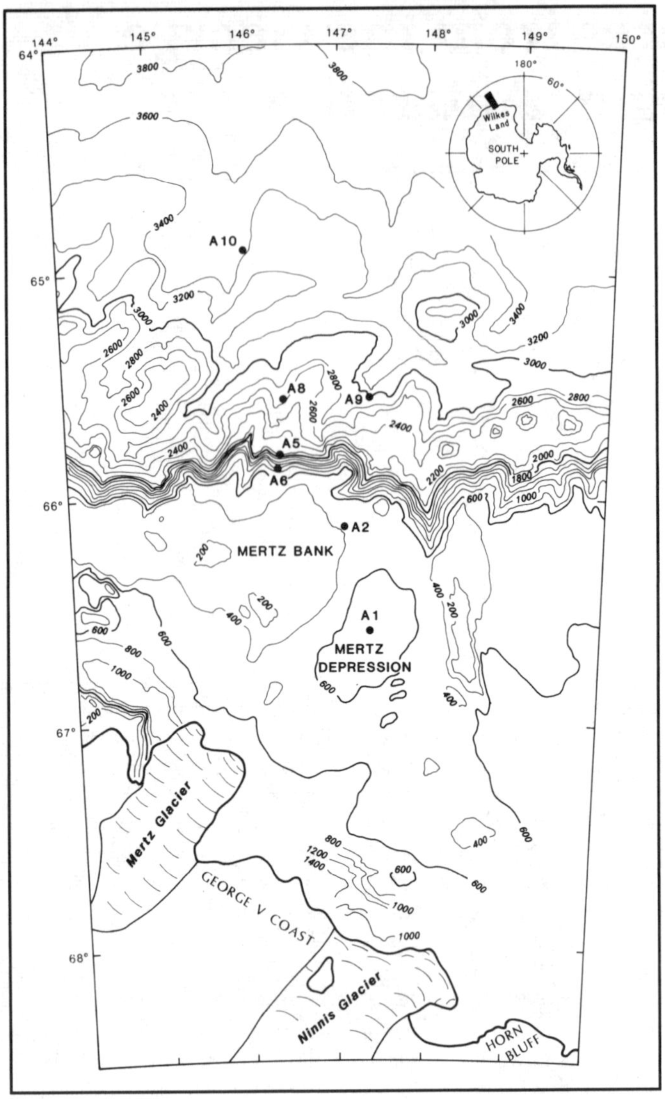

Figure 1—Map of the George V–Adelie continental margin showing location of sediment cores. Bathymetric contours in meters.

is minor on the shelf, probably because strong katabatic winds cause icebergs to transit rapidly across the area and because surface water is so cold that relatively little melting occurs (Domack and Anderson, 1983).

Domack (1980, 1982), in his study of cores collected on the George V–Adelie continental shelf during Operation Deep Freeze 1979, identified two source terrains for the glacial sediment that correlate well with a major geologic boundary originally discovered by Mawson (1940) in the vicinity of Ninnis Glacier. The distribution of distinctive heavy-mineral assemblages from the two terrains records a northwest flow direction of Pleistocene ice across the shelf (Domack, 1982).

Domack (1980, 1982) also proposed a scheme for distinguishing glacial till diamictons from glacial-marine diamictons on the George V–Adelie continental margin. The glacial-marine species exhibits crude stratification in x-radiographs, contains marine fossils, and has relatively low shear strength (as measured with laboratory vane shear devices). Glacial diamicton is characterized by substantially higher shear strength caused by ice-loading and a lack of stratification and first-cycle marine fossils.

The sedimentary history of the continental slope and rise is less well known than that of the shelf. The slope, as defined here for the area in which the cores were collected, is unusually steep and narrow, extending from ~400 to 600-m water depth at the shelf-break to ~2000-m water depth (Fig. 1). The continental rise consists of two segments: the steep upper rise (approximate depth range 2000–3000 m) and the less steep lower rise (approximate depth range 3000–4000 m). Judging from the physiography (Fig. 1), the rise appears to consist of a series of submarine fans, sourced and cut by many submarine canyons. Several canyons extend to the shelf-break, but some are situated entirely seaward of the shelf-break on the upper rise; none has a physiographic connection with the far-distant landmass. Seismic-reflection profiles suggest that the fans on the upper rise are relict and are dissected by younger canyons that feed or fed fans on the lower rise (Hampton et al., this volume). Cores from the continental slope and rise collected during Operation Deep Freeze 1979 have not been interpreted in the open literature, but unpublished core descriptions (Anderson, 1981) imply that glacial-marine processes, traction currents, and sediment gravity flows all have deposited sediment there.

Piper and Brisco (1975) studied DSDP Leg 28 drill cores taken from the base of the continental rise and the adjacent abyssal plain in the George V margin area. They concluded that the deposits in the cores accumulated by hemipelagic sedimentation, ice-rafting, turbidity currents, and contour currents. They also developed criteria for distinguishing deposits of turbidity currents from those of contour currents (see also Piper, 1978; Stow, 1979).

To summarize, the George V continental shelf is undergoing shallow-water reworking of sediment and deep-water biogenic sedimentation, and sediment gravity flows may be active locally on the slopes of depressions. Pleistocene sedimentation was dominated by glacial and associated marine processes. The continental slope is steep and narrow, apparently being shaped by strong geostrophic currents. The continental rise consists of turbidite fans where modern activity appears to be destructional in places and perhaps constructional in others.

The following discussion addresses several objectives. Sedimentary processes are inferred using a broad combination of geological observations, including sediment texture, clay mineralogy, heavy mineralogy, stratigraphy, sedimentary structures, and quartz grain microtextures. The physical properties of the seven cores collected from these polar sedimentary deposits are compared with those of sediment collected elsewhere to determine if major differences exist. Data from consolidation tests are analyzed to discern if they record past ice-loading events or significant erosional episodes, and the slope stability of the cored sediment types is evaluated in terms of the measured shear strength.

METHODS

Sediment samples were collected with a gravity corer, in plastic liners of 8.5-cm diameter. Maximum core length was 375 cm. Immediately after retrieval, the cores were cut into 1-m lengths, and a 10- to 20-cm length was cut from the bottom of each section for determination of hydrocarbon content (Kvenvolden et al., this volume). The remainder of each core was taken to the ship's laboratory, where shear strength was measured on the ends of core sections using a motorized torque-cell miniature vane device at a vane rotation rate of 90°/min. Subsamples were taken at each measurement location for determination of water content and grain-specific gravity. Core sections were then capped, wrapped in cheesecloth, and coated with microcrystalline wax to retard moisture loss. Cores were stored vertically at ~4°C onboard ship and later shipped by refrigerated air freight to the shorebase laboratory.

In the shorebase laboratory, x-radiographs were made of the full-round core sections and inspected to determine the locations for geotechnical testing. The stratigraphy in the radiographs also was described. Water content was determined by drying and weighing of the subsamples taken onboard ship (ASTM, 1982, standard D2216-80). Correction for salt content was made assuming a concentration of 35 ppt for sea water. Water content is reported as a percentage of dry sediment weight; therefore, values in excess of 100% are possible. Grain-specific gravity was measured on the dried subsamples, following ASTM (1982) standard D854. Organic carbon and carbonate carbon also were measured on these subsamples using an induction furnace and acid digestor carbon analyzer.

Consolidation tests were performed using a back-pressured triaxial cell and the constant rate of strain technique (Wissa et al., 1971). Test data were used to calculate the maximum past overburden stress (Casagrande, 1936) and the compression index (a measure of the amount of consolidation in response to an increase of overburden pressure).

Static triaxial strength testing also was performed in the back-pressured triaxial cell (Bishop and Henkel, 1964). Cylindrical test specimens were hand trimmed to 3.6-cm diameter, wrapped with strips of filter paper to aid pore-fluid drainage, and encased in a thin latex membrane. Specimens first were subjected to an isotropic consolidation pressure of four times the calculated maximum past overburden stress and allowed to come to equilibrium by permitting pore-fluid drainage. This high consolidation stress removes some of the effects of disturbance experienced during sampling and handling (Ladd and Foott, 1974). Shear stress was induced in the specimen with an axial ram advanced at a constant rate of 0.4 cm/hour, without allowing pore-fluid drainage (undrained test). Isotropic confining stress, axial load, axial strain, and pore pressure were monitored during the tests. Shear strength was calculated from the peak stress recorded over a 20% axial strain of the specimen. Atterberg limits (measurement of plasticity, ASTM, 1982 standards D423, D424, and D2217) were determined on trimmings from some of the triaxial specimens.

After geotechnical testing was completed, the remaining core sections were split longitudinally, then described macroscopically and with smear slides. Gravel-sized clasts were picked from the split core sections and identified with the aid of a binocular microscope. Thin-slab (5 mm thick) x-radiographs were made of some of the laminated portions of the cores. Monocrystalline quartz grains (typically coarser than 1ϕ but coarser than 4ϕ in some very fine sediment) were separated from some subsamples and inspected with a scanning electron microscope (SEM). Heavy minerals were separated from other subsamples using tetrabromoethane (S.G. = 2.96) (Carver, 1971) and identified using a petrographic microscope. Clay minerals were identified using an x-ray diffractometer with scintillation counter and nickel-filtered copper $\kappa\alpha$ radiations (Hein et al., 1976). Peak-area measurements and Biscaye's (1965) weighting factors were used to semiquantitatively calculate relative percentages of clay minerals.

DESCRIPTION OF THE SEDIMENT CORES

Stratigraphy and Sediment Type

Descriptions of the sediment cores and brief interpretations of their origin are given in Table 1. Cores A1G1 and A2G2 were collected from the continental shelf in the Mertz Depression and on the adjacent Mertz Bank, respectively, cores A5G1 and A6G1 were collected from the continental slope, cores A8G1 and A9G1 were collected from the upper continental rise, and core A10G1 was collected from the lower continental rise (Fig. 1).

The top 75 cm of core A1G1 is diatom mud, reflecting the dominantly biogenic sedimentary environment on the shelf during Holocene time (Domack, 1982; Domack and Anderson, 1983). Patches and thin stringers of the diatom mud extend across the otherwise sharp lower contact and down into the upper several centimeters of the diamicton. A concentration of angular outsize clasts occurs at the base of the diatom mud; they occur sparsely above the basal concentration. The subunit from 38 to 66 cm in the diatom mud is distinctive in its content of abundant small, dark, and typically ovoid cells (Fig. 2).

Both cores contain diamicton (below 75 cm in A1G1 and in all of A2G2) that is characterized by scattered faceted and scratched gravel clasts. Crude stratification is manifest in x-radiographs of the cores, either as gradational variations in the concentration of coarse clasts or as a variation in the x-ray density of the matrix that implies a variation in grain size (Fig. 3A). The stratification is subtle and not detectable in split core sections. The grain-size distributions determined for one sample from each core are listed in Table 2. Both samples contain a wide range of grain sizes, with large amounts of gravel and clay. The sample from core A1G1 has a fairly uniform distribution of grain sizes throughout the measured range, whereas the sample from core A2G2 has a mode in the fine sand range.

Cores A5G1 and A6G1 were taken from the steep continental slope. The small sample collected in core A5G1 contains gravel- to clay-sized grains and has a mode of medium to coarse sand. The sample is depleted in mud compared to the diamicton samples from cores A1G1 and A2G2 (Table 2). Foraminifers and carbonate fragments are significantly more abundant in this core than in the others.

The diamicton of core A6G1 has complex layering with some zones of clast-supported gravel (Table 1). The grain-size distribution from a clast-supported layer is similar to that of the sample from core A5G1, with a relatively low mud content (Table 2). Unpublished descriptions of cores from the Deep Freeze 79 expedition (Anderson, 1981) indicate that all samples collected from the steep continental slope consist predominantly of sandy sediment.

Cores A8G1 and A9G1 were collected on the upper continental rise, from the side of submarine canyons that incise either the same fan or two adjacent submarine fans (Fig. 1).

Table 1—Description and interpretation of sediment cores.

Core: A1G1 Location: 66°35.3′S, 147°21.7′E Water depth: 611 m Physiography: Mertz depression

Depth	Visual	X-radiograph	Smear slide	Interpretation
0-32 cm	Grayish olive (10Y 4/2) mud with faint black streaks. A few open voids to 1 cm size. A few black, 1-4 mm, angular clasts.	Medium x-ray density matrix with a few (~2 mm size) high-density clasts. Some mottling and a few faint individual burrows. Subtle irregular contact at 32 cm to slightly lower x-ray density material.	Diatom ooze with ~10% terrigenous silty clay.	Holocene biogenic sedimentation with a relatively small amount of ice-rafted terrigenous sedimentation. Warming conditions during deposition of "resting-spore(?)" subunit. Infaunal mixing across subunit boundaries.
32-38 cm	Transition zone with burrows of unit above within unit below.	Medium to low x-ray density matrix with mottling and a few high density clasts to 6 mm size. Subtle x-ray density change at lower contact.		
38-66 cm	Uniformly colored moderate olive brown (5Y 4/4) slightly sandy silty clay with a few open voids to 2 mm size.	Low x-ray density homogenous layer with a few small (~2 mm size) high x-ray density clasts. A slight concentration of clasts from ~55-66 cm. Some very faint mottling. Gradational lower contact.	Diatom ooze with abund dant (~1-2%) silica-filled resting spores(?). Less than 10% terrigenous grains; less silt size than above, finer mean grain size.	
66-75	Grayish olive (5Y 3/2) slightly sandy mud. Transitional from unit above to unit below. Common flat, black (~1-2 mm size) rock clasts. One 6 mm angular rock fragment observed.	Medium low x-ray density matrix with common (<6 mm size) high density clasts. A few irregular high x-ray density stringers from 68-78 cm.		
75-283 cm	Olive gray (5Y 3/2 sandy mud with scattered black clasts of varied size and abundance. Distinct patches and stringers of grayish olive (5Y 3/2) material from overlying unit in upper 10 cm.	Medium to low x-ray density matrix with abundant angular to subangular high x-ray density clasts (mostly 1-20 mm size), all matrix supported. Matrix is homogeneous with no indication of bioturbation except for horizontal burrow(?) at 134 cm. Some vague stratification, due to increased concentration of clasts, with gradational contacts.	Terrigenous mud with 5% siliceous tests (nearly all are broken diatoms).	Pleistocene glacial-marine sedimentation with a varied supply of large ice-rafted clasts.

Clast types identified in core: 2-Granite, distinctive salmon and black crystals (2, 2 cm), 1-Dacite (1 cm), 3-Black slate, one with pyrite cyrstals (3, 2, 1 cm), 1-Friable mudstone (1 cm).

Core: A2G2 Location: 66°08′S, 147°05′E Water depth: 458 m Physiography: Mertz bank

Depth	Visual	X-radiograph	Smear slide	Interpretation
1-139 cm	Grayish olive green (5GY 3/2) to dark greenish gray (5GY 4/1) diamicton (sandy mud with common larger clasts <2 mm size). Some black patches. Slightly less sandy below 28 cm.	Numerous angular to subangular granule- and pebble-size high x-ray density clasts (1-40 mm size) scattered throughout a low to medium x-ray density matrix. Clasts apparently are matrix supported. Horizontal, 2-cm-thick, lower x-ray density layer at 89-91 cm. Gravel particles show some preferred horizontal orientation and layering from 91-111 cm, with subtle thin layers of low and high x-ray density matrix.	Terrigenous sandy mud with 1-2% siliceous tests (mostly fragmented diatoms).	Glacial marine sedimentation.

Clast types identified in core: 3-Dacite (1, 1 1/2, 2 1/2 cm), 1-Quartz diorite (1 cm), 3-Gray-black slate, one with feldspar porphyroblasts (1, 1 1/2, 3 cm), 1-arkosic sandstone (3 cm).

Table 1 (Continued)

Core: A5G1 Location: 65°47.9′S, 146°23.3′E Water depth: 2200 m Physiography: Continental slope

Depth	Visual	X-radiograph	Smear slide	Interpretation
Core catcher (20 m)	Light olive gray (5Y 5/2) slightly muddy pebbly sand.		Muddy sand with abundant (~3%) opaques, Trace of carbonate shell material (forams). Only a trace of spicules, diatom and radiolarian fragments.	Traction deposit or winnowed lag deposit.

Core: A6G1 Location: 65°52.5′S, 146°21.1′E Water depth: 994 m Physiography: Continental slope

Depth	Visual	X-radiograph	Smear slide	Interpretation
0-4 cm	light olive gray (5Y 5/2) to dusky yellow (5Y 6/4) diamicton with relatively small amount of mud. Abundant angular to subangular, faceted and scratched gravel clasts (to 6 cm size). Coarse sand to gravel lens at 68-71 cm. Slight vertical textural varition.	Granule-size high x-ray density particles in low density matrix. Gradational lower contact.	98% terrigenous muddy sand with ~1% forams and other carbonate fragments. Siliceous biogenic material is mostly spicules and fragmented and whole diatoms and radiolarians (more abundant than in nearby A5G1. Fewer opaque grains than in A5G1.	Glacial-marine deposit, reworked.
4-27 cm		Clast-supported, angular to subrounded granules and pebbles (to 20 mm size). Gradational lower contact.		
27-38 cm		Relatively low x-ray density with less abundant granules and pebbles (to 10 mm size), some of which appear to be clast supported. Gradational lower contact.		
38-70 cm		Matrix-supported granules and pebbles, most ~5 mm in size. Sharp lower contact.		
70-78 cm		Large (to 6 cm size), clast-supported pebble-size high x-ray density particles.		

Clast types identified in core: 2-Quartz diorite (1, 5 cm size), 1-Salmon and black granite (3 cm size), 1-pink, fine-grained "granitic" (1 cm size), 2-Quartz-biotite "hornfels" (3 cm size), 1-Quartz-biotite schist (7 cm size), 5-gray aplite (1, 1, 1, 1-1/2, 4 cm size). NB-Clasts are faceted and scratched.

Core: A8G1 Location: 65°34.0′S, 146°25.4′E Water depth: 2635 m Physiography: Upper continental rise

Depth	Visual	X-radiography	Smear slide	Interpretation
0-193 cm	Olive gray (5Y 4/1) to light olive gray (5Y 5/2) mud to sandy mud with abundant circular to irregular mottles either lighter or darker than surrounding matrix, whose texture varies gradationally. Upper 20 cm has areas of brown color. Sparse angular black clasts.	Moderately mottled, mostly medium x-ray density but with local variations from low to high. Some faint laminations disturbed by bioturbation. A few high x-ray density particles (to 10 mm size). Highly varied density and crude layering between 103-140 cm.	Fine sandy mud with varied diatom flora, spicules, radiolarians, and silicoflagellates (~25% abundance); broken and whole forms, staining of many diatoms.	Hemipelagic marine sedimentation with slight to moderate input of glacial debris and with moderate infaunal activity. Perhaps some occasional bottom current action to produce laminations.

Table 1 (Continued)

Core A8G1 (continued)

Depth	Visual	X-radiograph	Smear slide	Interpretation
193-207 cm	Light olive gray (5Y 5/2), light olive brown (5Y 5/6), and moderate olive brown (5Y 4/4) mud. Circular to diffuse mottled areas.	Transition zone between unit above and unit below. Laminations disturbed by individual burrows. Burrows decrease in abundance downward. Laminations end abruptly upward.	Fewer siliceous tests (~10%), especially whole, circular varieties (e.g., Coacinodiacus sp.). Slightly sandy mud; finer than overlying unit.	Burrowing by early invading infauna into underlying unit.
207-356 cm	Gradational color variations, grayish olive (10Y 4/2) to olive gray (5GY 3/2), mud. Homogenous, structureless appearance except at some places where careful smoothing of core surface shows slight coarsening of grain size, suggesting few-mm thick laminations.	Alternating high and low x-ray density, thin and thick laminations and a few low x-ray density beds. High density laminations up to 3-mm thick; most are horizontal and many pinch out laterally. Some microfaults. Variation in number, thickness, clarity, and boundary sharpness of laminations, with no apparent pattern or cyclicity. Most high x-ray density laminations have sharp boundaries; some grading is apparent. Some lateral variations in x-ray density. Possible microcross-laminations at 353 cm. Sharp lower contact.	Terrigenous mud with ~1% siliceous tests (nearly all broken 5 superjacent smear slides show coarser sediment in lowest one and essentially identical texture in other 4. All slides distinctly finer grained (mud vs. sandy mud) compared to upper unit	Repeated bottom current action with traction current depositions, alternating with hemipelagic sedimentation and very little infaunal activity. Abrupt beginning and end of this episode.
356-369 cm	Olive gray (5GY 3/2) homogeneous mud.	Moderately high x-ray density with light to moderate bioturbation and some faint, high x-ray density thin laminations at base of unit.	Very fine sandy mud with ~1%, mostly broken siliceous tests. Trace broken calcite shell material.	Hemipelagic sedimentation with moderate infaunal activity. Weak bottom current action.

Core: A9G1 Location: 65°33.7′S, 147°21.6′E Water depth: 3037 m Physiography: Upper continental rise

Depth	Visual	X-radiography	Smear slide	Interpretation
0-78 cm	Olive gray (5Y 4/1) at top, grading to medium dark gray (N4) to olive gray (5Y 3/2) mud, with round to irregular patches (burrows) of slightly different color and consistency (texture).	Varied x-ray density and mottled, with several individual burrows. Below 60 cm the burrows are relatively abundant.	Terrigenous mud with trace of broken siliceous tests	Hemipelagic sedimentation with active infauna.
78-90 cm	Olive gray (5Y 3/4) mud with circular to irregular patches of slightly different color and texture from matrix.	Low x-ray density patches superimposed on vaguely laminated material. Transitional between overlying and underlying units.		Burrowing into underlying unit by infauna living in environment of overlying one
90-143 cm	Olive gray (5Y 3/4) mud with more-or-less distinct laminations of slightly coarser (fine sand to coarse silt) texture than mud interbeds, particularly visible in dried core section.	High x-ray density laminations and lower density interbeds. Laminations up to 10 mm thick and show lateral and vertical variation in texture, some lenses, and sharp to diffuse contacts. Some burrowing and small (<2 mm) high x-ray density particles.	Lamination: coarse silt to fine sand, very well sorted (no clay size), trace of broken siliceous tests. Interbed: mud, distinctly finer grained and more poorly sorted than lamination. Much clay-size material and a trace of broken siliceous tests.	Repeated traction current deposition, alternating with hemipelagic sedimentation with limited infaunal activity.
143-154 cm	Empty.			
154-368 cm	Olive gray (5Y 3/2) mud with circular to elongate to irregular patches and swirls of slightly different color (grayer) and texture. Light olive gray (5Y 5/2) swirled patches of sandy mud.	Medium x-ray density and moderately mottled, with numerous individual burrows. Some irregular and disturbed high x-ray density laminations. Burrows become more abundant and high x-ray density clasts more common below 194 cm.	Very fine sandy mud with 1-5% mostly broken but commonly whole siliceous tests, many having stain.	Hemipelagic sedimentation with input of glacial sediment. Moderate infaunal activity and some bottom current action.

Table 1 (Continued)

Core: A10G1 Location: 64°54.3′S, 145°59.9′E Water depth: 3379 m Physiography: Lower continental rise

Depth	Visual	X-radiograph	Smear slide	Interpretation
0-9.5 cm	Grayish olive (10Y 4/2) mud with abundant small burrows defined only by subtle color variations.	Medium x-ray density, lightly mottled, homogeneous. Sharp bottom contact.	Mud with ~3% whole and broken siliceous tests, mostly diatoms.	Hemipelagic sedimentation with low to moderate infaunal activity.
9.5-18 cm	Light olive gray (5Y 5/2) mud with more distinct burrows than above.	Several high x-ray density thick and thin, wavy to irregular laminations and low x-ray density interbeds. Moderate bioturbation. Gradational bottom contact.	Mud with 3-5% whole and broken siliceous tests (mostly diatoms). Slightly coarser (a few very fine sand grains) than overlying and underlying units.	Repeated bottom current activity interspersed with more quiet hemipelagic conditions. Moderate infaunal activity.
18-110 cm	Light olive gray (5Y 5/2) mud, strongly mottled with subtle color variations. A few large (to 5 cm size), angular, faceted clasts.	Scattered high x-ray density, angular to subangular particles (pebble size) in a lightly mottled matrix. Gradational bottom contact.	Mud with 3-5% whole and mostly broken silicecous (mostly diatoms). Fewer whole tests than in overlying units.	Hemipelagic sedimentation with input of large glacial particles; limited infaunal activity.
110-124 cm	Mud laminations of slightly coarser texture than clay interbeds. No lamination from 112-115 cm, which is mottled. Color from light olive gray (5Y 5/2) to nearly light olive brown (5Y 5/6).	High x-ray density, parallel laminations to 5-mm thick and lower density interbeds. Slight disturbance of lamination by burrows. Bottom contact distinct but slightly irregular, apparently due to erosion.	Lamination: well sorted medium to coarse silt with ~5% opaques and other heavy minerals; trace of broken siliceous tests. Interbed: clay-rich mud. (Maximum grain size < medium silt) with very few broken siliceous tests.	Repeated bottom current action, producing a lag deposit. Limited infaunal activity.
124-225 cm	Light olive gray (5Y 5/2) highly mottled mud with a few dark angular clasts to 2 cm size. Burrows circular to irregular shape; subtle texture and color differences.	Medium x-ray density and moderately mottled, with scattered high density angular to subangular particles from coarse sand to 2 cm size, generally increasing in abundance downward. Gradational bottom contact.	Very fine sandy mud with 3-5% mostly broken siliceous tests, mostly diatoms (some staining).	Hemipelagic and glacial sedimentation with moderate infaunal activity. Gradually varying level (and type?) of infaunal activity.

Table 1 (Continued)

Core: A10G1 (continued)

Depth	Visual	X-radiograph	Smear slide	Interpretation
225-238 cm	Light olive gray to slightly browner mud with coarser mottling than other units. More-or-less totally bioturbated.	Low x-ray density and highly mottled with many individual burrows and numerous angular to subangular high x-ray density particles, most < 5 mm but up to 1 cm size. Gradational bottom contact.	Coarse silty mud with abundant (~20%) whole and broken siliceous tests.	
238-257 cm	Similar to unit at 124-225 cm, but with slightly more distinct mottling. Totally bioturbated.	Medium x-ray density, with scattered high density gravel-size clasts. Gradational lower contact.	Coarse silty mud, similar to above, but with fewer (~2%) siliceous tests. Very few whole tests.	
257-267 cm		Thick and thin high x-ray density laminations with low density interbeds. Some laminations broken, bifurcated, or bioturbated. Some horizontal burrows in interbeds.		Repeated bottom current activity with intervening hemipelagic sedimentation. Limited infaunal activity.
267-284 cm	Empty			
284-294 cm	Light olive gray (5Y 5/2) mud with subtle mottling.	Low to medium x-ray density and homogeneous, with slight mottling. Irregular bottom contact.	Coarse silty mud with 3-5% broken and whole siliceous tests. Most whole diatoms are stained.	Hemipelagic sedimentation with variation in level (and type?) of infaunal activity. Occasional input of glacial sediment.
294-311 cm	Similar to overlying unit but with more distinct round to irregular mottles.	Similar to unit at 124-225 cm. High x-ray density angular to subangular particles, 2-5 cm size.	Very fine sandy mud (slightly coarser than above) with 3-5% whole and broken siliceous tests.	
311-327 cm	Slightly more brownish color than overlying unit, with less distinct but abundant mottling. More-or-less total bioturbation.	Low x-ray density with scattered high density particles and moderate to strong mottling. Gradational bottom contact.	Fine sandy mud (slightly coarser than above) with ~25% broken and whole siliceous tests.	
327-342 cm	Similar to overlying unit but with more distinct mottling. Several inclined to vertical burrows. More-or-less total bioturbation.	Higher x-ray density than above, with abundant burrowing that has obvious criss-cross pattern. Scattered high x-ray density particles, 2-5 mm size.		
342-370 cm	Light olive gray to slightly yellower color. Mud laminations (to 2 mm thick) and bioturbated clay interbeds.	Thick and thin, slightly wavy to highly irregular laminations with low x-ray density interbeds. Some bioturbation of laminations, moderate mottling of interbeds. Sharp lower contact.	Lamination: terrigenous silt with small amount of very fine sand and with virtually no clay-size sediment or fossils. Interbed: clay with a few % silt and a trace of broken siliceous tests.	Repeated traction current deposition, interspersed with hemipelagic sedimentation with infaunal activity.
370-375 cm		Moderate to high x-ray density and homogeneous, with indication of faint laminations.	Mud with 1-2% broken siliceous tests.	Hemipelagic sedimentation.

Clast types identified in core: 1-Diabase in contact with basalt (5 cm size), 2-Granitics (1/2, 1 cm size), 2-Black slate (1, 2 cm size), 1-Arkosic sandstone (1/2 cm size), 1-Coal fragment (1/2 cm size).

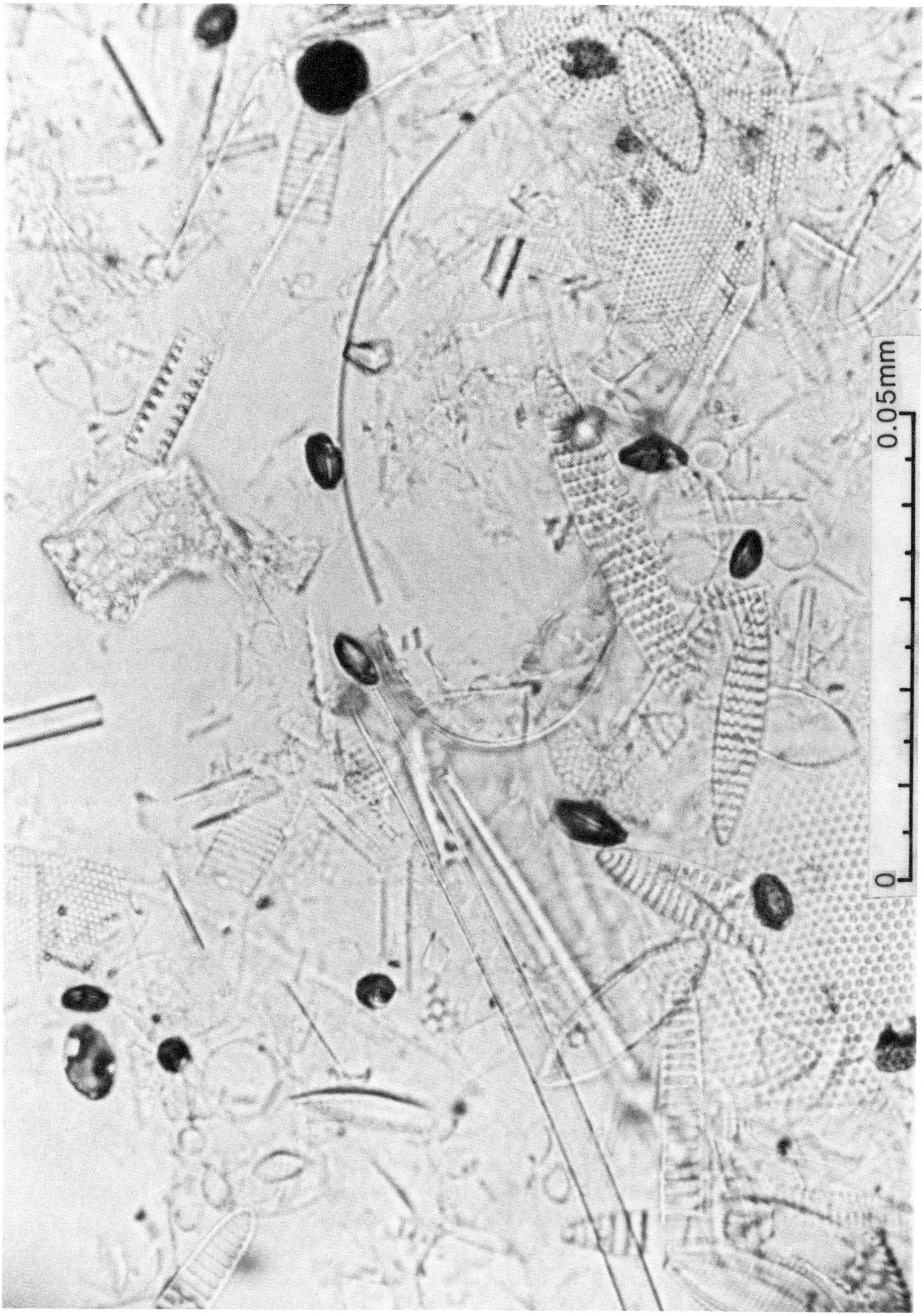

Figure 2—Photomicrograph of diatom mud showing dark ovoid cells, core A1G1 55 cm.

Figure 3—X-radiographs of sediment cores. A. Diamicton, core A1G1. B. Massive mud (above) and laminated mud (below), core A8G1. C. Heterogeneous mud (above) and laminated mud (below), core A10G1.

Table 2. Grain size of sediment samples, in percent.

Grain size (phi)	A1G1 173 cm. diamicton	A2G2 136 cm. diamicton	A5G1 cc pebbly muddy sand	A6G1 77 cm. diamicton	A8G1 67 cm. massive mud	A8G1 167 cm. massive mud	A8G1 230.1 cm. laminated mud	A8G1 230.3 cm. laminated mud	A8G1 230.5 cm. laminated mud	A8G1 230.7 cm. laminated mud	A8G1 230.9 cm. laminated mud	A9G1 41 cm. massive mud	A9G1 90 cm. coarse lamina	A9G1 94 cm. fine lamina	A9G1 274 cm. massive mud	A10G1 172 cm. massive mud
> −1	15.2	10.9	33.7	35.6	0.0	0.0	0.0	0.0	0.0	0.0	0.0	0.0	0.0	0.0	0.0	0.0
−1−0	1.5	2.6	3.1	1.8	0.4	0.6	0.0	0.0	0.0	0.0	0.0	0.0	0.0	0.0	0.1	0.1
0−1	5.0	10.0	18.0	14.5	0.6	0.7	0.0	0.0	0.0	0.0	0.0	0.0	0.1	0.0	0.4	0.4
1−2	7.2	13.7	12.9	26.0	1.0	1.7	0.0	0.0	0.0	0.0	0.0	0.1	0.2	0.1	0.7	0.6
2−3	4.8	16.8	10.8	11.5	0.9	1.3	0.0	0.0	0.0	0.0	0.0	0.2	0.3	0.2	0.9	0.4
3−4	3.3	8.4	6.7	1.4	5.0	1.9	0.5	0.7	0.7	0.6	0.5	0.4	3.7	0.4	4.1	0.6
4−5	6.5	8.5	3.6	1.7	23.9	12.4	2.8	2.0	1.9	3.8	4.6	6.2	57.4	4.2	15.9	6.3
5−6	5.9	5.8	3.1	1.1	16.7	12.6	9.5	9.1	11.8	12.2	13.9	11.9	30.5	9.8	22.2	16.7
6−7	6.8	3.2	2.0	0.8	12.2	11.1	13.4	15.8	9.6	16.5	18.8	11.6	2.5	15.1	12.9	14.9
7−8	3.2	1.1	1.6	0.7	7.5	9.9	10.3	5.7	12.2	10.1	10.0	8.5	0.0.	14.0	7.4	10.9
8−9	5.5	4.0	1.6	1.2	6.5	8.0	7.2	7.4	4.1	7.8	4.6	9.0	0.0	10.0	5.8	8.3
<9	34.6	13.9	2.7	2.6	25.3	39.7	56.2	59.3	59.8	49.0	47.6	52.0	5.1	46.0	29.6	40.8
% gravel	15.2	10.9	33.7	35.6	0.0	0.0	0.0	0.0	0.0	0.0	0.0	0.0	0.0	0.0	0.0	0.0
% sand	21.8	52.6	51.6	56.2	7.9	6.2	0.5	0.7	0.7	0.6	0.5	0.7	4.4	0.7	6.2	2.2
% silt	22.5	18.6	10.3	4.3	60.4	46.0	36.1	32.7	35.4	42.6	47.3	38.2	90.5	43.2	58.4	48.7
% clay	40.5	17.9	4.3	3.8	31.8	47.8	63.4	66.7	63.8	56.8	52.2	61.0	5.1	56.1	35.4	49.1

The cores contain two types of sedimentary units, massive mud and laminated sand and mud (Fig. 3B). The massive mud has only vague indications of stratification; it is moderately mottled and contains siliceous microfossils and only a few small gravel clasts (Table 1). The massive mud occurs at the sea floor in both cores and at other levels down-core. The intervening units are laminated, and mottles are rare (Fig. 3B).

The grain-size distribution of four samples from the massive units in cores A8G1 and A9G1 indicates a sandy mud with a large proportion of the material finer than 9ϕ (Table 2). Two samples have a well-defined mode in the coarse silt grade, whereas the others have only a minor one. Although the x-radiographs show vague stratification in a few places (Fig. 3B), the sediment structure generally is massive.

The laminated intervals are distinctive in several ways, compared both with each other and with the massive units. Mottles and large, outsize grains are rare to absent. Individual laminae are distinguished by variations in grain size, although the difference can be small. For example, the laminae in the 207–356-cm interval in core A8G1 are obvious in x-radiographs (Figs. 3B, 4A) but can barely be detected in split core sections. A series of five superjacent horizontal slices 2 mm thick was taken from within this interval, with a distinct coarse-grained lamina in the bottom slice and a few more subtle coarse laminae, but mostly fine-grained laminae above (Table 2, A8G1 230.1–230.9 cm). The bottom slab was intended to sample the lower coarse-grained lamina only, but some contamination with the surrounding fine-grained laminae is possible. The higher slices certainly contain some mixing of coarse- and fine-grained laminae, but probably in unequal amounts because successively fewer coarse-grained laminae occur higher upward. The grain-size distribution for the lowest slice is coarser than for the overlying ones, but only slightly so (Fig. 5, Table 2). Nearly all siliceous microfossils are broken in the coarse-grained laminae.

Individual coarse-grained laminae tend to vary laterally in both thickness and x-ray density (Fig. 3B). The upper and lower boundaries of most coarse-grained laminae are sharp, but gradational boundaries also are present. Lateral termination of laminae is common. Internal structures such as grading or cross-bedding are not evident.

A thin-slab x-radiograph of a part of the laminated sequence in core A8G1 is shown in Figure 4A. The thickest coarse-grained lamina is 2 mm, and most are less than 1 mm. Note that the top of the laminated unit terminates abruptly at the arrow in Figure 3B and that the upper several centimeters of the sequence are mottled.

Laminae in core A9G1 are coarser grained and thicker than in A8G1, and the coarse-grained laminae are readily visible in split core sections. The coarse-grained laminae consist of very well sorted silt- and sand-sized quartz grains with a few broken siliceous tests (Tables 1, 2). The fine-grained laminae are mud, and a grain-size measurement indicates that nearly half of the grains are finer than 9ϕ. Some relatively small outsize clasts and several small mottles appear in the x-radiographs. The high x-ray density portions of the radiograph are relatively thick packets of several coarse-grained laminae; thin and discontinuous fine-grained laminae (low x-ray density) are present within the packets (Fig. 4B). Furthermore, many of the coarse-grained laminae are discontinuous and wavy.

Core A10G1, taken farthest from land on the lower continental rise, contains both types of units present in A8G1 and A9G1. The several massive units in this core have significant variation in the amount and types of mottles and in the amount and size of outsize clasts (Table 1). Boundaries between adjacent units range from sharp to gradational (Fig.

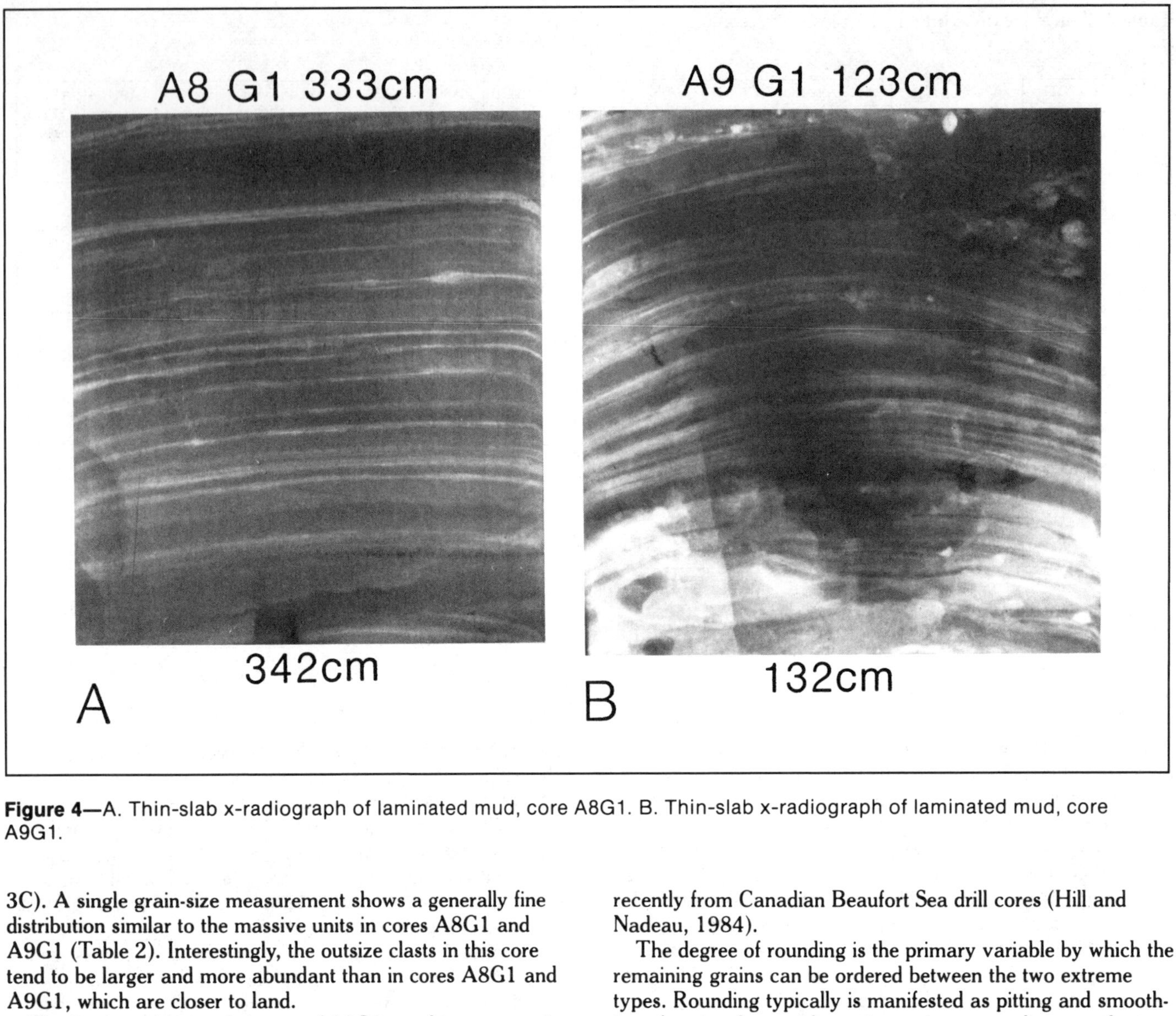

Figure 4—A. Thin-slab x-radiograph of laminated mud, core A8G1. B. Thin-slab x-radiograph of laminated mud, core A9G1.

3C). A single grain-size measurement shows a generally fine distribution similar to the massive units in cores A8G1 and A9G1 (Table 2). Interestingly, the outsize clasts in this core tend to be larger and more abundant than in cores A8G1 and A9G1, which are closer to land.

The laminated intervals in core A10G1 are thinner, contain more outsize clasts, and are more disturbed by mottling compared to the upper-rise cores. The coarse-grained laminae are very well sorted, as in core A9G1, and packets of laminae are present. Siliceous fossils are rare to absent.

Quartz Grain Surface Textures

Quartz sand grains examined with an SEM have a variety of surface textures. One extreme morphologic type has perfectly sharp edges, conchoidal fractures, and arc-step fractures that are characteristic of glacially transported grains (Fig. 6A, 6B) (Whalley and Krinsley, 1974). The other extreme type is highly rounded, in many instances spherical, and has abundant upturned plates and some shallow concavities that characterize eolian grains (Fig. 6C, 6D; Krinsley and Doornkamp, 1973). Many of the highly rounded grains are broken in half (Fig. 6E). The breakage surface has a very fresh appearance, and its boundary edge is sharp (Fig. 6F) but in some places shows a small amount of chipping. Highly rounded to spherical grains have been reported elsewhere from the Antarctic margin (Rex et al., 1970), and they have been described recently from Canadian Beaufort Sea drill cores (Hill and Nadeau, 1984).

The degree of rounding is the primary variable by which the remaining grains can be ordered between the two extreme types. Rounding typically is manifested as pitting and smoothing of grain edges, with continuous curvature between the two dihedral faces (Fig. 6G). However, some grain edges have unevenly spaced conchoidal spalls that make a jagged edge out of one that originally was knife-sharp (Fig. 6H).

The occurrence of surface textures was not treated quantitatively in this study, but distinctive aspects of some of the samples were noted. In particular, the average amount of rounding on grains with a glacial foundation morphology differs from sample to sample, perhaps related to the amount of current transport to which they have been subjected. For example, the edges on glacial grains from the shelf cores A1G1 and A2G2 vary from perfectly sharp to chipped to moderately rounded. The glacial grains from the slope cores A5G1 and A6G1 generally are more rounded than those in cores A1G1 and A2G2. Grains from the massive and laminated units of the rise cores show similar variations, some of the latter having more highly rounded grains. These differences are perceptible but not major, and a variety of surface textures is typical. For example, some of the most angular glacial grains occur in the laminated units (Fig. 6B).

Another commonly observed surface texture consists of large depressions in moderately rounded grains (Fig. 6I). The

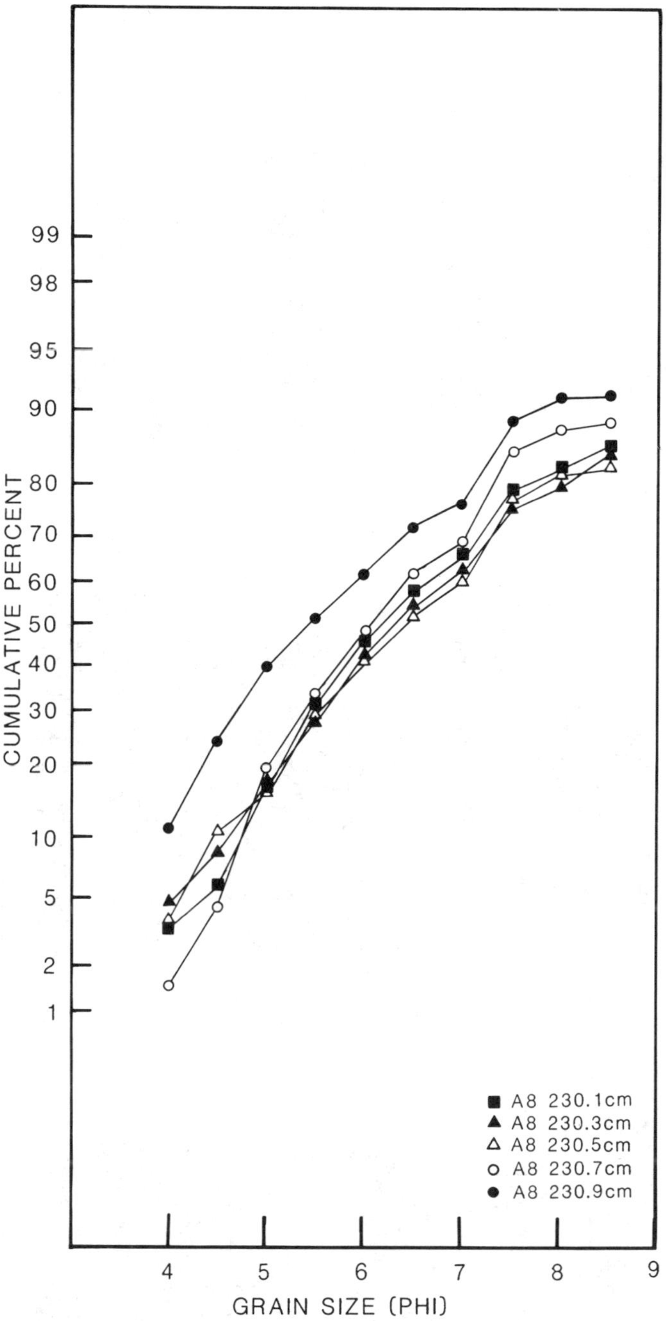

Figure 5—Cumulative grain size curves of five 2 mm thick horizontal sections from the laminated mud, core A8G1.

depressions have an irregular outline and a rough, corroded appearance.

Heavy Mineralogy, Clay Mineralogy, and Lithology of Clasts

The heavy mineralogy of the sand and coarse-silt fraction, the clay mineralogy of the <2-μm fraction, and the rock types in the gravel fraction were determined in order to detect relative changes in provenance or sedimentary processes, both within and between cores. The heavy-mineral assemblages are similar within and between cores, except for core A6G1 (Table 3). The typical assemblage is most abundant in clinopyroxene, followed by major amounts of orthopyroxene, epidote, garnet, and hornblende. Core A6G1 is richest in the same five minerals, but garnet is most abundant and clinopyroxene is less abundant relative to the other samples. Three subsamples from core A6G1 all have the same distinctive heavy-mineral assemblage.

The presence of orthopyroxene has not been reported previously from shelf sediment samples in this area, but only from farther west (von der Borch and Oliver, 1968; Domack, 1980, 1982). Most of the orthopyroxene is nonpleochroic enstatite, but a small to moderate percentage is weakly to distinctly pleochroic hypersthene.

The clay mineralogy data are summarized in Table 4. A plot of the data on a ternary diagram, using illite, kaolinite+chlorite, and smectite as end members, defines a principal trend of smectite variation, with relatively minor variation in illite and kaolinite+chlorite (Fig. 7). A line from the smectite corner of the diagram to the opposite baseline at a relative percentage of 78% illite and 22% kaolinite+chlorite best defines this trend.

Identifications of the gravel-sized clasts are listed in Table 1. The markedly diverse clast types include igneous, metamorphic, and sedimentary rocks. Although some rock types occur in more than one core, distinct suites or trends are not obvious. The total number of clasts is too small for generalized conclusions.

Physical Properties

Index physical properties measured on five of the cores are summarized in Figure 8. Some correlations with lithology are evident. The diatom mud in core A1G1 (0–75 cm) has a high value of water content and organic carbon and a low value of grain-specific gravity and wet unit weight, distinctly different from the underlying diamicton. Atterberg limit tests indicated that the diatom mud is nonplastic.

The diamictons in cores A1G1 and A2G1 have similar index properties. However, core A2G1, taken near core A2G2 and used only for geotechnical tests, has significantly lower water content values than the other two shelf cores (Table 5). The laminated units in cores A8G1, A9G1, and A10G1 have some distinctive properties. For example, the laminated interval in core A8G1, the thickest of all, has low shear strength and slightly higher organic carbon and carbonate content than does the overlying sediment. Also, it shows a consistent decrease of water content with depth (Fig. 8).

Our consolidation tests indicate a wide range of compression index (Table 5). Part of the variation is due to an anomalous feature of the laboratory compression curve, which is a plot of void ratio versus the logarithm of the consolidation pressure (Fig. 9). The "virgin" portion of the curve represents the phase of testing when consolidation has progressed beyond the level that existed under the maximum past stress (Fig. 9A). The virgin curve tends to be a straight line for most soil types (Lambe and Whitman, 1969, p. 297), and the slope of this line is the compression index. In most of our tests the virgin curve is concave upward or perhaps defines two straight-line segments (Fig. 9B), and therefore a unique value for compression index cannot be determined. We report two extreme values for compression index (Table 5) and consider the value for the lower part of the curve to be the most representative because the effects of initial cementation or disturbance are overcome at the higher stress levels. The continental shelf

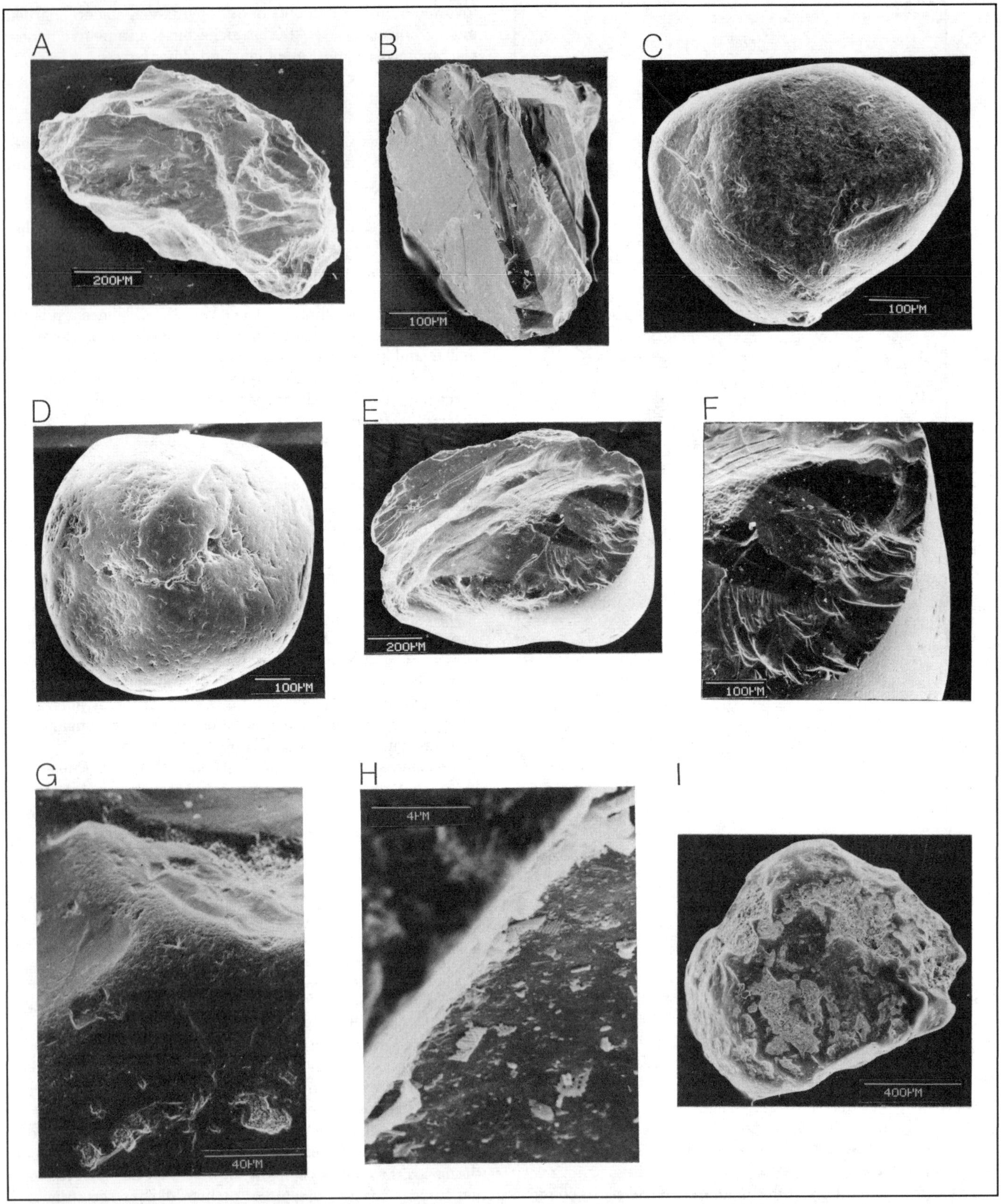

Figure 6—SEM photomicrographs of quartz sand grains. A. Angular grain with probable glacially induced surface texture, core A2G2 132 cm. B. Angular grain with probable glacially induced surface texture, core A10G1 359 cm. C. Rounded grain with probable eolian-induced surface texture, core A2G2 132 cm. D. Rounded grain with probable eolian-induced surface texture, Deep Freeze 79 core. E. Broken eolian grain, core A2G2 132 cm. F. Detail of fresh fracture shown in E. G. Smoothly rounded edge on grain, core A8G1 165 cm. H. Chipped edge on grain, core A1G1 57 cm. I. Large depressions on grain, core A9G1 139 cm.

Table 3. Heavy minerals in sediment samples, in percent.

Mineral type	A1G1 173 cm. diamicton	A2G2 136 cm. diamicton	A5G1 cc pebbly muddy sand	A6G1 17 cm. diamicton	A6G1 53 cm. diamicton	A6G1 77 cm. diamicton	A8G1 67 cm massive mud	A8G1 167 cm. massive mud	A8G1 224 cm. laminated mud	A9G1 41 cm. massive mud	A9G1 274 cm. massive mud	A10G1 172 cm. massive mud
orthopyroxene	23	15	23	10	17	16	19	26	24	25	24	17
clinopyroxene	45	33	41	30	31	29	37	38	38	31	39	32
hornblende	7	8	5	5	4	4	10	8	8	11	8	9
basaltic hornblende	-	-	-	-	T	P	-	-	-	P	P	T
epidote & clinozoisite	13	21	12	13	13	11	16	15	17	21	19	28
garnet	10	16	15	36	28	36	8	8	11	9	8	8
sphene	T	T	-	T	T	T	T	-	-	P	T	-
kyanite	T	1	T	1	T	T	8	3	-	T	T	2
apatite	T	1	T	T	2	T	T	T	1	3	T	3
tourmaline	T	2	T	T	T	T	T	T	T	T	T	1
zircon	T	P	1	1	T	T	-	T	T	T	-	T
rutile	-	T	T	1	1	2	P	T	T	P	-	P
aegerine-augite	T	-	T	-	-	-	-	-	-	-	-	-
glaucophane	-	P	-	-	-	-	-	-	T	-	-	-
andalusite	-	-	-	-	-	-	-	-	-	T	-	-
uvarovite garnet?	P	-	-	P	-	-	-	-	-	-	-	-
piedmontite?	-	-	-	-	-	-	P	-	-	-	-	-
chrome spinel	-	-	-	-	T	-	-	-	-	-	-	-
chloritoid	-	-	-	-	T	-	-	-	-	-	-	-

T: present in trace amount (< 1%). P, present in the sample, but not counted.

cores are less compressible than the continental rise cores (Table 5). Richards (1962) reported values of compression index from 0.20 to 0.87 for many tests of marine sediment. The George V margin sediment spans this range, although core A2G1 has unusually low compressibility (Table 5).

Concave virgin curves are by no means uncommon and have been reported elsewhere for sediment with high water content or high sensitivity (Quigley and Thompson, 1966; Bryant et al., 1974). No correlation between the degree of concavity and water content, plasticity index, or sensitivity exists for our data, but there is a qualitative correlation with the degree of lamination (Fig. 9). Samples from intervals with more laminations have a more highly concave virgin curve (Fig. 9). The only exception is core A1G1, in which no laminations were detected in the test intervals, although the virgin curves are perceptibly concave. The lower values of compression index do not vary significantly between the massive and laminated units in the continental rise cores.

Only a few static triaxial strength tests were conducted, because of the difficulty of obtaining a representative sample of acceptable grain size in these typically heterogeneous cores. The test results are summarized in Table 6. The important parameters are the effective angle of internal friction, ϕ', whose magnitude is an indication of the strength behavior under slow (drained) loading conditions, and the ratio of the undrained shear strength to the confining pressure, s_u/σ_c', whose magnitude is an indication of the strength behavior during rapid (undrained) loading. The values of ϕ' from 32° to 39° are at the high end of the normal range for marine sediment (Lambe and Whitman, 1969, p. 307) but are by no means extreme. These data imply that the sediment would be stable on relatively steep slopes under long-term loading conditions. The values of s_u/σ_c', 0.32–0.37, are in the midrange of values reported for marine sediment (Lambe and Whitman, 1969, p. 452). The laminated intervals tested (A9G1 100 cm and 109 cm) have lower values for the drained and undrained strength parameters than the massive sediment, but the difference is slight.

INTERPRETATION AND DISCUSSION

The cores collected on the 1984 *S.P. Lee* cruise to the George V continental margin span a range of marine environments from shelf to lower rise and record a history of sedimentation that has been influenced by ice-rafting, ocean currents, and biological activity in widely varying levels of importance. Interpretations of sedimentary processes are areally and temporally constrained by the sparse coverage and short length of the cores. Some extensions of the conclusions can be made using data from the Deep Freeze 79 and DSDP Leg 28 cruises, but a comprehensive regional sedimentation model can only be attempted when more numerous and longer cores become available.

Core A1G1 from the Mertz Depression contains 75 cm of diatom mud that Domack (1980, 1982) found at several places on the sea floor of closed depressions on the shelf, in water depth greater than 500 m. The diatom mud appears on high-resolution seismic-reflection profiles as an acoustically transparent sediment blanket up to 10 m thick (Barnes, this volume). These deposits consist of dominantly biogenic material that has settled directly through the water column in addi-

Table 4. Clay mineralogy of sediment samples, in percent.

Core	Depth (cm)	Lithology	Relative percent				
			illite	smectite	chlorite	kaolinite	chlorite + kaolinite
A1G1	0	diatom mud	64	14	14	8	22
	20	diatom mud	64	16	14	6	20
	50	diatom mud (ovoid cells)	69	6	19	6	25
	80	diamicton	70	7	16	7	23
	200a	diamicton	75	6	13	6	19
	200b	diamicton	74	6	13	7	20
	293	diamicton	66	14	13	7	20
A2G1	0	diamicton	75	5	14	6	20
	70	diamicton	73	5	15	7	22
A2G2	0	diamicton	69	6	16	9	25
	148	diamicton	71	5	16	8	24
A6G1	0	diamicton	61	18	12	9	21
	19	diamicton	54	30	11	7	16
	30	diamicton	56	28	10	6	16
	50	diamicton	58	26	11	5	16
	97	diamicton	52	35	8	5	13
A8G1	0	massive mud	68	9	15	8	23
	183	massive mud	71	15	9	5	14
	220	laminated mud	41	40	11	8	19
	240	laminated mud	54	24	12	10	22
	340	laminated mud	52	30	10	8	18
	378	massive mud	57	21	12	10	22
A9G1	0	massive mud	68	11	16	5	21
	41	massive mud	57	22	11	10	21
	90	laminated mud	54	20	15	11	26
	94	fine-grained lamina	60	3	21	15	37
	120	laminated sand, mud	54	21	14	11	25
	155a	massive mud	66	12	15	7	22
	155b	massive mud	68	14	12	6	18
	377	massive mud	62	18	14	6	20
A10G1	0	massive mud	63	16	14	7	21
	15	laminated mud	67	16	12	5	17
	60a	massive mud	69	15	12	4	16

tion to having been swept from the surrounding banks. Domack speculated that sinking of high-salinity shelf water, produced during the winter months, may sweep sediment from the banks and concentrate it in the troughs. The significance of the subunit from 38 to 66 cm in core A1G1, containing the dark, typically ovoid cells, presently is unknown (Fig. 2).

The sharp contact between the diatom mud and the underlying diamicton suggests a relatively sudden domination of biogenic over terrigenous sedimentation, probably at the beginning of Holocene time. The patches and thin stringers of diatom mud that extend down into the diamicton record biological mixing by early Holocene infauna. This also might explain

Table 4. Continued.

Core	Depth (cm)	Lithology	Relative percent				
			illite	smectite	chlorite	kaolinite	chlorite + kaolinite
	60b	massive mud	73	12	11	4	15
	115	laminated silt, mud	61	23	9	7	16
	124	fine-grained lamina	63	19	9	9	18
	190	massive mud	65	20	12	3	15
	240	massive mud	70	13	12	5	17
	264a	laminated silt, mud	74	9	11	6	17
	264b	laminated silt, mud	69	12	12	7	19
	290	massive mud	68	19	9	4	13
	300	massive mud	69	17	10	4	14
	320	massive mud	68	20	9	3	12
	340	massive mud	71	13	10	6	16
	358	laminated silt, mud	60	21	10	9	19
	364	fine-grained lamina	76	3	12	9	21
	384	mud (faint laminae)	71	11	11	7	18

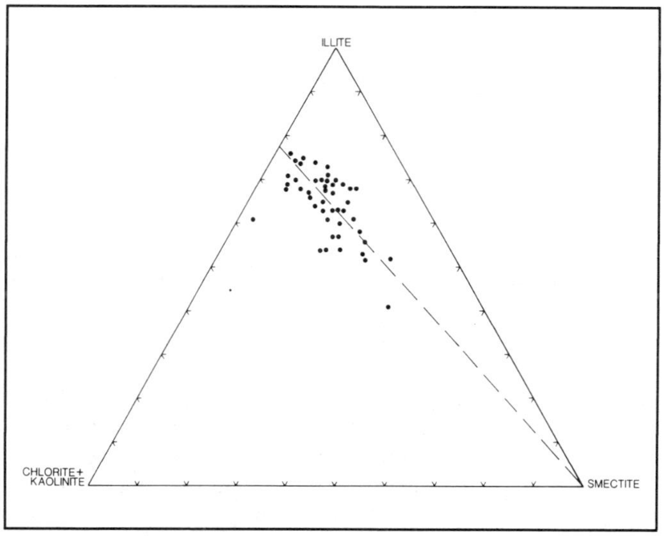

Figure 7—Ternary diagram of clay mineral relative abundances. Dashed line defines best fit line with constant illite-kaolinite+chlorite ratio of ~3.5.

the slight concentration of outsize clasts just above the contact, although just as likely it reflects a waning of post-Pleistocene glacial-marine deposition.

The index properties of the diatom mud are different from those of other cored sediment types and can be explained in terms of the sediment composition (Fig. 8). The low grain-specific gravity is due to the opaline composition of the diatoms and other siliceous tests, and the high water content and low wet unit weight are a consequence of the porous and highly nonspherical nature of the tests. The nonplastic behavior evidently reflects a low terrigenous clay abundance. The index properties are consonant with those reported for similar deposits elsewhere (Lee, 1973; Hampton, 1983).

The stratified, fossiliferous, muddy diamictons in cores A1G1 and A2G2 have the features of *compound glacial-marine* deposits (Anderson et al., 1980), although the characteristic silt-sized modes that result from traction-current deposition were not detected (Table 2). The diamictons are inferred to have been deposited from floating ice in a relatively quiescent environment. In spite of the abundance of striated and faceted gravel clasts that indicate transport in the basal zone of a glacier (Boulton, 1978), and in spite of the low measured water contents, we do not interpret the deposits as basal or lodgement till. The vane shear strengths are lower than those Domack (1982) reported for ice-loaded till deposits on the George V shelf (Fig. 8). Furthermore, the values of excess past pressure determined from the consolidation tests are less than would reasonably be produced by ice-loading, if the sediment was able to dewater during loading. The calculated values of ice thickness, in excess of neutrally buoyant thickness, that would produce the measured amount of excess pressure are only 7 m or less (Table 5), which is less than plausible for a fully developed grounded ice sheet (for example, Thomas, 1979). Moreover, the values of excess past pressure for sediment in deep-water cores A9G1 and A10G1, which certainly have not been ice-loaded, are similar to those mentioned above (Table 5). Another explanation of the excess past pressure values is that sediment overburden has been eroded by ocean currents. Calculated values of the thickness of eroded sediment are 1.6–10 m (Table 5). This also seems unlikely because no lag deposits appear in these cores. Most probably,

Figure 8—Index properties of cores A1G1, A2G2, A8G1, A9G1, and A10G1.

the excess past pressure is an artifact due to cementation or other structural bonding effects, which the consolidation test curve records as a false, high value of maximum past pressure (Richards and Hamilton, 1967).

The continental slope deposits recovered in cores A5G1 and A6G1 show evidence of stronger current activity compared to the shelf cores. The small amount of sandy sediment recovered in core A5G1 could be a traction-current deposit. The overall low mud content and the clast-supported gravel at some levels in core A6G1 imply reworking of a diamicton deposit by currents of various intensities. Sediment of this type is classified as *residual glacial marine* (Anderson et al., 1980). The pervasive occurrence of sandy sediment on the steep slope, as further documented in Deep Freeze 79 cores, suggests widespread reworking in water depths to greater than 2000 m. Domack and Anderson (1983) and Dunbar et al. (1985) speculated that reworked sediment they collected along the shelf-break was the product of an impinging geostrophic current with a velocity of 13–17 cm/sec. Evidently, this current is effective to the base of the continental slope. It is noteworthy that Deep Freeze 79 cores on the continental slope where it is broader and less steep west of our study area contain abundant muddy sediment; the current may not be as strong or persistent at the sea floor there.

Cores A8G1 and A9G1 contain distinct massive and laminated units that indicate abrupt changes in sedimentary environment from hemipelagic deposition for the former to fluctuating traction-current activity for the latter. The geomorphology and seismic stratigraphy suggest that the cores were collected from a submarine fan constructed during a prolonged period of deposition by turbidity currents (Hampton et al., this volume). Deep Freeze 79 cores contain graded sediment layers that clearly are turbidites, but cores A8G1 and A9G1 do not. The coarse-grained layers in the laminated sequences best fit the criteria for contourites as presented by Piper and Brisco (1975): common lateral terminations, wavy upper boundaries, gradational lower boundaries, and lack of grading. We prefer to use the more general term *traction-current deposit* as sug-

Table 5. Consolidation properties of sediment samples.

Core	Depth (cm)	Lithology	Water content (% dry wt.)	Overburden pressure (kPa)	Maximum past pressure (kPa)	Excess pressure (kPa)	Compression index*	Sediment removed† (m)	Ice removed†† (m)
A1G1	137	Diamicton	43	11	70	59	0.36-0.27	7	8
	220	Diamicton	41	17	58	41	0.30-0.22	5	5
	240	Diamicton	40	19	50	31	0.27-0.21	3	4
A2G1	24	Diamicton	23	2	16	14	0.09	1.6	1.4
	42	Diamicton	22	4	66	62	0.12	7	6
A8G1	153	Massive mud	76	9	72	63	0.66	7	11
	322	Laminated	90	15	60	45	0.82-0.58	5	8
	327	Laminated	92	15	45	30	1.31-0.61	3	5
	329	Laminated	104	15	54	39	1.42-0.68	4	7
	332	Laminated	105	15	68	53	0.80-0.69	6	10
A9G1	64	Massive mud	47.1	5	72	67	0.46-0.30	7	9
	115	Laminated	59	9	100	92	0.88-0.37	10	12
	187	Massive mud	71	12	75	63	0.92-0.57	7	9
A10G1	222	Massive mud	106	10	66	56	0.84	6	13

*Two values were calculated if the virgin curve is convex upward; one value for the upper curve and one value for the lower curve.

†Calculated from the excess burden pressure and the buoyant density of the sediment sample.

††Calculated from the excess overburden pressure and an assumed ice density of 9.0 kNm^3.

gested by Anderson et al. (1979). The well-sorted silt that constitutes the coarse-grained laminae in core A9G1 implies winnowing and bed-load transport. In contrast, the adjacent fine-grained laminae have a major proportion of clay-sized grains that suggest deposition from suspension (Table 2). Significant suspended-load deposition is indicated by the pervasive occurrence of silt and clay in all grain-size measurements of both the coarse- and fine-grained laminae in core A8G1. Surprisingly, these laminae, even the relatively coarse-grained basal one of the measured sequence, are finer grained than the massive units (Table 2). Perhaps this is because an ice-rafted component is present in the latter, although a modal class of such grains is not evident in the size distribution. Nevertheless, a few angular, gravel-sized grains that almost certainly were ice-rafted appear in x-radiographs (Table 1).

An active infauna partially reworked the sediment, producing the mottling, during deposition of the massive hemipelagic units in cores A8G1 and A9G1. In contrast, ice-rafted grains and biological burrows become insignificant when unsteady traction currents swept the area and deposited the alternate layers with subtle differences in grain size that typify the laminated units. Eittreim et al. (1972) measured geostrophic currents up to 15 cm/sec in the vicinity of cores A8G1 and A9G1 and detected a nepheloid layer. The currents and their entrained sediment are believed to derive from the formation of Antarctic Bottom Water (AABW) as far east as the Ross Sea, with some contribution from dense shelf water that spills across the shelf-break along the flow path (Gordon and Tchernia, 1972). Eittreim et al. (1972) commented that the strength of the current is likely to be highly variable, and probably it is this variability over long time spans that causes the stratigraphic changes observed in the cores. These environmental changes affect the ice-rafting and biological regimes as well as the sedimentological. This may be due to an increased rate of sedimentation during deposition of the laminated units, reducing the relative proportion of ice-rafted grains and limiting infaunal burrowing, or it may be a more complex combination of climate (cooling effects on the amount of ice-rafting) and oceanography (dissolved oxygen effects on biological populations). The invasion of infauna into previously barren territory when deposition of the massive units began is evidenced by the burrows that penetrate into the top of the laminated units.

Core A10G1 from the lower continental rise also contains both massive hemipelagic and laminated traction-current units, but the hemipelagic units are more diverse, especially in the degree and type of bioturbation and in the abundance of gravel-sized ice-rafted clasts. Also, some of the contacts between the massive and laminated units are gradational. The environment appears to have changed more gradually than at the locations of cores A8G1 and A9G1, perhaps because of a smoother continuum between times when ice-rafted grains and infaunal burrows were relatively abundant and times when traction-current transport dominated. Furthermore, the times of traction-current deposition permitted more burrowing at the site of core A10G1 compared to the shallower A8G1 and A9G1.

The more common appearance of ice-rafted gravel in core A10G1, compared to cores A8G1 and A9G1, may reflect warmer surface water farther offshore, causing accelerated meltout of ice-rafted sediment, or it may be an indicator of ice-

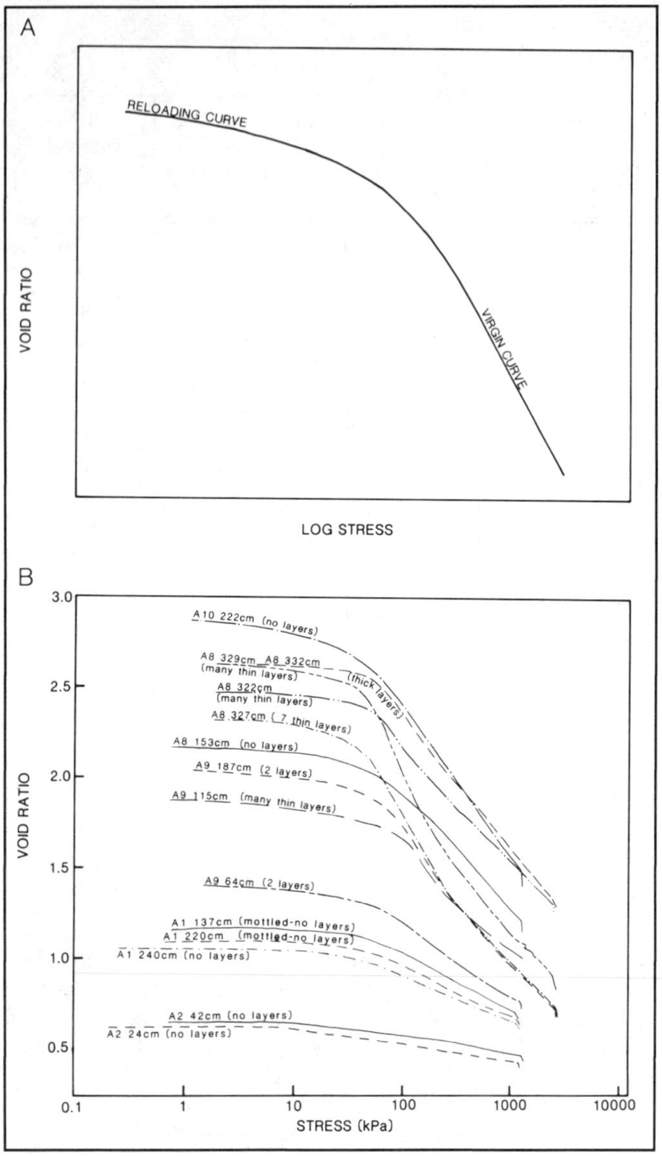

Figure 9—Compression curves. A. Ideal laboratory compression curve. B. Laboratory compression curves for sediment from the continental shelf and the continental rise.

berg tracks. Alternatively, there may be correlative units that contain more ice-rafted gravel at the upper-rise core sites, but at greater depth than was cored.

The laminated intervals in the three rise cores are further distinguished from the massive units by their physical properties, as discussed previously (Tables 5, 6; Fig. 8). The relatively thick laminated unit in core A8G1 is the only sediment that shows evidence of compaction under the in-place gravitational loads (Fig. 8); the water content is relatively high at the top of this unit and decreases downward. No other unit shows evidence of in-place compaction. The common lack of compaction in shallow-marine sediment has been attributed to structural bonding effects (Richards and Hamilton, 1967). Inconsistencies arise when this idea is applied to the George V sediment. Structural bonding produces strength in sediment that resists deformation. This strength should be manifest in the vane shear tests as high values of sensitivity and in the consolidation tests as a steep inclination of the upper portion of the virgin curve. Moderate to high values of sensitivity are recorded for the George V cores, and many of the compression curves are steep in the appropriate place (i.e., the concave-upward curves in Fig. 9), but there is no correlation to in-place compaction. In fact, compression curves for the laminated unit in core A8G1 have pronounced concavity, ultimately implying a resistance to compaction.

Turbidites have been recovered from the continental shelf depressions, the continental slope, and the continental rise, and slope instability probably was involved in the generation of the flows. The strength parameters from the triaxial tests can be used in an infinite-slope stability analysis to determine if the cored sediment is prone to gravitational failure. The effective angle of internal friction indicates the maximum stable slope for slow (drained) loading conditions. The minimum value of friction angle is 32° (Table 6), steeper than any slope displayed on bathymetric charts (Chase et al., this volume). The stability under rapid (undrained) loading conditions is determined from the relation derived by Morgenstern (1967):

$$S_u/\sigma = 0.5 \sin 2\alpha$$

where the term on the left is the strength to overburden ratio, and α is the maximum stable slope angle. Using the smallest strength to overburden ratio of 0.32, the maximum stable slope for the weakest material is 20°. There are no indicated slopes of this magnitude, either, although some on the continental slope approach this value. These analyses apply to static loading conditions only, and they indicate general stability of the tested sediment types on the George V margin.

The heavy-mineral identifications made in this study indicate a single provenance for the sand-sized sediment at all sampling stations, from the shelf to the lower rise, probably including the slightly anomalous assemblages in core A6G1. The pyroxene-rich assemblage is similar to the one identified by Domack (1980, 1982) in his study of shelf cores and attributed to the Ferrar Dolerite, which has been mapped onshore in the vicinity of Horn Bluff, east of Ninnis Glacier (Fig. 1). The previously unreported orthopyroxene is a common constituent of the Ferrar Dolerite (Gunn, 1962). The garnet that is abundant in all heavy-mineral counts and dominant in the three anomalous counts in core A6G1 likely was derived from the Beacon Group that underlies the Ferrar Dolerite (Domack, 1980, 1982). In light of the known distribution of these rocks, at Horn Bluff and in the Transantarctic Mountains (e.g., Gunn, 1962), the net transport direction of the sand-sized grains was inferred by Domack (1980, 1982) to be from the east, consistent with the inferred Pleistocene ice-flow directions and the modern regional current patterns (Domack, 1980, 1982; Gordon and Tchernia, 1972; Dunbar et al., 1985). The sediment of core A6G1 probably has been winnowed (Table 1), and because unique heavy-mineral types do not appear, we attribute the anomalous assemblage to hydraulic sorting rather than to a separate source. This idea could be tested by examining minerals of the coarse-sand fraction; Domack (1980, 1982) was able to distinguish mineralogical differences in this size fraction, less susceptible to the effects of hydraulic sorting because of smaller density variations, that suggest a distinct provenance for cores taken west of our study area. The conclusions reached in our study regarding prove-

Table 6. Triaxial strength properties of sediment samples.

Core	Depth (cm)	Lithology	Sheared water content (% dry wt.)	Initial effective confining pressure (kPa)	Maximum shear stress at failure (kPa)	Effective confining pressure at failure (kPa)	Effective angle of internal friction (degrees)	Strength/ overburden ratio
A9G1	32	Massive mud	44	277	98	214	36	0.35
A9G1	100	Laminated	37	362	117	239	32	0.32
A9G1	109	Laminated	37	356	120	232	33	0.34
A9G1	179	Massive mud	44	317	113	236	34	0.36
A10G1	215	Massive mud	68	260	96	164	39	0.37

nance are admittedly tentative in light of the cursory nature of the heavy-mineral observations and the poorly known distribution of rock types beneath the ice cap.

The gravel clasts listed in Table 1 indicate a provenance more diverse than the Ferrar Dolerite and Beacon Group—one that includes granitic and metamorphic rocks. Rocks of this varied nature have been identified along the George V–Adelie coast, especially west of Ninnis Glacier and the exposures of the Ferrar Dolerite and the Beacon Group (for example, Kleeman, 1940; Stillwell, 1940; Summers and Edwards, 1940; Craddock, 1972). On the basis of the gravel clasts, the entire area along the George V–Adelie coast must be considered a possible provenance of the ice-transported sediment. Furthermore, the palynological study by Truswell (this volume) indicates that pollen were derived from marine to marginally marine sediment of probable Cretaceous age, possibly from the same sequence of strata cored offshore in George V Basin by Domack et al. (1980). A diverse and complex source lithology is indicated for the coarse sediment along this section of the Antarctic continental margin, but multiple source terrains are not indicated.

The relative abundance of illite–smectite–kaolinite+chlorite clay minerals varies over a large range, with some association to the stratigraphic units in the cores (Table 4). In particular, smectite commonly has a relatively low percentage in the shelf diamictons. However, the clay-mineral proportions plot along a continuous trend rather than as separate assemblages (Fig. 7), and we attribute the variation to size-sorting processes rather than to differences in source. Gibbs (1977) and Knebel et al. (1977) proposed that clay minerals are physically sorted during transport as a result of different size distribution among the various mineral types, and smectite, being the smallest, generally is the most mobile. On the George V margin, samples of the laminated units from the continental rise cores have a greater variation in smectite abundance (range: 3–40%; standard deviation: 10%) relative to samples from the massive units (range: 9–22%; standard deviation: 4%). This is reasonable because variations in current strength (and physical sorting) probably were greater during deposition of the laminated units. However, two of the three analyses from distinct fine-grained laminae within the laminated units have a low smectite abundance of 3%, which clouds the issue because smectite (with a small grain size) should be relatively plentiful during tranquil deposition.

Eolian sand grains, similar to those discovered during the SEM examination of samples from our sediment cores, have been reported from other Antarctic and Arctic sediment cores (Rex et al., 1970; Hill and Nadeau, 1984; Hodel, 1985), suggesting that they are not an anomaly in polar environments; indeed, eolian deposits are a common feature of polar deserts (for example, Carter, 1983). The source of the dune grains recovered off George V Land is problematical, but they may have formed under arid and windy conditions in front of the Antarctic ice cap before it reached the edge of the continent, then carried out to sea as the ice advanced (J.B. Anderson, Rice University, personal communication). Most of the grains did not suffer breakage after removal from the eolian environment, which perhaps means that they were blown onto the ice rather than overridden by it. Eolian grains that are broken typically have a single large fracture surface that is unaltered (Figs. 6E, 6F). There has been speculation that these large fractures may form by a littoral freeze-thaw process (Hill and Nadeau, 1984), but if that were the case, abrasion of the fracture surface by littoral processes would be expected. Perhaps these grains experienced glacial crushing and then were dropped through the water column with no further significant abrasion by currents.

The abrasional features on the intermediate-type (rounded glacial) quartz grains probably were acquired subaerially for the most part. A recent study by Hodel (1985) demonstrated that grains in a sequence of samples down Alaska's Colville and Sagavanirktok drainage systems and onto the adjacent Beaufort Sea continental shelf exhibit mostly glacial features in the upper reach of the rivers and a trend of generally greater numbers of abrasional features downstream. Eolian grains are common, probably derived from relict Pleistocene dunes in this instance (Carter, 1983). A similar scenario can be envisioned for George V Land. When relatively temperate polar conditions existed in front of the advancing ice cap, meltwater from wet-base glaciers created an outwash plain, and grains released from the ice cap were abraded in this environment. The eolian grains formed subsequently, when cooler, arid conditions developed and wind reworked the outwash deposits. We conclude that some abrasion occurred in the ocean because grains from the slope cores, from laminated sections of the rise cores, and from turbidite and winnowed beds in Deep Freeze 79 cores are more rounded, on the average, than those from other units that were less affected by currents.

The sharp- or jagged-edge glacial grains that appear to some extent in virtually all samples are tempting to interpret as strictly ice-rafted grains, dropped directly over the location where they were cored. Sharp edges are extremely fragile and fracture under application of small stresses (Goodale and Hampton, in press), apparently ruling out current transport. However, the photograph in Figure 6B is of a grain taken from a coarse-grained lamina in core A10G1. It may be a directly deposited, ice-rafted grain, but unless it was dropped during the final stage of the lamina's accretion, it is difficult to envision how it escaped abrasion by the traction currents that deposited the remaining sediment.

We suspect that the large irregular depressions on grains from several samples (Fig. 6I) were created by pressure solution. If so, this implies a sedimentary source rock (Beacon Group) for these grains.

CONCLUSIONS

Sediment cores collected during the 1984 *S.P. Lee* cruise can be grouped into shelf, slope, and rise types based on common geologic features. A general picture of the regional sedimentation can be knit from the available geologic and oceanographic data.

Domack (1980, 1982) presented criteria for distinguishing between the Holocene diatom-rich and Pleistocene glacial and glacial-marine sediment types on the continental shelf and described the conditions under which each type was deposited. Our work confirms Domack's contention that glacial-marine diamictons can be identified on the basis of physical properties. We had hoped to take cores from sites where Domack identified glacial till, in order to see if we could detect high levels of ice-loading with consolidation tests, but were prevented from reaching these sites by adverse ice conditions. Edwards et al. (in press) successfully detected past ice-loading in Ross Sea tills, which have high values of excess pressure, up to 370 kPa (equivalent to loading by 41 m of ice).

Geostrophic currents strongly influence sedimentation on the continental slope and rise. Seismic-reflection profiles show that the continental slope is the front of a seaward-prograding sediment body (Hampton et al., this volume). The paucity of mud in sediment cores from the slope testifies to transport and reworking of sediment by a strong current whose effect has been documented at the edge of the adjacent continental shelf (Domack and Anderson, 1983; Dunbar et al., 1985). Our work indicates that it continues to at least a 2000-m water depth. The sediment transported along the depth contours may be intercepted by submarine canyons and eventually generate turbidity currents. Deep Freeze 79 cores from the less steep section of the continental slope west (downstream) of our study area contain intervals of mud, suggesting that the strength of the current at the sea floor is areally variable.

High-resolution seismic-reflection profiles show that sediment is accreting on most intercanyon areas of the upper continental rise, and erosion is occurring over broad regions of the lower rise sea floor (Hampton et al., this volume). Most of the canyons are characterized in the profiles by an opaque seafloor reflector that implies the presence of a dense, coarse-grained sediment layer. Canyon floors may be erosional or winnowed; indeed, some of the canyons are being actively downcut. We infer that the modern environment on the rise is dominated by the westward-flowing AABW, whose structure and sediment content, respectively, have been studied by Gordon and Tchernia (1972) and Eittreim et al. (1972). The erosional features on the lower rise mark the position of the high-velocity core of the current in recent time, and the sediment blanket on the upper rise occurs where the current generally is weaker and the regime is depositional. The alternation of massive and laminated units in both upper and lower rise cores documents the temporal variation in the strength of the current at a particular locality and an alternation between hemipelagic and traction-current deposition. Turbidity currents, laden with sediment either trapped by the heads of canyons on the continental slope or derived from slumping of canyon walls, may occasionally flow down the canyons, but heavy shelf water that repeatedly surges through a dynamic barrier near the shelf-break (Jacobs et al., 1970) also may form saline density currents that scour the canyon floors (Hampton et al., this volume).

Gravel-sized clasts were transported by ice to sites where they were cored; sand- and coarse silt-sized grains were transported either by ice, by contour-parallel traction currents, or by turbidity currents; and the smallest grains were transported by ice or in suspension by a possible variety of current processes. Therefore, the provenance of each size grade may be different. Our data do not indicate specific provenance areas, but they can be explained in terms of a single source terrain for each size grade. The down-core and between-core variations are small or vary in a manner that can be explained by current sorting. The gravel-clast lithologies could have been derived from rocks exposed along the entire George V–Adelie coast, in particular rocks farther west than the Beacon group and Ferrar Dolerite outcrops that were deduced on the basis of sand mineralogy by Domack (1980, 1982) to be the source of shelf glacial sediment in our study area.

Most quartz sand grains examined with the SEM have a blocky, angular foundation morphology indicative of glacial transport. We propose that most abrasional features on these grains were imparted by subaerial processes in front of the advancing ice sheet, but some grains also were abraded in the marine environment. Glacial grains with few or no abrasional features apparently were ice-rafted and released directly over their site of deposition. We infer that the highly spherical grains were rounded by eolian process in an arid polar-desert environment in front of the advancing ice cap, during a cooling event when meltwater did not form in significant quantity.

The physical properties of sediment cores collected for this study have some unusual features, such as the typically concave-upward nature of the virgin compression curves and the compaction characteristics of the laminated units. Some sediment types have distinctive physical properties—e.g., the high water content of the diatom mud—but most values of our laboratory measurements are within typical ranges for marine sediment. The cored sediment types are gravitationally stable on the seafloor slopes depicted on bathymetric charts (Chase et al., this volume). The slumps that are indirectly documented by the existence of turbidites in shelf basins and on the continental rise may have been triggered in weaker sediment types, on locally steep slopes, or by dynamic loads (for example, earthquakes).

ACKNOWLEDGMENTS

We thank Captain Alan McClenaghan and the crew of the *S.P. Lee* for their cooperation in achieving the goals of this study under the difficult operating conditions in Antarctic waters. Discussions with John Anderson, Homa Lee, and William Winters aided our interpretations of the data. Kris Johnson and Robert Kayen performed many of the laboratory tests. Dennis Cassidy kindly supplied descriptions and subsamples of Deep Freeze 79 cores. We gratefully acknowledge manuscript reviews by John Anderson, Eugene Domack, and Anders Elverhoi.

REFERENCES CITED

American Society for Testing and Materials (ASTM), 1982, 1982 Annual Book of ASTM Standards. Part 19, Natural Building Stones; Soil and Rock: Philadelphia, ASTM, 710 p.

Anderson, J. B., 1981, Marine sediment core descriptions—IWSOE 68, 69, 70; Deep Freeze 79: Houston, Rice University, Geology Department, unpublished report, 60 p.

Anderson, J. B., D. D. Kurtz, and F. M. Weaver, 1979, Sedimentation on the Antarctic continental slope, *in* L. J. Doyle and O. H. Pilkey, eds., Geology of continental slopes: Society of Economic Paleontologists and Mineralogists Special Publication 27, p. 265–283.

Anderson, J. B., D. D. Kurtz, E. W. Domack, and K. M. Balshaw, 1980, Glacial and glacial marine sediments of the Antarctic continental shelf: Journal of Geology, v. 88, p. 399–414.

Biscaye, P. E., 1965, Mineralogy and sedimentation of Recent deep sea clay in the Atlantic Ocean and adjacent sea and oceans: Geological Society of America Bulletin, v. 76, p. 803–832.

Bishop, A. W., and D. J. Henkel, 1964, The measurement of soil properties in the triaxial test: London, Edward Arnold Ltd., 228 p.

Boulton, G. S., 1978, Boulder shapes and grain-size distributions of debris as indicators of transport paths through a glacier and till genesis: Sedimentology, v. 25, p. 773–800.

Bryant, W. R., A. P. Deflache, and P. K. Trabant, 1974, Consolidation of marine clays and carbonates, *in* A. L. Inderbitzen, ed., Deep-sea sediments: New York, Plenum Press, p. 209–244.

Carter, L. D., 1983, Fossil sand wedges on the Alaskan arctic coastal plain and their paleoenvironmental significance: Permafrost: Fourth International Conference, Proceedings. Washington, DC, National Academy Press, p. 109–114.

Carver, R. E., 1971, Heavy-mineral separation, *in* R. E. Carver, ed., Procedures in sedimentary petrology: New York, Wiley-Interscience, p. 427–452.

Casagrande, A., 1936, The determination of the preconsolidation load and its practical significance: Proceedings of the 1st International Conference on Soil Mechanics and Foundation Engineering, p. 60.

Craddock, C., 1972, Geologic map of Antarctica: New York, American Geographical Society, 1:5,000,000.

Domack, E. W., 1980, Glacial marine geology of the George V–Adelie continental shelf, East Antarctica: M. A. Thesis, Rice University, Houston, Texas, 142 p.

———, 1982, Sedimentology of glacial and glacial marine deposits on the George V–Adelie continental shelf, East Antarctica: Boreas, v. 11, p. 79–97.

Domack, E. W., and J. B. Anderson, 1983, Marine geology of the George V continental margin: Combined results of Deep Freeze 79 and the 1911–14 Australasian expedition, *in* R. L. Oliver, P. R. James, and J. B. Jago, eds., Antarctic earth science: Canberra, Australian Academy of Science, p. 402–406.

Domack, E. W., W. W. Fairchild, and J. B. Anderson, 1980, Lower Cretaceous sediment from the East Antarctic continental shelf: Nature, v. 287, p. 625–626.

Dunbar, R. B., J. B. Anderson, E. W. Domack, and S. S. Jacobs, 1985, Oceanographic influences on sedimentation along the Antarctic continental shelf, *in* S. S. Jacobs, ed., Oceanology of the Antarctic continental shelf: American Geophysical Union, Antarctic Research Series, v. 43, p. 291–312.

Edwards, B. D., H. J. Lee, H. A. Karl, E. Reimnitz, and L. A. Timothy, in press, Geology and physical properties of Ross Sea, Antarctica, continental shelf sediment, *in* A. K. Cooper and F. J. Davey, eds., The Antarctic continental margin: Geology and geophysics of the western Ross Sea: Circum-Pacific Council for Energy and Mineral Resources Earth Science Series, v. 5B.

Eittreim, S., A. L. Gordon, M. Ewing, E. M. Thorndike, and P. Bruchhausen, 1972, The nepheloid layer and observed bottom currents in the Indian–Pacific Antarctic sea, *in* A. L. Gordon, ed., Studies in physical oceanography: London, Gordon and Breach, p. 19–35.

Gibbs, R. J., 1977, Clay mineral segregation in the marine environment: Journal of Sedimentary Petrology, v. 47, p. 237–243.

Goodale, J. L., and M. A. Hampton, in press, An experimental and theoretical study of some quartz grain surface features, *in* J. R. Marshall, ed., Clastic particles: Scanning electron microscopy and shape analysis of sedimentary and volcanic clasts: Stroudsberg, PA, Dowden, Hutchinson and Ross.

Gordon, A. L., and P. Tchernia, 1972, Waters of the continental margin off Adelie coast, Antarctica, *in* D. E. Hayes, ed., Antarctic oceanology. II. The Australian–New Zealand sector: American Geophysical Union, Antarctic Research Series, v. 19, p. 59–69.

Gunn, B. M., 1962, Differentiation in Ferrar Dolerites, Antarctica: New Zealand Journal of Geology and Geophysics, v. 5, p. 820–863.

Hampton, M. A., 1983, Geotechnical framework study of the Kodiak Shelf, Alaska: U.S. Geological Survey Open-File Report 83–171, 86 p.

Hein, J. R., D. W. Scholl, and C. E. Gutmacher, 1976, Neogene clay minerals of the far NW Pacific and southern Bering Sea: Sedimentation and diagenesis, *in* S. W. Bailey, ed., Proceedings of the International Clay Conference: Wilmette, IL, Applied Publishers, p. 71–80.

Hill, P. R., and O. C. Nadeau, 1984, Grain-surface textures of Late Wisconsinan sands from the Canadian Beaufort Shelf: Journal of Sedimentary Petrology, v. 54, p. 1349–1357.

Hodel, K. L., 1985, Microtextures of quartz grains from modern terrestrial and subaqueous environments: North Slope Alaska: M. S. Thesis, University of California, Berkeley, 139 p.

Jacobs, S. S., A. F. Amos, and P. M. Bruchhausen, 1970, Ross Sea oceanography and Antarctic Bottom Water formation: Deep-Sea Research, v. 17, p. 935–962.

Kleeman, A. W., 1940, Schists and gneisses from the moraines, Cape Denison, Adelie Land: Australasian Antarctic Expedition 1911–1914, Scientific Reports, Series A, v. IV, p. 201–292.

Knebel, H. J., T. J. Conomos, and J. A. Commeau, 1977, Clay-mineral variability in the suspended sediments of the San Francisco Bay system, California: Journal of Sedimentary Petrology, v. 47, p. 229–236.

Krinsley, D. H., and J. C. Doornkamp, 1973, Atlas of quartz sand surface textures: London, Cambridge University Press, 91 p.

Ladd, C. C., and R. Foott, 1974, New design procedure for stability of soft clays: Journal of the Geotechnical Engineering Division, American Society of Civil Engineers, v. 100, p. 763-786.

Lambe, T. W., and R. V. Whitman, 1969, Soil mechanics: New York, John Wiley & Sons, 553 p.

Lee, H. J., 1973, Engineering properties of some north Pacific and Bering Sea soils: Naval Civil Engineering Laboratory Technical Note N-1283, 26 p.

Mawson, D., 1940, Sedimentary rocks: Australasian Antarctic Expedition 1911-14, Scientific Reports, Series A, v. IV, p. 347-367.

Morgenstern, N. R., 1967, Submarine slumping and the initiation of turbidity currents, *in* A. F. Richards, ed., Marine geotechnique: Urbana, University of Illinois Press, p. 189-220.

Piper, D. J. W., 1978, Turbidite muds and silts on deepsea fans and abyssal plains, *in* D. J. Stanley, and G. Kelling, eds., Deposition in submarine canyons, fans, and trenches: Stroudsburg, PA, Dowden, Hutchinson and Ross, p. 163-176.

Piper, D. J. W., and C. D. Brisco, 1985, Deep water continental margin sedimentation, Leg 28, Antarctica, *in* D. E. Hayes, and L. A. Frakes, et al., eds., Initial reports of the Deep-Sea Drilling Project: Washington, DC, U.S. Government Printing Office, v. 28, p. 727-755.

Quigley, R. M., and C. D. Thompson, 1966, The fabric of anisotropically consolidated sensitive marine clay: Canadian Geotechnical Journal, v. 3, p. 61-73.

Rex, R. W., S. V. Margolis, and B. Murray, 1970, Possible interglacial dune sands from 300 meters water depth in the Weddell Sea, Antarctica: Geological Society of America Bulletin, v. 81, p. 3465-3472.

Richards, A. F., 1962, Investigation of deep-sea sediment cores, II. Mass physical properties: U.S. Navy Hydrographic Office Technical Report 106, 146 p.

Richards, A. F., and E. L. Hamilton, 1967, Investigations of deep-sea sediment cores. III. Consolidation, *in* A. F. Richards, ed., Marine geotechnique: Urbana, University of Illinois Press, p. 93-117.

Stillwell, F. L., 1940, Metamorphic rocks of Adelie Land: Australasian Antarctic Expedition 1911-1914, Scientific Reports, Series A, v. III, p. 3-230.

Stow, D. A. V., 1979, Distinguishing between fine-grained turbidites and contourites on the Nova Scotian deep water margin: Sedimentology, v. 26, p. 371-388.

Summers, H. S., and A. B. Edwards, 1940, Granites of King George Land and Adelie Land: Australasian-Antarctic Expedition 1911-1914, Series A, v. IV, p. 89-113.

Thomas, R. H., 1979, The dynamics of marine ice sheets: Journal of Glaciology, v. 24, p. 167-177.

Vanney, J. R., and G. L. Johnson, 1979, The sea floor morphology seaward of Terre Adelie (Antarctica): Deutsche Hydrographische Zeitschrift, v. 32, p. 965-978.

von der Borch, C. C., and R. L. Oliver, 1968, Comparison of heavy minerals in marine sediments with mainland rock outcrops along the coast of Antarctica between longitudes 40° E and 150° E: Sedimentary Geology, v. 2, p. 77-80.

Whalley, W. B., and D. H. Krinsley, 1974, A scanning electron microscope study of surface textures of quartz grains from glacial environments: Sedimentology, v. 21, p. 87-105.

Wissa, A. E., J. T. Christian, E. H. Davis, and S. Heiberg, 1971, Consolidation at constant rate of strain: Journal of the Soil Mechanics and Foundation Division, American Society of Civil Engineers, v. 97, p. 1393-1410.

Wright, R., J. B. Anderson, and P. P. Fisco, 1983, Distribution and association of sediment gravity flow deposits and glacial/glacial-marine sediments around the continental margin of Antarctica, *in* B. F. Molnia, ed., Glacial-marine sedimentation: New York, Plenum Press, p. 265-300.

Eittreim, S. L., and Hampton, M. A., 1987, *The Antarctic Continental Margin: Geology and Geophysics of Offshore Wilkes Land*, CPCEMR Earth Science Series, v. 5A: Houston, Texas, Circum-Pacific Council for Energy and Mineral Resources

Morphologic Studies of the Wilkes Land Continental Shelf, Antarctica—Glacial and Iceberg Effects

P. W. Barnes
U.S. Geological Survey
Menlo Park, California

Seismic-reflection data were obtained from the Wilkes Land continental margin in Antarctica. In addition, sidescan sonar data were obtained on a track parallel to the coast from the top of Mertz Bank (200 m) to the floor of Mertz Trough (600 m). Sparse ice gouge features occur to depths of over 500 m and typically form multiple-grooved incisions a few meters deep and tens of meters wide. The dominant bed forms on both the bank top and slope are circular to subcircular depressions 30–150 m in diameter. Locally these features are associated with ice gouges. They probably represent modern ice-resting sites indicating that modern sediment is being reworked by iceberg keels to depths of 500+ m. Deep draft ice is suggested by observed freeboards and extrapolation of measured iceberg drafts. The high concentration of depressions on the slope, their absence in the trough, and their subdued appearance on the bank top probably result from modern sediment reworking of older deposits on the bank top and sedimentation in the troughs. The reworking of the bank seems to be forming a modern ice-turbate deposit that may be widespread on the shallow portions of the Antarctic shelf. No morphologic evidence was found for sediment gravity flows or turbidites, as suggested by earlier studies.

Ridges and grooves perpendicular to the coast, draped with acoustically transparent sediment, are spaced about 100 m apart on the floor of Mertz trough. Their origin can best be related to Pleistocene glacial advances onto the shelf forming a fluted surface. During this same ice advance, asymmetric ridges up to 50 m high were developed on the southern and eastern rim of Mertz Bank suggesting that ice detoured around this feature. The ridges are remnants from lateral glacial ridges developed during the expansion of the Antarctic ice sheet onto the shelf.

INTRODUCTION

Geologic and morphologic data were gathered by the U.S. Geological Survey from the Wilkes Land continental margin during the 1984 austral summer (Eittreim et al., 1984). Ice and ice-related geologic processes are important in forming seabed morphology on high-latitude continental shelves. Previous studies of the Antarctic shelf have outlined the glacial domination of major sedimentary processes and large topographic features. The present study examines the morphologic and shallow seismic-reflection information from a small portion of common environments of the mid- to outer-Antarctic shelf (Fig. 1). The observations are related to the glacial and hydrologic processes responsible for the morphologies.

Vanney and Johnson (1979) described the morphology of the Antarctic continental shelf off Wilkes Land between long 135° and 147° E. Within this area, Domack (1982) incorporated additional data to make a more detailed bathymetric map for the area between long 140° and 150° E. These data show a shelf 100–200 km wide with a basin 1000 m or more deep paralleling the coastline. The central and outer shelf is characterized by cross-shelf troughs about 500 m deep that are separated by banks commonly having 200- to 300-m depths, with minimum depths less than 200 m (Fig. 1). The present study examines the morphology of one of these banks, Mertz Bank, and a cross-shelf trough, Mertz Trough.

Sedimentary Environment

Studies have shown that sedimentation in this area is related to modern and ancient glacial and glacial-marine processes (Domack, 1982). Surface sediment on Mertz Bank is moderately to poorly sorted sand and muddy sand. The slopes and floor of Mertz Trough are covered with very poorly sorted mud and sandy mud composed of ice-rafted debris and are rich in diatoms and sponge spicules (Domack, 1982).

Cores taken from the floor of Mertz Trough show a surficial layer of this Holocene diatom, spicule, and radiolaria ooze about 10 cm thick. This ooze overlies a stratified fossiliferous muddy diamicton that has been deposited from floating ice (Domack, 1982). Hampton et al. (this volume) confirm these observations. Farther to the west in another shelf trough, the presence of up to 40 m of siliceous oozes has been inferred from 12-kHz records (Domack and Anderson, 1983).

Only a few samples of sediment from the slopes of the depressions have been obtained, but they can be inferred to

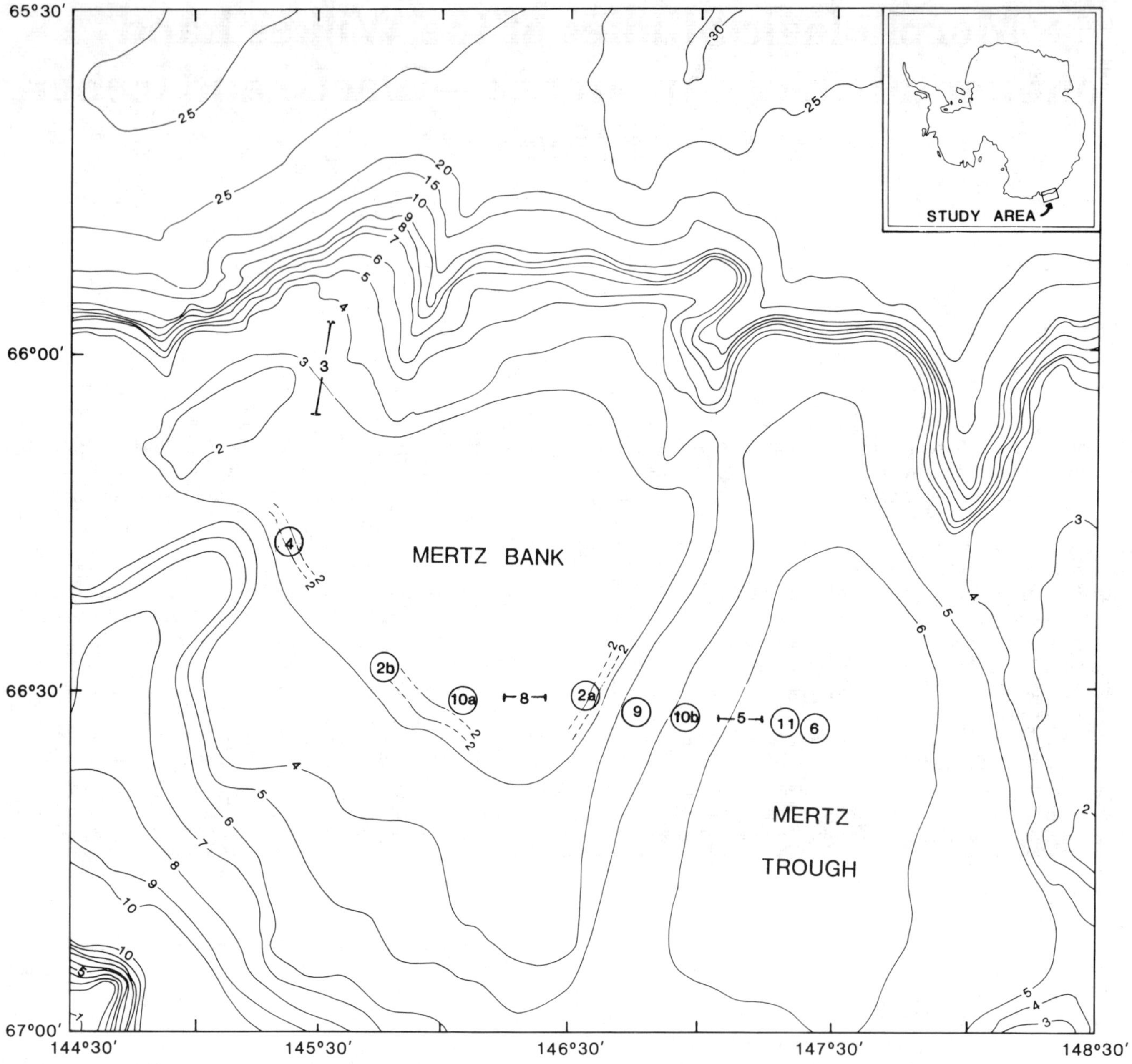

Figure 1—Location of study area and regional bathymetry modified after Vanney and Johnson (1979) and Domack (1982). Numbers in circles refer to the locations of other figures in the text to follow. Bathymetry is in hundreds of meters. Mertz Trough as used here is to be distinguished from the Mertz-Ninnis Trough, which is deeper and parallels the coast inshore from the study area (Defense Mapping Agency, 1981).

consist of massive to stratified diamictons that may or may not contain fossils (Domack et al., 1980; Anderson et al., 1980b). These are interpreted as having been derived from either basal tills or floating ice, which presently may be influenced by sediment gravity flows on the steeper slopes (Wright et al., 1983).

Sand prevails in the few surface samples collected from Mertz Bank. There are no cores from the top of the bank. However, a core from a bank top 200 km to the west recovered graded sand thought to be a turbidite (Domack, 1982). Elsewhere off Wilkes Land, cores from water depths less than 400 m contained sand and muddy sand, the distribution of which was limited to regions of "shallow" irregular shelf topography. Marine currents, aided by biologically active bottom fauna, have reworked glacial deposits on the banks and outer shelf leaving residual glacial-marine deposits (Dunbar et al., 1985). In the Weddell Sea, studies suggest that a lack of fine-grained sediments on the bank tops is related to winnowing of ice-rafted debris during descent through the water column (Orheim and Elverhoi, 1981).

In summary, previous studies have suggested that the shallow-water sediment consists of current-reworked glacial deposits, winnowed ice-rafted debris, and basal tills, whereas the deeper basin and trough sediments are primarily a result of biogenic deposition.

Seabed Morphology

Lien (1981), in his study of a small portion of the shelf of the Weddell Sea, used sidescan sonar to delineate four types of seabed morphology in water depths of less than 400 m: ice gouges, a washboard pattern, hummocky seabed, and undisturbed sea floor. Ice gouging was prevalent on the bank tops and down to about 300 m—maximum gouge widths reaching 250 m and maximum gouge depths 25 m—below the surrounding sea floor. The gouges at 300 m are difficult to explain, because the nearby ice shelf extends only 200 m below sea level. A washboard pattern of parallel ridges is associated with the gouging that extends down to at least 400 m. This pattern is attributed to wobbling icebergs. The gouging is more recent than the washboard pattern, as determined by superposition. Furthermore, the fresh appearance of the gouges suggests their recency; however, slow rates of sedimentation and/or erosion could account for the crisp morphology.

Icebergs and Glaciers

Data on the draft, underwater shape, and movement of icebergs in the Antarctic are sparse. Orheim (1980) used radio echo sounding techniques to measure tabular icebergs as much as 386 m thick. Radio echo sounding revealed very little detail of the underside of the bergs except that the thickness of any one iceberg varied an average of 18%. Orheim (1980) also found a higher ice density for drifting icebergs than the density of the ice shelves from which they were derived. Thus, floating icebergs should float deeper for an equivalent freeboard than ice shelves. He found the ratio of thickness to freeboard for tabular icebergs was about 5:1, or a draft of 284 m for the 386 m thick iceberg. Sonar studies indicate that the underwater shape of tabular icebergs varies greatly depending on the amount of exposure to waves and currents (Klepsvik, 1980).

Keys (1983) studied the distribution of upright icebergs (tabular and tilted tabular) versus overturned icebergs (irregular and rounded) in the Wilkes Land region. He found that 57% of the icebergs near the coast classify as upright, whereas farther north the bulk of the iceberg population (79%) was made up of overturned icebergs. Tabular icebergs make up 17% of the population in the Ross Sea but only 5% of the population north of there.

Glaciers are not presently active in the study area, except as coastal ice tongues of the Mertz and Ninnis glaciers. During the last glaciation, glaciers and ice shelves extended across the entire continental shelf of Wilkes Land continental margin (Stuvier et al., 1981) as elsewhere on Antarctica (Elverhoi and Maisey, 1983; Drewry, 1979). Controversy exists as to whether the expanded ice was in the form of a grounded glacier (Stuvier et al., 1981) or floating ice shelves (Drewry, 1979, 1980; Orheim and Elverhoi, 1981). In either case, ice would have expanded to the shelf-break and grounded on the top and flanks of Mertz Bank, aided by lowered sea levels.

In summary, the potential draft of Antarctic tabular ice is in excess of 284 m when the upper and lower surfaces of the icebergs are essentially parallel. Nontabular and tilted tabular ice blocks make up the major portion of the Antarctic iceberg population. These icebergs can be expected to have even deeper drafts (up to 50%), because when tabular icebergs are unstable and rotate they may increase their draft (Bass and Peters, 1984). Thus, ice and seafloor interaction in the Antarctic might be expected to occur to depths considerably greater than 284 m, perhaps in excess of 400 m. During the most recent glacial expansion, the ice extended to the shelf-break as either a glacier or ice shelf grounding on the offshore shoals.

Arctic Studies

Ice interaction with the sea floor is common on Arctic continental shelves; thus, studies of Arctic processes may apply to the Antarctic continental shelves. These Arctic studies have included sea ice (Pelletier and Shearer, 1972; Reimnitz and Barnes, 1974; Lewis, 1978), icebergs, and modern and ancient glacial ice (Belderson and Wilson, 1973; Moign, 1976; Barrie, 1980; Vorren et al., 1983; Lien, 1983; Solheim and Pfirman, 1985; Woodworth-Lynas et al., 1985). The latter studies are the most applicable to the present investigation. Barrie (1980) reports modern gouge features from Greenland icebergs on the sea floor off Labrador to depths of nearly 200 m, and Woodworth-Lynas and his coworkers demonstrate modern iceberg groundings to at least 175 m. Ice groundings penetrate the sea floor to a typical depth of 5 m and width of 30 m, but depths of more than 10 m, widths of hundreds of meters, and lengths in excess of 3 km have been observed. Barrie (1980) believes that the seafloor gouge record has accumulated over the past 20,000 years, many of the gouges being relict from a lower stand of sea level.

Studies on the Norwegian continental shelf (Vorren et al., 1983; Lien, 1983) suggest that the gouges on the northern European shelf are relict. The subdued and weathered seafloor features can be more than 25 m deep, over 250 m wide, and several kilometers long. They were formed at the end of the last ice age (10,000 to 13,000 years B. P.). Iceberg gouges cover about 55% of the shelf and are concentrated in water depths between 100 and 500 m on the continental slope and on bank margins rather than in the depressions. The 500-m depth limit for observed iceberg gouges probably is related to the maximum draft of icebergs of the last ice age. This draft limit would suggest that the icebergs had a draft of about 400 m below the reduced sea level at glacial maximum. Gouges tend to parallel the slopes but have random orientations where the bottom is flat. Soil conditions were found to influence the size and length of the gouges, because plowing resistance varied with soil type and consolidation.

RESULTS AND OBSERVATIONS

Data-Gathering Techniques

The data utilized in this study were gathered during January 1984 (Eittreim et al., 1984). Locations were determined with an integrated doppler sonar and satellite system with a probable accuracy on the order of a few hundred meters.

The geophysical data on which this report is based include 12-kHz bathymetry, 3.5-kHz bathymetry, 1-kHz seismic-boomer records, and a 7-kHz profiler coupled with a digital sidescan sonar fish. The 0.5- to 1.0-sec bathymetry records resolve morphologic features on the order of 1 m, partly because the mild sea conditions at the time of observation provided a stable platform. The seismic-reflection records and the profiler were recorded at 0.25 sec, and resolution of less than a meter was routine. Because the profiler was located in the sidescan fish, the relief observed with this instrument is directly comparable with the sonograph data; however, the profiler does not provide true water-depth information.

The sidescan was operated on a 200-m scan (400-m swath width) in shallow water and 300-m scan (600-m swath width) in deeper water. The digital records were taped and processed to remove slant-range and ship-speed variations, thus providing an anamorphically correct sonograph image. Subsequent to data acquisition, the tapes were replayed to produce "inverse images" (echoes appear as lighter areas) that proved more useful than normal images in visualizing the seabed morphology. Inverse images are used in this report.

Bathymetry

The bathymetry of the study area (Fig. 1) was modified from the earlier work of Vanney and Johnson (1979) and Domack (1982). The top of Mertz Bank has a general depth of about 250 m or about 300 m above the floor of Mertz Trough. The bank forms a smooth surface with microrelief of less than 2 m. Three bathymetric features stand out. On the eastern rim of the bank, a shoal abruptly rises 150 m above the regional bank surface, to a depth of 115 m (Fig. 2A). The shallowest portion of this shoal is almost flat for a width of 1.5 km.

On the south-central portion of the bank in water depths of 150–300 m, a pair of smooth asymmetric topographic highs 50–70 m in height form shoals 3–4 km wide on several profiles (Fig. 2B). The crest of these shoals ranges from less than 150 to more than 220 m. The steeper slopes of these broad features face seaward onto the bank top. The highest features are located on the western part of the bank. Although the lateral continuity of these features is unknown, if they are linear and parallel to the bathymetry, they would form a rim along the southwest-facing edge of the bank as we suggest in Figure 1.

A smooth 15-km ramp on the seaward edge of the bank was noted at the shelf-break at 600–700 m and rising up to 300 m (Fig. 3)—a more gradual slope from 300 m reaching the general elevation of the shoal crest at 250 m. No microrelief was noted on this ramp.

Seismic-Reflection Observations

Preliminary results from multichannel seismic-reflection data suggest that Mertz Bank is underlain by a few hundred meters of horizontal or subhorizontal beds of unknown age that cross beneath the shoals (Eittreim and Smith, this volume, Line 14B). An unconformity at the base of these units could mark the base of a sequence similar to that of the youngest sedimentary units described by Stagg (1985) and by Tanahashi et al. (this volume) more than 2000 km to the west.

The high-resolution seismic reflection systems were also equivocal in unraveling the shallow geology of the bank. The 3.5-kHz and boomer seismic-reflection systems commonly returned a seismic signal that indicates subbottom reflectors in the upper 10–20 m, but these reflectors generally are not laterally coherent (Figs. 2A, 4). This strong "hashy" seismic character for surface reflectors on high-resolution records commonly is associated with coarse sediment such as sand and gravel. Fine-grained material commonly allows more seismic penetration. Occasional irregular reflectors and stratigraphy can be traced within the near-surface seismic-reflection section.

The shallow seismic reflectors seen in the shoals at the southern and eastern edge of the bank indicate discontinuous reflectors and rough unconformable (?) relief. A reflector at 20–40 m subsurface can be traced on the slopes of the southern shoals for several hundred meters, although the reflectors never clearly outcrop (Fig. 4). The surficial rough relief, the discontinuous nature of the subbottom reflectors, and their common rough relief suggest that the present surface of the bank is being reworked and may be an erosional or nondepositional surface. In contrast, subbottom reflectors can be traced for a short distance beneath the shoal on the western edge of the bank suggesting that this feature was constructional (Fig. 2A). However, the rough surficial relief and discontinuous reflector suggest this shoal may presently have an erosional surface.

The slope to Mertz Trough east of the shoal is irregular—the hummocks and hyperbolic reflectors suggesting steep slopes. Microrelief of 5 m and more is common. Poor seismic-reflection records from the western flank of Mertz Trough preclude a definitive description of the stratigraphic character; however, from the records obtained of the surface morphology, the slope is believed to have a sediment cover that presently is being reworked.

The floor of the Mertz Trough at about 610 m is smooth with undulations of about 10 m and microrelief of less than 2 m. The sediments on the floor of the trough are characterized by ponds of acoustically transparent sediments visible on the 7-kHz pinger, 3.5-kHz, and boomer records. Ponds less than 10 m in depth occur on the flank of the trough at a water depth of 530 m. The surfaces of the ponds are commonly inclined from the horizontal (Fig. 5). Toward the central part of the trough, the transparent sediment is ubiquitous, commonly 10 m thick, and draped over an irregular subsurface reflector (Fig. 6).

In summary, the seismic-reflection information suggests to us that the shoals and the top and flanks of Mertz Bank are reworked surfaces of sand and coarser material. The irregular discontinuous horizontal reflectors suggest ice-related deposition or postdepositional disruption. The surficial sediments are reworked to a depth of 10–20 m, suggesting that the surface of the bank may become an unconformity. Mertz Trough has ponded transparent sediment about 10 m thick over a surface of rough relief.

Detailed Morphology—Sidescan Sonar and Microprofiler

Specific morphologic information from the study area is available from the combined sidescan, microprofiler, and 3.5-kHz data gathered from a single survey line extending east from the top of Mertz Bank down the bank slope to the center of Mertz Trough (Fig. 7). Four distinct morphologic environments were observed on this profile: (1) the essentially flat bank top with depths of 230–250 m and micro relief of up to 3 m; (2) the abrupt shoal at the edge of the bank with small-relief features of several meters and a flat smooth top; (3) the western slope of Mertz Trough with rugged microrelief com-

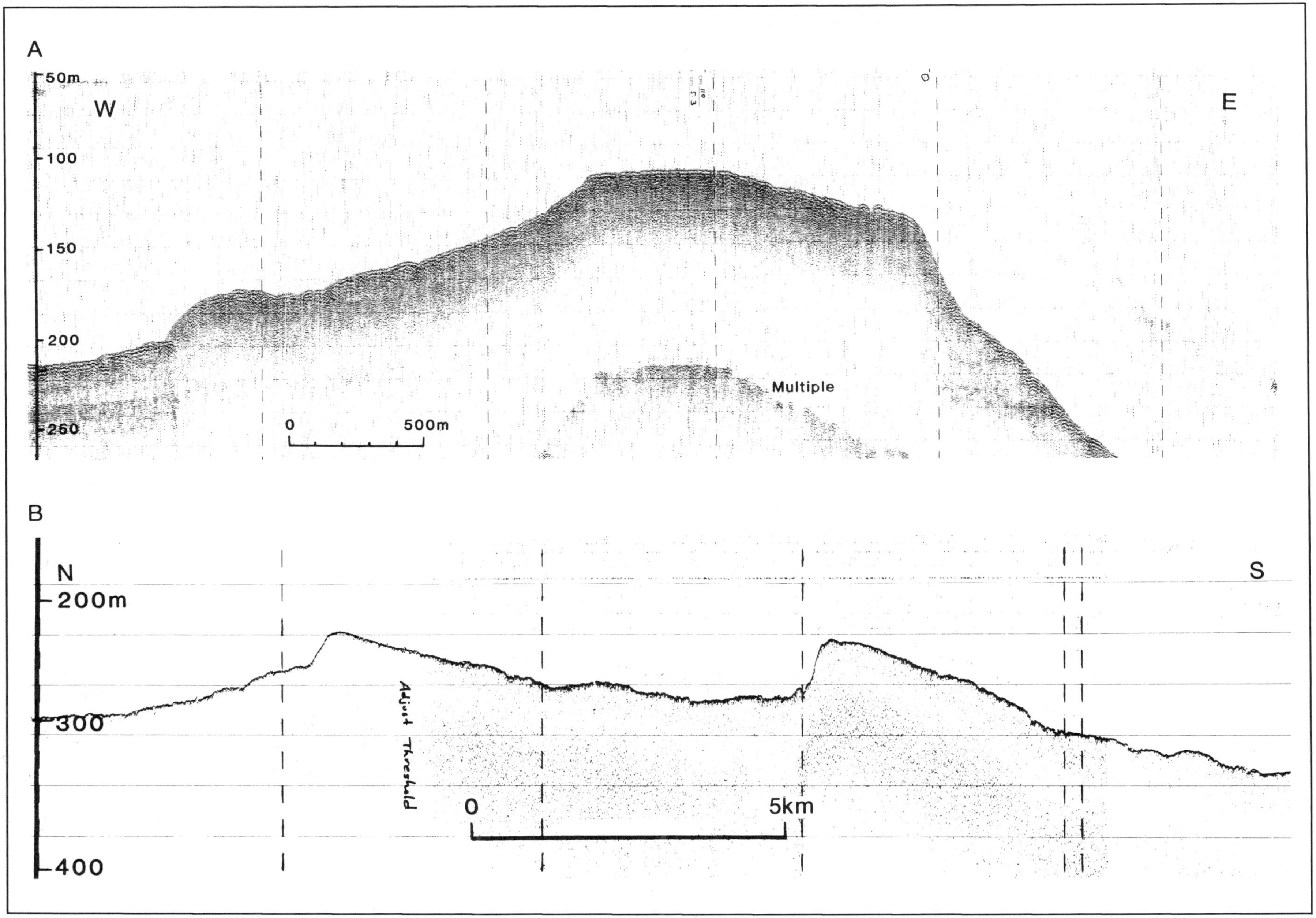

Figure 2—A. Ridge or shoal on the eastern flank of Mertz Bank; Mertz Trough is to the right. Note the lack of lateral continuity to any of the subbottom reflectors on this 1-kHz boomer record. See Figure 1 for location. B. Asymmetric ridges on the southern flank of Mertz Bank from 3.5-kHz record. South is to the right. See Figure 1 for location.

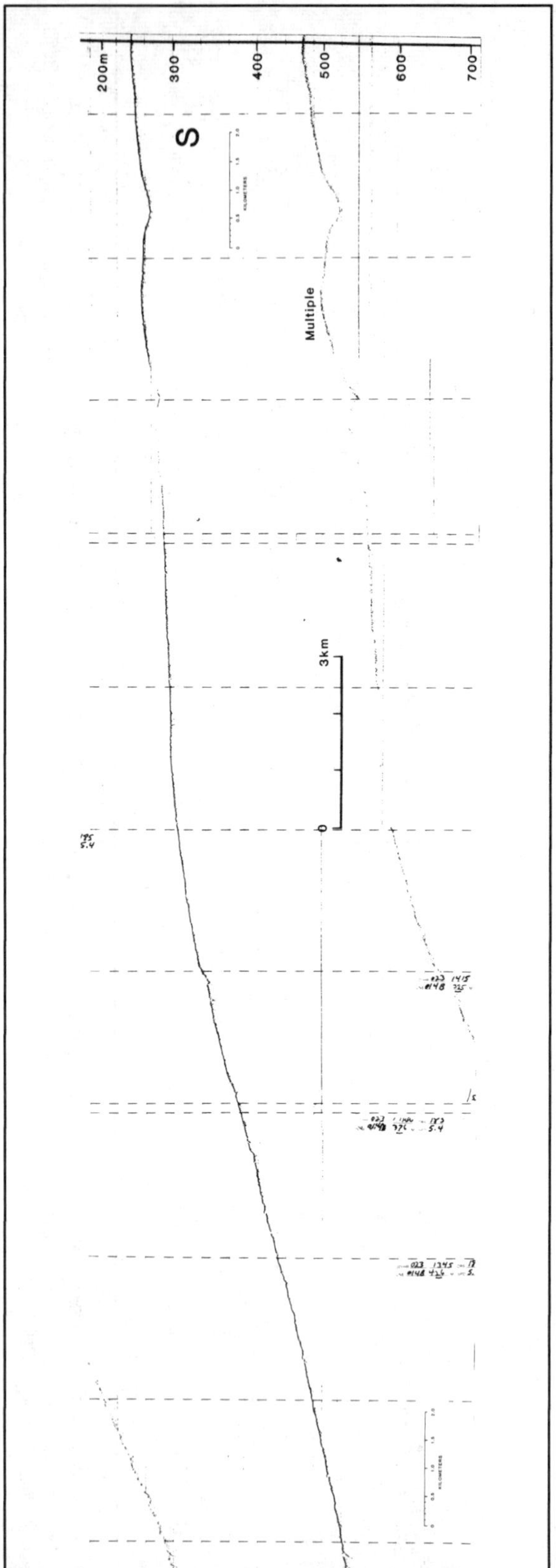

Figure 3—3.5-kHz profile of the shelf break and ramp from north (N) to south (S) across Mertz Bank. See Figure 1 for location.

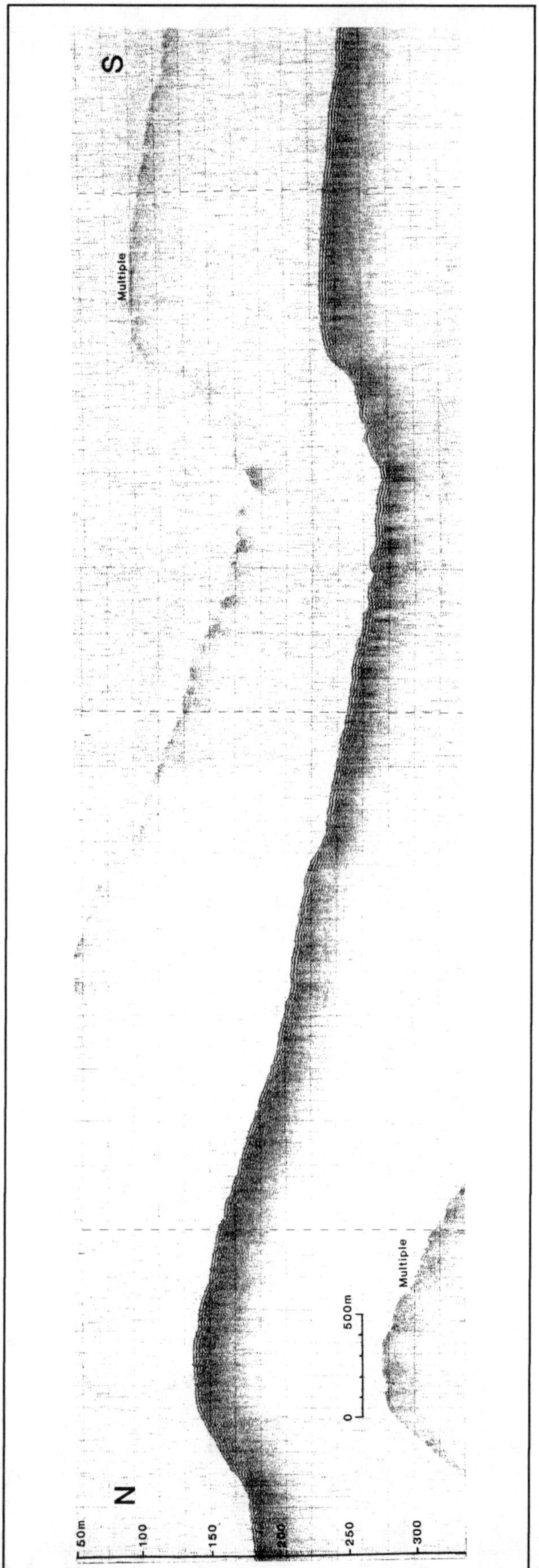

Figure 4—1-kHz seismic-reflection boomer record of asymmetric ridges along the southern margin of Mertz Bank. Note the lack of lateral continuity to the subbottom reflectors, although short correlations can be seen. See Figure 1 for location. N, north.

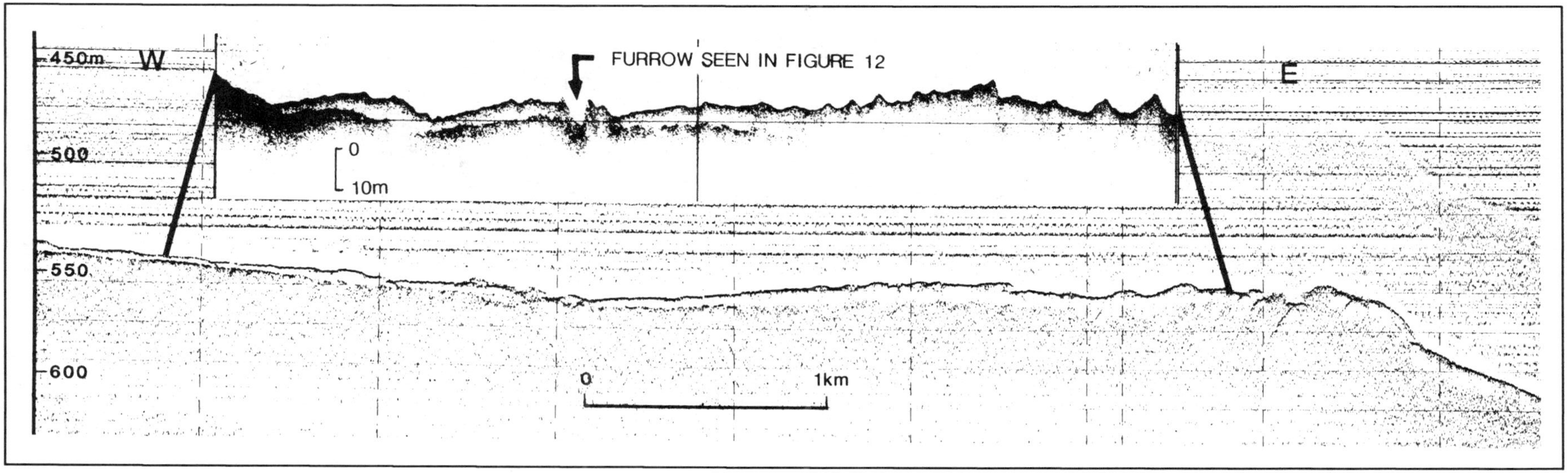

Figure 5—3.5-kHz profile of sediment drape at the base of the slope from Mertz Bank to Mertz Trough. Inset is 7-kHz pinger record of the same section of the sea floor obtained from the sidescan sonar fish. Note the sharper relief indicated on the pinger record. See Figure 1 for location.

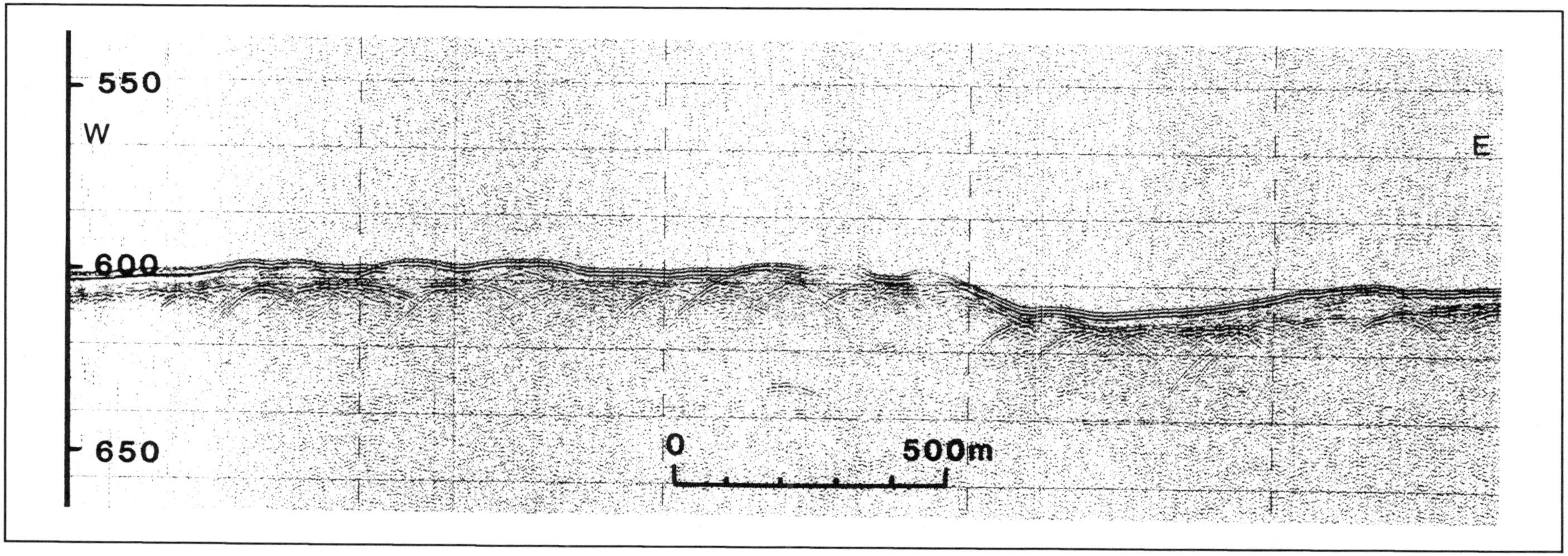

Figure 6—1-kHz profile of the sediment drape overlying a surface with rough relief in the northern part of Mertz Trough. See Figure 1 for location.

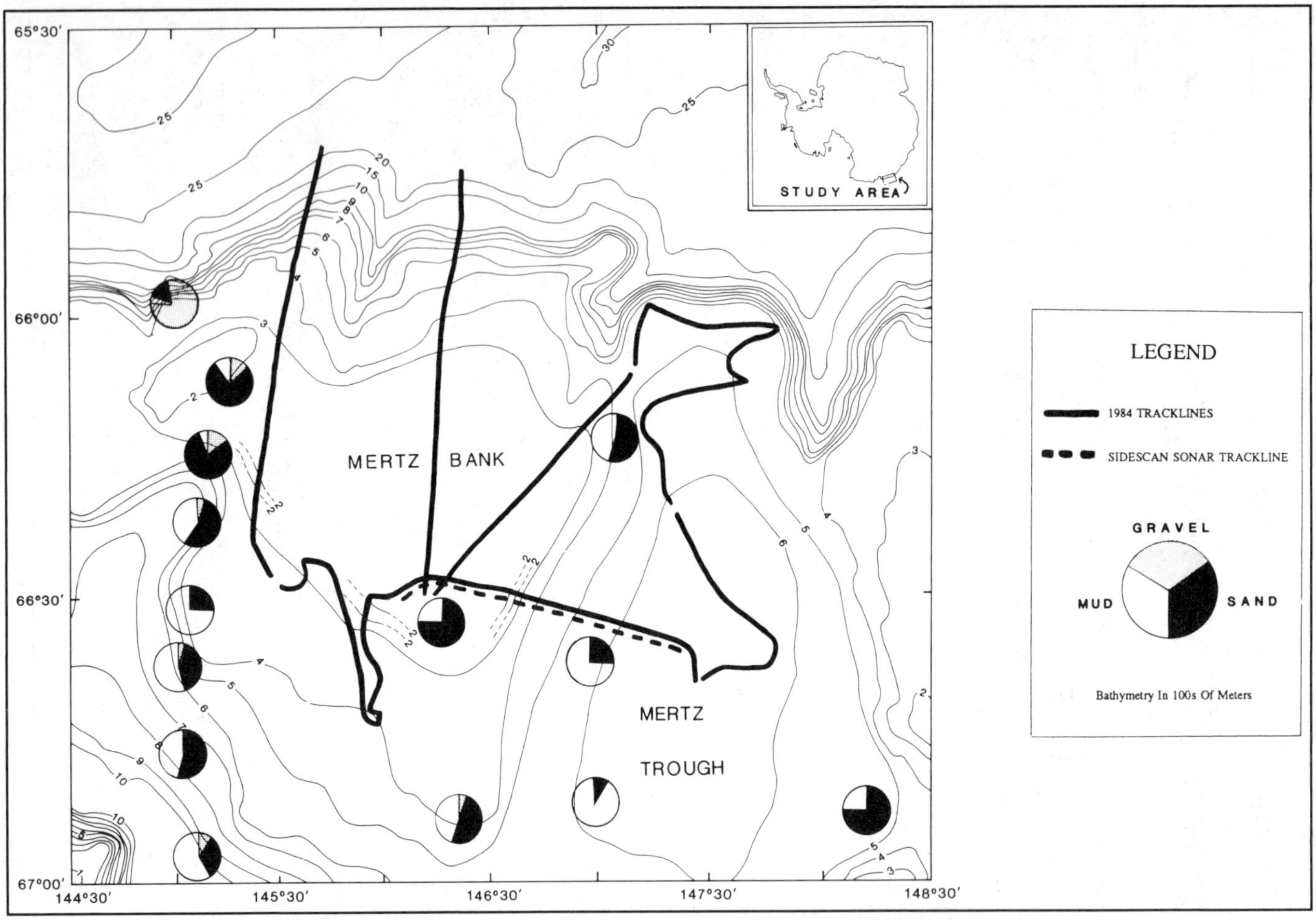

Figure 7—Ship's tracklines and surficial sediment textures. Most sediment texture data was obtained from Domack (1982, and written communication).

monly in excess of 5 m; (4) the floor of the trough at 600+ m water depth containing smooth areas several hundred meters across, separated by ridges with relief of about 10 m. The character of the bank top, slope, and trough floor environments is summarized in Table 1 and discussed in detail below.

Bed Forms

Four bed forms commonly occurred in more than one of the above morphologic environments. In addition, a sea floor with little relief and no discernible bed forms is common both on the top of the bank and in the floor of the depression; however, the flat sea floor does not persist for more than a few hundred meters, before being broken by one or more of the positive- or negative-relief features discussed below.

By far the most common bed form on both the bank top and slope are circular and semicircular depressions, commonly in an overlapping superimposed sequence. Their distribution is patchy on the bank top (Fig. 8) in contrast to the slope of the bank where they virtually saturate the sea floor (Fig. 9). In some cases, the features appear conical with well-developed rims and a sloping floor. The depressions more commonly lack or have poorly developed rims and flat floors and range from 50 to 100 m in diameter. A characteristic association of several depressions often occurs where several of about equal size are superimposed, en echelon, to form a single linear feature. The preferred trend of the linearity is northwest–southeast, the direction of superposition being to the northwest. As linear features, these depressions are as much as 100 m wide and several hundred meters long (Fig. 9). The largest feature is 350 m wide and 700 m long in water depth of 400 m.

The depth of the flat-floored depressions is less than 2 m, whereas ridge height was estimated from shadow lengths on the sonographs to be less than 0.5 m (Figs. 8, 9). A few of the superimposed depressions on the slope of Mertz Bank exhibited lineations parallel to their long dimension, either within the depression or as a short tail at their western end (Fig. 8A). These features may be related to the bed form described below.

Multiple parallel linear furrows occur on both the bank top and the slope of the bank (Fig. 10A). These features resemble the ice gouges described in studies from the Arctic (Pelletier and Shearer, 1972; Reimnitz and Barnes, 1974) and Antarctic (Lien, 1981). They are as much as 30 m wide and commonly have four or more parallel grooves. Their length ranges up to 400 m, the width of the sea floor recorded on the sonographs (Fig. 10A). The orientation of the gouge features is dominated by northwest–southeast trends. The gouge depth could not be resolved on the fathometer or profiler, and thus the depth is estimated to be less than 1 m below the sea floor. Gouges are few in number and are most clearly defined on the

Table 1—Summary characteristics of Antarctic seabed morphologies.

Bedform	Occurrence	Shape Characteristics	Sedimentology	Stratigraphy	Origin(s)
Shoals or Ridges	Southern and western rim of Mertz Bank at about 200 m	Asymmetric with steep slope toward the bank top, 30 to 50 m high and several km wide	Unknown	Discontinuous reflectors, sub-parallel to seafloor	Terminal or lateral ice sheet or ice shelf moraine- Drumlin(?)
Sea floor grooves	Floor of Mertz Trough	Ridges several meters high about 100 m apart oriented coast perpendicular(north-south)	Diamicton overlain by diatomaceous ooze in grooves	Ooze draped over ridges	Subglacial erosional or depositional grooves
Ice gouges	Sparse on bank top and trough slope to depths of 500+ m	Crisp, linear, multiple grooves ten's of meters in width, about a meter in depth and hundred's of meters in length, oriented east-west.	Muddy sands on bank top, diamictons suspected on trough slope	Disrupted surface deposits- ice turbate	Modern iceberg keels
Depressions	Sparse on banktop, saturate western slope of Mertz Trough to depths of 500+ m	Crisp, flat-floored, sub-circular, rimless, overlapping depressions tens of meters in diameter, about a meter in depth- overlapped east to west	Muddy sands on bank top, diamictons suspected on trough slope	Disrupted surface deposits- ice turbate	Modern iceberg keels

bank top. They are also found on the slope of the bank to depths of over 500 m (Fig. 10B). Gouges are also found in association with the circular depressions, often forming a tail at one end (Figs. 8A, 10B).

The rough relief of the slope of Mertz Bank at a depth of 500 m is interspersed with areas of smooth sea floor. Coincident with this change is the appearance of the thin transparent reflector on the 3.5-kHz and 7-kHz pinger records (Figs. 5, 6). Farther east, the morphology of the bottom of Mertz Trough at depths of 500–550 m is marked by parallel ridges about 1 m high and 10 m wide, spaced 50–100 m apart and oriented about 020°T parallel to the isobaths (Fig. 11). The ridge lineations seen on the sonograph appear to coincide with high points in topography underlying the blanketing transparent reflector. The ridges were poorly developed in the central part of the trough; however, indistinct positive-relief features less than 10 m in diameter are spaced a few hundred meters from one another. Another striking feature of the floor of Mertz Trough is a pronounced V-shape channel (gouge?) that crossed the record as an irregular path. This feature is several meters deep and 50 m wide and has clearly defined lateral ridges or levees (Figs. 5, 12). These lateral ridges appear to extend below the surrounding sediment drape (Fig. 5).

Sediment

Few sediment samples are available from the study area. However, a compilation of nearby surficial sediment samples from previous studies (Domack, 1982) shows the variability of the surface textures (Fig. 7). The troughs and depressions are characterized by mud and sandy muds, whereas slopes and bank tops are covered with sand—the amount of gravel increasing in a seaward direction.

A chain-bag dredge designed to retain clasts larger than 2 cm was used on the top of Mertz Bank. Although the dredge was towed approximately 2 km, only a mass of biologic material was recovered along with two small pebbles being used as holdfasts. This sampling suggests that the predominant texture of the bank top is finer than 2 cm and that muds, if present, are unconsolidated or coherent blocks of sediment that would have been recovered. The lack of cobbles and boulders also suggests that ice rafting of clasts larger than 2 cm has not occurred recently in this area and is probably a rare event. The bank top sediments seem to be composed of sand and mud as evidenced by the handful of sandy mud retrieved during an unsuccessful attempt to take a gravity core at this same site.

Grounded Ice Observations

A series of 15–20 ice-shelf and/or glacial icebergs were observed to be grounded along a north–south line corresponding to the eastern margin of Mertz Bank (S. Eittreim, written communication). As shown by fathograms, the icebergs were grounded in water depths of 250–320 m, perhaps on a shoal similar to that observed along several crossings of the southern

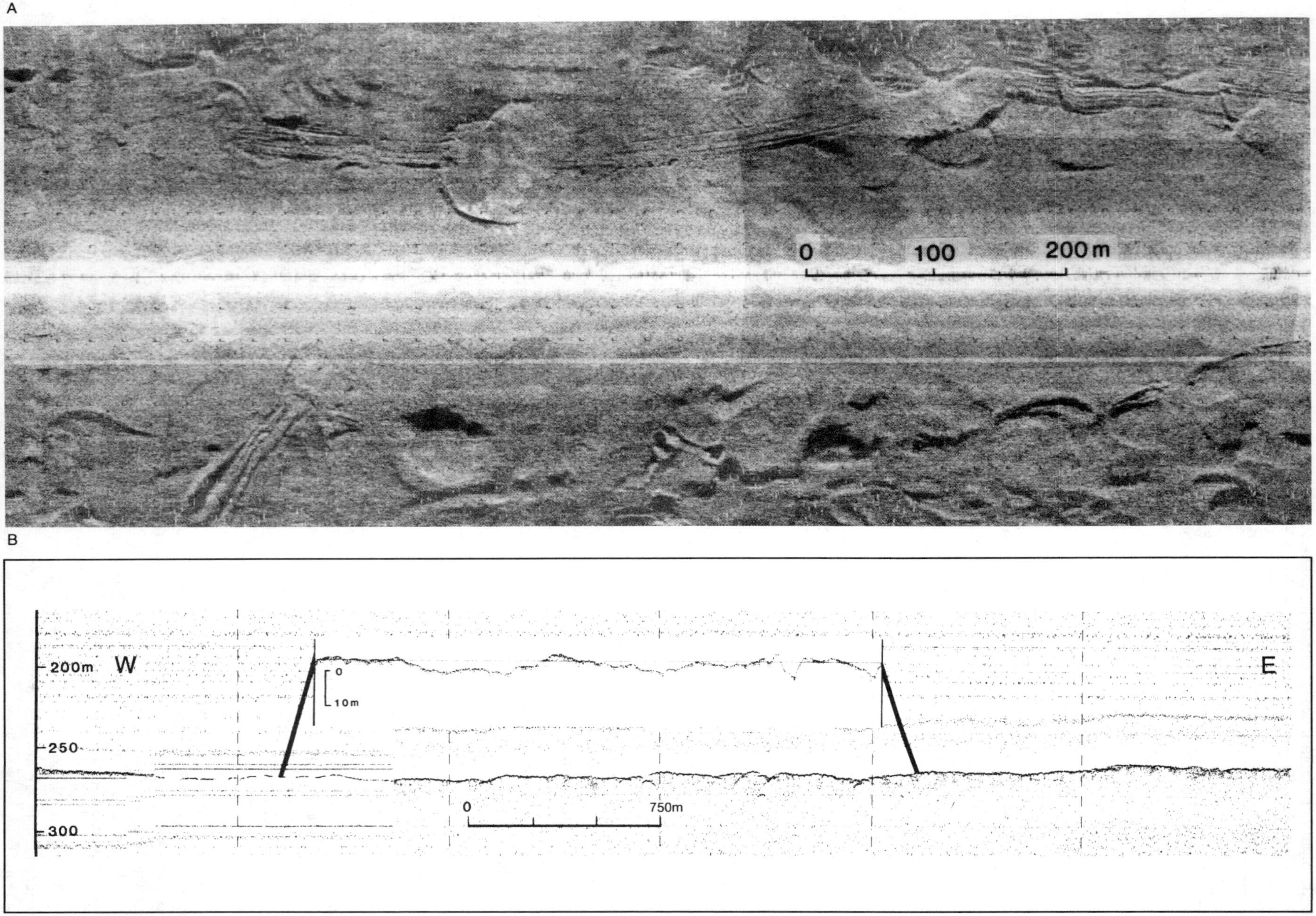

Figure 8—A. Sonograph illustrating circular depressions and linear features on the top of Mertz Bank. Note the overlapping pair of depressions and the associated gouge track in the upper center of the illustration. B. 3.5-kHz profile of the surface relief on the bank top in the vicinity of Figure 8A. Inset is a 7-kHz profile obtained from the sidescan fish towed near the bottom. Note the lack of bottom penetration by either profiling system. See Figure 1 for location.

and southeastern margin of the bank (Figs. 1, 2). The heights of three icebergs were measured by using vertical sextant angles and radar ranges in nautical miles to an estimated accuracy of about 5 m. The largest was 98 m high, whereas two others were 56 and 58 m high, respectively. The work of Orheim (1980) indicates these icebergs could easily have been grounded (Fig. 13).

DISCUSSION

The sedimentology, regional and detailed bathymetry, and high-resolution seismic-reflection data of this study generally support the earlier interpretations of Domack (1982) that the major topographic characteristics are related to glacial and glacial-marine processes (Table 1). For example, shallow seismic reflectors on Mertz Bank are discontinuous and poorly defined as might be expected in areas of till and unstructured glacial-marine sedimentation (Figs. 2A, 4). The discernible bedding is largely subparallel to the sea floor and probably reflects areas of hydraulic reworking of glacial-marine sediments either during or after initial deposition.

The acoustically indistinct surficial strata could also be interpreted as an ice turbate (Vorren et al., 1983). An ice turbate is developed from the tilling action of iceberg keels with the sea floor. The formation of an ice turbate could occur during a lower stand of sea level and thus be relict (Vorren et al., 1983), or ice keels may presently be furrowing the sea floor and reworking the surficial sediments, as is occurring off eastern Canada (Barrie, 1980). The repeated tilling action of ice keels would be expected to form surficial sedimentary strata several meters in thickness that could cover large areas composed of a heterogeneous mix of seabed, ice-rafted, and marine biogenic material. These sedimentary units would be difficult to distinguish from glacial marine deposits and perhaps even marine tills.

The large asymmetric ridges crossed at three sites on the southern and eastern rim of Mertz Bank (Figs. 2, 4, 7) are believed to be related to glacial or ice shelf advances onto the shelf producing a linear deposit of glacial material rimming Mertz Bank at around 200 m. The possibility that these features are fortuitous crossings of isolated, drumlin-like deposits cannot be completely ruled out. Descriptions of linear depositional ridges as submarine features have not been located in the Antarctic literature. Subbottom reflectors that are discontinuous but subparallel to the sea floor (Figs. 2A, 4) suggest that the shoals are constructional. These shoals are believed to be remnant depositional features such as a lateral moraine left from the advance of an ice shelf or glacier across the continental shelf, perhaps similar to the ice-shelf moraine described by Sugden and Clapperton (1981) from the Antarctic peninsula. A similar terminal feature formed by a surging glacier in Svalbard created an asymmetric ridge with the steeper slopes facing the glacier (Solheim and Pfirman, 1985). The steeper slopes of the Mertz Bank shoals face the top of the bank (Figs. 2B, 4).

These shoals are indicative of the age and nature of the Antarctic ice mass expansion. If the continental ice sheet expanded to the shelf edge during the last glacial period overriding the topography of the inner shelf, as suggested by Stuvier et al. (1981), preservation of shoals or moraine-like features inshore of the shelf edge seems unlikely. This suggests that the latest ice advance may have been in the form of ice shelves (Drewry, 1979) or expanded ice tongues of the Mertz and Ninnis glaciers. The cross-shelf expansion of ice during the last glaciation agreed upon by most workers (Drewry, 1979; Stuvier et al., 1981; Elverhoi and Maisey, 1981), coupled with concurrent lowered sea levels, suggests that the observed shoals at 150–250 m were created during this latest ice advance.

The faint ridges perpendicular to the coast on the floor of Mertz Depression (Fig. 11) are believed to be composed of the diamictons reported from cores in Mertz Trough by Domack (1982) and Hampton et al. (this volume). These diamictons underlie diatomaceous oozes that compose the surficial layer in cores (Domack, 1982). The ooze drapes and subdues the original diamicton morphology in the trough (Figs. 5, 6). The draped nature of the ooze and the lack of ice-rafted detritus and clasts in this sediment (Domack, 1982) indicate that hemipelagic sedimentation presently dominates over gravity mechanisms such as turbidites and debris flows.

The observed ridges in Mertz Trough are high points of a prominent subbottom acoustic reflector that can be traced beneath the blanket of ooze (Figs. 5, 6). I believe the ridges and grooves were formed subglacially, parallel to ice motion, at a time when sea level was lower and glaciers covered this area. Similar features have been described by Boulton (1976), who ascribed the ridges to the intrusion of subglacial sediments into subglacial openings in the lee of obstructions such as boulders. The regular spacing (Fig. 11) and the height of several meters above the subsurface diamicton reflector (Figs. 5, 6) of the observed ridges and grooves does not mesh well with a random distribution of obstructions that might be expected in subglacial materials. However, the ridges and grooves are interpreted to be a reflection of subglacial erosion or deposition and thus a reflection of ice motion indicating an onshore–offshore component. This northerly ice motion is at approximately right angles to the coast and parallel to the northwest ice motion suggested by Domack (1982) as indicated by mineral provenance studies.

The lack of gravel and large clasts on Mertz Bank and other bank tops (Fig. 7; Domack, 1982) is difficult to explain. Given the postulated glacial source for bank-top sediment and subsequent winnowing of these deposits by currents and by ice gouging, I would expect the lag deposits to be coarser than muddy sand (Fig. 7). Domack and Anderson (1983) suggest that the bank-top sand is winnowed from till exposed along the seaward margin of the shelf (Figs. 1, 2) and then carried onshore by currents. However, the presence of mud along with the sand and the lack of current-derived bed forms on the top of Mertz Bank (Figs. 3, 8, 10A) argue against this mechanism. Currents over the shoals rimming the bank, perhaps intensified around the grounded portion of icebergs, may winnow sediments from these shallow features to the deeper bank top. Another source for the sands could be wind. Eolian grains are reported from nearby samples (Hampton et al., this volume). This material may have been derived from polar deserts in front of the ice cap during a lower stand of sea level when the area was being swept by strong katabatic winds (Weyant, 1967). However, the limited number of eolian grains argue against wind as the sole source for the bank-top sand. Although coarse, ice-rafted debris have been observed on ice-

A

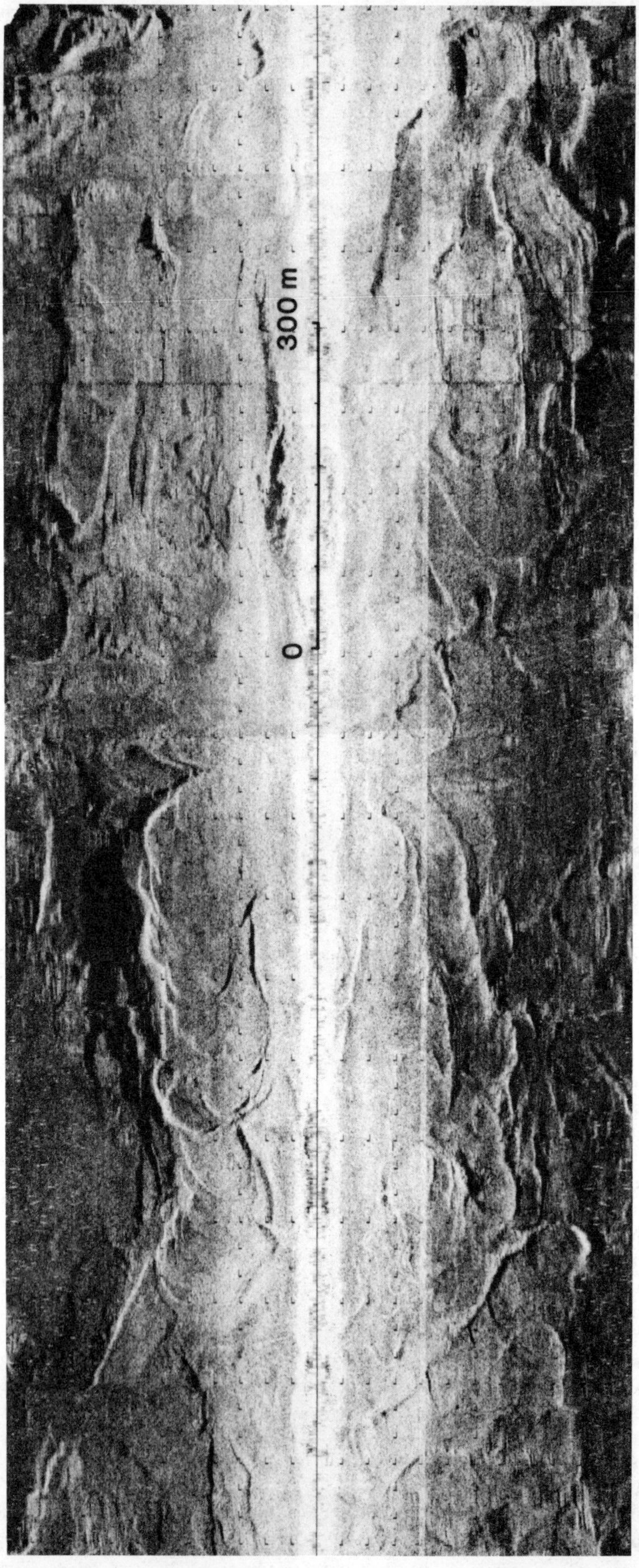

B

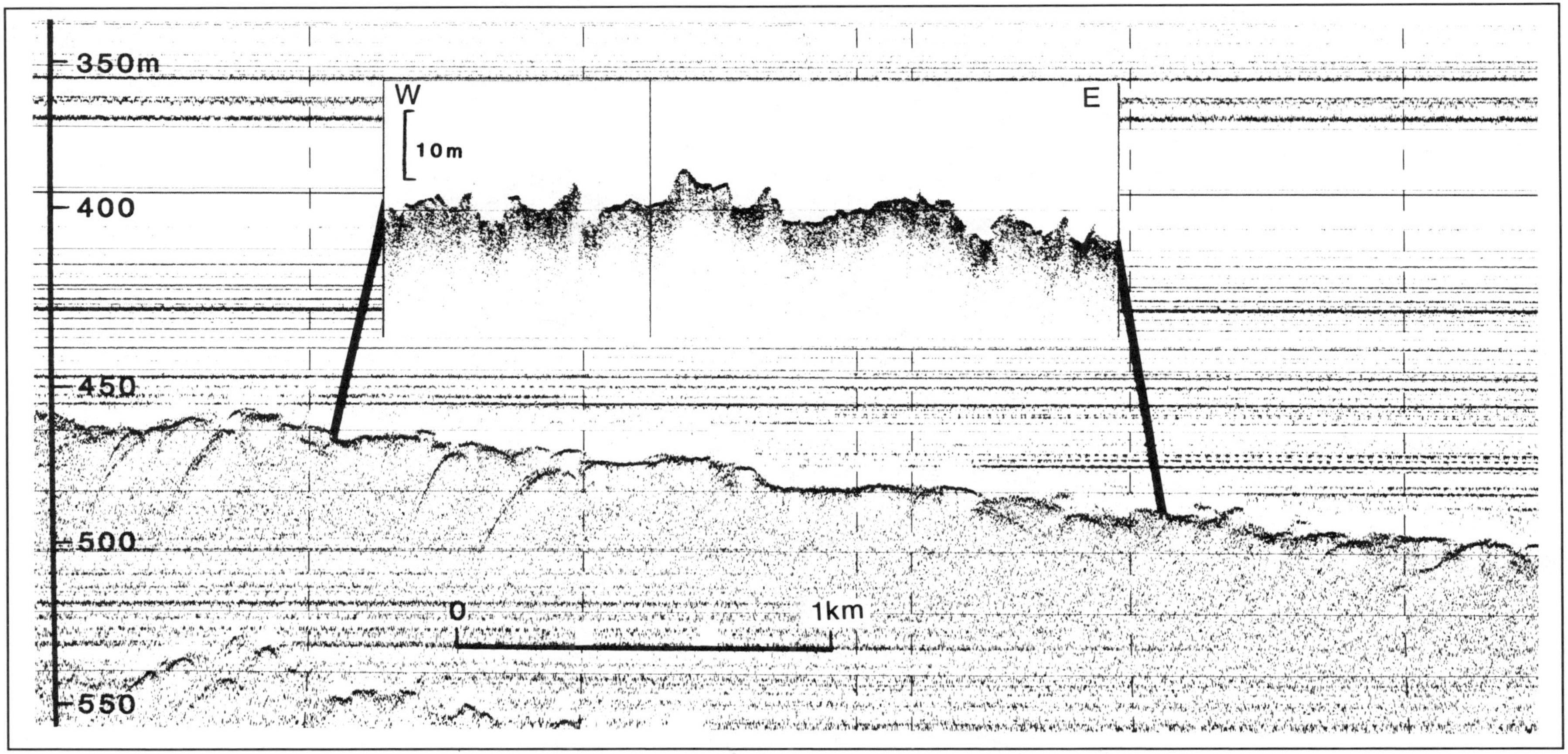

Figure 9—(opposite and above) A. Sonograph of circular depressions on the slope from Mertz Bank to Mertz Trough. Upslope is to the left. Note the uphill overlapping superposition of depressions in the right part of the figure. See Figure 1 for location. B. 3.5-kHz and 7-kHz (inset) profiles of the relief on the slope from Mertz Bank to Mertz Trough showing the sharp relief and little sediment ponding associated with the depressions. See Figure 1 for location.

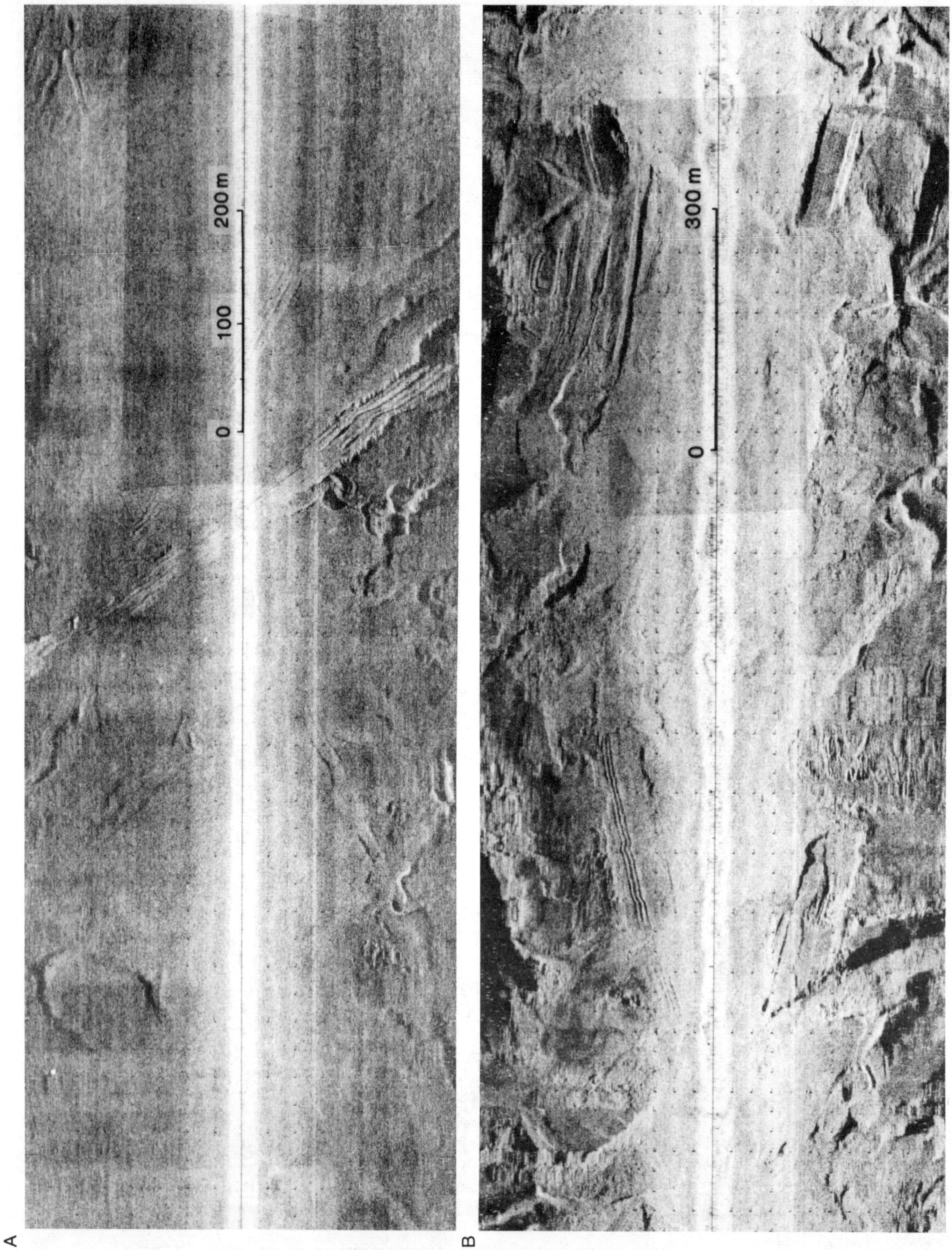
A
0
100
200 m
B
0
300 m

bergs in the area (Anderson et al., 1980a), it appears from the lack of gravel size material that the contribution from modern ice-rafting is also limited.

The multiple parallel furrows in the sea floor of the Antarctic (Figs. 8A, 10) are similar to gouge features generated by sea ice and glacial ice in the Arctic (Pelletier and Shearer, 1972; Belderson and Wilson, 1973; Reimnitz and Barnes, 1974). In the Arctic, multiple ice keels projecting from the base of ice canopy are responsible for the multiple gouges in the sea floor. Multiple gouges formed by sea ice keels are reported from water depths less than 60 m at latitudes around 70° N (Barnes et al., 1984). Assuming similar or less extreme sea ice conditions for the 67° latitude of the study area, the observed gouges occur in water depths too great to be formed by modern sea ice. Glacial ice, on the other hand, gouges the seabed to at least 200 m in the Arctic off eastern Canada (Barrie, 1980). Since Antarctic glacial and ice-shelf ice can be several hundred meters thick (Fig. 13; Barrie, 1980), we believe terrestrially derived ice is responsible for the observed gouges on the bank top and shoals and that the gouges are modern features. The modern age is based on the fresh appearance of the gouges and the presence of a modern sediment drape of the floor of Mertz Trough that is absent on the gentle slopes of Mertz Bank (Figs. 9A, 10B). Similarly, the bank-top sands are thought to be modern deposits (Domack and Anderson, 1983) that do not appear to be infilling the gouges on the banks (Figs. 8A, 10A).

Because there are more shallow-draft than deep-draft icebergs (Orheim, 1980), we would expect to observe higher densities of ice gouges on the upper slopes and top of Mertz Bank, paralleling the observations reported from the Weddell Sea (Lien, 1981). However, this is not the case. Not only are there fewer gouges on the bank top, but everywhere the gouges are shallow and only tens of meters wide. These dimensions are modest by Arctic standards (Barnes et al., 1984), but minuscule when compared to the size and masses of Antarctic ice of sufficient draft to interact with Mertz Bank. Thus, the forces and keels creating the Antarctic gouges must be small compared to the mass of the ice creating the gouge.

The shape of the gouges indicates that the ice creating them has sharp irregular relief a few meters wide (Fig. 10), in contrast to the broad undulating relief observed at the base of tabular Antarctic icebergs (Orheim, 1980). The sharp relief may be created by thermal erosion of an iceberg at or above the sea surface that comes in contact with the sea floor when the unstable iceberg rolls to a new orientation and contacts the sea floor (Lewis and Bennett, 1984); or the gouge shapes may reflect the underwater configuration of outlet glacier icebergs. The absence of broad gouges is puzzling, unless the footprint of a broad flat surface is too slight to be resolved with the sidescan sonar.

Figure 10—(opposite) A. Sonograph of the top of Mertz Bank illustrating the parallel linear features interpreted as ice gouges. See Figure 1 for location. B. Sonograph from 530 m water depth on the slope from Mertz Bank to Mertz Trough illustrating the parallel linear features and associated bottom morphology. See Figure 1 for location.

The origin of the circular depressions on the western slope of Mertz Trough is problematical. The salient features are shown in Figures 8, 9, and 10 and include the following:

1. Subcircular shapes on the order of 100 m in diameter.
2. General "fresh" appearance with sharp near-vertical scarps.
3. Shallow flat floors and no or slight rimming ridges.
4. Common occurrence on bank top and dominant morphology on the slope to Mertz Trough.
5. Common lateral extension of depressions by repeated superposition.
6. Lack of sediment infilling or ponding.

Postulated mechanisms for forming the depressions must include iceberg and seafloor interaction, subglacial processes, permafrost collapse, slope failure, and mass movement. The occurrence of the depressions on the slope to the trough, their general slope-parallel orientation, and their general circular shape (Fig. 9) suggest they could result from slope failures or slumping. I believe, rather, that their occurrence on the flat bank top (Fig. 9), their uniform size, and the lack of downslope sediment accumulations is strong evidence that the depressions are not related to slumping.

Subglacial processes are discounted as agents in forming the depressions because the features appear fresh, equidimensional, and recently formed, whereas glaciers probably have not been in contact with the sea floor in this area for many thousands of years (Kellogg et al., 1979). Although frozen substrate can be produced without subaerial exposure (Menzies, 1981; Orheim and Elverhoi, 1981), the incorporation of excess pore fluids—a prerequisite to forming thermal collapse depressions—in a salt-water environment has not been reported and seems unlikely at 300–500 m below sea level where the depressions are best developed (Figs. 1, 10). Furthermore, the freezing and thawing of Arctic marine sediments have been postulated as a mechanism of sediment overconsolidation (Chamberlin, 1981).

Depressions have been reported from other marine environments (King and MacLean, 1970; Whiticar and Werner, 1981; Flood and Johnson, 1984; Barrie and Collins, 1985). Although all features described are from glacial environments, their origin is often reported as complex or problematic. Rapid dewatering of glacial sediments, gas and water percolation, freshwater seeps, and iceberg impacts are postulated as mechanisms. None of the reported features matches the Wilkes Land continental margin features; they are usually smaller, deeper, and more conical in shape when compared to the shallow, flat-floored depressions reported here. In addition, the overlapping character of the Antarctic depressions are unique, thereby minimizing the usefulness of previous studies in explaining their origin.

Modern iceberg interaction with the sea floor is proposed as a mechanism to account for the fresh appearance of the depressions, their orientation relative to the movement of the ice from east to west in the east-wind drift, and their occurrence on both the slope of Mertz Trough and on Mertz Bank (Figs. 8, 9, 10). The occurrence of the same features in close association with ice gouging and ice wobbling in the Weddell Sea (Lien, 1981) also suggests that the depressions may be related to iceberg action. Modern circular depressions asso-

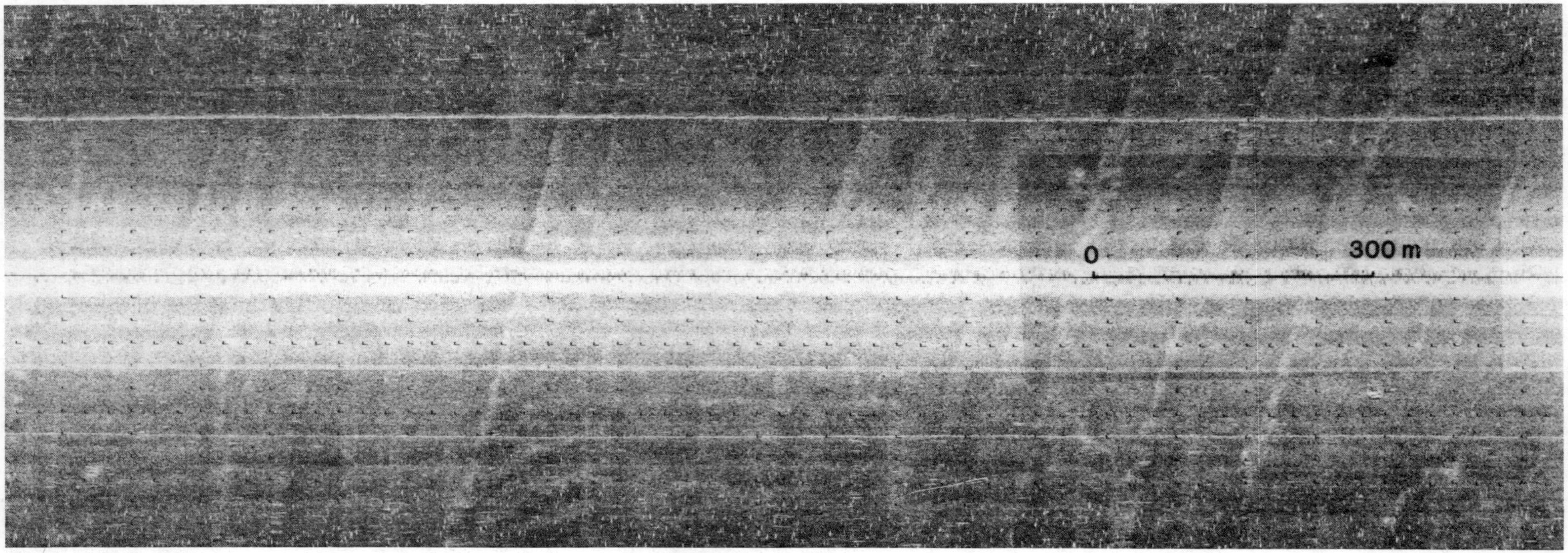

Figure 11—Sonograph of ridges normal to the coast on the floor of Mertz Trough. See Figures 5 and 6 for the relief and subbottom character of the trough floor. Figure 1 indicates the location.

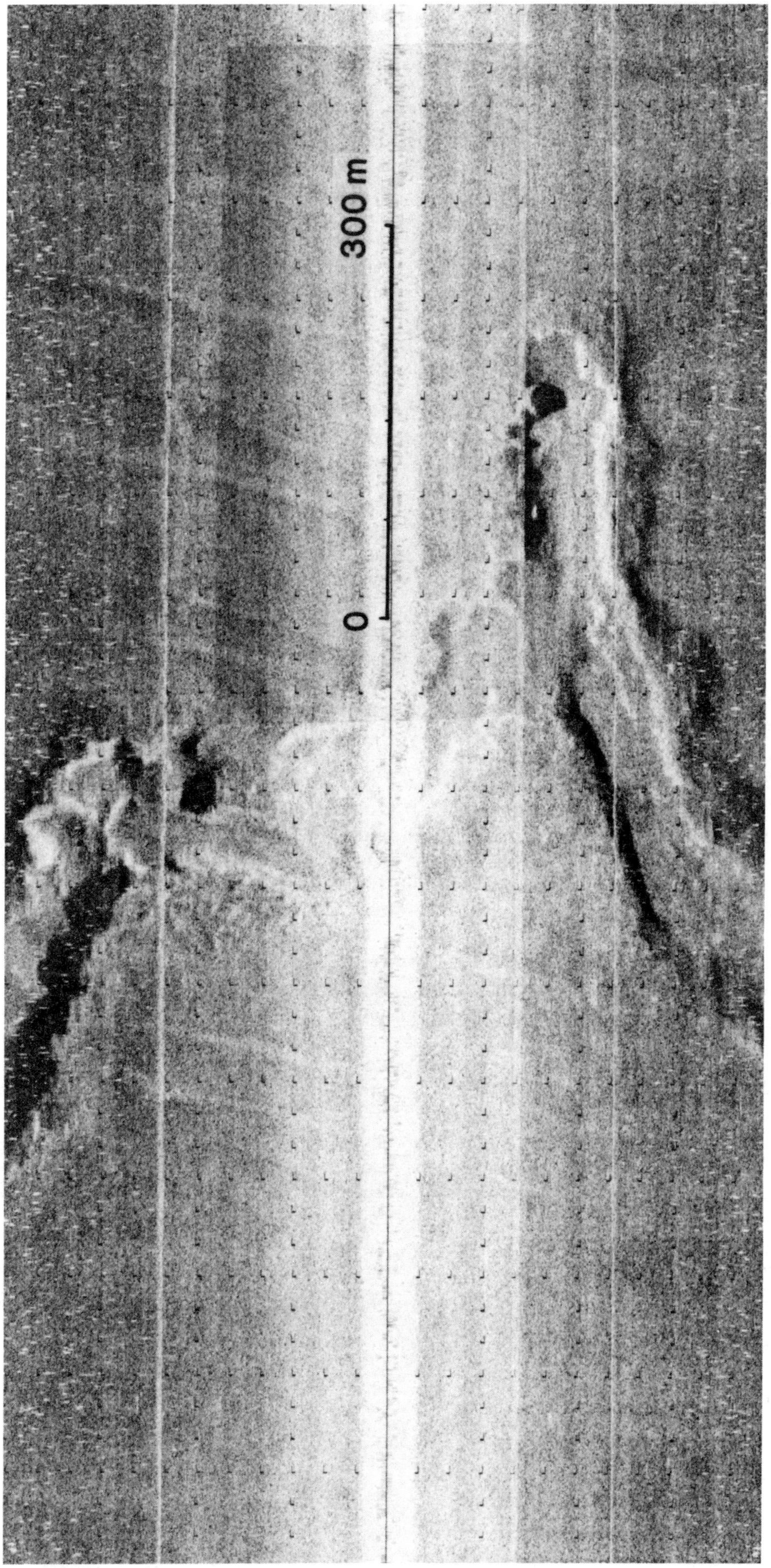

Figure 12—Sonograph of the irregular furrow in the diatom oozes on the floor of Mertz Trough from a water depth of 540 m. Figure 5 shows a profiler crossing of this furrow. Location is the same as Figure 5 shown in Figure 1.

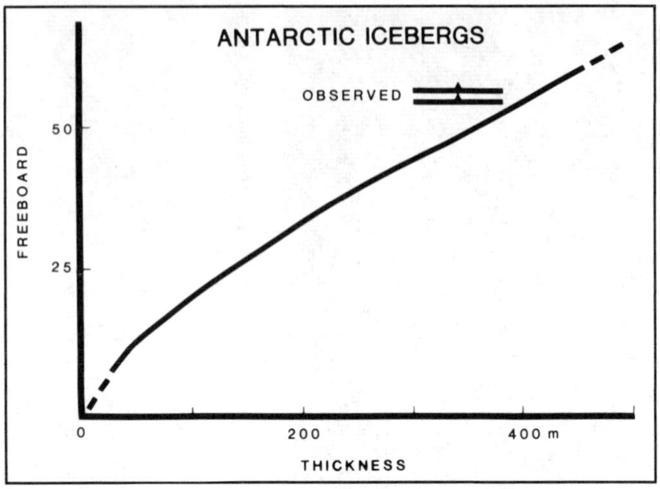

Figure 13—Distribution of iceberg thickness versus freeboard for tabular Antarctic icebergs after Orheim (1980). Two additional observations are based on water depth (bar of uncertainty) and measured freeboard in January 1984.

ciated with icebergs have been described from the Arctic (Moign, 1976), although these depressions are only a few meters across—much smaller than the features reported here. Miller and Barnes (1985) described current-driven rotational motion of a grounded Antarctic iceberg and suggest several scenarios for the formation of "rotational depressions." With a slight rise and fall of sea level resulting from storms or even a change of tide, icebergs may become grounded on shallower parts of the shelf, thereby becoming readily available to make keel marks on the sea floor. The swinging of ice in a steady current, as a leaf often oscillates in the wind, could create the flat-floored circular depressions. Rotational motion may also be imparted to grounded ice blocks when free-drifting icebergs impact and "shunt" grounded icebergs laterally, generating circular ice motions (H. Keys, written communication, 1986). The presence of a pivot point in the center of the seafloor depressions is problematical, since there is no reason to expect the keel of an iceberg to be pointed.

If modern icebergs are responsible for both the circular depressions and the multiple gouges, then their small size and shallow depth can be used to infer ice conditions and currents. The concentration and maximum relief of the features (Fig. 9) on the east flank of Mertz Bank suggest that the major direction of impact is from the east. However, these impacts are not driven with the large forces observed in the Arctic nor in the Weddell Sea (Reimnitz and Barnes, 1974; Lien, 1981), otherwise much deeper and wider gouges and depressions in this readily cored sediment (Domack, 1982) would be observed. The small forces may indicate generally weak westerly currents. The ice keels responsible for creating the gouges and depressions are remarkably small in comparison to the ice masses needed to reach the observed depths of occurrence, and their underwater shape has sharper relief than is indicated by radio echo-sounding (Orheim, 1980).

Domack and Anderson (1983, p. 404) suggest that, in the absence of waves and wind-driven currents, slumping, debris-flow transport, and turbidity-current transport are important on the continental shelf. Wright et al. (1983) indicate that the margins of bank tops and the depressions should be the location of the most numerous sediment gravity-flow deposits. Dunbar et al. (1985) outline the role of marine currents in shaping the sedimentary environment of the shallow banks and outer shelf and discount the role of ice below depths of 150–200 m. Observations reported here suggest that, at a minimum, iceberg gouging is also important in moving sediment and influencing the surficial sedimentary stratigraphy of the continental shelf at depths shallower than about 500 m, or over half the Wilkes Land continental margin. The data presented here, as well as the data of Lien (1981), suggest that slumping, debris flows, and turbidity currents are not important processes at shallow depths on the Antarctic continental shelf in the areas studied.

CONCEPTUAL MODEL

Based on the seabed morphologies and the earlier work of Domack (1982) detailing the sedimentology and sedimentary history, a conceptual model of the geologic processes responsible for the observed morphologies and sediment is presented in Figure 14. The most recent advance of the Antarctic ice sheet onto the shelf from Mertz and Ninnis glaciers was parallel to the isobaths at the seabed, diverging around Mertz Bank in a slight modification of Domack's (1982) flow directions. Subglacial ridges were formed in Mertz Trough parallel to the flow direction of the ice. The glaciers to the south and east of Mertz Bank developed lateral ridges (moraines?) that form the shoals rimming the southern and eastern edge of Mertz Bank (Figs. 1, 2). Waves and currents, perhaps intensified around grounded icebergs, are transporting finer grained sediment from the fringing shoals to the inner bank top, leaving a blanket of muddy sand without coarse clasts. Eolian transport on the exposed shelf during the lower stand of sea level contributed to the surficial sediment blanket.

Subsequent to the retreat of the ice sheet, icebergs drifting westward in the east-wind drift gounded on the east flank of Mertz Bank, creating gouges and circular rotational depressions as they moved vertically and horizontally with the winds, currents, and tides. The seafloor sediment is being reworked to a depth of a few meters by these ice processes and may be forming an ice turbate, especially on the slope of Mertz Trough. Only rare icebergs, perhaps drifting from the west or northeast with sufficient draft to interact with the sea floor, encroach onto the bank top, because the rimming shoal on the south and east blocks their passage. Therefore, relatively few ice-related features occur on the bank top.

Hemipelagic and biogenic sedimentation has dominated in Mertz Trough (Domack, 1982) since the most recent glacial regression, draping the older subglacially formed ridges and furrows with a muddy diatom ooze.

CONCLUSIONS

The following conclusions can be drawn from these studies:

1. Glacial ice was active in shaping the major morphologic features of Mertz Bank and Mertz Trough and in shaping the bank-top shoals and the ridges in the trough. The gla-

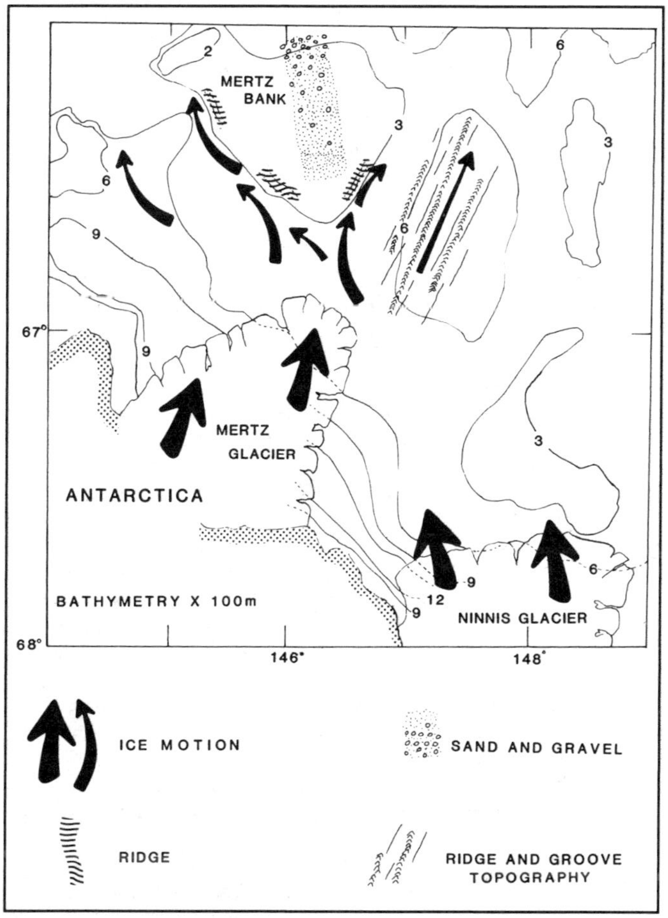

Figure 14—Conceptual diagram of glacial ice motion suggested for developing the basic morphologies observed on Mertz Bank and in Mertz Trough.

cial activity has resulted in a lack of lateral continuity in the near-surface seismic reflectors on the bank top.

2. Icebergs are presently gouging tops of the outer-shelf banks and the slopes (to depths over 500 m), creating shallow multiple-gouge tracks. The distribution of linear gouge features on the bank top indicates either rare gouge events or episodic reworking by currents.
3. Shallow flat-floored circular depressions occur on the bank top and are ubiquitous on the slope. The origin of these features is related to "rotation" or "wobbling" of icebergs, although they may be formed by relict subglacial processes.
4. Biogenic sediment, draping the glacially carved grooves that are perpendicular to the coast at the base of Mertz Trough, is acoustically transparent.

ACKNOWLEDGMENTS

I would like to thank the officers and crew of the *S.P. Lee* and especially Captain Alan McClenaghan for perseverance under the trying conditions of the voyage. I would also like to thank Doug Rearic and Ed Kempema for helpful observations and general assistance during preparation of this manuscript. Insightful comments and observations by Eugene Domack and Harry Keys greatly improved the manuscript.

REFERENCES CITED

Anderson, J. B., E. W. Domack, and D. D. Kurtz, 1980a, Observations of sediment-laden icebergs in Antarctic waters: Implications to glacial erosion and transport: Journal of Glaciology, v. 25, p. 387–396.

Anderson, J. B., D. D. Kurtz, E. W. Domack, and K. M. Balshaw, 1980b, Glacial and glacial marine sediments of the Antarctic continental shelf: Journal of Geology, v. 88, p. 399–414.

Barnes, P. W., D. M. Rearic, and E. Reimnitz, 1984, Ice gouge characteristics and processes, *in* P. W. Barnes, D. M. Schell, and E. Reimnitz, eds., The Alaskan Beaufort Sea—ecosystems and environments: New York, Academic Press, p. 159–184.

Barrie, J. V., 1980, Iceberg-seabed interaction (Northern Labrador Sea): Annals of Glaciology, v. 1, p. 71–76.

Barrie, J. V., and W. T. Collins, 1985, Iceberg pits: Description and postulated origin: Arctic land-sea interaction, Abstract: 14th Arctic Workshop, Dartmouth, Nova Scotia, Canada, Bedford Institute of Oceanography, p. 90.

Bass, D., and G. R. Peters, 1984, Computer simulation of iceberg instability: Cold Regions Science and Technology, v. 9, p. 163–169.

Belderson, R. H., and J. B. Wilson, 1973, Iceberg plough marks in the vicinity of the Norwegian Trough: Norsk Geologisk Tidsskrift, v. 533, p. 323–328.

Boulton, G. S., 1976, The origin of glacially fluted surfaces—observations and theory: Journal of Glaciology, v. 17, p. 287–309.

Chamberlin, E. J., 1981, Overconsolidation effects of ground freezing: Engineering Geology, v. 18, p. 97–110.

Defense Mapping Agency, 1981, Gazetteer of Undersea Features, 3rd Ed.: Washington, DC, Defense Mapping Agency, 250 p.

Domack, E. W., 1982, Sedimentology of glacial and glacial marine deposits of the George V–Adelie continental shelf, East Antarctica: Boreas, v. 11, p. 79–97.

Domack, E. W., and J. B. Anderson, 1983, Marine geology of the George V continental margin: Combined results of Deep Freeze 79 and the 1911–14 Australian expedition, *in* R. L. Oliver, P. R. Jones, and J. B. Jago, eds., Antarctic earth science: Canberra, Australian Academy of Science, p. 402–406.

Domack, E. W., J. B. Anderson, and D. D. Kurtz, 1980, Clast shape as an indicator of transport and depositional mechanisms in glacial marine sediments: George V continental shelf, Antarctica: Journal of Sedimentary Petrology, v. 50, p. 813–820.

Drewry, D. J., 1979, Late Wisconsin reconstruction for the Ross Sea region, Antarctica: Journal of Glaciology, v. 24, p. 231–244.

———, 1980, Pleistocene bimodal response of Antarctic ice: Nature, v. 287, p. 214–216.

Dunbar, R. B., J. B. Anderson, E. W. Domack, and S. S. Jacobs, 1985, Oceanographic influences on sedimentation along the Antarctic continental shelf: Oceanology of the Antarctic continental shelf: American Geophysical Union, Antarctic Research Series, v. 43, p. 291–312.

Elverhoi, A., and G. Maisey, 1983, Glacial erosion and morphology of the eastern and southeastern Weddell Sea shelf, *in* R. L. Oliver, P. R. James, and J. B. Jago, eds., Antarctic earth science: Canberra, Australia, Australian Academy of Science, p. 483–487.

Eittreim, S. L., A. K. Cooper, and the Scientific Staff, 1984, Marine geological and geophysical investigations of the Antarctic continental margin, 1984: U.S. Geological Survey Circular 935, 12 p.

Flood, R. G., and T. C. Johnson, 1984, Side-scan targets in Lake Superior—evidence for bedforms and sediment transport: Sedimentology, v. 31, p. 311–333.

Kellogg, T. B., R. S. Truesdale, and L. E. Osterman, 1979, Late Quaternary extent of the West Antarctic ice sheet: New evidence from Ross Sea cores: Geology, v. 7, p. 249-253.

Keys, J. R., 1983, Iceberg quantities, shapes and sizes in western Ross and D'Urville Seas: Antarctic Journal of the United States, v. 18, p. 125-127.

King, L. H., and B. MacLean, 1970, Pockmarks on the Scotian Shelf: Geological Society of America Bulletin, v. 81, p. 3141-3148.

Klepsvik, J. O., 1980, Studies of icebergs, ice fronts and ice walls using side-scanning sonar: Annals of Glaciology, v. 1, p. 31-36.

Lewis, C. F. M., 1978, The frequency and magnitude of drift-ice groundings from ice-scour tracks in the Canadian Beaufort Sea, *in* POAC 77: The Fourth International Conference on Port and Ocean Engineering under Arctic Conditions: St. John's, Newfoundland, Memorial University 1977, Proceedings, v. 1, p. 568-579.

Lewis, J. C., and G. Bennett, 1984, Monte Carlo calculations of iceberg draft changes caused by roll: Cold Regions Science and Technology, v. 10, p. 1-10.

Lien, R., 1981, Sea bed features in the Blaaenga area, Weddell Sea, Antarctica: POAC 81, Proceedings, v. 2, p. 706-716, Quebec, Canada, University of Laval.

______, 1983, Ploymerker etter isfjell pa norsk kontinentalsokkel. (Iceberg scouring on the Norwegian continental shelf.): Trondheim, Continental Shelf Institute of Norway Publication 109, 147 p.

Menzies, J., 1981, Freezing fronts and their possible influence on processes of subglacial erosion and deposition: Annals of Glaciology v. 2, p 52-56.

Miller, R. G., and P. W. Barnes, 1985, formation of iceberg keel marks on the Antarctic seafloor: Iceberg Research, v. 11, p. 10-12.

Moign, A., 1976, L'action des glaces flottantes sur le lettoral et les fonds marins du Spitsberg central and Nord-occidental: La Revue de Geographic de Montreal, v. 30, p. 51-64.

Orheim, O., 1980, Physical characteristics and life expectancy of tabular Antarctic icebergs: Annals of Glaciology, v. 1, p. 11-18.

Orheim, O., and A. Elverhoi, 1981, Model for submarine glacial deposition: Annals of Glaciology, v. 2, p. 123-128.

Pelletier, B. R., and J. M. Shearer, 1972, Sea bottom scouring in the Beaufort Sea of the Arctic Ocean: 24th session of the International Geological Congress, Montreal, Sect. 8, pp. 251-261.

Reimnitz, E., and P. W. Barnes, 1974, Sea ice as a geologic agent, *in* J. Reed, and J. Sater, eds., The coast and shelf of the Beaufort Sea: Arlington, Virginia, Arctic Institute of North America, p. 301-353.

Solheim, A., and S. Pfirman, 1985, Sea-floor morphology outside a grounded surging glacier; Brasvellbreen, Svalbard: Marine Geology, v. 65 p. 127-143.

Stagg, H. M. J., 1985, The structure and origin of Prydz Bay and MacRoberston shelf, East Antarctica: Tectonophysics, v. 114, p. 315-340.

Stuvier, M., G. H. Denton, T. J. Hughes, and J. L. Fastook, 1981, History of the marine ice sheet in West Antarctica during the last glaciation: A working hypothesis, *in* G. H. Denton, and T. J. Hughes, The last great ice sheets: New York, John Wiley and Sons, p. 319-436.

Sugden, D. E., and C. M. Clapperton, 1981, An ice-shelf moraine, George VI Sound, Antarctica: Annals of Glaciology, v. 2, p. 135-141.

Vanney, J. R., and G. L Johnson, 1979, The seafloor morphology seaward of Terre Adelie (Antarctica): Deutsch Hydrographische Zeitschrift, v. 32, p. 39-88.

Vorren, T. O., M. Hald, M. Edvardsen, and O. Lind-Hansen, 1983, Glaciagenic sediments and sedimentary environments on continental shelves: General principles with a case study from the Norwegian shelf, *in* J. Ehlers, ed., Glacial deposits of northwest Europe: Rotterdam, A. A. Balkema, p. 61-73.

Weyant, W. S., 1967, The Antarctic atmosphere: Climatology of the surface environment: American Geographical Society, Antarctic Map Folio Series, Folio 8.

Whiticar, M. J., and F. Werner, 1981, Pockmarks: Submarine vents of natural gas or freshwater seeps?: Geo-Marine Letters, v. 1, p. 193-199.

Woodworth-Lynas, C. M. T., A. Simms, and C. M. Rendell, 1985, Iceberg grounding and scouring on the Labrador continental shelf: Cold Regions Science and Technology, v. 10, p. 163-186.

Wright, R., J. B. Anderson, and P. P. Fisco, 1983, Distribution and association of sediment gravity flow deposits and glacial/glacial marine sediments around the continental margin of Antarctica, *in* B. F. Molnia, ed., Glacial-marine sedimentation: New York, Plenum Press, p. 265-300.

Eittreim, S. L., and Hampton, M. A., 1987, *The Antarctic Continental Margin: Geology and Geophysics of Offshore Wilkes Land*, CPCEMR Earth Science Series, v. 5A: Houston, Texas, Circum-Pacific Council for Energy and Mineral Resources

Preliminary Stratigraphy for a Portion of the Wilkes Land Continental Shelf, Antarctica: Evidence from Till Provenance

E. W. Domack
Hamilton College
Clinton, New York

Mineralogic variation between glacial sediments within the Mertz Trough is compared to the direction of ice flow inferred from recently discovered seafloor morphology. Systematic changes in till provenance parallel to the direction of glacial transport indicate a varied pre- to late glacial stratigraphy consisting of Early Cretaceous siltstone, Triassic(?)-early Tertiary(?) sandstone, basaltic intrusive, Quaternary(?) diamictite, and arkosic sandstone. These units were partially eroded during the most recent glacial expansion and are recognized from gravel lithologies and proportions of quartz, dolerite, diamictite, and arkose within the coarse sand fraction of glacial sediments. Proportions of pyroxene, garnet, and other heavy minerals also vary in a manner consistent with the stratigraphic model. Palynologic and organic maturation data along with the provenance indicators suggest that a significant preglacial sedimentary section exists on a portion of the Wilkes Land continental shelf.

INTRODUCTION

Marine geologic data were obtained during Operation Deep Freeze 1979 from the George V–Adelie continental shelf and slope (Fig. 1; Anderson et al., 1979). Subsequent studies of the data have been published previously (Anderson et al., 1980, 1983; Domack et al., 1980a, 1980b; Domack, 1982; Domack and Anderson, 1983; Wright et al., 1983; Dunbar et al., 1985). Poorly sorted sediments of glacial and glacial-marine origin were found to be the predominant subsurface deposit in the region (Domack, 1982). The lithologic and mineralogic characters of these sediments were related to derivation from both coastal outcrops and inferred local lithologies on the continental shelf. Such data suggest that during the most recent glacial expansion, regional ice-flow directions were to the northwest. This is based on the observation that provenance boundaries extend from coastal outcrops and run parallel to this direction (Domack, 1982).

Subsequent investigation of the seafloor morphology in a portion of this area (the Mertz Trough) has indicated that glacial ice flowed to the north–northeast (Barnes, this volume), a direction at odds with the earlier studies. The purpose of this paper is to suggest an alternative hypothesis that addresses the petrologic variation of the glacial sediments in relation to the newly proposed flow direction. It is believed that the variation provides clues as to the nature of the stratigraphic succession that lies beneath the most recent glacial unconformity on the continental shelf.

Bathymetry

The bathymetry of the area shown in Figure 1 is taken from Vanney and Johnson (1979) with modifications by Domack (1982). It differs slightly in detail from more recent compilations of Hampton et al. (this volume) and Barnes (this volume). The continental shelf is dominated by the Mertz–Ninnis Trough, a linear inner shelf feature that parallels the coast and reaches depths of over 1400 m. The Mertz Trough (a separate feature) extends at right angles to the Mertz–Ninnis Trough and reaches the shelf-break. Within this 500–600 m deep depression "ridge and groove" morphology, trending to the north–northeast, has been described and interpreted as a subglacially fluted surface (Barnes, this volume). The broad shelf is broken by prominent banks, the Mertz and Ninnis Banks, which reach depths of less than 200 m but average 300–400 meters in depth. The Mertz Bank is apparently encircled along its east, south, and southwest margin by a pair of linear ridges. These ridges reach depths of less than 200 m and are believed to represent lateral moraines of submarine origin (Barnes, this volume). This paper focuses on piston cores taken from the Mertz Trough and east end of the Mertz–Ninnis Trough (Fig. 1).

Seismic Stratigraphy

Seismic stratigraphic information from the Wilkes Land continental shelf and slope are limited to recently acquired reflection data of the Institut Francais du Petrole (Wannesson, 1985; Wannesson et al., 1985) and the U.S. Geological Survey (Eittreim et al., 1985). Although the French data do extend into the region of the Mertz–Ninnis Trough, publication to date has been limited to lines located approximately 400 km northwest (offshore Adelie Land coast).

The French survey revealed the presence of three major seismic sequences on the continental shelf. Landward of the continent–ocean boundary the sequences comprise fault-bound

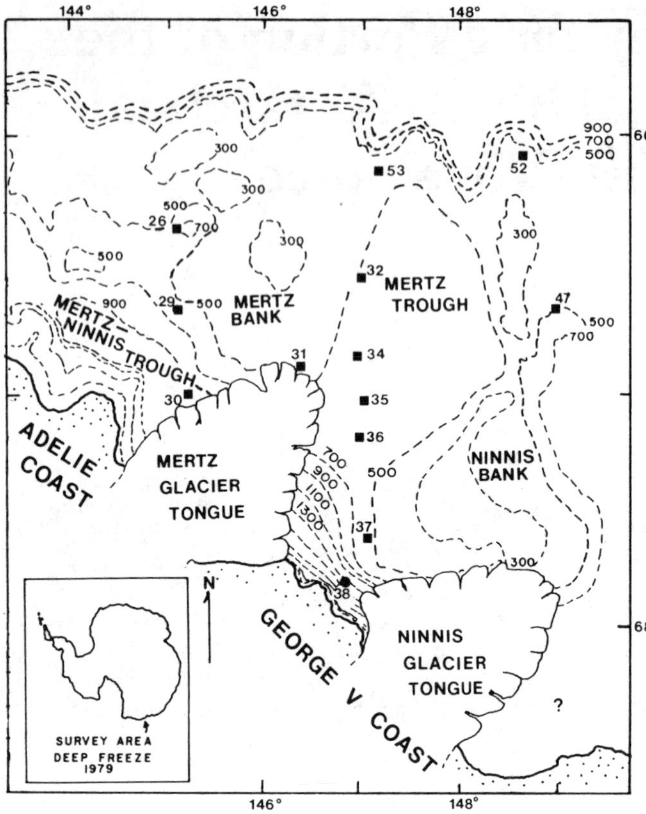

Figure 1—Bathymetry and core location on the George V portion of the Wilkes Land continental shelf (modified from Vanney and Johnson, 1979). Contours are dashed and the interval is 200 m. The Mertz Trough or depression is distinct from the Mertz–Ninnis Trough or George V Basin. Diamictons are shown by solid squares while a siltstone breccia is shown as a solid dot.

(synrift) basins and continental crust, onlapping and seaward-dipping (drift phase) reflectors, and prograding (synglacial) reflectors. Major unconformities that define the sequence boundaries are related to the onset of rifting, mid-Cretaceous continental breakup (110–90 m.y.; Cande and Mutter, 1982) and mid-Tertiary glaciation (Wannesson, 1985). The extent of the late Pleistocene(?) glacial erosion is quite extreme as Wannesson infers that lower Miocene units are exposed along the seaward flank of a broad shelf depression.

METHODS

Visual descriptions of stratification, structure, and color were made on freshly cut cores. Observations of stratification were also made from x-ray radiographs. In addition, the presence or absence of microfauna in the sand fraction, as revealed by binocular microscopic examination, was noted for each sample that was texturally analyzed (Table 1).

Textural data for selected samples were obtained using the Rice University Automated Sediment Analyser, RUASA (Anderson and Kurtz, 1979). Sand, silt, and clay ratios were obtained by a combination of sieving and hydrophotometer (Jordan et al., 1971).

Detailed lithologic analyses were conducted on selected piston core samples. Lithologies of the coarse sand fraction were determined following the procedure outlined by the Iowa Geologic Survey (Hallberg, 1978). Coarse sand was separated by dry sieving, and counts were made on all grains present in the coarse fraction. Only samples that contained over 100 grains were used in interpretation, since lithologic percentages for samples containing less than this number may not be representative. The total number of grains for each sample varied from 116 to 482.

Heavy minerals were separated from the 0.25–0.063 mm (fine and very fine sand) fraction of selected samples and are recorded as a weight percent. Heavy mineral separation procedures of Bates and Bates (1960) were followed, and tetrabromethane, specific gravity = 2.97, was used as a heavy liquid. Grains were mounted on glass slides coated with a very thin film of epoxy resin. Mineral assemblages were determined from 300 grain counts per sample. Minerals were identified using a Ziess combination reflected-transmitted light microscope equipped with a mechanical stage. Duplicates of two samples were also prepared and analyzed in the same manner, and reproducibility ranged from $\pm 7\%$ to $\pm 0.3\%$ for dominant mineral constituents.

OBSERVATIONS

Core Lithology

The location of piston cores discussed in this paper is shown on Figure 1. Except for core DF-79-38 (discussed later), the piston cores recovered unlithified diamictons of internally homogeneous character (Table 1). Diamictons are poorly sorted terrigenous sediments composed of a wide range of grain sizes (Flint et al., 1960a, 1960b). Those from the Mertz Trough contain various percents of gravel, sand, silt, and clay (Figs. 2, 3; Table 1).

Physical examination of freshly split cores and x-ray radiographs revealed only slight stratification within cores DF-79-36, 34, 32, and 53. Very thick to thin laminae are horizontal and consist of slight variations in sand, silt, and clay within the matrix (Domack et al., 1980, Fig. 3). Convolution of the laminae occur in DF-79-36 at two levels within the core, but these are bounded by horizontal laminae. Cores DF-79-37 and 35 lacked stratification. Clasts of unlithified clay occur near the base of core DF-79-34.

Hampton et al. (this volume) describe similar diamictons that were also obtained within the Mertz Trough. The diamictons are overlain by Recent siliceous oozes several centimeters to several meters thick (Domack, 1982; Hampton et al., this volume; Barnes, this volume).

Pebble Lithology

Pebble lithologies are listed in Table 1 and represent clasts greater than 1 cm in dimension. Because pebbles were collected in such small numbers for individual cores (<19), such data cannot be treated in a statistical manner. However, several observations are worth mentioning. Several argillite (indurated shale) clasts from DF-79-37 contain a Lower Cretaceous palynomorph assemblage similar to that found in core DF-79-38 (Domack, in press). The latter is believed to represent a submarine outcrop (Domack et al., 1980b). In addition, DF-

Table 1—Deep Freeze 1979 core descriptions.

DF-79-32

Lat: 66° 33'S, Long: 147° 00' E, Water Depth: 534 m

0-4 cm	Gravelly muddy sand, light olive gray (5Y 5/2) with very fine sand in discontinuous laminae, gradational contact with	radiolarians foraminifera common to rare
4-70 cm	Muddy diamicton greenish-gray (5GY 4/1)	common foraminifera

Clast lithologies: argillite (2), sandstone (2), dolerite (1), quartz (1)

DF-79-34

Lat: 66° 50'S, Long: 146°59'E, Water Depth: 595 m

0-3 cm	Mud, light olive gray (5Y 5/2) sharp contact with	radiolarians and diatoms common
3-9 cm	Siliceous ooze, light olive (10Y 5/4) gradational contact with	sparse foraminifera throughout
9-112 cm	Muddy diamicton, medium dark gray (N4) to light olive gray (5Y 6/1) clay clasts scattered below 100 cm	rare

Clast lithologies: diamictite (4), quartzite (1), granite-diorite (1)

DF-79-35

Lat: 67°03'S, Long: 147° 00'E, Water Depth: 540 m

0-138 cm	Muddy diamicton, light olive gray, (5Y 6/1) to greenish black, (5G 2/1)	no fauna or flora

Clast lithologies: quartzite (1), dolerite (1), diamictite (1)

Table 1 (Continued)

DF-79-36

Lat: 67° 17'S, Long: 147° 00'E, Water Depth: 503 m

0-5 cm	Gravelly sandy mud, olive gray (5Y 4/1) - olive black (5Y 2/1) sharp contact with	rare foraminifera
5-7 cm	Clay, brownish gray (5YR 4/1) sharp contact with	
7-188 cm	Muddy diamicton, olive black (5Y 2/1), indistinct laminations, contorted at 54 cm	rare foraminifera

Clast lithologies: siltstone (1), sandstone (1), schist (1), granite (1), granite-diorite (1), feldspar (1)

DF-79-37

Lat: 67°33'S, Long: 147°00'E, Water Depth: 582 m

0-402 cm	Sandy diamicton, light olive (5Y 3/2) to light olive gray (5Y 5/2). Water saturated and gradational contact with	no fauna or flora
402-467	Gravelly muddy sand, light olive brown (5Y 5/6) disturbed (?)	

Clast lithologies: dolerite (5), sandstone (3), argillite (2), basal (2), granite (2), granite-diorite (2), feldspar (1)

DF-79-38

Lat: 67°44'S, Long: 146°51'E, Water Depth: 1407 m

0-40 cm	Brecciated siltstone, olive black (5Y 2/1)	

Clast lithologies: organic rich siltstone, pollen & spores of Early Cretaceous age.

DF-79-53

Lat: 66°08'S, Long: 147°06'E, Water Depth: 445 m

0-381 cm	Sandy diamicton, light olive gray (5Y 5/2), below 2 cm olive gray (5Y 3/2). Possible core disturbance (flow-in) below 75 cm.	common radiolarians rare foraminifera

Clast lithologies: red granite (2), sandstone (2), dolerite (5), schist (2), argillite (2), quartzite (6)

79-37 contains several sandstone clasts, a dolerite type lithology, which appears to predominate, and two basalt clasts (Table 1).

Diamictite (lithified diamicton) clasts (Fig. 4) are quite abundant within cores from the central portion of the Mertz Trough (i. e., DF-79-36, 32, and 34). They contain a very immature mineralogy including plagioclase and pyroxene. The cements tend to be either organic (disaggregates in H_2O_2) or carbonate in character.

Coarse Sand

The lithologic variation of the coarse sand fraction (2.0–1.0 mm) is minimal within cores (Domack, 1982) but is pronounced between groups of cores (Fig. 5). The major components include quartz, dolerite, diamictite, with minor feldspar, and arkose (Fig. 5). Various crystalline lithologies inlcuding red granite, schist, gneiss, and granodiorite comprise a minor part of the remaining fraction (Domack, 1980).

Quartz is dominant in core DF-79-37 and less so in core DF-79-36. Quartz comprises a uniform (27–31%) percent of the fraction in cores DF-79-35, 34, 32, and 53. Various ratios of well-rounded to subrounded and glacially abraded angular grains can be found (Table 2).

Lithic grains consisting of interlocking phenocrysts of clear-greenish plagioclase and pyroxene are classified as dolerites based on their dominant mineralogy and ophitic texture. They at least represent an intrusive of basaltic composition. The dolerite grains gradually increase in abundance from core DF-79-37 (2%) to DF-79-53 (16%), but on average comprise 7–10% of the coarse sand fraction (Fig. 5).

Lithified grains resembling diamictite or tillite(?) matrix are rather abundant (11–26%) in cores DF-79-35, 34, and 32, but rare in cores DF-79-37, 36, and 53 (Fig. 5). They are identical to gravel clasts described in the previous section (Fig. 4).

Heavy Mineralogy

Heavy minerals in the very fine sand fraction (0.25–0.063 mm) are rather abundant (2–3%) within the diamictons and include garnet, pyroxene, opaques (chiefly magnetite and illmenite), zircon, epidote, hornblende, kyanite, and various other minor constituents (Domack, 1982; Fig. 6). Variation in the proportion of these minerals is minimal within a single core (Domack, 1980).

A stable assemblage of red and pink garnet, zircon, epidote, and opaques dominates in core DF-79-37 and to a lesser degree in core DF-79-36 (Fig. 6). Pale greenish pyroxene dominates a second, unstable assemblage in cores DF-79-35, 34, 32, and 53.

INTERPRETATION AND DISCUSSION

Diamicton Origin

The sedimentologic criteria of Anderson et al. (1980) were applied to the diamictons from the Mertz Trough with the result that both basal tills (DF-79-37 and 35) and glacial-marine, or ice-rafted, diamictons (DF-79-36, 34, 32, and 53) were recognized (Domack, 1982). Both cores DF-79-37 and 35 are massive, texturally homogeneous (Fig. 3), unfossiliferous and have shear strengths in excess of 2.5 kg/cm^2 (Domack, 1980, 1982; Table 1). These observations are consistent with the criteria of Anderson et al. (1980) for basal till and are in contrast to the stratification, textural variation (Fig. 3), fossil content, and lower shear strengths observed in cores DF-79-36, 34, 32, and 53 (Domack, 1980, 1982; Table 1). Detailed study by Hampton et al. (this volume) of similar diamictons, also from the Mertz Trough, led to the conclusion that such sediments were glacial-marine in origin. The depositional model of Domack (1982) suggests that the ice-rafted diamictons were deposited beneath a retreating ice shelf and onto basal tills.

Barnes (this volume) interprets the "ridge and groove" morphology of the Mertz Trough as a subglacially formed surface consisting of relict diamicton, namely the sediments discussed in this paper. It is apparent from morphologic and shallow seismic data that the diamictons represent a continuous deposit within the floor of the Mertz Trough (Barnes, this volume). The basal tills in cores DF-79-37 and 35 strongly support the interpretation of Barnes (this volume). However, the glacial-marine deposits recovered only a short distance from "ridge and groove" relief are not entirely consistent with a subglacial origin. This is because deposition of 3–4 m of ice-rafted sediment would bury and mask the fluted surface, which has a relief of only 1 m. Thus, either the glacial-marine deposits represent only a local or patchy deposit, or their characteristics (i.e., fossil content, stratification, etc.) have been misinterpreted.

The presence of stratification and microfossils is not always a reliable indicator of deposition by ice-rafting, because tills derived from contemporaneous marine sediments also contain these properties (Domack and Lawson, 1985). The diamictons in cores DF-79-36, 34, 32, and 53 may represent previous ice-rafted deposits that have been overridden and remolded by grounded ice. Their mineralogic integrity most likely reflects derivation from basal debris zones and/or homogenization associated with the development of the "ridge and groove" morphology (Barnes, this volume). The low compaction indicated by sediment shear strengths (less than 2.0 kg/cm^2, Anderson et al., 1980; Domack, 1982) remains problematic (Hampton et al., this volume). However, if dewatering of the sediments was prevented, then consolidation may have been retarded even within the subglacial environment (Muller, 1983).

Provenance

Previous Interpretation

The mineralogic data discussed above were originally interpreted as an indication of northwest ice flow during the most recent glacial advance (Domack, 1982). Quartz and garnet-rich diamictons were thought to have been derived from outcrops of Triassic Beacon Sandstone found approximately 75 km to the southeast (Mawson, 1942). The dolerite grains were also related to exposures of Jurassic Ferrar Dolerite, which has intruded the Beacon clastics (Mawson, 1942). Diamictite lithologies were thought to have been derived from Ninnis Bank (Fig. 1), as ice flowed over, from the southeast, and eroded this feature. This model related the distinct provenance boundary between quartz–garnet units and diamictite–pyroxene units to a parallel direction of paleo-ice flow (Domack, 1982). The boundary extends along the seaward flank of the

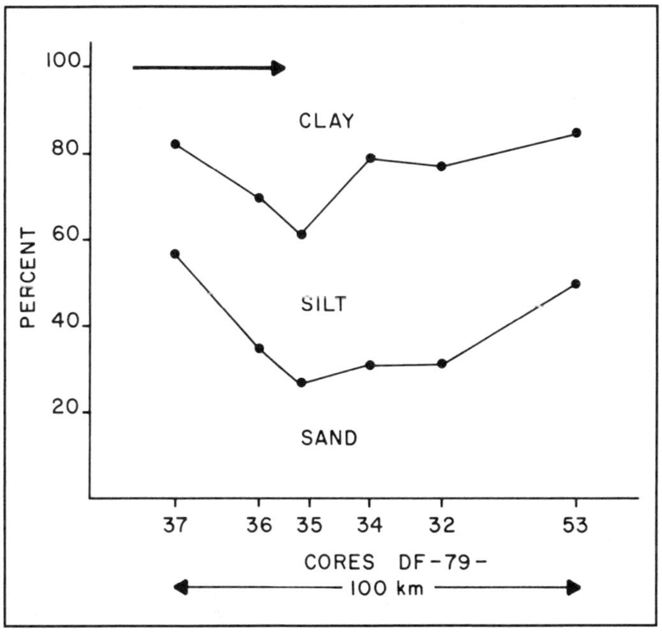

Figure 2—Variation in sand, silt, and clay proportions between diamicton cores from the Mertz Trough. The arrow shows direction of glacial flow according to Barnes (this volume).

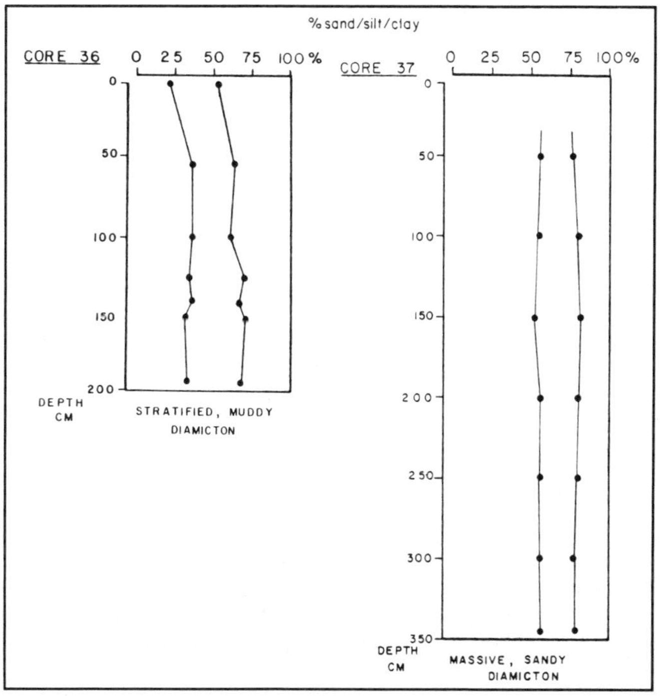

Figure 3—Textural variation within a stratified (DF-79-36) and massive (DF-79-37) diamicton.

Figure 4—Photograph of diamictite [tillite(?)] clasts recovered from cores DF-79-32, 35, 52, and 47.

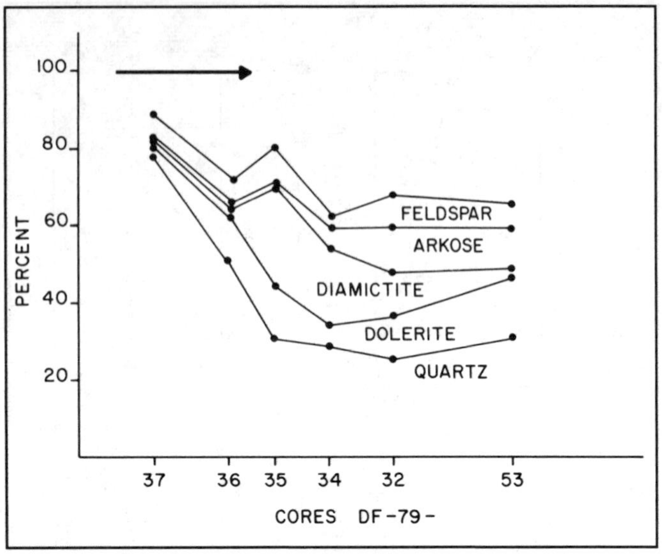

Figure 5—Variation in coarse sand lithology between diamictons from the Mertz Trough. The arrow shows direction of glacial flow according to Barnes (this volume).

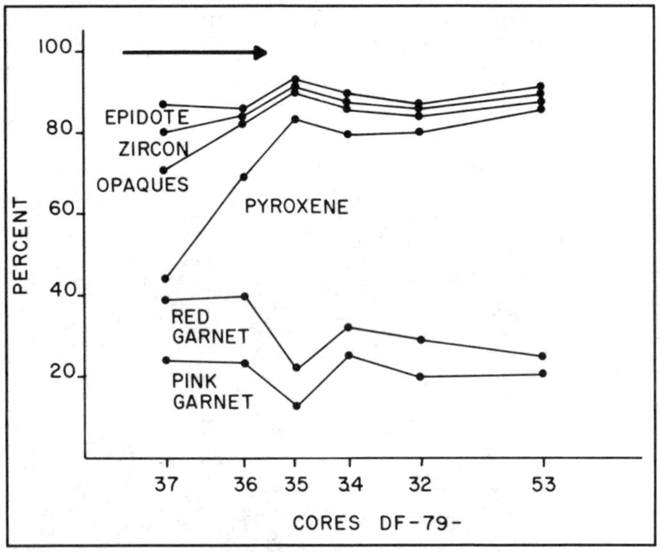

Figure 6—Variation in heavy mineralogy (fine and very fine sand fraction) between diamicton cores from the Mertz Trough. The arrow shows direction of glacial flow according to Barnes (this volume).

Mertz-Ninnis Trough and is well defined by the available core coverage (Domack, 1982). However, the north-northeast flow direction within the Mertz Trough (Barnes, this volume) demonstrates that the provenance boundary actually trends about 90° to the actual flow direction (Figs. 4, 5). Therefore, the significance of the mineralogic boundary needs to be reconsidered.

Glacial Processes

A new interpretation for the mineralogic data is suggested herein based on our present understanding of glacial erosion and depositional processes. It is known that as glacial ice

Table 2—Quartz grain shape.

	Coarse Sand/Medium Sand		
Roundness[1]	DF-79-36, 80-82cm;	DF-79-37, 80-81cm;	250-253cm
1	9.5/0	9/10.6	5 /13.3
2	19/12	27/24.4	22 /20.5
3	28.5/40	29/26.8	36 /28.1
4	28/34	18.5/22	22 /23.3
5	10.5/10	6/9.8	9 /10.9
6	4.5/4	10.5/6.4	6/3.8
	100%/100%	100%/100%	100%/100%
Mean Roundness	3.25/3.54	3.17/3.07	3.26/3.10

[1]Roundness Scale from Powers (1953). Percentages based on 200 grains/sample.

advances across geologic contacts the composition of basal debris and till change, as local bedrock is incorporated by various entrainment processes (Shilts, 1977; Linden, 1975; Dilabio and Shilts, 1979; Humlum, 1985). The rapidity and extent to which the local bedrock or substrate comes to dominate the basal debris, and thus the resulting till, depends to a large degree on the thermal character of the basal ice, subglacial relief, comminution, and erodability of the substrate (Dreimanis and Vagners, 1971; Boulton, 1972, 1978; Sugden and John, 1976). For example, the coarser (i.e., gravel) fractions of till often may reflect local sources while the matrix, due to significant transport and comminution, reflects a distant source terrain (Dreimanis and Vagners, 1971). The opposite relationship may also occur particularly when the local bedrock is composed of easily eroded and rapidly comminuted lithologies (i.e., sedimentary) and the distant source is crystalline rock (Dreimanis and Vagners, 1972; Dilabio and Shilts, 1979).

Stratigraphic Models

With the above considerations in mind, two stratigraphic models for the Mertz Trough area are shown in Figures 7 and 8. It is proposed that as ice flowed up and out of the Mertz-Ninnis Trough and into the Mertz Trough, it encountered and partially eroded a succession of units starting with siltstones of Early Cretaceous age (i.e., core DF-79-38). Till deposited at the site of core DF-79-37 contains argillite pebbles, also of Early Cretaceous age (Domack, in press). The presence of a significant stratigraphic section above the Lower Cretaceous unit (core DF-79-38) is supported by vitrinite reflectance (R_o) values of 0.51 obtained from the Lower Cretaceous siltstone (Domack, in press). This degree of thermal maturation would have occurred at burial depths of 1500-2200 m assuming a reasonable geothermal gradient (Domack, in press).

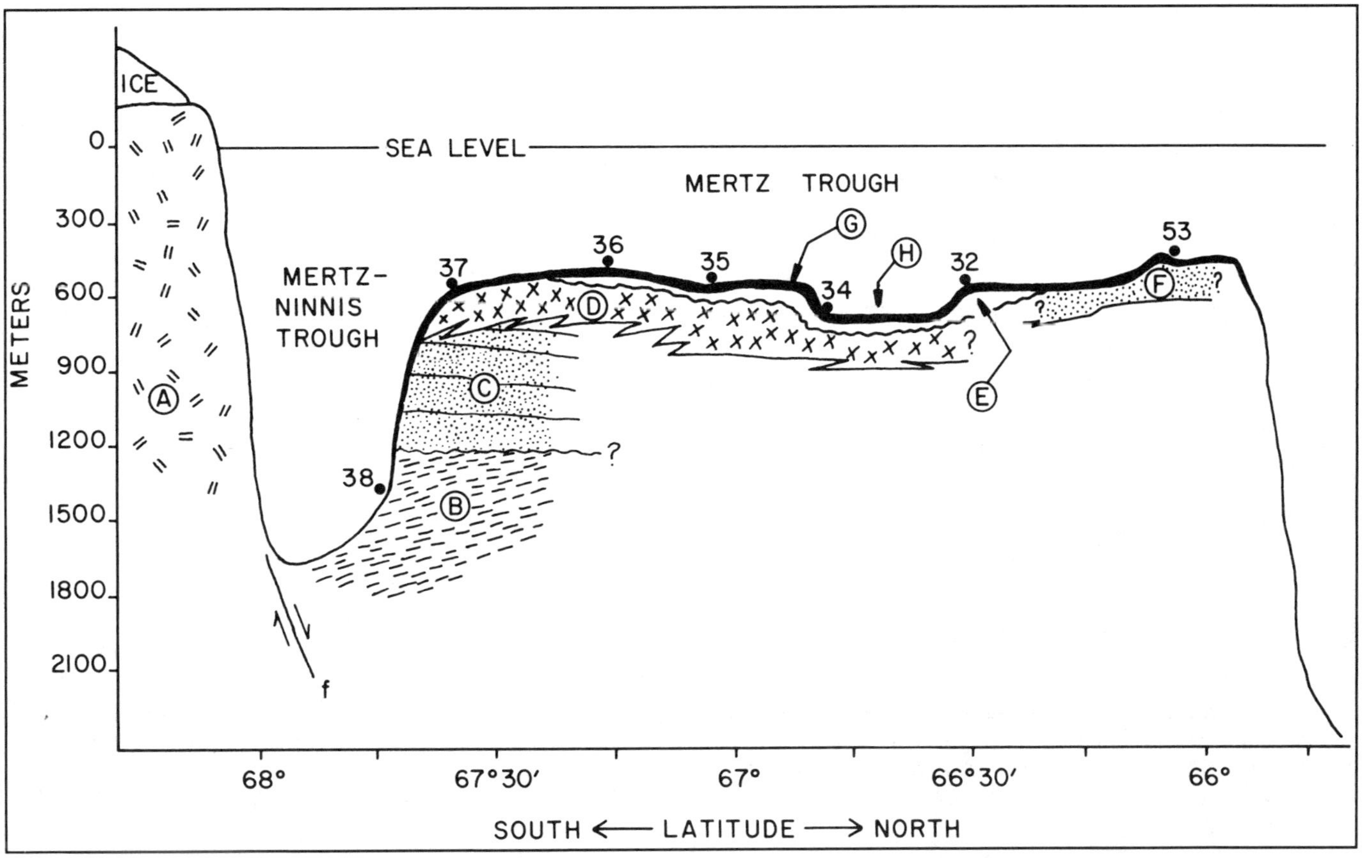

Figure 7—Stratigraphic model for the Mertz Trough area of the Wilkes Land continental shelf. The bathymetric profile is from Domack (1982). Stratigraphic units are: A, Precambrian granite in coastal outcrop (Mawson, 1942); B, Lower Cretaceous (Aptian) siltstone in submarine outcrop (Domack et al., 1980b); C, Lower Cretaceous-lower Tertiary(?) sandstone; D, dolerite or basaltic intrusive; E, Quaternary(?) diamictite; F, arkosic sandstone; G, Pleistocene-Holocene(?) diamicton; H, Holocene siliceous ooze. Core locations are also indicated (dots). Profile is parallel to long 147° E.

The Lower Cretaceous siltstone must comprise part of a synrift sequence since its age (Aptian) predates breakup by at least 3 million and at most 29 million years (Cande and Mutter, 1982; Domack, in press). Therefore, beds within this unit are assumed to dip toward a major boundary fault and the Precambrian basement (Figs. 7, 8). Next to the siltstone, a sandstone of undetermined thickness (but at most 600 m) likely exists. The abundance of quartz, a stable heavy mineral fraction (garnet, opaques, zircon, and epidote), and the sandy texture within core DF-79-37 are strong evidence for the presence of this sandstone. Sandstone clasts were also found within the till at site 37 (Table 1). The sandstone unit subcrops along most of the seaward flank of the Mertz-Ninnis Trough as diamictons in cores DF-79-26, 29, and 31 (Fig. 1) contain a coarse sand and heavy mineral assemblage identical to DF-79-37 (Domack, 1982). Sandstone pebbles are also abundant in core DF-79-29 (Domack, 1980). Lignite and thin volcanics (basalt) may also be part of the section, since these components have also been found in nearby surficial sediments (Mawson, 1942) and cores (Table 1; Domack, 1982).

The age of the sandstone is unknown, but two possibilities are suggested: Early Cretaceous and younger or Triassic. Sparse palynomorphs of Late Cretaceous and early Tertiary age have been found within nearby surface sediments, suggesting a source in proximity to the Mertz-Ninnis Trough (Truswell, 1982, 1983a). The arenaceous nature of the sandstone unit may explain the paucity of pollen and spores of this age recovered by Truswell (1982, 1983b). If the sandstone and related units are indeed younger than Early Cretaceous then they would comprise part of the drift phase sequence recognized by Wannesson (1985) and Eittreim et al. (1985). Figure 7, therefore, shows these units to lie above the breakup unconformity and with seaward dips.

It is also possible that the sandstone is Triassic in age and correlates with exposures of Beacon sandstone found 75 km to the southeast (Mawson, 1942). This is supported by the mineralogic and textural similarities between the till sands and Beacon Sandstone (Domack, 1982) and the association of the Beacon clastics with the Ferrar Dolerite (see below). If this is the case, then the Mertz-Ninnis Trough represents a half-graben feature with portions of the continental basement, or prerift, sequence (Wannesson, 1985; Eittreim et al., 1985) exposed along its seaward flank (Fig. 8).

A contact of the sandstone unit with intrusive rocks of basaltic composition (dolerites) likely occurs just south of site 37. This is proposed because the sandstone unit is represented primarily within the matrix fraction of the till, while the clast fraction at site 37 is dominated by dolerite (Table 1; Fig. 5). As ice flowed into the Mertz Trough, it encountered a more or less flat-lying unit of diamictite. It is at site DF-79-35 that diamictite first dominates the coarse sand fraction and is abundant within the gravel (Table 1; Figs. 4, 5). Therefore,

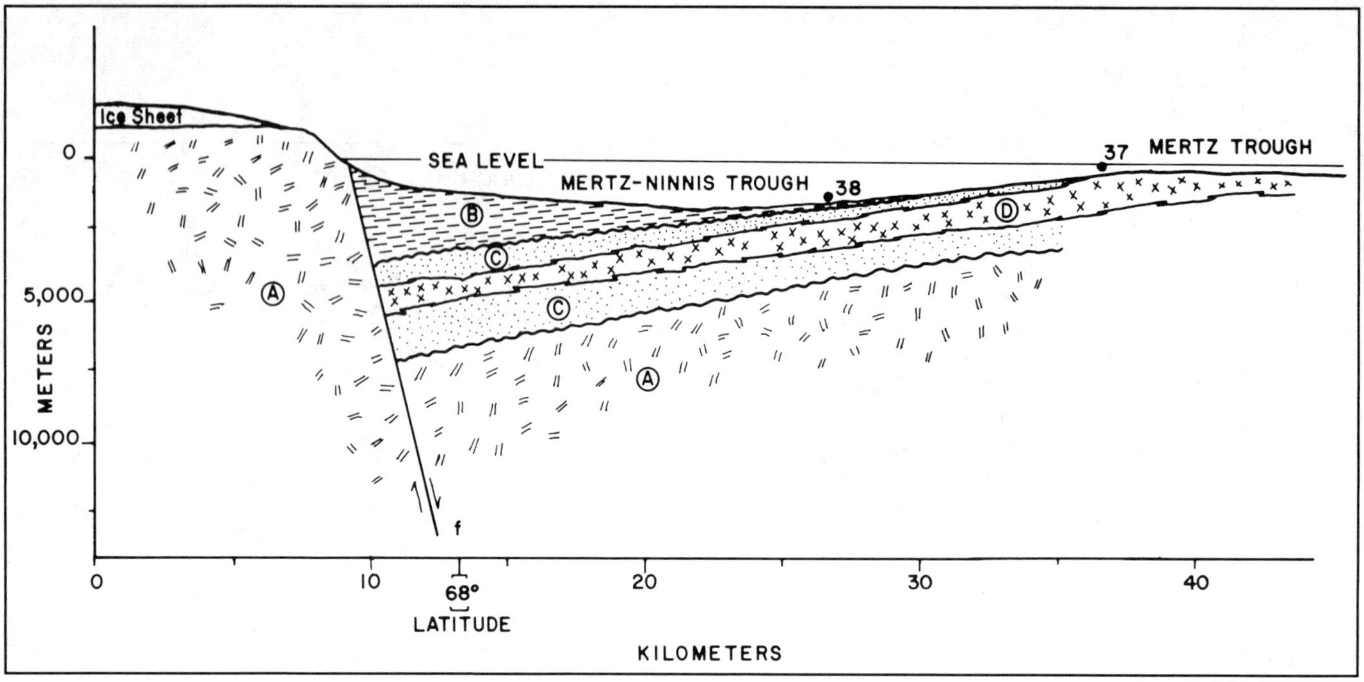

Figure 8—Alternative stratigraphic model for the Mertz-Ninnis Trough region of the continental shelf. The bathymetric profile is modified after Domack (1982) and is shown here without vertical exaggeration. Stratigraphic units are: A, Precambrian granite in coastal outcrop (Mawson, 1942); B, Lower Cretaceous (Aptian) siltstone in submarine outcrop (Domack et al., 1980b); C, Triassic sandstone; D, Jurassic dolerite or basaltic intrusive. Core locations are also indicated (dots). Profile is parallel to long 147° E.

the dolerite–diamictite contact lies just south of site 35 (Fig. 7). The age of the diamictite is unknown, but it is certainly much older than the overlying, unlithified, diamicton. Comminution of dolerite and diamictite grains along the transport direction would increase the proportion of fines (silt and clay) and release pyroxene into the heavy mineral fraction. The data in Figures 2 and 6 strongly support this view. The fact that the diamictite contains relatively abundant pyroxene and plagioclase suggests that it too was derived, in part, from the dolerite. The diamictite is therefore assumed to overlie the intrusive unit. Such a contact suggests that significant erosion may have occurred. Therefore, the contact might be recognized on seismic profiles of the shelf as a distinct, unconformable reflector similar to the oldest glacial unconformity recognized by Wannesson (1985).

The coarse sand lithologies suggest that the subcrop limit of the diamictite lies in the vicinity of core DF-79-32 and that it is perhaps succeeded by an arkosic sandstone. However, gravel clasts from these cores are so few in number that the presence of the arkose cannot be strongly supported with the available data.

CONCLUDING REMARKS

Though the interpretations in Figures 7 and 8 are speculative, they do explain the mineralogic variation observed in the diamicton units. They are also compatible with paleo-ice flow directions, which are interpreted from geomorphologic evidence (Barnes, this volume). Identical changes in lithology within glacial sediments have been observed in other areas where ice sheets have crossed geologic contacts along the flow direction (Linden, 1975; and others). Both the distances involved and rates of lithologic variation are similar between these studies and those reported in this paper.

The mineralogic data reexamined in this study suggest that observed provenance boundaries within glacial sediments on the shelf may be related to contacts of several previously unrecognized stratigraphic units. The proposed, seaward, succession consists of siltstone, sandstone, basaltic intrusive, diamictite, and arkose. Further study should be done on age dating of the intrusive unit as well as detailed petrographic characteristics of the sedimentary lithologies.

ACKNOWLEDGMENTS

This research was originally supported by National Science Foundation grants DPP-77-26407 and DPP-80-80242 to Dr. John Anderson (Rice University). Gratitude is extended to D. Cassidy at Florida State University, for his help in providing samples, and K. Sanchirico, for providing the quartz grain shape data. I would also like to thank D. Kurtz, J. Anderson, A. Elverhoi, and R. Powell, for their helpful reviews, and S. L. Eittreim for his encouragement.

REFERENCES CITED

Anderson, J. B., and D. D. Kurtz, 1979, RUASA: An automated sediment analysis system: Journal of Sedimentary Petrology, v. 49, p. 625–627.

Anderson, J. B., K. Balshaw, E. Domack, D. Kurtz, R. Milam, and R. Wright, 1979, Geological survey of East Antarctic continental margin aboard USCGC Glacier: Antarctic Journal of the United States, v. 14, p. 142–144.

Anderson, J. B., D. D. Kurtz, E. W. Domack, and K. Balshaw, 1980, Glacial and glacial marine sediments from the Antarctic continental shelf: Journal of Geology, v. 88, p. 399-414.

Anderson, J. B., C. Brake, E. Domack, N. Myers, and R. Wright, 1983, Development of a polar glacial-marine sedimentation model from Antarctic Quaternary deposits and glaciological information, *in* B. F. Molnia, ed., Glacial-marine sedimentation: New York, Plenum Press, p. 233-264.

Bates, J. D., and B. H. Bates, 1960, Evaluation of heavy mineral separation using artificial samples: Journal of Sedimentary Petrology, v. 30, p. 148-153.

Boulton, G. S., 1972, The role of thermal regime in glacial sedimentation, *in* R. J. Price and D. E. Sugden, eds., Polar geomorphology: Institute of British Geography Special Publication 4, p. 1-19.

______, 1978, Boulder shapes and grain-size distributions of debris as indicators of transport paths through a glacier and till genesis: Sedimentology, v. 25, p. 773-799.

Cande, S. C., and J. C. Mutter, 1982, A revised identification of the oldest sea-floor spreading anomalies between Australia and Antarctica: Earth and Planetary Science Letters, v. 58, p. 151-160.

Dilabio, R. N. W., and W. W. Shilts, 1979, Composition and dispersal of debris by modern glaciers, Bylot Island, Canada, *in* C. Schluchter, ed., Moraines and varves: Rotterdam, A. A. Balkema Press, p. 145-155.

Domack, E. W., 1980, Glacial marine geology of the George V-Adelie continental shelf, East Antarctica: M. A. Thesis, Rice University, Houston, Texas, 142 p.

______, 1982, Sedimentology of glacial and glacial marine deposits on the George V-Adelie continental shelf, East Antarctica: Boreas, v. 11, p. 79-97.

______, in press, Glacial erosion on the George V-Adelie continental margin, East Antarctica: Antarctic Journal of the United States.

Domack, E. W., and J. B. Anderson, 1983, Marine geology of the George V continental margin: Combined results of Deep Freeze 1979 and the 1911-14 Australasian Expedition, *in* R. L. Oliver, P. R. James, and J. B. Jago, eds., Antarctic earth science: Canberra, Australian Academy of Science, p. 402-406.

Domack, E. W., and D. E. Lawson, 1985, Pebble fabric in an ice-rafted diamicton: Journal of Geology, v. 93, p. 577-591.

Domack, E. W., J. B. Anderson, and D. D. Kurtz, 1980a, Clast shape as an indicator of transport and depositional mechanisms in glacial marine sediments: George V continental shelf, Antarctica: Journal of Sedimentary Petrology, v. 50, p. 813-820.

Domack, E. W., W. W. Fairchild, and J. B. Anderson, 1980b, Lower Cretaceous sediment from the Antarctic continental shelf: Nature, v. 287, p. 625-626.

Dreimanis, A., and U. J. Vagners, 1971, Bimodal distribution of rock and mineral fragments in basal tills, *in* R. P. Goldthwaite, ed., Till, a symposium: Ohio State University Press, Columbus, p. 237-250.

______, 1972, The effect of lithology upon texture of till, *in* E. Yatsu and A. Falconer, eds., Research methods in Pleistocene geomorphology: Proceedings of second Guelph symposium on geomorphology, 1971, p. 86-102.

Dunbar, R. B., J. B. Anderson, E. W. Domack, and S. S. Jacobs, 1985, Oceanographic influences on sedimentation along the Antarctic continental shelf, *in* S. S. Jacobs, ed., Oceanology of the Antarctic continental shelf: American Geophysical Union, Antarctic Research Series, v. 43, p. 291-312.

Eittreim, S. L., M. A. Hampton, and J. R. Childs, 1985, Seismic-reflection signature of Cretaceous continental breakup on the Wilkes Land margin, Antarctica: Science, v. 229, p. 1082-1084.

Flint, R. F., J. E. Sanders, and J. Rodgers, 1960a, Symmictite: A name for unsorted terrigenous sedimentary rocks that contain a wide range of particle sizes: Geological Society of America Bulletin, v. 71, p. 507-510.

______, 1960b, Diamictite: A substitute term for symmictite: Geological Society of America Bulletin, v. 71, p. 1809-1810.

Jordan, Jr., C. F., G. E. Fryer, and E. Hemman, 1971, Size analysis of silt and clay by hydrophotometer: Journal of Sedimentary Petrology, v. 41, p. 489-496.

Hallberg, G. R., 1978, Standard procedures for evaluation of Quaternary materials in Iowa: Iowa Geological Survey Technical Information Series, Publication 8.

Humlum, O., 1985, Changes in texture and fabric of particles in glacial traction with distance from source, Myrdalsjökull, Iceland: Journal of Glaciology, v. 31, p. 150-156.

Linden, A., 1975, Till petrographical studies in an Archean bedrock area in southern central Sweden: Striae, v. 1, p. 1-57.

Mawson, D., 1942, Sedimentary rocks, *in* D. Mawson, ed., Australasian Antarctic Expedition 1911-14, Scientific Reports, Series A, v. 4, Geology, pt. 2.

Muller, E. H., 1983, Dewatering during lodgement of till, *in* E. B. Evenson, C. Schluchter, and J. Rabassa, eds., Tills and related deposits: Rotterdam, A. A. Balkema, p. 13-18.

Powers, M. C., 1953, A new roundness scale for sedimentary particles: Journal of Sedimentary Petrology, v. 23, p. 117-119.

Shilts, W. W., 1977, Glacial till and mineral exploration, *in* R. F. Legget, ed., Glacial till, an interdisciplinary study: Ottawa, Royal Society of Canada, p. 205-224.

Sugden, D. E., and B. S. John, 1976, Glaciers and landscape: A geomorphological approach: London, Halsted Press, 376 p.

Truswell, E. M., 1982, Palynology of seafloor samples collected by the 1911-14 Australasian Antarctic Expedition: implications for geology of coastal East Antarctica: Journal of the Geological Society of Australia, v. 29, p. 343-356.

______, 1983a, Geological implications of recycled palynomorphs in continental shelf sediments around Antarctica, *in* R. L. Oliver, P. R. James, and J. B. Jago, eds., Antarctic earth science: Canberra, Australian Academy of Sciences, p. 394-399.

______, 1983b, Recycled Cretaceous and Tertiary pollen and spores in Antarctic marine sediments: A catalogue: Palaeontographica, Abt. B, v. 186, p. 121-174.

Vanney, J. R., and G. L. Johnson, 1979, The sea floor morphology seaward of Terre Adélie (Antarctica): Deutsche Hydrographische Zeitschrift, v. 32, p. 39-88.

Wannesson, J., 1985, Structure de la marge continental de Terre-Adélie interprétation de la campagne géophysique 1982: Actes du collogue sur la recherche francaise dans l'Antarctique, Grenoble, Comité National Francais des Recherches Antarctiques, p. 124-128.

Wannesson, J., M. Pelras, B. Petitperrin, M. Perret, and J. Segoufin, 1985, A geophysical transect of the Adélie margin, East Antarctica: Marine Petroleum Geology, v. 2, p. 192-201.

Wright, R., J. B. Anderson, and P. P. Fisco, 1983, Distribution and association of sediment gravity flow deposits and glacial/glacial-marine sediments around the continental margin of Antarctica, *in* B. F. Molnia, ed., Glacial marine sedimentation: New York, Plenum Press, p. 265-300.

Eittreim, S. L., and Hampton, M. A., 1987, *The Antarctic Continental Margin: Geology and Geophysics of Offshore Wilkes Land*, CPCEMR Earth Science Series, v. 5A: Houston, Texas, Circum-Pacific Council for Energy and Mineral Resources

Hydrocarbon Geochemistry of Sediments Offshore from Antarctica: Wilkes Land Continental Margin

K. A. Kvenvolden
M. Golan-Bac
J. B. Rapp
U.S. Geological Survey
Menlo Park, California

Near-surface sediments of the Wilkes Land continental margin off the George V coast and north of the Mertz Glacier contain both gaseous and heavy hydrocarbons (*n*-alkanes and isoprenoids, pristane and phytane) in low concentrations. Analyses of hydrocarbon gases with one through four carbon atoms show that methane is the most abundant. The hydrocarbon gases likely represent "background" from biologic and early diagenetic processes. Heavy hydrocarbons with 13 to 36 carbon atoms are dominantly of odd-carbon number, especially molecules with more than 24 carbon atoms. The molecular distribution of *n*-alkanes suggests that many of these compounds came from land plants. Because there are no vascular plants in Antarctica at present, the sediment *n*-alkanes may represent plant material that was transported from other continents, or, more likely, the compounds could be from eroded ancient sediments of Antarctica. Information from this study contributes little to understanding possible petroleum resources in the offshore area of Antarctica. Geophysical evidence, however, suggests that gas hydrates may be present in sediments of the Wilkes Land margin.

INTRODUCTION

Interest in the petroleum potential of Antarctica is increasing as more nations explore this remote part of the world. There is a consensus among current Antarctic researchers that petroleum likely occurs in some of the sedimentary basins both onshore and offshore, but economics and politics will inhibit active exploration and exploitation for some time. Information about the petroleum potential of Antarctica comes from interpretation of geophysical studies (Behrendt, 1982, 1983; Cook and Davey, 1984), the results of drilling by the Deep-Sea Drilling Project (DSDP) in the Ross Sea (McIver, 1975), and on gross comparisons and analogies with other continents and continental margins (Cameron, 1981). Direct geologic observations of the rocks in the basins of greatest interest have not been possible because these basins are located either on the continent beneath the ice sheet or on the continental margin beneath the water.

Hydrocarbon geochemistry has thus far played only a minor role in ascertaining the petroleum potential of this continent. Most cited are the observations of McIver (1975), who analyzed the hydrocarbon gases in sediments cored at three DSDP sites in the Ross Sea. High concentrations of methane were found along with small but significant quantities of ethane. The carbon isotopic composition of this methane (relative to PDB standard) ranged from −68 to −79 o/oo (Claypool and Kvenvolden, 1983, Table 3), values indicative of a microbial source. The source of ethane is not known, but ethane could result from in situ diagenesis or could have migrated from deeper sources. Hydrocarbons from inferred thermogenic sources were recently observed in surface sediments of the Bransfield Strait of the Antarctic Peninsula (Whiticar et al., 1985). Volcanic sills and dikes in this rifted basin may have provided local sources of heat to accelerate the thermal decomposition of sediment organic matter, thereby producing thermogenic hydrocarbon gases. Other hydrocarbon geochemistry studies have been directed toward possible environmental contamination of a site in the sub-Antarctic near South Georgia (Mackie et al., 1978; Platt and Mackie, 1979, 1981).

Our work considers the hydrocarbon geochemistry of sediments from two localities offshore from Antarctica: (1) the continental margin north of the Mertz Glacier, which is adjacent to the George V coast of Wilkes Land (this paper), and (2) the western Ross Sea, offshore Victoria Land (Rapp et al., in press). Our geochemical studies are part of the U.S. Geological Survey's 1984 marine geological and geophysical investigation of the Antarctic continental margin (Eittreim et al., 1984).

During Leg 1 of this investigation, nine gravity cores of bottom sediment were obtained from various depositional environments on the shelf, slope, and rise off of Wilkes Land. From eight of these cores, samples were obtained for analyses of hydrocarbons, and the results of the analyses are discussed here. On Leg 2, 15 gravity cores were recovered from bottom sediment of the Ross Sea, and samples from nine of these

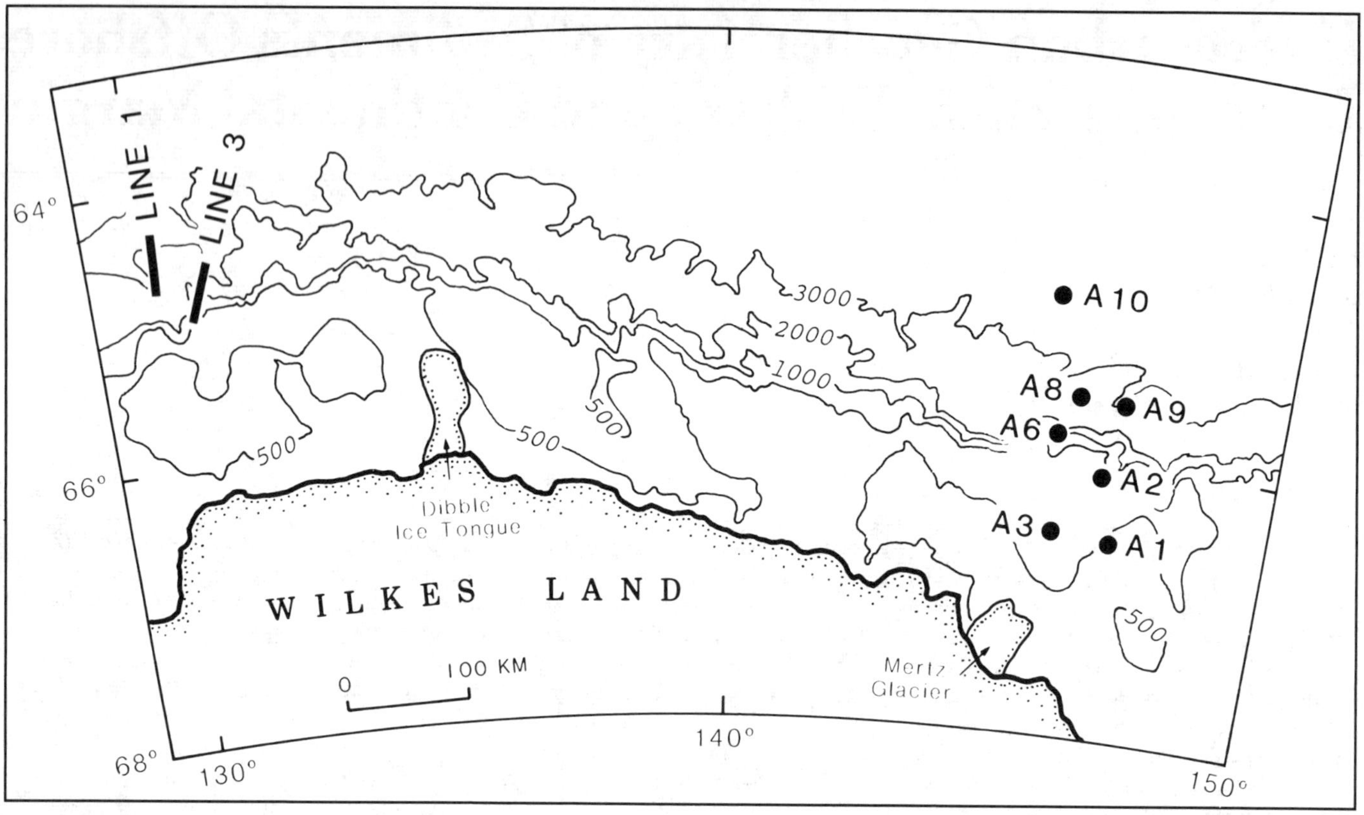

Figure 1—Map showing the location of seven stations on the Wilkes Land margin where gravity coring recovered sediments for hydrocarbon geochemistry studies. Heavy lines indicate segments of multichannel seismic Lines 1 and 3 (see Fig. 3). Bathymetry is in meters.

cores were examined for hydrocarbons; results from these are described by Rapp et al. (in press).

Our main objective was to obtain preliminary information on the concentration, distribution, and possible sources of specific hydrocarbons in near-surface sediments sampled during Legs 1 and 2. We measured (1) the gaseous hydrocarbons, methane (C_1), ethane (C_2), ethene ($C_{2:1}$), propane (C_3), propene ($C_{3:1}$), isobutane (i-C_4), and n-butane (n-C_4) in samples from 17 cores; and (2) the heavy hydrocarbons (n-C_{13} to n-C_{36} plus pristane and phytane) in 15 samples.

PROCEDURES

Sampling

The Leg 1 sampling program followed an extensive geophysical survey along the Wilkes Land continental margin (Eittreim et al., 1984). Sample stations were located on a traverse north of Mertz Glacier (Fig. 1). Sediments were obtained from several physiographic features of the margin: Mertz Depression, Mertz Bank, the upper continental slope, and the upper and lower continental rise; depths of water ranged from 253 to 3379 m (Table 1). A gravity core (8-cm internal diameter core liner) from each station was subsampled for this study. For hydrocarbon gases (C_1–C_4), 9 cm long segments of the core (one sample was 8 cm long) were cut at intervals ranging in depth from 13 to 384 cm. For heavy hydrocarbon studies (C_{13}–C_{36}), a single sample, 8 cm in length, was taken at each site. The geochemical samples were generally fine-grained, diatomaceous mud with variable amounts of pebbles.

Hydrocarbon Gas Analyses

The procedures used for gas analyses have been discussed previously by Kvenvolden and Redden (1980). As soon as the 9 cm long segments of core liner had been cut, the sediment was extruded from the liner into cans (approximately 1-l volume), each of which had two septa-covered entry ports for the later removal of gas. Each can was filled with distilled water that had been previously purged with helium to remove dissolved gases. One hundred milliliters of water were removed from the can, which was then sealed with a double-friction seal top, and the resulting 100-ml headspace was purged with helium via the septa. The can was shaken for 10 min to extract the gases from the sediment into the headspace. About 5 ml of the headspace gas mixture was removed in a gas-tight syringe, and exactly 1 ml of this mixture was injected into a gas chromatograph equipped with both flame-ionization and thermal conductivity detectors. The chromatograph was calibrated with standard mixtures of hydrocarbon gases. Concentrations of gases in our samples were calculated from integrated peak areas or peak heights on the resulting chromatograms and are reported as nanoliters/liter of wet sediment (nl/l). Partition coefficients, based on averages from previous studies, were used to correct for differences in gas solubilities: $C_1 = 0.8$; C_2, C_3, C_4, $= 0.7$; $C_{2:1}$, $C_{3:1} = 0.6$.

Heavy Hydrocarbon Analyses

The procedure for the extraction and identification of heavy hydrocarbons (C_{13}–C_{36}) was modified from Rapp et al. (1982) and Kvenvolden et al. (1986). Immediately upon recovery,

Table 1—Location of stations on the Wilkes Land margin where samples were recovered for hydrocarbon geochemistry studies.

Core	Latitude	Longitude	Water Depth (m)	Physiography
A1G1	66° 35.4'S	147° 21.7'E	611	Mertz Depression
A2G1	66° 08.1'S	147° 05.5'E	460	Mertz Bank
A2G2	66° 08.1'S	147° 05.4'E	458	Mertz Bank
A3G1	66° 30.8'S	146° 10.9'E	253	Mertz Bank
A6G1	65° 52.5'S	146° 21.1'E	994	Upper Slope
A8G1	65° 34.0'S	146° 25.4'E	2635	Upper Rise
A9G1	65° 34.2'S	147° 21.6'E	3037	Upper Rise
A10G1	64° 54.3'S	145° 59.9'E	3379	Lower Rise

each sediment sample was transferred to a precleaned glass jar, and the jar was sealed with an aluminum-foil-lined lid. The sample was frozen immediately and kept in this state until analyzed in our shore-based laboratory. The entire sample was freeze dried in the original container, gently pulverized with a mortar and pestle, and sieved through a 32-mesh screen. Only the less than 32-mesh size fraction was analyzed for hydrocarbons. About 10 g of this size fraction were used to determine the organic carbon content by wet combustion, following the procedure of Bush (1970).

For extraction of hydrocarbons, 100 g of each sample was placed in a glass-stoppered centrifuge bottle and extracted with 200 ml dichloromethane (pesticide quality) by shaking for 2 hours on a wrist-action shaker. The extract solution was recovered by centrifugation and filtered through glass wool. Fresh solvent (100 ml) was added to the sediment, and the extraction procedure was repeated twice. The combined filtered extracts were concentrated to 5 ml by rotary evaporation. Sulfur was removed by filtration through activated copper. The filtrate volume was adjusted to 5 ml, and a portion was removed for determination of the weight of extract. The dichloromethane solvent was then exchanged with *n*-hexane.

The *n*-hexane solutions (5 ml) were sonicated and fractionated by liquid–solid chromatography using a 10-mm ID column (25 cm long) packed sequentially from bottom to top with activated silica gel (5 g Davidson No. 923 and 2.5 g Davidson No. 62) and 2.5 g deactivated (5% water) alumina. The column was eluted in sequence with 20 ml *n*-hexane, 25 ml benzene, and 25 ml methanol to yield fractions containing aliphatic hydrocarbons, medium polarity organic compounds including aromatic hydrocarbons, and polar organic compounds, respectively. Portions of the first two fractions were removed for weight determinations.

Gas chromatography of the hexane fraction was done using a 30 m × 0.3 mm DB-1 bonded phase, fused-silica, capillary column. Conditions for gas chromatography were: column temperatures, 90°C for 3 min followed by 4°C/min temperature increase to 310°C and held for 20 min; injection port and detector temperatures, 300°C; column pressure, 10 psi helium; splitless injection. Retention times and concentrations of compounds were determined by electronic integration of the chromatographic peaks.

RESULTS

Hydrocarbon Gases

The concentrations of hydrocarbon gases recovered in eight cores from seven sites offshore Wilkes Land are listed in Table 2. Water depths at the sites ranged from 253 to 3379 m providing samples from the shelf, slope, and rise. The depths of penetration of the cores within the sediment were variable. Core samples from the Mertz Depression (core A1G1) and from the rise (cores A8G1, A9G1, and A10G1) ranged in subbottom depth from 43 to 384 cm. Four samples were analyzed for hydrocarbon gases from each of these cores. Core samples from the Mertz Bank (cores A2G1, A2G2, A3G1) and upper slope (core A6G1) came from subbottom depths less than 150 cm; hydrocarbon gases were determined in only one sample from each of these cores.

Hydrocarbon gases are present in all samples analyzed. As expected from previous work elsewhere (summarized by Claypool and Kvenvolden, 1983), C_1 is the most abundant hydrocarbon gas, with concentrations ranging from 53 to 16,000 nl/l. The other hydrocarbon gases are present in much lower concentrations, and when detected, range from 4 to 78 nl/l. C_1 constitutes from 40% to 99% of the hydrocarbon gases measured. At three of four stations, where profiles of gas concentrations could be determined, C_1 concentrations increase with depth; the other hydrocarbons show no obvious trends.

Table 2—Concentrations (nl/l of wet sediment) of hydrocarbon gases in sediments from the Wilkes Land margin.

Core	Interval (cm)	Water depth (m)	C1	C2	C2:1	C3	C3:1	iC4	nC4
A1G1	87-96	611	1800	n.d.[1]	13	10	8	n.d.	n.d.
A1G1	176-184		3400	10	14	12	5	n.d.	n.d.
A1G1	283-292		5100	14	16	8	8	n.d.	n.d.
A1G1	336-345		5400	14	26	13	23	n.d.	n.d.
A2G1	61-70	460	7600	60	20	53	10	n.d.	8
A2G2	139-148	458	16000	63	25	78	11	9	9
A3G1	13-22	253	900	20	33	22	8	n.d.	n.d.
A6G1	88-97	994	2800	50	20	21	13	8	6
A8G1	70-79	2635	220	8	14	4	8	n.d.	n.d.
A8G1	170-179		740	n.d.	n.d.	n.d.	6	n.d.	n.d.
A8G1	270-279		2500	10	14	n.d.	4	n.d.	n.d.
A8G1	369-378		3500	n.d.	n.d.	n.d.	4	n.d.	n.d.
A9G1	43-52	3037	100	n.d.	n.d.	n.d.	n.d.	n.d.	n.d.
A9G1	143-152		610	n.d.	8	n.d.	n.d.	n.d.	n.d.
A9G1	285-294		1300	23	18	16	12	n.d.	n.d.
A9G1	368-377		1500	10	18	10	6	n.d.	n.d.
A10G1	76-85	3379	85	13	20	7	72	n.d.	n.d.
A10G1	175-184		75	10	16	8	61	n.d.	n.d.
A10G1	275-284		53	8	14	5	54	n.d.	n.d.
A10G1	375-384		57	10	18	6	52	n.d.	n.d.

[1]n.d., not detected.

The distribution of hydrocarbon gases in samples from the deepest water site (3379 m) is most unusual in that the amounts of C_1 and $C_{3:1}$ are present at comparable concentrations between 52 and 85 nl/l. For several samples, the hydrocarbon gases C_2 and larger were not detected because shipboard motion increased the noise level on the gas chromatograph and obscured the peaks representing these compounds.

Heavy Hydrocarbons

Heavy hydrocarbons were extracted from six samples from six cores. In addition, the total organic carbon content of each of these samples was measured. Results are summarized in Table 3. The total organic-carbon content of these samples is low, ranging from 0.02% to 0.33% of the dry weight of sediment. Likewise, extractable organic matter is low, ranging from about 1 to 23 $\mu g/g$ of dry sediment. Of the extractable organic matter, the hexane fraction (mainly aliphatic hydrocarbons) constitutes 0.2–2.7 $\mu g/g$. The benzene fraction (mainly aromatic hydrocarbons) contains from 0.1 to 4.5 $\mu g/g$.

The dominant components of the aliphatic hydrocarbons are *n*-alkanes (n-C_{13} to about n-C_{36}), and the concentrations of these *n*-alkanes are given in Table 4. Also, the isoprenoid hydrocarbons, pristane and phytane, were identified, and their concentrations are listed in Table 4. Figure 2 shows the relative distribution of the *n*-alkanes. The distribution of *n*-alkanes in all samples is characterized by an odd-carbon-number dominance, particularly for compounds having 24 or more carbon atoms. In fact, the carbon preference index (CPI) from n-C_{24} to n-C_{34} ranges from 1.8 to 3.4 (Table 3). In general, the *n*-alkanes from C_{13} to C_{24} do not show a significant odd-carbon-number dominance. The *n*-alkanes in samples A6G1, A9G1, and A10G1 are strongly bimodal, with the primary mode cen-

Table 3—Geochemical analyses—Wilkes Land margin.

Core	Interval (cm)	OC (%)	EOM (μg/g)	Hexane Fraction (μg/g)	Benzene Fraction (μg/g)	CPI	Pristane / Phytane
A1G1	184-193	0.15	11	0.6	1.8	3.0	1.5
A2G2	40-49	0.16	6	0.2	0.6	2.9	1.5
A6G1	80-88	0.02	1	0.3	0.2	1.8	1.0
A8G1	262-270	0.33	23	2.7	4.5	3.4	2.5
A9G1	277-285	0.12	1	0.2	0.1	2.9	1.0
A10G1	267-275	0.16	1	0.2	0.1	2.9	0.7

[1]OC, organic carbon.
[2]EOM, extractable organic material.
[3]μg/g = μg of fraction/g of dry sediment.
[4]CPI, carbon preference index $= \frac{1}{2}\left[\frac{C_{25}+C_{27}+C_{29}+C_{31}+C_{33}}{C_{24}+C_{26}+C_{28}+C_{30}+C_{32}}+\frac{C_{25}+C_{27}+C_{29}+C_{31}+C_{33}}{C_{26}+C_{28}+C_{30}+C_{32}+C_{34}}\right]$.

tered about n-C_{19}. However, the measured distributions of n-alkanes smaller than about n-C_{20} have been affected by losses due to evaporation when the extracts went to dryness before final analysis. Thus, the measured distributions probably only approximate the in situ distributions.

Pristane and phytane are present in all samples. The pristane/phytane ratios range from 0.7 to 2.5 (Table 3). The three samples (A6G1, A9G1, and A10G1) with the least amount of extractable organic material and the bimodal n-alkane distributions had the smallest pristane/phytane ratios (1.0 or less).

DISCUSSION

Our results show that gaseous (C_1–C_4) and heavy (C_{13}–C_{36}) hydrocarbons are ubiquitous components of near-surface sediment of a small segment of the continental margin offshore the George V coast and north of the Mertz Glacier. Gaseous hydrocarbons were extracted from a total of 20 samples from seven sites, and heavy hydrocarbons were extracted from six samples from six sites. The maximum subbottom depth of sampling was 384 cm. Thus, the small number of samples analyzed and the shallow depths of sampling limit the extent to which our data can be extrapolated. Our study should be considered a preliminary survey of sediment hydrocarbons based on samples of opportunity. That is, selection of sampling sites was based on sedimentological considerations and the timing and progress of the multichannel-seismic survey of Leg 1 (Eittreim et al., 1984). The sample stations were not targeted on geophysical anomalies caused by, for example, faulting or gas-charged sediment. Thus, the probability of our finding hydrocarbons related to deep-source processes was very low. Although we had hoped to contribute information regarding the petroleum potential of offshore Antarctica, our results do not add significantly to this topic.

Hydrocarbon Gases

Of the hydrocarbon gases found in these sediments, methane (C_1) is most abundant. Other gaseous hydrocarbons through C_4 are usually present but much less abundant. The concentrations of gaseous hydrocarbons observed in these sediments are in the same general range as found on Leg 2 and can be considered "background," and are at approximately the same concentration as in sediment of the Bering Sea (Kvenvolden and Redden, 1980). The sources of these low concentrations of hydrocarbon gases have not been documented, but the fact that they are found in near-surface oceanic sediment wherever they have been sought suggests that they result from in situ biologic and/or early diagenetic processes (Claypool and Kvenvolden, 1983).

Possible Gas Hydrates

Our geochemical survey of hydrocarbon gases does not independently suggest the possibility of gas hydrates in the sediments offshore from Wilkes Land. Nevertheless, on Lines 1 and 3 of the Leg 1 multichannel survey (Fig. 1), an anomalous acoustic reflector (bottom-simulating reflector or BSR) was observed on seismic records (Fig. 3). This reflector has characteristics that suggest it could represent the base of a zone of gas hydrate (Shipley et al., 1979): (1) The reflector is at a depth consistent with the pressure-temperature stability field of methane hydrates; and (2) the reflector apparently shows a polarity reversal (Fig. 4). This possible gas hydrate is the first to be inferred from seismic records obtained from the continental margins of Antarctica.

A third criterion, i.e., that the subbottom depth of the reflector increases with increasing depth of water, is not met. In fact, the depth of the reflector off of Wilkes Land decreases with

Table 4—Concentration (ng/g)[1] of aliphatic hydrocarbons in sediments of the Wilkes Land margin.

Core	A1G1	A2G2	A6G1	A8G1	A9G1	A10G1
Interval (cm)	184-193	40-49	80-88	262-270	277-285	267-275
Hydrocarbons						
n-C13	0.4	0.0	0.0	0.8	0.0	0.0
n-C14	1.3	0.0	0.0	2.6	0.0	0.1
n-C15	4.2	0.8	0.1	6.9	0.2	0.6
n-C16	5.7	2.1	0.5	9.4	0.6	1.7
n-C17	7.9	3.5	1.1	14.0	1.2	3.9
n-C18	8.0	3.7	1.5	16.0	1.4	6.6
n-C19	9.4	4.8	1.7	20.0	1.5	9.7
n-C20	9.4	3.9	1.3	27.0	1.2	7.4
n-C21	12.0	4.5	0.9	36.0	0.9	5.1
n-C22	9.0	3.4	0.6	33.0	0.7	2.2
n-C23	16.0	5.6	0.5	64.0	0.7	1.8
n-C24	8.5	3.0	0.4	36.0	0.5	1.2
n-C25	18.0	5.8	0.4	85.0	0.8	2.0
n-C26	6.7	2.3	0.3	35.0	0.5	1.2
n-C27	15.0	5.2	0.4	110.0	1.1	2.9
n-C28	4.3	1.6	0.2	30.0	0.4	1.1
n-C29	11.0	4.4	0.4	96.0	1.2	3.4
n-C30	1.7	1.2	0.2	18.0	0.3	0.8
n-C31	7.1	3.6	0.4	64.0	1.4	2.9
n-C32	1.6	0.3	0.1	7.7	0.2	0.5
n-C33	3.3	1.2	0.2	16.0	0.5	1.0
n-C34	0.6	0.3	0.1	2.9	0.1	0.2
n-C35	0.8	0.3	0.1	4.7	0.3	0.3
n-C36	0.5	0.3	0.0	1.2	0.0	0.1
Pristane	8.8	2.3	0.5	26.0	0.5	1.8
Phytane	5.8	1.6	0.5	10.0	0.5	2.4

[1]ng/g = nanograms of hydrocarbon/gram of dry sediment.

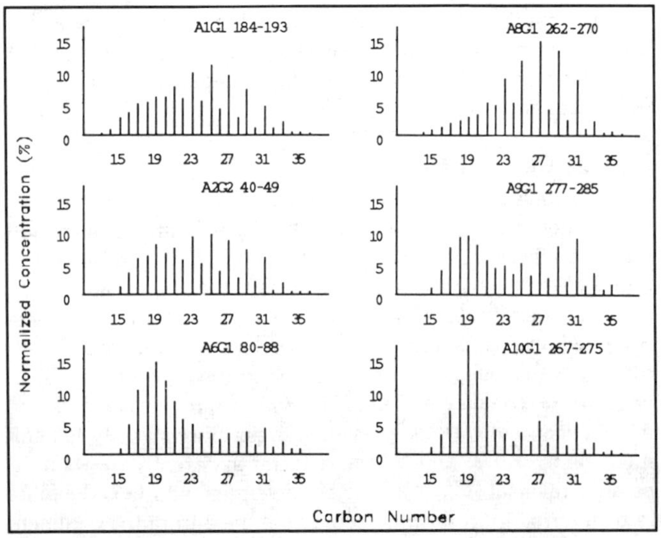

Figure 2—Graphs showing the relative distributions (concentrations normalized to 100%) of *n*-alkanes in six sediment samples from the Wilkes Land margin.

increasing water depth (Fig. 3). However, studies by Yamano et al. (1982) of BSRs on marine seismic records from the Nankai Trough also show that the reflectors become more shallow with increasing water depth. These authors attribute the BSRs to the base of the gas hydrate, and they postulate oceanward-increasing heat flow to explain the shallowing seismic reflection with increasing water depth. If the BSR on Leg 1, Lines 1 and 3, indicates the presence of gas hydrates, then the geothermal gradient can be estimated (Shipley et al., 1979; MacLeod, 1982). By measuring the depth to the reflector at several locations and using the pressure–temperature requirements for gas-hydrate stability, we can determine that the average geothermal gradient in the region of Line 1 is 41 ± 6°C/km and in the region of Line 3 is 37 ± 5°C/km (Fig. 5).

Heavy Hydrocarbons

The six samples examined for heavy hydrocarbons came from subbottom depths ranging from 40 to 285 cm. We purposely avoided very near-surface samples because they can be greatly disturbed in gravity coring and may be contaminated by modern pollution. The aliphatic hydrocarbons were found

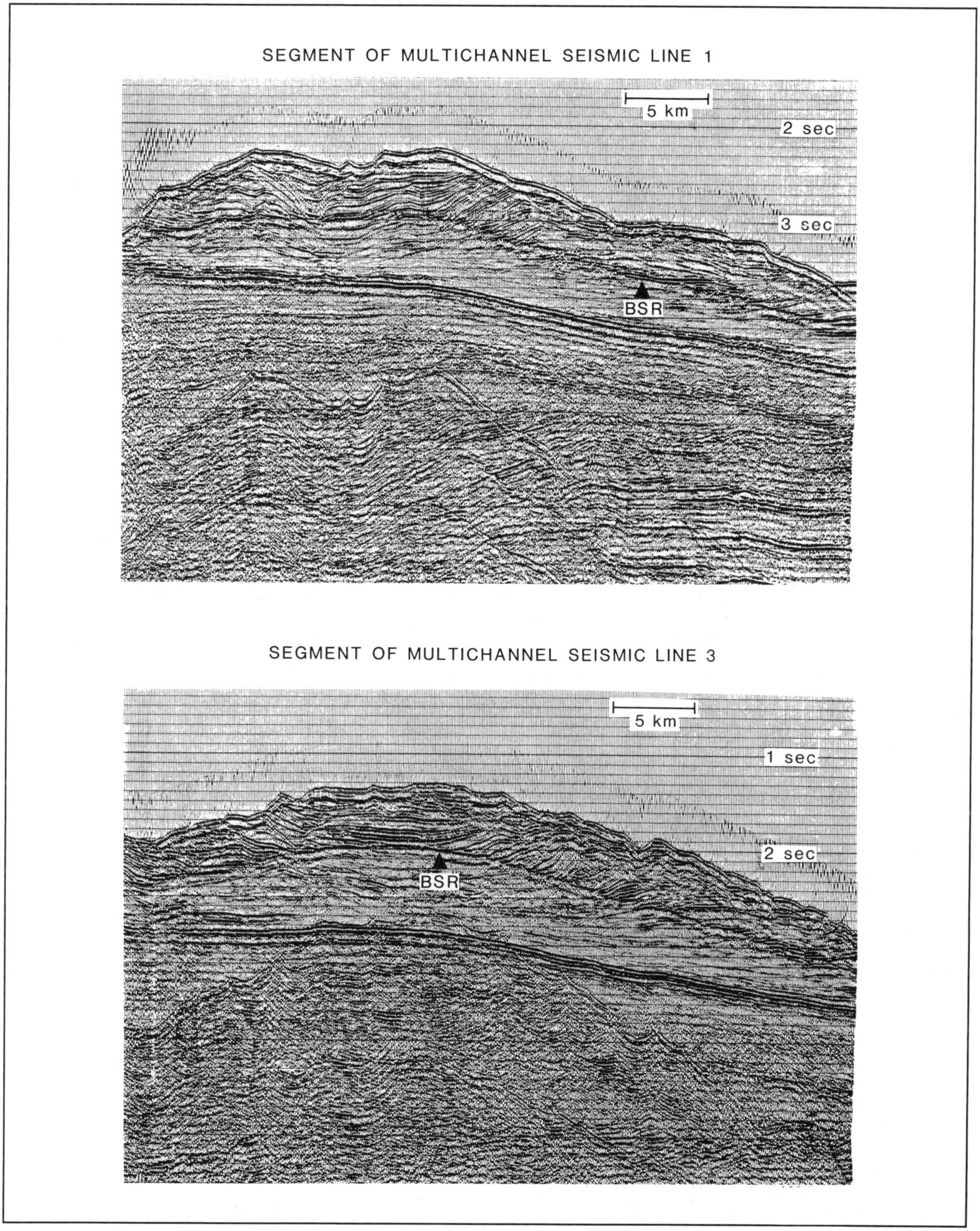

Figure 3—Segments of multichannel seismic Lines 1 and 3 showing a bottom-simulating reflector (BSR) that may mark the base of the zone of gas hydrates.

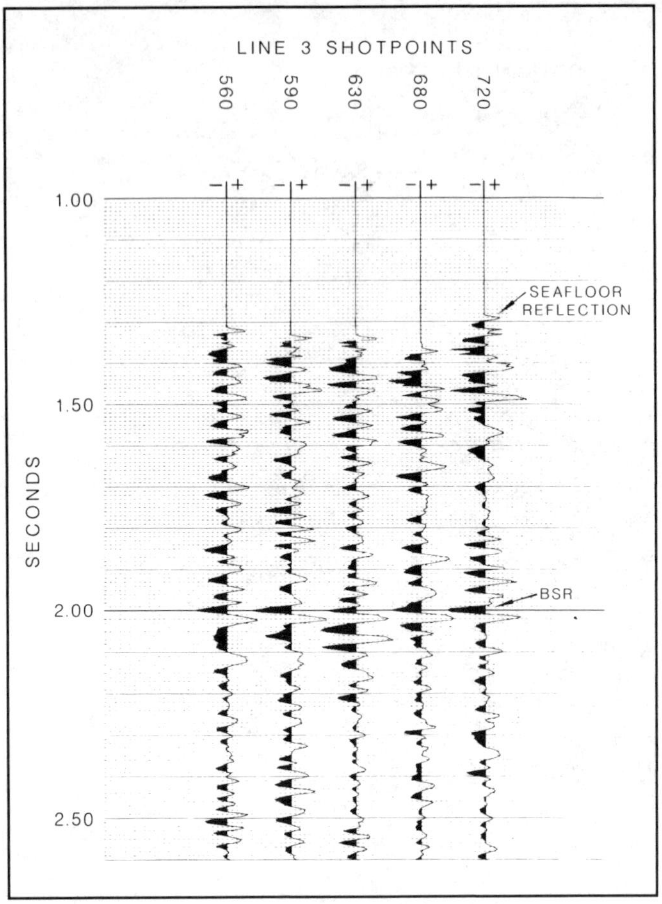

Figure 4—Five single-trace seismic reflection records from Line 3 showing reversal of reflection polarity at the inferred base of gas hydrate. Dark peaks are negative polarity. Note that the seafloor reflection and the BSR are of opposite polarity.

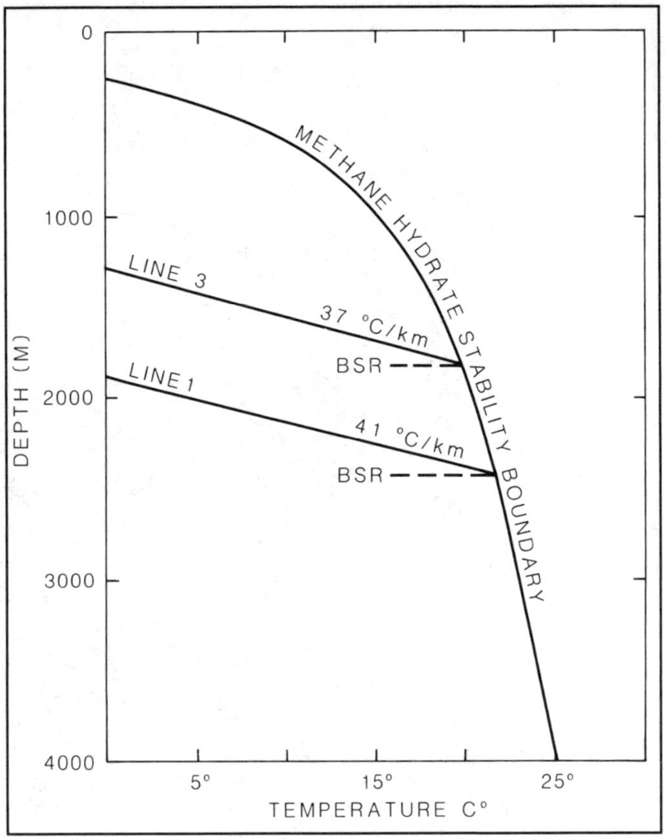

Figure 5—Average geothermal gradients of the Wilkes Land margin determined from depths of the BSR on seismic Lines 1 and 3 and from the depth-temperature stability field of methane hydrates. Subbottom BSR depths of occurrence were averaged for each of lines 1 and 3.

to be present in low concentrations (weight of hexane fraction ranged from 0.2 to 2.7 $\mu g/g$) in sediments with low organic-carbon contents (0.12–0.33%). A striking characteristic of the *n*-alkane distributions (*n*-C_{13} to *n*-C_{36}) in these and the Leg 2 sediments is the dominance of odd-carbon-numbered molecules (Fig. 2), especially in the range *n*-C_{24} and larger where the CPI ranges from 1.8 to 3.4 (Table 3). The *n*-alkanes with this range and distribution likely represent products from vascular plant material of continental origin (Eglinton et al., 1962). Because Antarctica today does not have vascular plants, these *n*-alkanes may represent either plant material that has been transported great distances from other continents, for example, by eolian transport (Simoneit and Eglinton, 1977), or these compounds are very old plant materials from ancient sediments eroded from Antarctica. Palynomorphs found in Leg 1 samples support the latter suggestion in that these microfossils may result from recycling of older sedimentary deposits (Truswell, this volume). Further evidence for erosion and redeposition is found elsewhere in Antarctica particularly in the sediments of the Ross Sea (Sackett et al., 1974; Wrenn and Beckman, 1982).

The three samples from the upper slope, the upper rise, and the lower rise have the lowest amounts of extractable organic material. The *n*-alkanes in these samples (A6G1, A9G1, and A10G1) are bimodally distributed (Fig. 2). One mode centers about C_{19} and the other at *n*-C_{29} or *n*-C_{31}. The in situ distribution of *n*-alkanes centered at C_{19} is uncertain, however, because of the losses that occurred during evaporation of our extracts. The center of this mode may actually be at a lower carbon number (*n*-C_{17} or *n*-C_{18}). The *n*-alkanes in the range of *n*-C_{13} to *n*-C_{24} in these three samples do not show a large odd-carbon dominance. These molecules may represent autochthonous compounds. Similar distributions of *n*-alkanes were noted in sub-Antarctic sediments of King Edward Cove on South Georgia (Mackie et al., 1978). These *n*-alkanes are biomodally distributed, but the mode at lower molecular weights centers usually at *n*-C_{17} or *n*-C_{18}.

The isoprenoid hydrocarbons, pristane and phytane, are present in all samples (Table 4). The pristane/phytane ratios are equal to or less than 1.0 in the samples with biomodal distributions. In the other three samples, this ratio is greater than 1.0 (1.5 to 2.5). Samples with higher pristane/phytane ratios may represent samples from more oxidizing environments than samples with lower ratios.

SUMMARY

Our study represents the first in which hydrocarbon gases and heavy hydrocarbons have been measured in this part of the continental margin of Antarctica; we found that the oceanic, near-surface sediments north of the Mertz Glacier, from the banks to the continental rise, contain aliphatic hydrocarbons that have a wide range of molecular weights (C_1 to C_4 and C_{13} to C_{36}). It is likely that the intermediate molecular-weight hydrocarbons (C_5 to C_{12}) are also present, but they were not sought. The gaseous hydrocarbons that are reported represent "background" and likely result from biologic and early diagenetic processes. The heavier *n*-alkanes, especially C_{24} and larger, probably represent terrestrial plant material. Although hydrocarbons are ubiquitous in the sediments sampled, their concentrations are low.

ACKNOWLEDGMENTS

We thank the staff and crew of the R/V *S.P. Lee* for their role in helping us accomplish our offshore Antarctica studies. This manuscript benefited from reviews by G.L. Claypool, G. W. Moore, R. D. McIver, and W. M. Sackett.

REFERENCES CITED

Behrendt, J. C., 1982, Geophysical and geological studies relevant to assessment of the petroleum resources of Antarctica, *in* R. L. Oliver, P. R. James, and J. B. Jago, eds., Antarctic earth science: Cambridge, Cambridge University Press, p. 423–428.

———, 1983, Are there petroleum resources in Antarctica?, *in* J. C. Behrendt, ed., Petroleum and mineral resources of Antarctica: U.S. Geological Survey Circular 909, p. 3–24.

Bush, P. R., 1970, A rapid method for the determination of carbonate carbon and organic carbon: Chemical Geology, v. 6, p. 59–62.

Cameron, P. J., 1981, The petroleum potential of Antarctica and its continental margin: Australian Petroleum Exploration Association Journal, v. 21, pt. 1, p. 99–111.

Claypool, G. E., and K. A. Kvenvolden, 1983, Methane and other hydrocarbon gases in marine sediment: Annual Reviews of Earth and Planetary Sciences, v. 11, p. 299–327.

Cook, R. A., and F. J. Davey, 1984, The hydrocarbon exploration of the basins of the Ross Sea, Antarctica, from modeling of the geophysical data: Journal of Petroleum Geology, v. 7, p. 213–226.

Eglinton, G., A. G. Gonzalez, R. J. Hamilton, and R. A. Raphall, 1962, Hydrocarbon constituents of the wax coatings of plant leaves: A taxonomic survey: Phytochemistry, v. 1, p. 89–102.

Eittreim, S. L., A. K. Cooper, et al., 1984, Marine geological and geophysical investigations of the Antarctic continental margin, 1984: U.S. Geological Survey Circular 935, 12 p.

Kvenvolden, K. A., and G. D. Redden, 1980, Hydrocarbon gas in sediment from the shelf, slope and basin of the Bering Sea: Geochimica et Cosmochimica Acta, v. 44, p. 1145–1150.

Kvenvolden, K. A., J. B. Rapp, and J. Bourell, 1986, Comparison of molecular markers in crude oils and rocks from the North Slope of Alaska, *in* L. B. Magoon, and G. E. Claypool, eds., Alaska North Slope oil/rock correlation study: Analysis of North Slope crude: American Association of Petroleum Geologists Studies in Geology 20, p. 593–617.

Mackie, P. R., H. M. Platt, and R. Hardy, 1978, Hydrocarbons in the marine environment. II. Distribution of *n*-alkanes in the fauna and environment of the sub-Antarctic island of South Georgia: Estuarine and Coastal Marine Science, v. 6, p. 301–313.

MacLeod, M. K., 1982, Gas hydrates in ocean bottom sediments: Bulletin of the American Association of Petroleum Geologists, v. 66, p. 2649–2662.

McIver, R. D., 1975, Hydrocarbon gases in canned core samples from Leg 28, Sites 271, 272, and 273, Ross Sea, *in* D. E. Hayes, L. A. Frakes, et al., Initial reports of the Deep Sea Drilling Project, Washington, DC, U.S. Government Printing Office, v. 28, p. 815–817.

Platt, H. M., and P. R. Mackie, 1979, Analyses of aliphatic and aromatic hydrocarbons in Antarctic marine sediment layers: Nature, v. 280, p. 576–578.

———, 1981, Sources of Antarctica hydrocarbons: Marine Pollution Bulletin, v. 12, p. 407–409.

Rapp, J. B., K. A. Kvenvolden, and H. E. Clifton, 1982, Aliphatic hydrocarbons in surface sediments of Willapa Bay and Grays Harbor, Washington: U.S. Geological Survey Open-File Report 82-609, 13 p.

Rapp, J. B., K. A. Kvenvolden, and M. Golan-Bac, in press, Hydrocarbon geochemistry of sediment offshore Antarctica, *in* A. Cooper, and F. Davey, eds., The Antarctic continental margin: Geology and geophysics of the Western Ross Sea: Circum-Pacific Council for Energy and Mineral Resources Earth Science Series, v. 5B.

Sackett, W. M., C. W. Poag, and B. J. Eadie, 1974, Kerogen recycling in the Ross Sea, Antarctica: Science, v. 185, p. 1045–1047.

Shipley, T. H., M. H. Houston, R. T. Buffler, F. J. Shaub, K. J. McMillen, J. W. Ladd, and J. L. Worzel, 1979, Seismic evidence for widespread possible gas hydrate horizons on continental slopes and rises: Bulletin of the American Association of Petroleum Geologists, v. 63, p. 2204–2213.

Simoneit, B. R. T., and G. Eglinton, 1977, Organic matter of eolian dust and its input to marine sediment, *in* R. Campos, and J. Goni, eds., Advances in organic geochemistry 1975: Madrid, Enadimsa, p. 415–430.

Whiticar, M. J., E. Suess, and H. Wehner, 1985, Thermogenic hydrocarbons in surface sediments of the Bransfield Strait, Antarctica Peninsula: Nature, v. 314, p. 87–90.

Wrenn, J. H., and S. W. Beckman, 1982, Maceral, total organic carbon and palynological analyses of Ross Ice Shelf Project Site J9 cores: Science, v. 216, p. 187–189.

Yamano, M., S. Uyeda, Y. Aoki, and T. H. Shipley, 1982, Estimates of heat flow derived from gas hydrates: Geology, v. 10, p. 339–343.

Eittreim, S. L., and Hampton, M. A., 1987, *The Antarctic Continental Margin: Geology and Geophysics of Offshore Wilkes Land*, CPCEMR Earth Science Series, v. 5A: Houston, Texas, Circum-Pacific Council for Energy and Mineral Resources

The Palynology of Core Samples from the *S.P. Lee* Wilkes Land Cruise

E. M. Truswell
Bureau of Mineral Resources
Canberra, Australia

Of nine core catcher samples collected on the Antarctic continental shelf and slope during the *S.P. Lee* Cruise L1-84-AN off Wilkes Land, seven yielded palynomorphs after maceration. The organic-walled microfossils are of mixed age; Early Cretaceous spores and pollen predominate, with a minor component of early Tertiary pollen, and a single Permian pollen grain. None of the assemblages suggest that in situ sequences of sediments were sampled; all the palynomorphs are believed to result from recycling of older deposits. The source is probably the same sequence of strata penetrated during earlier coring in the George V Basin. The recycled suites do not permit an extension of the Aptian age assigned to that sequence; the presence of rare Cretaceous dinocysts does, however, suggest that part of the sequence may be marine or marginally marine. The highest density of recycled material occurs in clays from sites on the continental slope.

INTRODUCTION

Palynological examination was undertaken on nine core samples collected during the *S.P. Lee* Cruise L1-84-AN off Wilkes Land, Antarctica (Fig. 1). All samples were from core catchers and represent the oldest sediment at the sampled sites. The purpose was to determine whether or not palynomorphs—spores, pollen, and dinocysts—were present that would indicate the presence of in situ sedimentary sequences, or suggest that such sequences were being eroded nearby. Palynomorphs of recycled origin are a common constituent of marine sediments around Antarctica (see Truswell and Drewry, 1984, for a review). They derive from the erosion of sequences occurring below the ice on the continent, or cropping out on the continental shelf. From sediments onshore they are transported seaward by processes of glacial entrainment and subsequent meltout. Once deposited on the sea floor, they may be redistributed by bottom currents, but quantitative studies in Ross Sea sediments (Truswell and Drewry, 1984) suggest that this process is only of minor importance in controlling their distribution. The recycled microfloras can be used to pinpoint, or at least hint at, the position of hidden sedimentary rocks that are the source of the microfossils.

In the area covered by the present study, a sedimentary sequence is known to occur in situ in the George V Basin, between the Mertz and Ninnis glaciers close to the George V–Adelie coast. There, siltstones recovered from piston core DF79-38 contain pollen and spores which, on the basis of comparisons with well-known Australian sequences, were determined to be of Aptian age (Domack et al., 1980). The assemblages recovered from this site are diverse and entirely nonmarine in character. The only other palynological data from the area were obtained during analyses of dredge samples collected from the continental shelf to the west of the Mertz Glacier by Douglas Mawson, during the 1911–1914 Australian Antarctic Expedition (Truswell, 1982). These dredge samples (Fig. 1) yielded sparse palynomorph assemblages containing a mixture of Early Cretaceous and Tertiary forms. Because of the sparseness of palynomorphs and the mixture of ages they represent, they are considered to be of recycled origin and may represent material deposited during a Quaternary expansion of ice shelves in the area, an expansion that probably involved the grounding of ice and the deposition of basal tills on topographically high regions of the continental shelf and of glacial-marine sediments in depressions (Domack, 1982).

PALYNOMORPH RECOVERY FROM SAMPLES EXAMINED

Sediments submitted for examination in the present investigation came from sites on both the continental shelf and the continental slope seaward of the shelf break. They derive from a variety of sediment depths at the sites cored, a factor which may mean that sedimentary units of different ages are represented. The yield of palynomorphs and organic debris is given below; the species identified in each sample are listed in Table 1.

Sample A1G1: 66°35.4′ S, 147°21.7′ E. From a water depth of 611 m. 292–293 cm. Sandy silty clay. Yielded a sparse but well-preserved suite of Early Cretaceous palynomorphs, mostly nonmarine, but including rare dinoflagellate cysts suggestive of a marine influence. A single Tertiary pollen grain was recorded.

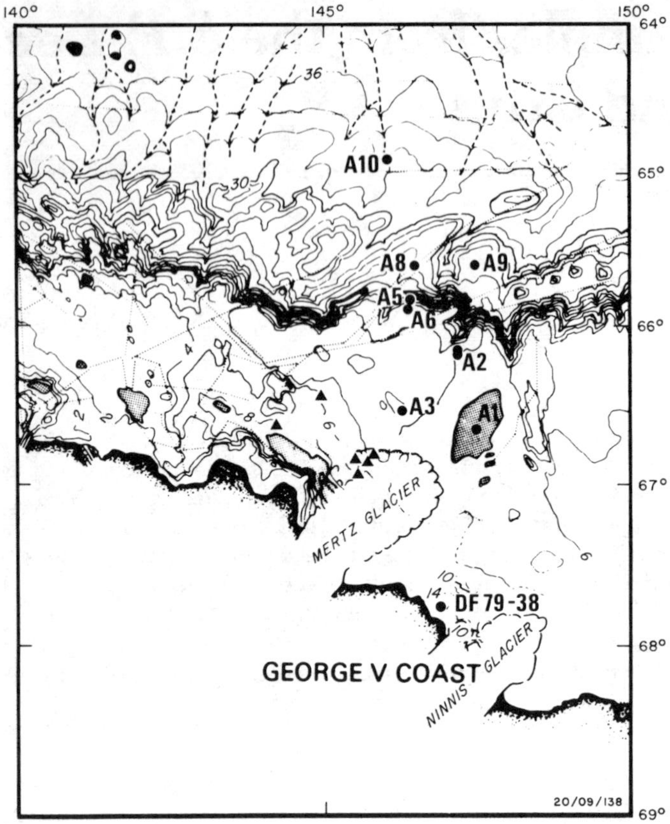

Figure 1—Locality map showing sampled sites. Bathymetry from Gebco Bathymetric map (1980). *S.P. Lee* cruise sites prefixed A; sites marked by a triangular symbol were sampled during the Australasian Antarctic Expedition of 1911–1914.

Sample A2G1: 66°08.1′ S, 147°05.5′ E. Water depth 460 m. 70–71 cm. Sandy silty clay. Abundant finely divided organic debris, a few tracheids. No identifiable palynomorphs.

Sample A2G2: 66°08.1′ S, 147°05.4′ E. Water depth 458 m. 148–149 cm. Pebbly sandy silt. Yielded very rare woody fragments: two pollen grains only were observed, both long-ranging Cretaceous–Tertiary forms.

Sample A3G1: 66°30.9′ S, 146°10.9′ E. Water depth 253 m. 22–23 cm. Spiculiferous fine sand. Yielded a fragment of plant cuticle and a single Cretaceous dinocyst.

Sample A5G1: 65°47.9′ S, 146°23.3′ E. Water depth 2200 m. 19–20 cm. Silty sand. Yielded a small quantity of woody debris and rare Cretaceous palynomorphs, mostly long-ranging saccate forms and featureless trilete spores. All of these were mid-brown in color, suggesting mature levels of organic metamorphism.

Sample A6G1: 65°52.5′ S, 146°21.1′ E. Water depth 994 m. 97–98 cm. Pebbly clayey sand. A sparse yield of Mesozoic palynomorphs, all nonmarine types, and all of a brown color indicating a mature organic facies.

Sample A8G1: 65°34.0′ S, 146°25.4′ E. Water depth 2635 m. 378–379 cm. Clay. Yielded abundant pollen, including spores, pollen, and dinoflagellates. The majority are of Early Cretaceous age, but early Tertiary pollen and dinoflagellates are present, and a single pollen grain of Permian age was recorded.

Sample A9G1: 65°34.2′ S, 147°21.6′ E. Water depth 3037 m. 377–378 cm. Silty clay. Yielded sparse but well-preserved spores and pollen, dominantly of Early Cretaceous age, but including some early Tertiary forms. Fine woody fragments are common.

Sample A10G1: 64°54.3′ S, 145°59.9′ E. Water depth 3379 m. 384–385 cm. Silty clay. Fine organic debris only, no palynomorphs.

DISCUSSION

There is no indication, from the distribution of palynomorphs described above and in Table 1, that any in situ Mesozoic or Tertiary sediments have been sampled. In those samples where palynomorphs are abundant, they are a mixture of forms of Cretaceous and early Tertiary age, with, in the case of Sample A8, a single Permian pollen grain. This mixed aspect suggests that all of the microfossils have been derived by the reworking of older sediments. In those samples where only Mesozoic (probably Early Cretaceous) pollen and spores were recorded, the recovery was too low to suggest an in situ origin.

Mesozoic forms predominate in the recycled suites. From the stratigraphic ranges shown in Table 1, it is apparent that these are mostly long-ranging spores and pollen that are common in the Early Cretaceous, although not confined to it. Many of these forms were reported by Domack et al. (1980) from the apparently in situ Aptian siltstones in the George V Basin; there is no evidence, from the present study, to suggest derivation of these recycled palynomorphs from sequences of a different age. However, rare Early Cretaceous dinoflagellate cysts in samples A1, A3, and A5 suggest that Cretaceous sequences in the region may contain marine or near-marine facies. The Aptian siltstones identified in the George V Basin were entirely nonmarine (Domack et al., 1980).

In those samples taken on the continental shelf during the *Lee* cruise, the densities of palynomorphs are low. A similar situation is evident for samples collected during the 1911–1914 Mawson Expedition (Truswell, 1982). This may reflect winnowing of the silt-sized fraction, into which most palynomorphs fall. The fact that the highest densities of palynomorphs seen in this study occur in sediments on the continental slope is noteworthy: Samples A8 and A9 are much richer in microfossils than any other samples. Two explanations are possible. The first is that the sediments containing these rich suites—predominantly of Early Cretaceous age but with a minor early Tertiary component—represent material derived from the continental shelf by slumping, which has concentrated palynomorphs in the fine fraction. Sample A8 appears to lie seaward of a channel or minor canyon feature, which may reinforce the argument that sediment from the shelf has been concentrated in this area. Alternatively, the high density of palynomorphs at these two sample sites may result from the erosion of sedimentary strata cropping out nearby, in the steeper sections of the continental slope. This seems less likely in view of the mixed ages of microfossils contained in the sediments.

Table 1—Distribution of pollen, spores, and dinoflagellate cysts in samples examined. Species occurring in common with suites from George V Basin (Domack et al., 1980) are shown in first sample columns. Stratigraphic ranges shown in second column are as these are known in Australia.

		Core 38 George V Basin	A1-G1	A2-G2	A3-G1	A5-G1	A6-G1	A8-G1	A9-G1
1. Permian palynomorphs									
Striatopodocarpidites cancellatus (Balme & Hennelly)	Early-late Permian							•	
2. Jurassic-Early Cretaceous palynomorphs									
A. Spores and pollen									
Aequitriradites spinulosus (Cookson & Dettmann)	Late Jurassic-Albian	•						•	
Araucariacites australis Cookson	Jurassic-Recent		•				•		
Callialasporites dempieri (Balme)	Early Jurassic-Neocomian							•	
Cicatricosisporites australiensis (Cookson)	Late Jurassic-Senonian	•						•	
Classopollis spp.	Late Triassic-Cenomian	•	•						
Contignisporites cooksonae (Balme)	Mid-Jurassic-Aptian								•
Cyathidites australis Cookson	Jurassic-Recent	•		•	•	•	•	•	
C. asper (Bolkhovitina)	Neocomian-Albian								
C. minor Couper	Jurassic-Recent		•			•			•
Foveotriletes sp.	Unknown							•	
Ginkgocycadophytus sp.	Permian-Recent							•	
Gleicheniidites circiniidites (Cookson)	Jurassic-Recent							•	•
Gleicheniidites sp.	Jurassic-Recent						•		
Klukisporites sp.	Jurassic-Neocomian							•	•
Leptolepidites verrucatus Couper	Late Jurassic-Albian	•						•	
Lycopiacidites dettmannae Burger	Late Jurassic-Albian							•	
Lycopodiumsporites austroclavatidites (Cookson)	Jurassic-Tertiary	•						•	•
L. eminulus Dettmann	Mid-Jurassic-Albian		•					•	•
L. reticulumsporites (Rouse)	Mid-Jurassic-Neocomian							•	
Microcachryidites antarcticus Cookson	Mid-Jurassic-Recent	•		•	•		•	•	•
Neoraistrickia truncata (Cookson)	Mid-Jurassic-Late Cretaceous	•						•	
Osmundacidites wellmanii Couper	Late Triassic-Tertiary	•	•				•	•	
Pilosisporites notensis Cookson & Dettmann	Neocomian-Albian	•						•	
Podocarpidites ellipticus Cookson	Jurassic-Tertiary							•	•
Podocarpidites spp.	Jurassic-Recent		•			•	•		
Podosporites microsaccatus (Couper)	Late Jurassic-Tertiary		•						•
Stereisporites antiquasporites (Wilson & Webster)	Early Jurassic-Recent							•	
B. Dinocysts									
Apteodinium sp. cf. *A. granulatum* Eisenack	Aptian		•						
Diconodinium sp. cf. *D. davidii* Morgan	Aptian-Albian		•		•	•			
3. Late Cretaceous-early Tertiary palynormorphs									
A. Spores and pollen									
Gambierina rudata Stover	Late Cretaceous-Paleocene							•	
Malvacipollis diversus Harris	Paleocene-Miocene							•	
Nothofagidites brachyspinulosus (Cookson)	Paleocene-Miocene							•	•
N. flemingii (Couper)	Paleocene-Oligocene		•					•	
N. lachlanae (Couper)	Late Cretaceous-Recent							•	
Nothofagidites sp.	Unknown							•	
Phyllocladidites mawsonii Cookson	Late Cretaceous-Miocene							•	
Proteacidites sp. cf. *P. amolosexinous* Dettman & Playford	Unknown							•	
B. Dinocysts									
Deflandrea sp. cf. *D. flounderensis* Stover	Probably early Tertiary							•	•
Spinidinium macmurdoense (Wilson)	Eocene							•	
Spiniferites ramosus (Ehrenberg)	Cretaceous-Tertiary							•	

CONCLUSION

Because of the sparse scatter of sample sites in the region and the likelihood that they may represent different stratigraphic levels, it is not possible, with the data available, to closely pinpoint possible source sites for the recycled palynomorphs. It seems likely that the dominant component, spores and pollen of Early Cretaceous age, derive from the same sequence of rocks that was penetrated in the George V Basin. There is insufficient evidence available to indicate any extension of the Aptian age originally assigned to that sequence. Nor can possible outcrop areas be identified; data from the present study admit a slight possibility that there are areas of outcrop on the continental slope in the area surveyed. The presence of rare Early Cretaceous dinocysts among the recycled microfossils suggests that there may be a marine or marginal marine component in the sequence. The source of the Tertiary pollen in the sediments is unknown; specimens are rare and may have been transported some distance.

REFERENCES CITED

Domack, E. W., 1982, Sedimentology of glacial and glacial marine deposits on the George V–Adelie continental shelf, East Antarctica: Boreas, v. 11, p. 79–97.

Domack, E. W., W. W. Fairchild, and J. B. Anderson, 1980, Lower Cretaceous sediment from the East Antarctic continental shelf: Nature, v. 287, p. 625–626.

Truswell, E. M., 1982, Palynology of seafloor samples collected by the 1911–1914 Australasian Antarctic Expedition: Implications for the geology of coastal East Antarctica: Journal of the Geological Society of Australia, v. 29, p. 343–356.

Truswell, E. M., and D. J. Drewry, 1984, Distribution and provenance of recycled palynomorphs in surficial sediments of the Ross Sea, Antarctica: Marine Geology, v. 59, p. 187–214.

Eittreim, S. L., and Hampton, M. A., 1987, *The Antarctic Continental Margin: Geology and Geophysics of Offshore Wilkes Land*, CPCEMR Earth Science Series, v. 5A: Houston, Texas, Circum-Pacific Council for Energy and Mineral Resources

Provenance of Glacial Stones Dredged off Wilkes Land, Antarctica

G. W. Moore
U.S. Geological Survey
Menlo Park, California

Ten stones, some with glacial striations, were collected in a dredge haul from the continental slope at a depth of about 1600 m about 160 km off Wilkes Land. Most of the rocks are similar to rocks cropping out on the continent, and their inferred sources are: Robertson Bay Group of Cambrian and Ordovician age; Granite Harbour Intrusives of Cambrian or Ordovician age; Beacon Supergroup of Devonian to Jurassic age; Ferrar Dolerite of Jurassic age; and possible unnamed synrift deposits of Early Cretaceous age.

INTRODUCTION

During its 1984 survey of Antarctica, the U.S. Geological Survey's research vessel *S.P. Lee* collected ten rocks along a 2-km dredge line from the upper continental slope off Wilkes Land (Fig. 1). The location of the sample was from 65°48.9′ S, 146°19.3′ E, to 65°49.9′ S, 146°17.3′ E, and the depth was from 1785 to 1570 m. Rifting had separated Australia from this part of Antarctica during the Cretaceous, left pre-Cretaceous basement rocks on the continent, and led to synrift and postrift deposits on the shelf and slope (Eittreim et al., 1985).

The dredge site was selected because of the steep local slope of about 7° up to the south. Seismic-reflection profiles indicate that erosion may have exposed Lower Cretaceous synrift deposits at the site of the dredge haul but not the prerift basement (Eittreim and Smith, this volume).

Several specimens in the collection are flatiron shaped and have well-developed glacial striations. Therefore, glaciers or icebergs probably transported these specimens to the dredge site.

Glacial transport in this area, now and during the last glacial maximum, has been suggested by Stuiver et al. (1981) to have been roughly northward perpendicular to the coast. In contrast, because of prevailing wind and current directions, iceberg drift has been from the southeast (Keys et al., 1984). Hence the transported stones are most likely to have come from pre-Cretaceous outcrops either on the adjacent coast or from areas to the southeast (Fig. 1).

DESCRIPTION OF ROCK SPECIMENS

The ten rock specimens from the dredge line off the coast of Wilkes Land have a total weight of about 20 kg. A chain dredge with a mesh opening of 50 mm recovered the rocks, so smaller clasts were not collected. The rocks, studied in thin section supplemented by x-ray diffraction of mineral separates, are described as follows.

L1-84-AN-A7-1

Granite (60 × 100 × 140 mm, 1234 g): Coarse-grained, medium light gray; 35% quartz (average 5 mm), with undulatory extinction, containing tourmaline needles; 40% orthoclase (5 mm), partly altered to sericite, containing inclusions of biotite, zircon, and muscovite; 15% reddish brown biotite (5 mm), somewhat bent and partly altered to chlorite, containing zircons with well-developed (old) pleochroic halos; minor plagioclase and microcline. Probable provenance: Granite Harbour Intrusives of Cambrian or Ordovician age.

L1-84-AN-A7-2

Diabase (110 × 160 × 190 mm, 5544 g): Ophitic, intersertal, medium grained, medium gray, deeply striated; 50% plagioclase (labradorite), 1–2 mm; 20% augite, average 1 mm, more altered than the plagioclase; interstitial glass that contains apatite needles and has altered to quartz and chlorophaeite; minor magnetite; fine secondary biotite. Probable provenance: Ferrar Dolerite of Jurassic age.

L1-84-AN-A7-3

Diabase (30 × 60 × 90 mm, 355 g): Subophitic, intersertal, medium grained, olive gray; 50% plagioclase (laboradorite), 1 mm; 20% augite, 1 mm; interstitial glass that contains apatite needles and has altered partly to quartz; 2% magnetite; calcite infillings in cavities. Probable provenance: Ferrar Dolerite of Jurassic age.

L1-84-AN-A7-4

Sandstone (60 × 100 × 120 mm, 907 g): Medium grained, subangular, pale red with 1 cm pinkish gray reduced spots; 80% quartz, 0.5 mm, some contain biotite inclusions, rare polycrystalline grains; 5% microcline; minor amphibole

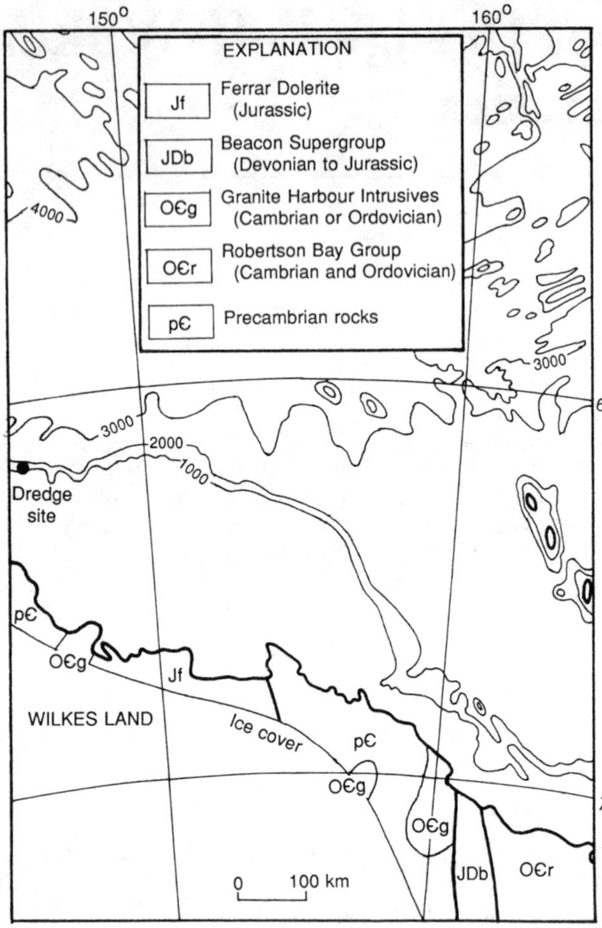

Figure 1—Location of dredge haul L1-84-AN-A7 off Wilkes Land with respect to the simplified onshore bedrock geology (after Craddock, in press).

(pargasite?), metaquartzite, phyllite, and sandstone grains; sericite cement and hematite grains (0.01 mm). Probable provenance: Beacon Supergroup of Devonian to Jurassic age.

L1-84-AN-A7-5

Sandstone (50 × 90 × 140 mm, 1280 g): Fine-grained, subangular, light olive gray with grayish orange stains; 70% quartz, 0.2 mm, 10% polycrystalline; 10% microcline; rare muscovite flakes, white amphibole (actinolite?), olive amphibole (hornblende?), plagioclase, pyrite; ankerite and clay cement. Probable provenance: Beacon Supergroup of Devonian to Jurassic age.

L1-84-AN-A7-6

Sandstone (100 × 170 × 190 mm, 5336 g): Foliated, fine grained, subangular, olive gray; 60% quartz (0.1 mm); 10% potassium feldspar; 10% chlorite; 10% muscovite; 10% biotite (pleochroic); 5% siderite rhombs with iron oxide rinds; minor plagioclase, opaque minerals, zircon, clay; contains a 0.5 mm quartz vein with a chlorite sheath. Probable provenance: Robertson Bay Group of Cambrian and Ordovician age.

L1-84-AN-A7-7

Phyllite (80 × 150 × 190 mm, 3216 g): Medium gray; 60% quartz, 0.1 mm; 20% muscovite foliated in two directions; 10% chlorite; rare opaque minerals; contains phacoids of quartz sandstone. Probable provenance: Robertson Bay Group of Cambrian and Ordovician age.

L1-84-AN-A7-8

Slate (10 × 180 × 200 mm, 1105 g): Kink banded, medium greenish gray; 60% quartz, 0.01 mm; 10% chlorite, some coarse grained but mostly fine; 10% carbonate (detrital?); rare opaque minerals and a detrital rutile rod. Probable provenance: Robertson Bay Group of Cambrian and Ordovician age.

L1-84-AN-A7-9

Conglomerate (50 × 50 × 90 mm, 200 g): Grayish black, encrusted by siliceous sponges, bryozoans, tube worms, and forams; pebbles of 1–3 cm size, diabase and grayish pink sandstone; coarse-grained quartz sandstone matrix, subangular, moderately indurated, manganese coated (birnessite). Probable provenance: Lower Cretaceous synrift deposits or younger.

L1-84-AN-A7-10

Concretion (20 × 80 × 120 mm, 450 g): Brownish black rind with black coating; grayish orange siltstone core; siltstone composed of clay, quartz, rock fragments, minor plagioclase; rind cement is goethite, and coating is birnessite. Possible provenance: Lower Cretaceous synrift deposits.

CONCLUSIONS

Comparison with the descriptions of rocks from onshore outcrops suggests that the most likely sources for the samples dredged off Wilkes Land are those stated with the descriptions above. The first rocks taken from Antarctica during the discovery of the continent came from ice about 1000 km west of our site and consisted of our site and consisted of "red sandstone and basalt" (Wilkes, 1845). Bradshaw et al. (1985) published a modern synthesis of the geology of this part of Antarctica. Also, Wong et al. (in press) prepared a detailed description of a large suite of rocks dredged from the Ross Sea, 1500 km east of the present site, that contains many specimens similar to those described here.

The conglomerate and the concretion are the only rocks in the collection that are possibly local. Domack et al. (1980) cored fossiliferous Lower Cretaceous rocks within about 150 km of this dredge site, and Truswell (1983, this volume) found reworked fossil pollen of the same age in nearby surficial seafloor sediment. Therefore, the conglomerate and concretion samples may have come from Lower Cretaceous synrift deposits laid down during the first phases of the breakup of Antarctica and Australia and subsequently exposed by late Cenozoic erosion at the top of the continental slope.

ACKNOWLEDGMENTS

Walter B. Friesen provided useful suggestions in this study. The manuscript was reviewed by Florence L. Wong and Arthur B. Ford.

REFERENCES CITED

Bradshaw, J. D., S. D. Weaver, and M. G. Laird, 1985, Suspect terranes and Cambrian tectonics in northern Victoria Land, Antarctica: Circum-Pacific Council for Energy and Mineral Resources Earth Science Series, v. 1, p. 467–479.

Craddock, C., in press, Geologic map of the circum-Pacific region, Antarctica sheet: American Association of Petroleum Geologists. Scale 1:10,000,000.

Domack, E. W., W. W. Fairchild, and J. B. Anderson, 1980, Lower Cretaceous sediment from the East Antarctic continental shelf: Nature, v. 287, p. 625–626.

Eittreim, S. L., M. A. Hampton, and J. R. Childs, 1985, Seismic-reflection signature of Cretaceous continental breakup on the Wilkes Land margin, Antarctica: Science, v. 229, p. 1082–1084.

Keys, J. R., M. J. McDonnell, and K. W. Piddington, 1984, Immediate report of the McMurdo iceberg pilot study: Wellington, Commission for the Environment, 52 p.

Stuiver, M., G. H. Denton, T. J. Hughes, and J. L. Fastook, 1981, History of the marine ice sheets in West Antarctica during the last glaciation: A working hypothesis, *in* G. H. Denton and T. J. Hughes, eds., The last great ice sheets: New York, Wiley-Interscience, p. 319–436.

Truswell, E. M., 1983, Geological implications of recycled palynomorphs in continental shelf sediments around Antarctica, *in* R. L. Oliver, P. R. James, and J. B. Jago, eds., Antarctic earth science: Canberra, Australian Academy of Science, p. 394–399.

Wilkes, C., 1845, Narrative of the United States exploring expedition during the years 1838, 1839, 1840, 1841, 1842: Philadelphia, Sherman, p. 325.

Wong, F. L., P. J. Barrett, J. Gamble, and D. G. Howell, in press, Petrography of rock samples dredged from Iselin Bank, Ross Sea, Antarctica, *in* A. L. Cooper and F. J. Davey, the Antarctic continental margin: Geology and geophysics of the Western Ross Sea, CPCEMR Earth Science Series, V. 5B: Houston, Texas, Circum-Pacific council for Energy and Mineral Resources.